Springer Series in
Computational
Mathematics

10

Yu. Ermoliev R. J-B Wets (Eds.)

Numerical Techniques for Stochastic Optimization

With 62 Figures

Springer-Verlag
Berlin Heidelberg New York
London Paris Tokyo

Yuri Ermoliev
Glushkov Institute of Cybernetics
Ukrainian Academy of Sciences
252207 Kiev 207
USSR

Roger J-B Wets
Department of Mathematics
University of California
Davis, CA 95616
USA

Mathematics Subject Classification (1980):
90C15, 90C06, 90C30, 90C50, 49D27, 49D45, 49D35, 65U05
Library of Congress Cataloging-in-Publication Data
Numerical techniques for stochastic optimization / Yuri Ermoliev,
    Roger J-B Wets, editors.
        p.   cm. — (Springer series in computational mathematics ; 10)
    Bibliography: p.
    ISBN-13:978-3-642-64813-7   e-ISBN-13:978-3-642-61370-8
    DOI: 10.1007/978-3-642-61370-8
    1. Stochastic programming.   I. Ermol 'ev, IUriĭ Mikhaĭlovich.
    II. Wets, Roger J.-B.   III. Series.
    T57.79.N86   1988
    003—dc19   88–12189

Camera-ready text prepared by the editors.

9 8 7 6 5 4 3 2 1

THE INTERNATIONAL INSTITUTE FOR APPLIED SYSTEMS ANALYSIS

is a nongovernn ental research institution, bringing together scientists from around the world to work cn problems of common concern. Situated in Laxenburg, Austria, IIASA was founded in October 1972 by the academies of science and equivalent organizations of twelve countries. Its founders gave IIASA a unique position outside national, disciplinary, and institutional boundaries so that it might take the broadest possible view in pursuing its objectives:

To promote international cooperation in solving problems arising from social, economic, technological, and environmental change

To create a network of institutions in the national member organization countries and elsewhere for joint scientific research

To develop and formalize systems analysis and the sciences contributing to it, and promote the use of analytical techniques needed to evaluate and address complex problems

To inform policy advisors and decision makers about the potential application of the Institute's work to such problems

The Institute now has national member organizations in the following countries:

Austria
The Austrian Academy of Sciences

Bulgaria
The National Committee for Applied Systems Analysis and Management

Canada
The Canadian Committee for IIASA

Czechoslovakia
The Committee for IIASA of the Czechoslovak Socialist Republic

Finland
The Finnish Committee for IIASA

France
The French Association for the Development of Systems Analysis

German Democratic Republic
The Academy of Sciences of the German Democratic Republic

Federal Republic of Germany
Association for the Advancement of IIASA

Hungary
The Hungarian Committee for Applied Systems Analysis

Italy
The National Research Council

Japan
The Japan Committee for IIASA

Netherlands
The Foundation IIASA—Netherlands

Poland
The Polish Academy of Sciences

Sweden
The Swedish Council for Planning and Coordination of Research

Union of Soviet Socialist Republics
The Academy of Sciences of the Union of Soviet Socialist Republics

United States of America
The American Academy of Arts and Sciences

PREFACE

Rapid changes in today's environment emphasize the need for models and methods capable of dealing with the uncertainty inherent in virtually all systems related to economics, meteorology, demography, ecology, etc. Systems involving interactions between man, nature and technology are subject to disturbances which may be unlike anything which has been experienced in the past. In particular, the technological revolution increases uncertainty as each new stage perturbs existing knowledge of structures, limitations and constraints. At the same time, many systems are often too complex to allow for precise measurement of the parameters or the state of the system. Uncertainty, nonstationarity, disequilibrium are pervasive characteristics of most modern systems.

In order to manage such situations (or to survive in such an environment) we must develop systems which can facilitate our response to uncertainty and changing conditions. In our individual behavior we often follow guidelines that are conditioned by the need to be prepared for all (likely) eventualities: insurance, wearing seat-belts, savings versus investments, annual medical check-ups, even keeping an umbrella at the office, etc. One can identify two major types of mechanisms: the short term *adaptive* adjustments (defensive driving, marketing, inventory control, etc.) that are made after making some observations of the system's parameters, and the long term *anticipative* actions (engineering design, policy setting, allocation of resources, investment strategies, etc.). The main challenge to the system analyst is to develop a modeling approach that combines both mechanisms (adaptive and anticipative) in the presence of a large number of uncertainties, and this in such a way that it is computationally tractable.

The technique most commonly used, *scenario analysis*, to deal with long term planning under uncertainty is seriously flawed. Although it can identify "optimal" solutions for each scenario (that specifies some values for the unknown parameters), it does not provide any clue as to how these "optimal" solutions should be combined to produce merely a reasonable decision.

As uncertainty is a broad concept, it is possible—and often useful—to approach it in many different ways. One rather general approach, which has been successfully applied to a wide variety of problems, is to assign explicitly or implicitly, a probabilistic measure—which can also be interpreted as a measure of confidence, possibly of subjective nature—to the various unknown parameters. This leads us to a class of stochastic optimization problems, conceivable with only partially known distribution functions (and incomplete observations of the unknown parameters), called *stochastic programming problems*. They can be viewed as extensions of the linear and nonlinear programming models to decision problems that involve random parameters.

Stochastic programming models were first introduced in the mid 50's by Dantzig, Beale, Tintner, and Charnes and Cooper for linear programs with random coefficients for decision making under uncertainty; Dantzig even used the name "linear programming under uncertainty". Nowadays, the term "stochastic programming" refers to the whole field—models, theoretical underpinnings, and in particular, solution procedures—that deals with optimization problems involving random quantities (i.e., with stochastic optimization problems), the accent being placed on the computational aspects; in the USSR the term "stochastic programming" has been used to designate not only various types of stochastic optimization problems but also stochastic procedures that can be used to solve deterministic nonlinear programming problems but which play a particularly important role as solution procedures for stochastic optimization problems, cf. Chapter 1, Section 9.

Although stochastic programming models were first formulated in the mid 50's, rather general formulations of stochastic optimization problems appeared much earlier in the literature of mathematical statistics, in particular in the theory of sequential analysis and in statistical decision theory. All statistical problems such as estimation, prediction, filtering, regression analysis, testing of statistical hypotheses, etc., contain elements of stochastic optimization; even Bayesian statistical procedures involve loss functions that must be minimized. Nevertheless, there are differences between the typical formulation of the optimization problems that come from statistics and those from decision making under uncertainty.

Stochastic programming models are mostly motivated by problems arising in so-called "here-and-now" situations, when decisions must be made on the basis of, existing or assumed, a *priori* information about the random (relevant) quantities, without making additional observation. The situation is typical for problems of long term planning that arise in operations research and systems analysis. In mathematical statistics we are mostly dealing with "wait-and-see" situations when we are allowed to make additional observations "during" the decision making process. In addition, the accent is often on closed form solutions, or on *ad hoc* procedures that can be applied when there are only a few decision variables (statistical parameters that need to be estimated). In stochastic programming, which arose as an extension of linear programming, with its sophisticated computational techniques, the accent is on solving problems involving a large number of decision variables and random parameters, and consequently a much larger place is occupied by the search for efficient solutions procedures.

Unfortunately, stochastic optimization problems can very rarely be solved by using the standard algorithmic procedures developed for deterministic optimization problems. To apply these directly would presuppose the availability of efficient subroutines for evaluating the multiple integrals of rather involved (nondifferentiable) integrands that characterize the system as functions of the decision variables (objective and constraint functions), and such subroutines are neither available nor will they become available short of a small upheaval in (numerical) mathematics. And that is why there is presently not software available which is capable of handling general stochastic optimization problems, very much for the same reason that there is no universal package for solving partial differential equations where one is also confronted by multidimensional integrations. A number of computer codes have been written to solve certain specific applications, but it is only now that we can reasonable hope to develop generally applicable software; generally applicable that is within well-defined classes of stochastic optimization problems. This means that we should be able to pass from the artisanal to the production level. There are two basic reasons for this. First, the available technology (computer technology, numerically stable subroutines) has only recently reached a point where the computing capabilities match the size of the numerical problems faced in this area. Second, the underlying mathematical theory needed to justify the computational shortcuts making the solution of such problems feasible has only recently been developed to an implementable level.

This book is a result of a project on "Numerical Methods for Stochastic Optimization Problems" of the Adaptation and Optimization Task of the International Institute for Applied Systems Analysis (IIASA). This project was started in 1982. IIASA's traditional role as a network coordinator between individual scientists as well as research institutes was a vital component of this collaborative network of researchers whose interactions contributed significantly to the advances made in this field during the last 2–3 years. Let this book serve as a testimony to this collaborative effort.

The book is divided in five parts. Part I is just an introduction to some general and particular stochastic programming problems as models for decision making under uncertainty. Part II consists of a number of chapters, each covering some of the numerical questions that must be dealt with when developing solution procedures for stochastic programming problems. This part is also meant to provide the background to the description of the implementation of a number of methods given in Part III. Part IV is a collection of selected applications and test problems. This volume, and a tape collecting the computer codes for stochastic programming problems developed either at IIASA or at other research institutions that have collaborated in this project, is the state-of-the-art of algorithmic development in this field. The main objective of the IIASA project was to demonstrate that software can be built which solves a wide variety of stochastic programming problems. For certain classes of problems the software now available is nearly of production-level quality, whereas for others only experimental codes have been included. This is a first step in software development; it should provide a solid base and serious encouragement for more ambitious endeavors in this area.

TABLE OF CONTENTS

PART I

Models, Motivation and Methods

CHAPTER 1

STOCHASTIC PROGRAMMING, AN INTRODUCTION

Yu. Ermoliev and R. Wets

The purpose of this introduction is to discuss the way to deal with uncertainties in a stochastic optimization framework and to develop this theme in a general discussion of modeling alternatives and solution strategies. We shall be concerned with motivation and general conceptual questions rather than by technical details. Most everything is supposed to happen in finite dimensional Euclidean space (decision variables, values of the random elements) and we shall assume that all probabilities and expectations, possibly in an extended real-valued sense, are well defined.

1.1 Optimization Under Uncertainty

Many practical problems can be formulated as optimization problems or can be reduced to them. Mathematical modeling is concerned with a description of various types of relations between the quantities involved in a given situation. Sometimes this leads to a unique solution, but more generally it identifies a set of possible states, a further criterion being used to choose among them a more, or most, desirable state. For example the "states" could be all possible structural outlays of a physical system, the preferred state being the one that guarantees the highest level of reliability, or an "extremal" state that is chosen in terms of certain desired physical property: dielectric conductivity, sonic resonance, etc. Applications in operations research, engineering, economics have focussed attention on situations where the system can be affected or controlled by outside decisions that should be selected in the best possible manner. To this end, the notion of an optimization problem has proved very useful. We think of it in terms of a set S whose elements, called the feasible solutions, represent the alternatives open to a decision maker. The aim is to optimize, which we take here to be minimize, over S a certain function g_0, the objective function. The exact definition of S in a particular case depends on various circumstances, but it typically involves a number of functional relationships among the variables identifying the possible "states". As prototype for the set S we take the following description

$$S := \{x \in R^n | x \in X, g_i(x) \le 0, \quad i = 1, \ldots, m\}$$

where X is a given subset of R^n (usually of rather simple character, say R^n_+ or possibly R^n itself), and for $i = 1, \ldots, m, g_i$ is a real-valued function on R^n.

The optimization problem is then formulated as:

$$\text{find} \quad x \in X \subset R^n$$
$$\text{such that} \quad g_i(x) \leq 0, \quad i = 1, \ldots, m, \tag{1.1}$$
$$\text{and} \quad z = g_0(x) \text{ is minimized.}$$

When dealing with conventional deterministic optimization problems (linear or nonlinear programs), it is assumed that one has precise information about the objective function g_0 and the constraints g_i. In other words, one knows all the relevant quantities that are necessary for having well-defined functions $g_i, i = 1, \ldots, m$. For example, if this is a production model, enough information is available about future demands and prices, available inputs and the coefficients of the input-output relationships, in order to define the cost function g_0 as well as give a sufficiently accurate description of the balance equations, i.e., the functions $g_i, i = 1, \ldots, m$. In practice, however, for many optimization problems the functions $g_i, i = 0, \ldots m$ are not known very accurately and in those cases, it is fruitful to think of the functions g_i as depending on a pair of variables (x, ω) with ω as vector that takes its values in a set $\Omega \subset R^q$. We may think of ω as the *environment*-determining variable that conditions the system under investigation. A decision x results in different outcomes

$$(g_0(x, \omega), g_1(x, \omega), \ldots, g_m(x, \omega))$$

depending on the uncontrollable factors, i.e. the environment (state of nature, parameters, exogenous factors, etc.). In this setting, we face the following "optimization" problem:

$$\text{find} \quad x \in X \subset R^n$$
$$\text{such that} \quad g_i(x, \omega) \leq 0, \quad i = 1, \ldots, m, \tag{1.2}$$
$$\text{and} \quad z(\omega) = g_0(x, \omega) \text{ is minimized.}$$

This may suggest a parametric study of the optimal solution as a function of the environment ω and this may actually be useful in some cases, but what we really seek is some x that is "feasible" and that minimizes the objective for all or for nearly all possible values of ω in Ω, or is some other sense that needs to be specified. Any fixed $x \in X$, may be feasible for some $\omega' \in \Omega$, i.e. satisfy the constraints $g_i(x, \omega') \leq 0$ for $i = 1, \ldots, m$, but infeasible for some other $\omega \in \Omega$. The notion of feasibility needs to be made precise, and depends very much on the problem at hand, in particular whether or not we are able to obtain some information about the environment, the value of ω, before choosing the decision x. Similarly, what must be understood by optimality depends on the uncertainties involved as well as on the view one may have of the overall objective(s), e.g. avoid a disastrous situation, do well in nearly all cases, etc. We cannot "solve" (1.2) by finding the optimal solution for every possible value of ω in Ω, i.e. for every possible environment, aided possibly in this by parametric analysis. This is the approach preconized by *scenario analysis*. If the problem is not insensitive to its environment, then knowing that $x^1 = x^*(\omega^1)$ is the best decision in environment ω^1 and $x^2 = x^*(\omega^2)$ is the best decision in environment ω^2 does not really tell us how to choose some x that will be a reasonably good decision whatever be the environment, ω^1 or ω^2; taking a (convex) combination of x^1 and x^2 may lead to an infeasible decision for both possibilities: problem (1.2) with $\omega = \omega^1$ or $\omega = \omega^2$.

In the simplest case of complete information, i.e. when the environment ω will be completely known before we have to choose x, we should, of course, simply select the optimal solution of (1.2) by assigning to the variables ω the known values of these parameters. However, there may be some additional restrictions on this choice of x in certain practical situations. For example, if the problem is highly nonlinear and/or quite large, the search for an optimal solution may be impractical (too expensive, for example) or even physically impossible in the available time, the required response-time being too short. Then, even in this case, there arises—in addition to all the usual questions of optimality, design of solutions procedures, convergence, etc.—the question of implementability. Namely, how to design a practical (implementable) decision rule (function)

$$\omega \mapsto x(\omega)$$

which is viable, i.e. $x(\omega)$ is feasible for (1.2) for all $\omega \in \Omega$, and that is "optimal" in some sense, ideally such that for all $\omega \in \Omega, x(\omega)$ minimizes $g_0(\cdot, \omega)$ on the corresponding set of feasible solutions. However, since such an ideal decision rule is only rarely simple enough to be implementable, the notion of optimality must be redefined so as to make the search for such a decision rule meaningful.

A more typical case is when each observation (information gathering) will only yield a partial description of the environment ω : it only identifies a particular collection of possible environments, or a particular probability distribution on Ω. In such situations, when the value of ω is not known in advance, for any choice of x the values assumed by the functions $g_i(x, \cdot), i = 1, \ldots, m$, cannot be known with certainty. Returning to the production model mentioned earlier, as long as there is uncertainty about the demand for the coming month, then for any fixed production level x, there will be uncertainty about the cost (or profit). Suppose, we have the very simple relation between x (production level) and ω (demand):

$$g_0(x, \omega) = \begin{cases} \alpha(x - \omega) & \text{if } \omega \geq x \\ \beta(\omega - x) & \text{if } x \geq \omega \end{cases} \tag{1.3}$$

where α is the unit surplus-cost (holding cost) and β is the unit shortage-cost. The problem would be to find an x that is "optimal" for all foreseeable demands ω in Ω rather than a function $\omega \mapsto x(\omega)$ which would tell us what the optimal production level should have been once ω is actually observed.

When no information is available about the environment ω, except that $\omega \in \Omega$ (or to some subset of Ω), it is possible to analyze problem (1.2) in terms of the values assumed by the vector

$$(g_0(x,\omega), g_1\{x,\omega\}, \ldots, g_m(x,\omega))$$

as ω varies in Ω. Let us consider the case when the functions $g_1, \ldots, g_m$ do not depend on ω. Then we could view (1.2) as a multiple objective optimization problem. Indeed, we could formulate (1.2) as follows:

$$\begin{aligned} \text{find} \quad & x \in X \subset R^n \\ \text{such that} \quad & g_i(x) \leq 0, \quad i = 1, \ldots, m \\ \text{and for each} \quad & \omega \in \Omega, z_\omega = g_0(x, \omega) \text{ is minimized.} \end{aligned} \qquad (1.4)$$

At least if Ω is a finite set, we may hope that this approach would provide us with the appropriate concepts of feasibility and optimality. But, in fact such a reformulation does not help much. The most commonly accepted point of view of optimality in multiple objective optimization is that of Pareto-optimality, i.e. the solution is such that any change would mean a strictly less desirable state in terms of at least one of the objectives, here for some ω in Ω. Typically, of course, there will be many Pareto-optimal points with no equivalence between any such solutions. There still remains the question of how to choose a (unique) decision among the Pareto-optimal points. For instance, in the case of the objective function defined by (1.3), with $\Omega = [\underline{\omega}, \overline{\omega}] \subset (0, \infty)$ and $\alpha > 0, \beta > 0$, each $x = \omega$ is Pareto-optimal, see Figure 1.1,

$$g_0(x, \omega) = g_0(\omega, \omega) = 0$$

$$g_0(\omega, \omega') > 0 \quad \text{for all } \omega' \neq \omega.$$

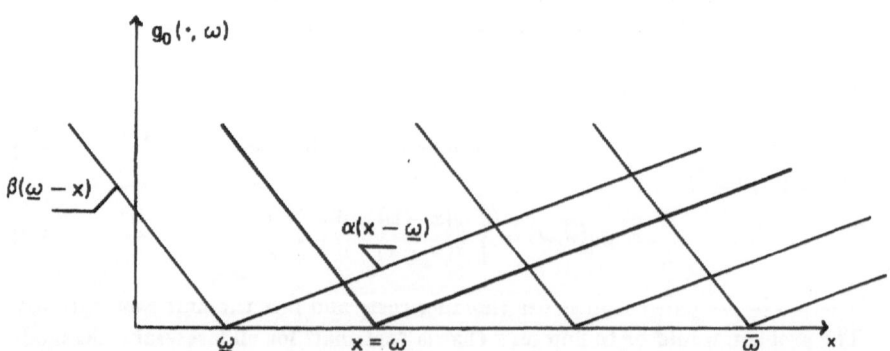

Figure 1.1 Pareto-optimality

One popular approach to selecting among the Pareto-optimal solutions is to proceed by "worst-case analysis". For a given x, one calculates the worst that could happen—in terms of all the objectives—and then choose a solution that minimizes the value of the worst-case loss; scenario analysis also relies on a similar approach. This should single out some point that is optimal in a pessimistic minimax sense. In the case of the example (1.3), it yields $x^* = \overline{\omega}$ which suggests a production level sufficiently high to meet every foreseeable demand. This may turn out to be a quite expensive solution in the long run!

1.2 Stochastic Optimization: Anticipative Models

The formulation of problem (1.2) as a stochastic optimization problem presupposes that in addition to the knowledge of Ω, one can rank the future alternative environments ω according to their comparative frequency of occurrence. In other words, it corresponds to the case when weights—an *a priori probability measure, objective* or *subjective*—can be assigned to all possible $\omega \in \Omega$, and this is done in a way that is consistent with the calculus rules for probabilities. Every possible environment ω becomes an element of a probability space, and the meaning to assign to feasibility and optimality in (1.2) can be arrived at by reasonings or statements of a probabilistic nature. Let us consider the here-and-now situation, when a solution must be chosen that does not depend on future observations of the environment. In terms of problem (1.2) it may be some $x \in X$ that satisfies the constraints

$$g_i(x, \omega) \leq 0, \quad i = 1, \ldots, m, \tag{1.2}$$

with a certain level of reliability:

$$\text{prob.}\{\omega | g_i(x, \omega) \leq 0, \quad i = 1, \ldots, m\} \geq \alpha \tag{1.5}$$

where $\alpha \in (0, 1)$, not excluding the possibility $\alpha = 1$, or in the average:

$$E\{g_i(x, \omega)\} \leq 0, \quad i = 1, \ldots, m. \tag{1.6}$$

There are many other possible probabilistic definitions of feasibility involving not only the mean but also the variance of the random variable $g_i(x, \cdot)$,

$$\text{Var } g_i(x, \cdot) := E[g_i(x, \omega) - E\{g_i(x, \omega)\}]^2,$$

such as

$$E\{g_i(x, \omega)\} + \beta(\text{Var } g_i(x, \cdot))^{\frac{1}{2}} \leq 0 \tag{1.7}$$

for β some positive constant, or even higher moments or other nonlinear functions of the $g_i(x, \cdot)$ may be involved. The same possibilities are available in definiting optimality. Optimality could be expressed in terms of the (feasible) x that minimizes

$$\text{prob.}\{\omega | g_0(x, \omega) \geq \alpha_0\} \tag{1.8}$$

for a prescribed level α_0, or the expected value of future cost

$$E\{g_0(x,\omega)\},\tag{1.9}$$

and so on.

Despite the wide variety of concrete formulations of stochastic optimization problems, generated by problems of the type (1.2) all of them may finally be reduced to the following rather general version given below, and for conceptual and theoretical purposes it is useful to study stochastic optimization problems in those general terms: Given a probability space (Ω, A, P), that gives us a description of the possible environments Ω and all possible events A with associated probability measure P, a *stochastic programming problem* is:

$$\text{find} \quad x \in X \subset R^n$$
$$\text{such that} \quad F_i(x) = E\{f_i(x,\omega)\}$$
$$= \int f_i(x,\omega) P(d\omega) \leq 0, \text{ for } i = 1,\ldots,m,$$
$$\text{and} \quad z = F_0(x) = E\{f_0(x,\omega)\}$$
$$= \int f_0(x,\omega) P(d\omega) \text{ is minimized,}$$

$$(1.10)$$

where X is a (usually closed) fixed subset of R^n, and the functions

$$f_i : R^n \times \Omega \to R, \quad i = 1,\ldots,m,$$

and

$$f_0 : R^n \times \Omega \to \overline{R} := R \cup \{-\infty, +\infty\},$$

are such that, at least for every x in X, the expectations that appear in (1.10) are well-defined.

For example, the constraints (1.5) that are called *probabilistic or chance constraints*, will be of the above type if we set:

$$f_i(x,\omega) = \begin{cases} \alpha - 1 & \text{if } g_\ell(x,\omega) \leq 0 \text{ for } \ell = 1,\ldots m, \\ \alpha & \text{otherwise} \end{cases}\tag{1.11}$$

The variance, which appears in (1.7) and other moments, are also mathematical expectations of some nonlinear functions of the $g_i(x,\cdot)$.

How one actually passes from (1.2) to (1.10) depends very much on the concrete situation at hand. For example, the criterion (1.8) and the constraints (1.5) are obtained if one classifies the possible outcomes

$$g_0(x,\omega), g_1(x,\omega),\ldots, g_m(x,\omega),$$

as ω varies on Ω, into "bad" and "good" (or acceptable and nonacceptable). To minimize (1.8) is equivalent to minimizing the probability of a "bad" event. The choice of the level α as it appears in (1.5), is a problem in itself, unless such a constraint is introduced to satisfy contractually specified reliability levels. The natural tendency is to choose the reliability level α as high as possible, but this may result in a rapid increase in the overall cost. Figure 1.2 illustrates a typical situation where increasing the reliability level beyond a certain level $\overline{\alpha}$ may result in enormous additional costs.

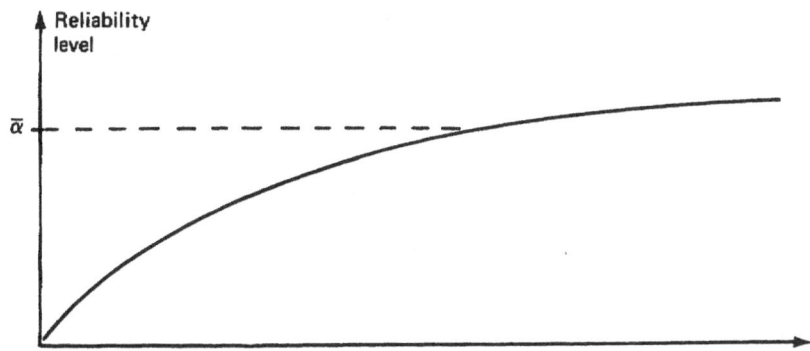

Figure 1.2 Reliability versus cost.

To analyze how high one should go in the setting of reliability levels, one should, ideally, introduce the loss that would be incurred if the constraints were violated, to be balanced against the value of the objective function. Suppose the objective function is of type (1.9), and in the simple case when violating the constraint $g_i(x,\omega) \leq 0$, it generates a cost:

$$q_i \cdot g_i(x,\omega), \qquad (q_i \geq 0)$$

proportional to the amount by which we violate the constraint, we are led to the objective function:

$$f_0(x,\omega) = g_0(x,\omega) + \sum_{i=1}^{m} q_i \left(\max[0, g_i(x,\omega)]\right), \qquad (1.12)$$

for the stochastic optimization problem (1.10). For the production (inventory) model with cost function given by (1.3), it would be natural to minimize the expected loss function

$$F_0(x) = \alpha \int_{\omega \leq x} (x - \omega) P(d\omega) + \beta \int_{x \leq \omega} (\omega - x) P(d\omega) = E\{g_0(x,\omega)\}$$

which we can also write as

$$F_0(x) = E\{\max[\alpha(x - \omega), \beta(\omega - x)]\}. \qquad (1.13)$$

A more general class of problems of this latter type comes with the objective function:

$$F_0(x) = E \max_{y \in Y} \rho(x, y, \omega) \qquad (1.14)$$

where $Y \subset R^P$. Such a problem can be viewed as a model for decision making under uncertainty, where the x are the decision variables themselves, the ω variables correspond to the states of nature with given probability measure P, and the y variables are there to take into account the worst case.

1.3   About Solution Procedures

In the design of solution procedures for stochastic optimization problems of type (1.10), one must come to grips with two major difficulties that are usually brushed aside in the design of solution procedures for the more conventional nonlinear optimization problems (1.1): in general, the exact evaluation of the functions F_i, $i = 1, \ldots, m$, (or of their gradients, etc.) is out of question, and moreover, these functions are quite often nondifferentiable. In principle, any nonlinear programming technique developed for solving problems of type (1.1) could be used for solving stochastic optimization problems. Problems of type (1.10) are after all just special case of (1.1), and this does also work well in practice if it is possible to obtain explicit expressions for the functions F_i, $i = 1, \ldots, m$, through the analytical evaluation of the corresponding integrals

$$F_i(x) = E\{f_i(x, \omega)\} = \int f_i(x, \omega) P(d\omega).$$

Unfortunately, the exact evaluation of these integrals, either analytically or numerically by relying on existing software for quadratures, is only possible in exceptional cases; for very special types of probability measures P and integrands $f_i(x, \cdot)$. For example, to calculate the values of the constraint function (1.5) even for $m = 1$, and

$$g_1(x, \omega) = h(\omega) - \sum_{j=1}^{n} t_j(\omega) x_j \tag{1.15}$$

with random parameters $h(\cdot)$ and $t_j(\cdot)$, it is necessary to find the probability of the event

$$\{\omega | \sum_{j=1}^{n} t_j(\omega) x_j \geq h(\omega)\}$$

as a function of $x = (x_1, \ldots, x_n)$. Finding an analytical expression for this function is only possible in a few rare cases, the distribution of the random variable

$$\omega \mapsto h(\omega) - \sum_{j=1}^{n} t_j(\omega) x_j$$

may depend dramatically on x; compare $x = (0, \ldots, 0)$ and $x = (1, \ldots, 1)$.

Of course, the exact evaluation of the functions F_i is certainly not possible if only partial information is available about P, or if information will only become available while the problem is being solved, as is the case in optimization systems in which the values of the outputs $\{f_i(x, \omega), i = 0, \ldots, m\}$ are obtained through actual measurements or Monte Carlo simulations.

In order to bypass some of the numerical difficulties encountered with multiples integrals in the stochastic optimization problem (1.10), one may be tempted to solve a substitute problem obtained from (1.2) by replacing the parameters by their expected values, i.e. in (1.10) we replace

$$E\{f_i(x,\omega)\} \text{ by } f_i(x,\overline{\omega}),$$

where $\overline{\omega} = E\{\omega\}$. This is relatively often done in practice, sometimes the optimal solution might only be slightly affected by such a crude approximation, but unfortunately, this supposedly harmless simplification, may suggest decisions that not only are far from being optimal, but may even "validate" a course of action that is contrary to the best interests of the decision maker. As a simple example of the errors that may derive from such a substitution let us consider:

$$f_0(x,\omega) = (\omega x)^2, x \in R, P[\omega = +1] = P[\omega = -1] = \tfrac{1}{2},$$

then

$$f_0(x,\overline{\omega}) \equiv 0, \text{ but } E\{f_0(x,\omega)\} = x^2.$$

Not having access to precise evaluation of the function values, or the gradients of the $F_i, i = 0, \ldots, m$, is the main obstacle to be overcome in the design of algorithmic procedures for stochastic optimization problems. Another peculiarity of this type of problems is that the functions

$$x \mapsto F_i(x), \quad i = 0, \ldots, m,$$

are quite often nondifferentiable—see for example (1.5), (1.7), (1.8), (1.13) and (1.14)—they may even be discontinuous as indicated by the simple example in Figure 1.3.

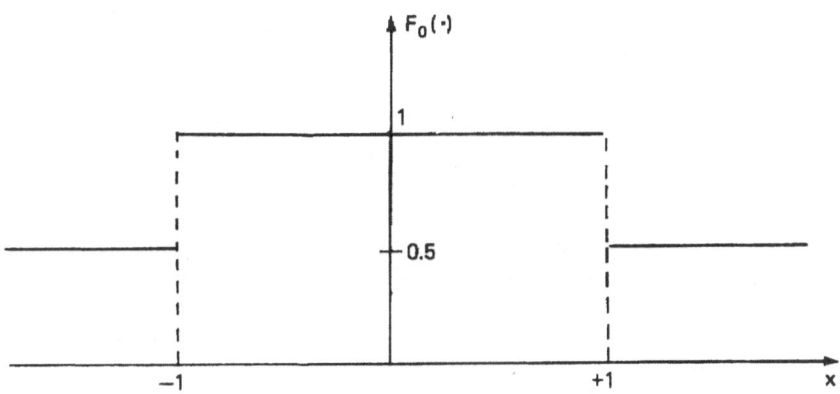

Figure 1.3 $F_0(x) = P\{\omega | \omega x \leq 1\}, P[\omega = +1] = P[\omega = -1] = \tfrac{1}{2}.$

The stochastic version of even the simplest linear problem may lead to a nondifferential problem as vividly demonstrated by Figure 1.3. It is now easy to imagine how complicated similar functions defined by linear inequalities in R^n might become. As another example of this type, let us consider a constraint of the type (1.2), i.e. a probabilistic constraint, where the $g_i(\cdot, \omega)$ are linear, and involve only one 1-dimensional random variable $h(\cdot)$. The set S of feasible solutions are those x that satisfy

$$P\{\omega | x + 3 \geq h(\omega), x \leq h(\omega)\} \geq \tfrac{2}{3},$$

where $h(\cdot)$ is equal to 0, 2, or 4 each with probability $\tfrac{1}{3}$. Then

$$S = [-1, 0] \cup [1, 2]$$

is disconnected.

The situation is not always that hopeless, in fact for well-formulated stochastic optimization problem, we may expect a lot of regularity, such as convexity of the feasibility region, convexity and/or Lipschitz properties of the objective function, and so on. This is well documented in the literature.

In the next two sections, we introduce some of the most important formulations of stochastic programming problems and show that for the development of conceptual algorithms, problem (1.10) may serve as a guide, in that the difficulties to be encountered in solving very specific problems are of the same nature as those one would have when dealing with the quite general model (1.10).

1.4 Stochastic Optimization: Adaptive Models

In the stochastic optimization model (1.10), the decision x has to be chosen by using an *a priori* probabilistic measure P without having the opportunity of making additional observations. As discussed already earlier, this corresponds to the idea of an optimization model as a tool for planning for possible future environments, that is why we used the term: anticipative optimization. Consider now the situation when we are allowed to make an observation before choosing x, this now corresponds to the idea of optimization in a learning environment, let us call it *adaptive optimization*.

Typically, observations will only give a partial description of the environment ω. Suppose B is a collection of sets that contains all the relevant information that could become available after making an observation; we think of B as a subset of A. The decision x must be determined on the basis of the information available in B, i.e. it must be a function of ω whose values are B dependent or equivalently is "B-measurable". The statement of the corresponding optimization is similar to (1.10), except that now we allow a larger class of solutions—the B-measurable functions—instead of just points in R^n (which in this setting would just correspond to the constant functions on Ω). The problem is to find a B-measurable function

$$\omega \mapsto x(\omega)$$

that satisfies: $x(\omega) \in X$ for all ω,

$$E\{f_i(x(\cdot),\cdot)|B\}(\omega) \le 0, \quad i = 1,\ldots,m,$$

and

$$z = E\{f_0(x(\omega),\omega)\} \text{ is minimized.} \tag{1.16}$$

where $E\{\cdot|B\}$ denotes the *conditional expectation* given B. Since x is to be a B-measurable function, the search for the optimal x, can be reduced to finding for each $\omega \in \Omega$ the solution of

$$\text{find} \quad x \in X \subset R^n$$
$$\text{such that} \quad E\{f_i(x,\cdot)|B\}(\omega) \le 0, \quad i = 1,\ldots,m \tag{1.17}$$
$$\text{and} \quad z_\omega = E\{f_0(x,\cdot)|B\}(\omega) \text{ is minimized.}$$

Each problem of this type has exactly the same features as problem (1.10) except that expectation has been replaced by conditional expectation; note that problem (1.16) will be the same for all ω that belong to the same elementary event of B. In the case when ω becomes completely known, i.e. when $B = A$, then the optimal $\omega \mapsto x(\omega)$ is obtained by solving for all ω, the optimization problem:

$$\text{find} \quad x \in X \subset R^n$$
$$\text{such that} \quad f_i(x,\omega) \le 0, \quad i = 1,\ldots,m, \tag{1.18}$$
$$\text{and} \quad z_\omega = f_0(x,\omega) \text{ is minimized,}$$

i.e. we need to make a parametric analysis of the optimal solution as a function of ω.

If the optimal decision rule $\omega \mapsto x^*(\omega)$ obtained by solving (1.16), is implementable in a real-life setting it may be important to know the distribution function of the optimal value

$$\omega \mapsto E\{f_0(x^*(\cdot),\cdot)|B\}(\omega)$$

This is known as the *distribution problem* for random mathematical programs which has received a lot of attention in the literature, in particularly in the case when the functions $f_i, i = 0,\ldots,m$, are linear and $B = A$; references can be found in Part V of this volume, consult the section on the distribution problem.

Unfortunately in general, the decision rule $x^*(\cdot)$ obtained by solving (1.17), and in particular (1.18), is much too complicate for practical use. For example, in our production model with uncertain demand, the resulting output may lead to highly irregular transportation requirements, etc. In inventory control, one has recourse to "simple", (s,S)-policies in order to avoid the possible chaotic behavior of more "optimal" procedures; an (s,S)-policy is one in which an order is placed as soon as the stock falls below a buffer level s and the quantity ordered will restore to a level S the stock available. In this case, we are restricted to a

specific family of decision rules, defined by two parameters s and S which have
to be defined before any observation is made.

More generally, we very often require the decision rules $\omega \mapsto x(\omega)$ to belong
to a prescribed family

$$\{x(\lambda, \cdot), \lambda \in \Lambda\}$$

of decision rules parametrized by a vector λ, and it is this λ that must be chosen
here-and-now before any observations are made. Assuming that the members
of this family are B-measurable, and substituting $x(\lambda, \cdot)$ in (1.16), we are led
to the following optimization problem

$$
\begin{aligned}
&\text{find} \quad \lambda \in \Lambda \\
&\text{such that} \quad x(\lambda, \omega) \in X \text{ for all } \omega \in \Omega \\
&\qquad H_i(\lambda) = E\{\phi_i(x(\lambda, \cdot), \cdot)\} \le 0, \quad i = 1, \ldots, m \\
&\text{and} \quad H_0(\lambda) = E\{f_0(x(\lambda, \omega), \omega)\} \text{ is minimized.}
\end{aligned}
\tag{1.19}
$$

This again is a problem of type (1.10), except that now the minimization is with
respect to λ. Therefore, by introducing the family of decision rules $\{x(\lambda, \cdot), \lambda \in \Lambda\}$ we have reduced the problem of adaptive optimization to a problem of
anticipatory optimization, no observations are made before fixing the values of
the parameters λ.

It should be noted that the family $\{x(\lambda, \cdot), \lambda \in \Lambda\}$ may be given implicitly.
To illustrate this let us consider a problem studied by Tintner. We start with
the linear programming problem (1.20), a version of (1.2):

$$
\begin{aligned}
&\text{find} \quad x \in R_+^n \\
&\text{such that} \quad \sum_{j=1}^n a_{ij}(\omega)x_j \ge b_i(\omega), \quad i = 1, \ldots, m \\
&\text{and} \quad z = \sum_{j=1}^n c_j(\omega)x_j \text{ is minimized,}
\end{aligned}
\tag{1.20}
$$

where the $a_{ij}(\cdot), b_i(\cdot)$ and $c_j(\cdot)$ are positive random variables. Consider the
family of decision rules: let λ_{ij} be the portion of the i-th resource to be assigned
to activity j, thus

$$\sum_{j=1}^n \lambda_{ij} = 1, \lambda_{ij} \ge 0 \text{ for } i = 1, \ldots, m; j = 1, \ldots, n,
\tag{1.21}$$

and for $j = 1, \ldots n$,

$$x_j(\lambda, \omega) \in \operatorname*{argmin}_{x \in R_+}\{c_j(\omega)x \,|\, a_{ij}(\omega)x \ge \lambda_{ij}b_i(\omega), i = 1, \ldots, m\}$$

i.e.

$$x_j(\lambda, \omega) = \max_{1 \le i \le m} \lambda_{ij}b_i(\omega)/a_{ij}(\omega).$$

This decision rule is only as good as the λ_{ij} that determine it. The optimal λ's are found by minimizing

$$\sum_{j=1}^{n} E\{c_j(\omega) \max_{1 \le i \le m} (\lambda_{ij} b_j(\omega)/a_{ij}(\omega))\}$$

subject to (1.21), again a problem of type (1.10).

1.5 Anticipation and Adaptation: Recourse Models

The (two-stage) recourse problem can be viewed as an attempt to incorporate both fundamental mechanisms of anticipation and adaptation within a single mathematical model. In other words, this model reflects a trade-off between long-term anticipatory strategies and the associated short-term adaptive adjustments. For example, there might be a trade-off between a road investment's program and the running costs for the transportation fleet, investments in facilities location and the profit from its day-to-day operation. The linear version of the *recourse problem* is formulated as follows:

$$\begin{aligned}
\text{find} \quad & x \in R_+^n \\
\text{such that} \quad & F_i(x) = b_i - A_i x \le 0, \quad i = 1, \ldots, m, \\
\text{and} \quad & F_0(x) = cx + E\{Q(x,\omega)\} \text{ is minimized}
\end{aligned} \tag{1.22}$$

where

$$Q(x,\omega) = \inf_{y \in R_+^{n'}} \{q(\omega)y \mid W(\omega)y = h(\omega) - T(\omega)x\}; \tag{1.23}$$

some or all of the coefficients of matrices and vectors $q(\cdot), W(\cdot), h(\cdot)$ and $T(\cdot)$ may be random variables. In this problem, the long-term decision is made before any observation of $\omega \sim (q(\omega), W(\omega), h(\omega), T(\omega))$. After the true environment is observed, the discrepancies that may exist between $h(\omega)$ and $T(\omega)x$ (for fixed x and observed $h(\omega)$ and $T(\omega)$) are corrected by choosing a recourse action y, so that

$$W(\omega)y = h(\omega) - T(\omega)x, \quad y \ge 0, \tag{1.24}$$

that minimizes the loss

$$q(\omega)y.$$

Therefore, an optimal decision x should minimize the total cost of carrying out the overall plan: direct costs as well as the costs generated by the need of taking correct (adaptive) action.

A more general model is formulated as follows. A long-term decision x must be made before the observation of ω is available. For given $x \in X$ and observed ω, the recourse (feedback) action $y(x,\omega)$ is chosen so as to solve the problem

$$\begin{aligned}
\text{find} \quad & y \in Y \subset R^{n'} \\
\text{such that} \quad & f_{2i}(x, y, \omega) \le 0, \quad i = 1, \ldots, m', \\
\text{and} \quad & z_2 = f_{20}(x, y, \omega) \text{ is minimized,}
\end{aligned} \tag{1.25}$$

assuming that for each $x \in X$ and $\omega \in \Omega$ the set of feasible solutions of this problem is nonempty (in technical terms, this is known as relatively complete recourse). Then to find the optimal x, one would solve a problem of the type:

$$\begin{aligned} &\text{find} \quad x \in X \subset R^n, \\ &\text{such that} \quad F_0(x) = E\{f_{20}(x, y(x,\omega), \omega)\} \text{ is minimized.} \end{aligned} \tag{1.26}$$

If the state of the environment ω remains unknown or partially unknown after observation, then

$$\omega \mapsto y(x, \omega)$$

is defined as the solution of an adaptive model of the type discussed in Section 1.4. Given B the field of possible observations, the problem to be solved for finding $y(x, \omega)$ becomes: for each $\omega \in \Omega$

$$\begin{aligned} &\text{find} \quad y \in Y \subset R^{n'} \\ &\text{such that} \quad E\{f_{2i}(x,y,\cdot)|B\}(\omega) \le 0, \quad i = 1, \ldots, m' \\ &\text{and} \quad z_{2\omega} = E\{f_{20}(x,y,\cdot)|B\}(\omega) \text{ is minimized.} \end{aligned} \tag{1.27}$$

If $\omega \mapsto y(x, \omega)$ yields the optimal solution of this collection of problems, then to find an optimal x we again have to solve a problem of type (1.26).

Let us notice that if

$$f_{20}(x, y, \omega) = cx + q(\omega)y$$

and for $i = 1, \ldots, m'$,

$$f_{2i}(x, y, \omega) = \begin{cases} 1 - \alpha & \text{if } T_i(\omega)x + W_i(\omega)y - h_i(\omega) \ge 0, \\ \alpha & \text{otherwise} \end{cases}$$

then (1.26), with the second stage problem as defined by (1.27), corresponds to the statement of the recourse problem in terms of conditional probabilistic (chance) constraints.

There are many variants of the basic recourse models (1.22) and (1.26). There may be in addition to the deterministic constraints on x some expectation constraints such as (1.7), or the recourse decision rule may be subject to various restrictions such as discussed in Section 1.4, etc. In any case as is clear from the formulation, these problems are of the general type (1.10), albeit with a rather complicated function $f_0(x, \omega)$.

1.6 Dynamic Aspects: Multistage Recourse Problems

It should be emphasized that the "stages" of a two-stage recourse problem do not necessarily refer to time units. They correspond to steps in the decision process, x may be a here-and-now decision whereas the y correspond to *all* future actions to be taken in different time period in response to the environment created by the chosen x and the observed ω in that specific time period. In another instance, the x, y solutions may represent sequences of control actions over a given time horizon,

$$x = (x(0), x(1), \ldots, x(T)),$$
$$y = (y(0), y(1), \ldots, y(T)),$$

the y-decisions being used to correct for the basic trend set by the x-control variables. As a special case we have

$$x = (x(0), x(1), \ldots, x(s)),$$
$$y = (y(s+1), \ldots, y(T)),$$

that corresponds to a mid-course maneuver at time s when some observations have become available to the controller. We speak of *two-stage dynamic* models. In what follows, we discuss in more detail the possible statements of such problems.

In the case of dynamical systems, in addition to the x, y solutions of problems (1.26)-(1.25), there may also be an additional group of variables

$$z = (z(0), z(1), \ldots, z(T))$$

that record the *state of the system* at times $0, 1, \ldots, T$. Usually, the variables x, y, z, ω are connected through a (differential) system of equations of the type:

$$\Delta z(t) = h(t, z(t), x(t), y(t), \omega), \quad t = 0, \ldots, T-1, \tag{1.28}$$

where

$$\Delta z(t) = z(t+1) - z(t), z(0) = z_0,$$

or they are related by an implicit function of the type:

$$h(t, z(t+1), z(t), x(t), y(t), \omega) = 0, \quad t = 0, \ldots, T-1. \tag{1.29}$$

The latter one of these is the typical form one finds in operations research models, economics and system analysis, the first one (1.28) is the conventional one in the theory of optimal control and its applications in engineering, inventory control, etc. In the formulation (1.28) an additional computational problem arises from the fact that it is necessary to solve a large system of linear or nonlinear equations, in order to obtain a description of the evolution of the system.

The objective and constraints functions of stochastic dynamic problems are generally expressed in terms of mathematical expectations of functions that we take to be:

$$g_i\big(z(0), x(0), y(0), \ldots, z(T), x(T), y(T)\big), \quad i = 0, 1, \ldots, m. \tag{1.30}$$

If no observations are allowed, then equations (1.28), or (1.29), and (1.30) do not depend on y, and we have the following one-stage problem

$$
\begin{aligned}
\text{find} \quad & x = (x(0), x(1), \ldots, x(T)) \\
\text{such that} \quad & x(t) \in X(t) \subset R^n, \quad t = 0, \ldots, T, \\
& \Delta z(t) = h\big(t, z(t), x(t), \omega\big), \quad t = 0, \ldots, T-1, \\
& E\big(g_i\big(z(0), x(0), \ldots, z(T), x(T), \omega\big) \le 0, \quad i = 1, \ldots, m \\
\text{and} \quad & v = E\{g_0\big(z(0), x(0), \ldots, z(T), x(T), \omega\big)\} \text{ is minimized}
\end{aligned}
\tag{1.31}
$$

or with the dynamics given by (1.29). Since in (1.28) or (1.29), the variables $z(t)$ are functions of (x, ω), the functions g_i are also implicit functions of (x, ω), i.e. we can rewrite problem (1.31) in terms of functions

$$f_i(x, \omega) = g_i\big(z(x, \omega), x, \omega\big),$$

the stochastic dynamic problem (1.31) is then reduced to a stochastic optimization problem of type (1.10). The implicit form of the objective and the constraints of this problem requires a special calculus for evaluating these functions and their derivatives, but it does not alter the general solution strategies for stochastic programming problems.

The two-stage recourse model allows for a recourse decision y that is based on (the first stage decision x and) the result of observations. The following simple example should be useful in the development of a dynamical version of that model. Suppose we are interested in the design of an optimal trajectory to be followed, in the future, by a number of systems that have a variety of (dynamical) characteristics. For instance, we are interested in building a road between two fixed points (see Figure 1.4) at minimum total cost taking into account, however, certain safety requirements. To compute the total cost we take into account not just the construction costs, but also the cost of running the vehicles on this road.

For a fixed feasible trajectory

$$z = (z(0), z(1), \ldots, z(T)),$$

and a (dynamical) system whose characteristics are identified by a parameter $\omega \in \Omega$, the dynamics are given by the equations, for $t = 0, \ldots, T-1$, and $\Delta z(t) = z(t+1) - z(t)$,

$$\Delta z(t) = h\big(t, z(t), y(t), \omega\big), \tag{1.32}$$

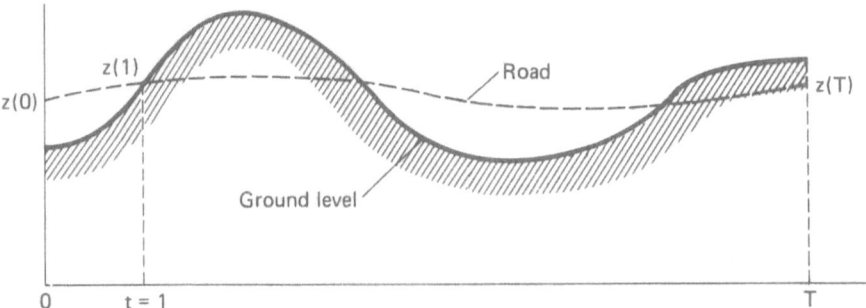

Figure 1.4   Road design problem.

and

$$z(0) = z_0, z(T) = z_T.$$

Here the variable t records position (between 0 and T). The variables

$$y = (y(0), y(1), \ldots, y(T))$$

are the control variables at $t = 0, 1, \ldots, T$ that determine the way a dynamical system of type ω will be controlled when following the trajectory z from 0 to T. The choice of the z-trajectory is subject to certain restrictions, that include safety considerations, such as

$$|\Delta z(t)| \leq d_1, |\Delta z(t) - \Delta z(t-1)| \leq d_2, \tag{1.33}$$

i.e. the first two derivatives cannot exceed certain prescribed levels.

For a specific system $\omega \in \Omega$, and a fixed trajectory z, the optimal control actions (recourse)

$$y(z, \omega) = (y(0, z, \omega), y(1, z, \omega), \ldots, y(T, z, \omega))$$

is determined by minimizing the loss function

$$g_0(z(0), y(0), \ldots, z(T-1), y(T-1), z(T), \omega)$$

subject to the system's equations (1.32) and possibly some constraints on y. If P is the *a priori* distribution of the systems parameters, the problem is to find a trajectory (road design) z that minimizes in the average the loss function, i.e.

$$F_0(z) = E\{g_0(z(0), y(0, z, \omega), \ldots, z(T-1), y(T-1, z, \omega), z(T), \omega)\} \tag{1.34}$$

subject to constraints of type (1.33).

In this problem the observation takes place in one step only. We have amalgamated all future observations that will actually occur at different time periods in a single collection of possible environments (events). There are situations when ω has the structure

$$\omega = (\omega(0), \omega(1), \ldots, \omega(T))$$

and the observations take place in T steps. As an important example of such a class, let us consider the following problem: the long term decision $x = (x(0), x(1), \ldots, x(T))$ and the corrective recourse actions $y = (y(0), y(1), \ldots, y(T))$ must satisfy the linear system of equations:

$$
\begin{array}{llll}
A_{00}x(0) & + & B_0 y(0) & \geq h(0) \\
A_{10}x(0) & + & A_{11}x(1) \quad\quad +B_1 y(1) & \geq h(1) \\
\cdots & & \cdots \quad\quad\quad \cdots & \\
A_{T0}x(0) & + & A_{T1}x(1) \quad +\cdots+ A_{TT}x(T) + B_T y(T) & \geq h(T), \\
x(0) \geq 0, & \cdots & , x(T) \geq 0; \quad y(0) \geq 0, \ldots, y(T) \geq 0 &
\end{array}
$$

where the matrices A_{tk}, B_t and the vectors $h(t)$ are random, i.e. depend on ω. The sequence $x = (x(0), \ldots, x(T))$ must be chosen before any information about the values of the random coefficients can be collected. At time $t = 0, \ldots, T$, the actual values of the matrices, and vectors,

$$A_{tk}, k = 0, \ldots, t; B_t, h(t), d(t)$$

are revealed, and we adapt to the existing situation by choosing a corrective action $y(t, x, \omega)$ such that

$$y(t, x, \omega) \in \operatorname{argmin}\Big[d(t)y \big| B_t y \geq h(t) - \sum_{k=0}^{t} A_{tk}x(k), y \geq 0\Big].$$

The problem is to find $x = (x(0), \ldots, x(T))$ that minimizes

$$F_0(x) = \sum_{t=0}^{T}[c(t)x(t) + E\{d(t)y(t, x, \omega)\}] \tag{1.35}$$

subject to $x(0) \geq 0, \ldots, x(T) \geq 0$.

In the functional (1.35), or (1.34), the dependence of $y(t, x, \omega)$ on x is nonlinear, thus these functions do not possess the separability properties necessary to allow direct use of the conventional recursive equations of dynamic programming. For problem (1.31), these equations can be derived, provided the functions $g_i, i = 0, \ldots, m$, have certain specific properties. There are, however, two major obstacles to the use of such recursive equations in the stochastic case: the tremendous increase of the dimensionality, and again, the more serious problem created by the need of computing mathematical expectations.

For example, consider the dynamic system described by the system of equations (1.28). Let us ignore all constraints except $x(t) \in X(t)$, for $t = 0, 1, \ldots, T$. Suppose also that

$$\omega = (\omega(0), \omega(1), \ldots, \omega(T))$$

where $\omega(t)$ only depends on the past, i.e. is independent of $\omega(t+1), \ldots, \omega(T)$. Since the minimization of

$$F_0(x) = E\{g_0(z(0), x(0), \ldots, z(T), x(T), \omega)\}$$

with respect to x can then be written as:

$$\min_{x(0)} \min_{x(1)} \ldots \min_{x(T)} E\{g_0\}$$

and if g_0 is separable, i.e. can be expressed as

$$g_0 := \sum_{t=0}^{T-1} g_{0t}(\Delta z(t), x(t), \omega(t)) + g_{0T}(z(t), \omega(T))$$

then

$$\min_x F_0(x) = \min_{x(0)} E\{g_{00}(\Delta z(0), x(0), \omega(0))\} + \min_{x(1)} E\{g_{01}(\Delta z(1), x(1), \omega(1))\}$$
$$+ \cdots + \min_{x(T-1)} E\{g_0, T-1(\Delta z(T-1), x(T-1), \omega(T-1))\} +$$
$$+ E\{g_{0T}(z(T), \omega(T))\}$$

Recall that here, notwithstanding its sequential structure, the vector ω is to be revealed in one global observation. Rewriting this in backward recursive form yields the Bellman equations:

$$v_t(z_t) = \min[E\{g_{0t}(h(t, z_t, x, \omega(t)), x, \omega(t)) \\ + v_{t+1}(z_t + h(t, z_t, x, \omega(t)))\} | x \in X(t)] \tag{1.36}$$

for $t = 0, \ldots, T-1$, and

$$v_T(z_T) = E\{g_{0T}(z_T, \omega(T))\}, \tag{1.37}$$

where v_t is the value function (optimal loss-to-go) from time t on, given state z_t at time t, that in turn depends on $x(0), x(1), \ldots, x(t-1)$.

To be able to utilize this recursion, reducing ultimately the problem to:

find $x \in X(0) \subset R^n$ such that v_0 is minimized, where
$$v_0 = E\{g_{00}(h(0, z_0, x, \omega(0)), x, \omega(0)) + v_1(z_0 + h(0, z_0, x, \omega(0)))\},$$

we must be able to compute the mathematical expectations

$$E\{g_{0t}(\Delta z(t), x, \omega(t))\}$$

as a function of the intermediate solutions $x(0), \ldots, x(t-1)$, that determine $\Delta z(t)$, and this is only possible in special cases. The main goal in the development of solution procedures for stochastic programming problems is the development of appropriate computational tools that precisely overcome such difficulties.

A much more difficult situation may occur in the (full) multistage version of the recourse model where observation of some of the environment takes place at each stage of the decision process, at which time (taking into account the new information collected) a new recourse action is taken. The whole process looks like a sequence of alternating: decision-observation-...-observation-decision.

Let x be the decision at stage $k = 0$, which may itself be split into a sequence $x(0), \ldots, x(N)$, each $x(k)$ corresponding to that component of x that enters into play at stage k, similar to the dynamical version of the two-stage model introduced earlier. Consider now a sequence

$$y = (y(0), y(1), \ldots, y(N))$$

of recourse decisions (adaptive actions, corrections), $y(k)$ being associated specifically to stage k. Let

$$B_k := \text{information set at stage } k,$$

consisting of past measurements and observations, thus $B_k \subset B_{k+1}$.

The *multistage recourse problem* is

$$
\begin{aligned}
\text{find} \quad & x \in X \subset R^n \\
\text{such that} \quad & f_{0i}(x) \leq 0, \quad i = 1, \ldots, m_0, \\
& E\{f_{1i}(x, y(1), \omega)|B_1\} \leq 0, \quad i = 1, \ldots, m_1, \\
& \cdots \qquad \cdots \\
& E\{f_{Ni}(x, y(1), \ldots, y(N), \omega)|B_N\} \leq 0, \quad i = 1, \ldots, m_N, \\
& y(k) \in Y(k), \quad k = 1, \ldots, N, \\
\text{and} \quad & F_0(x) \text{ is minimized}
\end{aligned}
\tag{1.38}
$$

where

$$F_0(x) = E^{B_0}\{\min_{y(1)} E^{B_1}\{\ldots \min_{y(N-1)} E^{B_{N-1}}\{f(x, y(1), \ldots, y(N), \omega)\}.\}\}$$

If the decision x affects only the initial stage $k = 0$, we can obtain recursive equations similar to (1.36) - (1.37) except that expectation E must be replaced by the conditional expectations E^{B_t}, which in no way simplifies the numerical problem of finding a solution. In the more general case when $x = (x(0), x(1), \ldots, x(N))$, one can still write down recursion formulas but of such (numerical) complexity that all hope of solving this class of problems by means of these formulas must quickly be abandoned.

1.7   Solving the Deterministic Equivalent Problem

All of the preceding discussion has suggested that the problem:

$$
\begin{aligned}
&\text{find} \quad x \in R^n \\
&\text{such that} \quad F_i(x) = \int f_i(x,\omega)P(d\omega) \leq 0, \quad i = 1,\ldots,m, \\
&\text{and} \quad z = F_0(x) = \int f_0(x,\omega)P(d\omega) \text{ is minimized,}
\end{aligned}
\tag{1.39}
$$

exhibits all the peculiarities of stochastic programs, and that for exploring computational schemes, at least at the conceptual level, it can be used as the canonical problem.

Sometimes it is possible to find explicit analytical expressions for an acceptable approximation of the F_i. The randomness in problem (1.39) disappears and we can rely on conventional deterministic optimization methods for solving (1.39). Of course, such cases are highly cherished, and can be dealt with by relying on standard nonlinear programming techniques.

One extreme case is when $\overline{\omega} = E\{\omega\}$ is a *certainty equivalent* for the stochastic optimization problem, i.e. the solution to (1.39) can be found by solving:

$$
\begin{aligned}
&\text{find} \quad x \in X \subset R^n \\
&\text{such that} \quad f_i(x,\overline{\omega}) \leq 0, \quad i = 1,\ldots,m, \\
&\text{and} \quad z = f_0(x,\overline{\omega}) \text{ is minimized,}
\end{aligned}
\tag{1.40}
$$

this would be the case if the $f_i, i = 0,\ldots,m$ are linear functions of ω. In general, as already mentioned in Section 1.3, the solution of (1.40) may have little in common with the initial problem (1.39). But if the f_i are convex functions, then according to Jensen's inequality

$$
E\{f_i(x,\omega)\} \geq f_i(x,\overline{\omega}), \quad i = 1,\ldots,m,
$$

This means that the set of feasible solutions in (1.40) is larger than in (1.39) and hence the solution of (1.40) could provide a lower bound for the solution of the original problem.

Another case is a stochastic optimization problem with simple probabilistic constraints. Suppose the constraints of (1.39) are of the type

$$
P\Big\{\omega \Big| \sum_{j=1}^{n} t_{ij}x_j > h_i(\omega)\Big\} \geq \alpha_i, \quad i = 1,\ldots,m,
\tag{1.41}
$$

with deterministic coefficients t_{ij} and random right-hand sides $h_i(\cdot)$. Then these constraints are equivalent to the linear system

$$
\sum_{j=1}^{n} t_{ij}x_j \geq h_i^*, \quad i = 1,\ldots,m,
$$

where
$$h_t^* = \inf\{t \mid P[\omega \mid h_t(\omega) < t] \geq \alpha_i\}$$
I' all the parameters t_{ij} and h_i in (1.41) are jointly normally distributed (and $c_i \geq .5$), then the constraints

$$x_0 = 1$$

$$\sum_{j=0}^{n} \bar{t}_{ij} x_j + \vartheta \left(\sum_{j=0}^{n} \sum_{k=0}^{n} \tau_{ijk} x_j x_k \right)^{\frac{1}{2}} \leq 0 \qquad i = 1, \ldots, m,$$

can be substituted for (1.41), where

$$t_{i0}(\cdot) = -h_i(\cdot)$$
$$\bar{t}_{ij} := E\{t_{ij}(\omega)\}, \quad j = 0, 1, \ldots, n,$$
$$\tau_{ijk} := \mathrm{cov}(t_{ij}(\cdot), t_{ik}(\cdot)), \quad j = 0, \ldots, n; k = 0, \ldots, n,$$

and β is a coefficient that identifies the α-fractile of the normalized normal distribution.

Another important class are those problems classified as stochastic programs with simple recourse (see Chapter 4), or more generally recourse problems where the random coefficients have a discrete distribution with a relatively small number of density points (support points), as discussed in Chapter 3. For the linear model (1.22) introduced in Section 1.5, where

$$\Omega = \{(q^1, W^1, h^1, T^1), \ldots, (q^N, W^N, h^N, T^N)\}$$

where for $k = 1, \ldots, N$, the point (q^k, W^k, h^k, T^k) is assigned probability p_k, one can find the solution of (1.22) by solving:

$$\text{find} \quad x \in R_+^n, (y^k \in R_-^{n'}, k = 1, \ldots, N)$$

such that

$$
\begin{array}{llllll}
Ax & & & & \geq & b, \\
T^1 x + W^1 y^1 & & & & = & h^1, \\
T^2 x & +W^2 y^2 & & & = & h^2, \qquad (1.42) \\
\vdots & & \ddots & & & \vdots \\
T^N x & & & +W^n y^n & = & h^N, \\
cx + p_1 q^1 y^1 & +p_2 q^2 y^2 & \cdots & +p_N q^N y^N & = & z,
\end{array}
$$

and z is minimized.

This problem has a (dual) block-angular structure. It should be noticed that the number N could be astronomically large, if only the vector h is random and each component of the vector

$$h = (h_1, h_2, \ldots, h_{m'})$$

has two independent outcomes, then $N = 2^{m'}$. A direct attempt at solving (1.42) by conventional linear programming techniques will only yield at each iteration very small progress in the terms of the x variables. Therefore, a special large scale optimization technique is needed for solving even this relatively simple stochastic programming problem.

1.8 Approximation Schemes

If a problem is too difficult to solve one may have to learn to live with approximate solutions. The question however, is to be able to recognize an approximate solution if one is around, and also to be able to assess how far away from an optimal solution one still might be. For this one needs a convergence theory complemented by (easily computable) error bounds, improvement schemes, etc. This is an area of very active research in stochastic optimization, both at the theoretical and the software-implementation level. These questions are studied in much more detail in Chapter 2, here we only want to highlight some of the questions that need to be raised and the main strategies available in the design of approximation schemes.

For purposes of discussion it will be useful to consider a simplified version of (1.39):

$$\text{find} \quad x \in X \subset R^n$$
$$\text{that minimizes} \quad F_0(x) = \int f_0(x,\omega)P(d\omega), \tag{1.43}$$

we suppose that the other constraints have been incorporated in the definition of the set X. We deal with a problem involving one expectation functional. Whatever applies to this case also applies to the more general situation (1.39), making the appropriate adjustments to take into account the fact that the functions

$$F_i(x) = \int f_i(x,\omega)P(d\omega), \qquad i = 1,\ldots,m,$$

determine constraints.

Given a problem of type (1.43) that does not fall in one of the nice categories mentioned in Section 1.7, one solution strategy may be to replace it by an approximation*. There are two possibilities to simplify the integration that appears in the objective function, replace f_0 by an integrand f_0^v or to replace P by an approximation P_v, and of course, one could approximate both quantities at once.

The possibility of finding an acceptable approximate of f_0 that renders the calculation of

$$\int f_0^v(x,\omega)P(d\omega) =: F_0^v(x),$$

sufficiently simple so that it can be carried out analytically or numerically at low-cost, is very much problem dependent. Typically one should search for a separable function of the type

$$f_0^v(x,\omega) = \sum_{j=1}^q \varphi_j(x,\omega_j),$$

---

* Another approach will be discussed in Section 1.9.

recall that $\Omega \subset R^q$, so that

$$F_0^\nu(x) = \sum_{j=1}^q \int \varphi_j(x, \omega_j) P(d\omega) = \sum_{j=1}^q \int \varphi_j(x, \omega_j) P_j(d\omega_j)$$

where the P_j are the marginal measures associated to the j-th component of ω. The multiple integral is then approximated by the sum of 1-dimensional integrals for which a well-developed calculus is available, (as well as excellent quadrature subroutines). Let us observe that we do not necessarily have to find approximates that lead to 1-dimensional integrals, it would be acceptable to end up with 2-dimensional integrals, even in some cases—when P is of certain specific types—with 3-dimensional integrals. In any case, this would mean that the structure of f_0 is such that the interactions between the various components of ω play only a very limited role in determining the cost associated to a pair (x, ω). Otherwise an approximation of this type could very well throw us very far off base. We shall not pursue this question any further since they are best handled on a problem by problem basis. If $\{f_0^\nu u, \nu = 1, \ldots\}$ is a sequence of such functions converging, in some sense, to f, we would want to know if the solutions of

$$x^\nu \in \arg\min F^\nu = \int f_0^\nu(\cdot, \omega) P(d\omega), \qquad \nu = 1, \ldots$$

converge to the optimal solution of (1.43) and if so, at what rate. These questions would be handled very much in the same way as when approximating the probability measure as will be discussed next.

Finding valid approximates for f_0 is only possible in a limited number of cases while approximating P is always possible in the following sense. Suppose P_ν is a probability measure (that approximates P), then

$$|F_0^\nu(x) - F_0(x)| \le \int |f_0(x, \omega)| \, |P_\nu - P|(d\omega), \qquad (1.44)$$

where now

$$F_0^\nu(x) := \int f_0(x, \omega) P_\nu(d\omega).$$

Thus if f_0 has Lipschitz properties, for example, then by choosing P_ν sufficiently close to P we can guarantee a maximal error bound when replacing (1.43) by:

$$\text{find} \quad x \in X \subset R^n$$
$$\text{that minimizes} \quad F_0^\nu(x) = \int f_0(x, \omega) P_\nu(d\omega). \qquad (1.45)$$

Since it is the multidimensional integration with respect to P that was the source of the main difficulties, the natural choice—although in a few concrete cases there are other possibilities—for P_ν is a discrete distribution that assigns to a finite number of points

$$\omega^1, \omega^2, \ldots, \omega^L$$

the probabilities

$$p_1, p_2, \ldots, p_L;$$

Problem (1.45) then becomes:

$$\text{find} \quad x \in X \subset R^n$$

$$\text{that minimizes} \quad F_0^\nu(x) = \sum_{\ell=1}^{L} p_\ell f_0(x, \omega^\ell) \tag{1.46}$$

At first glance it may now appear that the optimization problem can be solved by any standard nonlinear programming, the sum $\sum_{\ell=1}^{L}$ involving only a "finite" number of terms, the only question being how "approximate" is the solution of (1.46). However, if inequality (1.44) is used to design this approximation, to obtain a relatively sharp bound from (1.44), the number L of discrete points required may be so large that problem (1.46) is in no way any easier than our original problem (1.43). To fix the ideas, if $\Omega \subset R^{10}$, and P is a continuous distribution, a good approximation—as guaranteed by (1.44)—may require having $10^{10} \le L \le 10^{11}$! This is jumping from the fire into the frying pan.

This clearly indicates a need for more sophisticated approximation schemes. As background, we have the following convergence results. Suppose $\{P_\nu, \nu = 1, \ldots\}$ is a sequence of probability measures that converge in distribution to P, and suppose that for all $x \in X$, the function $f_0(x, \omega)$ is uniformly integrable with respect to all P_ν, and suppose there exists a bounded set D such that

$$D \cap \operatorname{argmin}\left[F_0^\nu(x) = \int f_0(x, \omega) P_\nu(d\omega) \,|\, x \in X \right] \neq \emptyset$$

for almost all ν, then

$$\inf_X F_0 = \lim_{\nu \to \infty} \left(\inf_X F_0^\nu \right)$$

and

$$if \, x^\nu \in \operatorname*{argmin}_X F_0^\nu, x = \lim_{k \to \infty} x^{\nu_k}$$

then

$$x \in \operatorname*{argmin}_X F_0.$$

The convergence result indicates that we are given a wide latitude in the choice of the approximating measures, the only real concern is to guarantee the convergence in distribution of the P_ν to P, the uniform integrability condition being from a practical viewpoint a pure technicality.

However, such a result does not provide us with error bounds, but since we can choose the P_ν in such a wide variety of ways, we could for example have P_ν such that

$$\inf_X F_0^\nu \leq \inf_X F_0 \tag{1.47}$$

and $P_{\nu+1}$ such that

$$\inf_X F_0 \leq \inf_X F_0^{\nu+1} \tag{1.48}$$

providing us with upper and lower bounds for the infimum and consequently error bounds for the approximate solutions:

$$x^\nu \in \operatorname*{argmin}_X F_0^\nu, \text{ and } x^{\nu+1} \in \operatorname*{argmin}_X F_0^{\nu+1}.$$

This, combined with a sequential procedure for redesigning the approximations P_ν so as to improve the error bounds, is very attractive from a computational viewpoint since we may be able to get away with discrete measures that involve only a relatively small number of points (and this seems to be confirmed by computational experience).

The only question now is how to find these measures that guarantee (1.47) and (1.48). There are basically two approaches: the first one exploits the properties of the function $\omega \mapsto f_0(x,\omega)$ so as to obtain inequalities when taking expectations, and the second one chooses P_ν in a class of probability measures that have characteristics similar to P but so that P_ν dominates or is dominated by P and consequently yields the desired inequality (1.47) or (1.48). A typical example of this latter case is to choose P_ν so that it majorizes or is majorized by P, another one is to choose P_ν so that for at least for some $\hat{x} \in X$:

$$P_\nu \in \operatorname{argmax}\left[\int f_0(\hat{x},\omega)Q(d\omega) | Q \in D\right] \tag{1.49}$$

where D is a class of probability measures on Ω that contains P, for example

$$D = \{Q| \int \omega Q(d\omega) = E\{\omega\}\}.$$

Then

$$F_0^\nu(\hat{x}) \geq F_0(x) \geq \inf_X F_0$$

yields an upper bound. If instead of P_ν in the argmax we take P_ν in the argmin we obtain a lower bound.

If $\omega \mapsto f_0(x,\omega)$ is convex (concave) or at least locally convex (locally concave) in the area of interest we may be able to use Jensen's inequality to construct probability measures that yield lower (upper) approximates for F_0 and probability measures concentrated on extreme points to obtain upper (lower) approximates of F_0. We have already seen such an example in Section 1.7 in connection with problem (1.40) where P is replaced by P_ν that concentrate all the probability mass on $\overline{\omega} = E\{\omega\}$.

Given an approximate measure P_ν, we also need a scheme to refine it so that the error bounds can be improved, if necessary. One cannot hope to have a unive·sal scheme since so much will depend on the problem at hand as well as the discretizations that have been used to build the upper and lower bounding problems. There is, however, one general rule that seems to work well, in fact surprisingly well, in practice: choose the region of refinement of the discretization in such a way as to capture as much of the nonlinearity of $f_0(x,\cdot)$ as possible.

It is, of course, not necessary to wait until the optimal solution of an approximate problem has been reached to refine the discretization of the probability measure. Conceivably, and ideally, the iterations of the solutions procedure should be intermixed with the sequential procedure for refining the approximations. Common sense dictates that as we approach the optimal solution we should seek better and better estimates of the function values and its gradients. How many iterations should one perform before a refinement of the approximation is introduced, or which tell-tale sign should trigger a further refinement, are questions that have only been scantily investigated, but are ripe for study at least for certain specific classes of stochastic optimization problems.

As to the rate of convergence this is a totally open question, in general and in particular, except on an experimental basis where the results have been much better than what could be expected from the theory. One open challenge is to develop the theory that validates the convergence behavior observed in practice.

1.9 Stochastic Procedures

Let us again consider the general formulation (1.10) for stochastic programs:

$$\text{find}\quad x \in X \subset R^n$$
$$\text{such that}\quad F_i(x) = \int f_i(x,\omega)P(d\omega) \leq 0, \qquad i = 1,\dots,m, \tag{1.50}$$
$$\text{and}\quad F_0(x) = \int f_0(x,\omega)P(d\omega)\ \text{is minimized.}$$

We already know from the discussion in Sections 1.3 and 1.7 that the exact evaluation of the integrals is only possible in exceptional cases, for special types of probability measures P and integrands f_i. The rule in practice is that it is only possible to calculate random observations $f_i(x,\omega)$ of $F_i(x)$. Therefore in the design of universal solution procedures we should rely on no more than the random observations $f_i(x,\omega)$. Under these premises, finding the solution of (1.50) is a difficult problem at the border between mathematical statistics and optimization theory. For instance, even the calculation of the values $F_i(\bar{x})$, $i = 0,\dots,m$, for a fixed $\bar{x}$ requires statistical estimation procedures: on the basis of the observations

$$f_i(\bar{x},\omega^0), f_i(\bar{x},\omega^1),\dots,f_i(\bar{x},\omega^s),\dots$$

one has to estimate the mean value

$$E\{f_i(\overline{x},\omega)\}.$$

The answer to the simplest question, whether or not a given $\overline{x} \in X$ is feasible, requires verifying the statistical hypothesis that

$$E\{f_i(\overline{x},\omega)\} \leq 0, \text{ for } i = 1,\ldots,m.$$

Since we can only rely on random observations, it seems quite natural to think of stochastic solution procedures that do not make use of the exact values of the $F_i(x)$, $i = 0,\ldots,m$. Of course, we cannot guarantee in such a situation a monotonic decrease (or increase) of the objective value as we move from one iterate to the next, thus these methods must, by the nature of things, be non-monotonic.

Deterministic processes are special cases of stochastic processes, thus stochastic optimization gives us an opportunity to build more flexible and effective solution methods for problems that cannot be solved within the standard framework of deterministic optimization techniquest. *Stochastic quasi-gradient methods* is a class of procedures of that type. They are described in more detail in Chapter 6, here we shall only sketch out their major features. We consider two examples in order to get a better grasp of the main ideas involved.

Example 1: *Optimization by simulation.* Let us imagine that the problem is so complicated that a computer based simulation model has been designed in order to indicate how the future might unfold in time for each choice of a decision x. Suppose that the stochastic elements have been incorporated in the simulation so that for a single choice x repeated simulation runs results in different outputs. We always can identify a simulation run as the observation of an event (environment) ω from a sample space Ω. To simplify matters, let us assume that only a single quantity

$$f_0(x,\omega)$$

summarizes the output of the simulation run ω for given x. The problem is to

$$\begin{aligned} &\text{find} \quad x \in R^n \\ &\text{that minimizes} \quad F_0(x) = E\{f_0(x,\omega)\}. \end{aligned} \tag{1.51}$$

Let us also assume that F_0 is differentiable. Since we do not know with any level of accuracy the values or the gradients of F_0 at x, we cannot apply the standard gradient method, that generates iterates through the recursion:

$$x^{s+1} := x^s - \rho_s \sum_{j=1}^n \frac{F_0(x^s + \Delta_s e^j) - F_0(x^s)}{\Delta_s} e^j, \tag{1.52}$$

where ρ_s is the step-size, Δ_s determines the mesh for the finite difference approximation to the gradient, and e^j is the unit vector on the j-th axis. A well-known procedure to deal with the minimization of functions in this setting is the so-called *stochastic approximation method* that can be viewed as a recursive Monte-Carlo optimization method. The iterates are determined as follows:

$$x^{s+1} := x^s - \rho_s \sum_{j=1}^n \frac{f_0(x^s + \Delta_s e^j, \omega^{sj}) - f_0(x^s, \omega^{s0})}{\Delta_s} e^j, \qquad (1.53)$$

where $\omega^{s0}, \omega^{s1}, \dots, \omega^{sn}$ are observations, not necessarily mutually independent one possibility is $\omega^{s0} = \omega^{s1} = \dots = \omega^{sn}$. The sequence $\{x^s, s = 0, 1, \dots\}$ generated by the recursion (1.53) converges with probability 1 to the optimal solution provided, roughly speaking, that the scalars $\{\rho_s, \Delta_s; s = 1, \dots\}$ are chosen so as to satisfy

$$\rho_s \geq 0, \sum_s \rho_s = \infty, \sum_s (\rho_s^2 + \rho_s \Delta_s) < \infty,$$

$(\rho_s = \Delta_s = 1/s$ are such sequences), the function F_0 has bounded second derivatives and for all $x \in R^n$,

$$E\{\|\Delta f_0(x, \omega)\|^2\} \leq d(1 + \|x\|^2), d > 0. \qquad (1.54)$$

This last condition is quite restrictive, it excludes polynomial functions $f_0(\cdot, \omega)$ of order greater than 3. Therefore, the methods that we shall consider next will avoid making such a requirement, at least on all of R^n.

Example 2: *Optimization by random search.* Let us consider the minimization of a convex function F_0 with bounded second derivatives and n a relatively large number of variables. Then the calculation of the exact gradient ∇F_0 at x requires calling up a large number of times the subroutines for computing all the partial derivatives and this might be quite expensive. The finite difference approximation of the gradient in (1.52) require $(n+1)$ function-evaluations per iteration and this also might be time-consuming if function-evaluations are difficult. Let us consider the following random search method: at each iteration $s = 0, 1 \dots$, choose a direction h^s at random, see Figure 1.5.

If F_0 is differentiable, this direction h^s or its opposite $-h^s$ leads into the region

$$\{x | F_0(x) \leq F_0(x^s)\}$$

of lower values for F_0, unless x^s is already the point at which F_0 is minimized. This simple idea is at the basis of the following random search procedure:

$$x^{s+1} := x^s - \frac{3}{2}\rho_s \frac{F_0(x^s + \Delta_s h^s) - F_0(x^s)}{\Delta_s} h_s, \qquad (1.55)$$

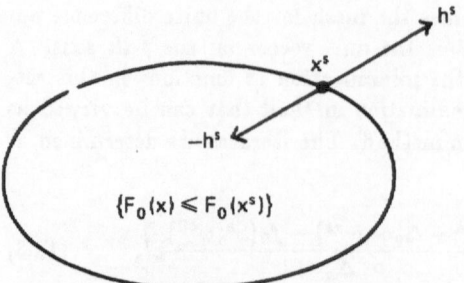

Figure 1.5 Random search directions $+-h^s$.

which requires only two function-evaluations per iteration. Numerical experimentation shows that the number of function-evaluations needed to reach a good approximation of the optimal solution is substantially lower if we use (1.55) in place of (1.52). The vectors $h^0, h^1, \ldots, h^s, \ldots$ often are taken to be independent samples of vectors $h(\cdot)$ whose components are independent random variables uniformly distributed on $[-1, +1]$.

Convergence conditions for the random search method (1.55) are the same, up to some details, as those for the stochastic approximation method (1.53). They both have the following feature: the direction of movement from each $x^s, s = 0, 1, \ldots$ are statistic estimates of the gradient $\nabla F_0(x^s)$. If we rewrite the expressions (1.53) and (1.55) as :

$$x^{s+1} := x^s - \rho_s \xi^s, s = 0, 1, \ldots \qquad (1.56)$$

where ξ^s is the direction of movement, then in both cases

$$E\{\xi^s | x^s\} = \nabla F_0(x^s) + O(\Delta_s) \qquad (1.57)$$

A general scheme of type (1.56) that would satisfy (1.57) combines the ideas of both methods. There may, of course, be many other procedures that fit into this general scheme. For example consider the following iterative method:

$$x^{s+1} := x^s - \rho_s \frac{f_0(x^s + \Delta_s h^s, \omega^{s1}) - f_0(x^s, \omega^{s0})}{\Delta_s} h^s, \qquad (1.58)$$

which requires only two observations per iteration, in contrast to (1.53) that requires $(n+1)$ observations. The vector

$$\xi^s = \frac{3}{2} \frac{f_0(x^s + \Delta_s h^s, \omega^{s1}) - f_0(x^s, \omega^{s0})}{\Delta_s} h^s$$

also satisfies the condition (1.57),

$$\frac{2}{3} E\{\xi^s | x^s\} = E\{\frac{F_0(x^s + \Delta_s h^s) - F_0(x^s)}{\Delta_s} h^s\}$$

$$= E\{\sum_{j=1}^{n} (\frac{\partial}{\partial x_j} F_0(x^s)) h_i^s\} + O(\Delta_s) = \nabla F_0(x^s) + O(\Delta_s).$$

The convergence of all these particular procedures (1.53), (1.55), (1.58) follow from the convergence of the general scheme (1.56)–(1.57). These questions are studied in detail in Chapter 6. The vector ξ^s satisfying (1.57) is called a *stochastic quasi-gradient* of F_0 at x_s, and the scheme (1.56)–(1.57) is an example of a stochastic quasi-gradient procedure.

Unfortunately this procedure cannot be applied, as such, to finding the solution of the stochastic optimization problem (1.50) since we are dealing with a constrained optimization problem, and the functions $F_i, i = 0,\ldots,m$, are in general nondifferentiable. So, let us consider a simple generalization of this procedure for solving the constrained optimization problem with nondifferentiable objective:

$$\text{find} \quad x \in X \subset R^n \tag{1.59}$$
$$\text{that minimzes} \quad F_0(x)$$

where X is closed convex set and F_0 is a real-valued (continuous) convex function. The new algorithm generates a sequence $x^0, x^1,\ldots,x^s,\ldots$ of points in X by the recursion:

$$x^{s+1} := \operatorname*{prj}_X[x^s - \rho_s\xi^s] \tag{1.60}$$

where prj_X means projection on X, and ξ^s satisfies

$$E\{\xi^s|x^0,x^1,\ldots,x^s\} \in \partial F_0(x^s) + \eta^s \tag{1.61}$$

with

$$\partial F_0(x^s) := \text{ the set of subgradients of } f_0 \text{ at } x^s,$$

and η^s is a vector, that may depend on $(x^0,\ldots,x^s)$, that goes to 0 (in a certain sense) as s goes to ∞. The sequence $\{x^s, s = 0,1,\ldots\}$ converges with probability 1 to an optimal solution, when the following conditions are satisfied with probability 1:

$$\rho_s \geq 0, \sum_s \rho_s = \infty, \sum_s E\{\rho_s\|\eta^s\| + \rho_s^2\} < \infty,$$

and

$$E\{\|\xi^s\|^2|x^0,\ldots,x^s\} \text{ is bounded whenever } \{x^0,\ldots,x^s\} \text{ is bounded.}$$

Convergence of this method, as well as its implementation, and different generalizations are considered in Chapter 6.

To conclude let us suggest how the method could be implemented to solve the linear recourse problem (1.22). From the duality theory for linear programming, and the definition (5.2) of Q, one can show that

$$\partial Q(x,\omega) := \{-uT(\omega)|u \in \operatorname*{argmax}_v[v(h(\omega - T(\omega)x)|vW(\omega) \leq q(\omega)]\}.$$

Thus an estimate ξ^s of the gradient of F_0 at x^s is given by

$$\xi^s = c - u^s T(\omega^s)$$

where ω^s is obtained by random sampling from Ω (using the measure P), and

$$u^s \in \operatorname*{argmax}_v[v(h(\omega^s) - T(\omega^s)x)|vW(\omega^s) \le q(\omega^s)]$$

The iterates could then be obtained by

$$x^{s+1} := \operatorname*{prj}_X[x^s + \rho_s u^s T(\omega^s) - \rho_s c]$$

where

$$X = \{x \in R_+^n | Ax \le b\}.$$

It is not difficult to show that under very weak regularity conditions (involving the dependence of $W(\omega)$ on ω),

$$E\{\xi^s|x^s\} \in \partial F_0(x^s).$$

1.10   Conclusion

In guise of conclusion, let us just raise the following possibility. The stochastic quasi-gradient method can operate by obtaining its stochastic quasi-gradient from 1 sample of the subgradients of $f_0(\cdot, \omega)$ at x^s, it could equally well—if this was viewed as advantageous—obtain its stochastic quasi-gradient ξ^s by taking a finite sample of the subgradients of $f_0(\cdot, \omega)$ at x^s, say L of them. We would then set

$$\xi^s := \frac{1}{L}\sum_{\ell=1}^{L} \nu^\ell \text{ where } \nu^\ell \in \partial f_0(x^s, \omega^\ell) \tag{1.62}$$

and $\omega^1, \ldots, \omega^L$ are random samples (using the measure P). The question of the efficiency of the method taking just 1 sample versus $L \ge 1$ should, and has been raised, cf. the implementation of the methods described in Chapter 16. But this is not the question we have in mind. Returning to Section 1.8, where we discussed approximation schemes, we nearly always ended up with an approximate problem that involves a discretization of the probability measures assigning probabilities $p_1, \ldots, p_L$ to points $\omega^1, \ldots, \omega^L$, and if a gradient-type procedure was used to solve the approximating problem, *the* gradient, or a subgradient of F_0 at x^s would be obtained as

$$\varsigma^s := \sum_{\ell=1}^{L} p_\ell \nu^\ell \text{ where } \nu^\ell \in \partial f_0(x^s, \omega^\ell). \tag{1.63}$$

The similarity between expressions (1.62) and (1.63) suggest possibly a new class of algorithms for solving stochastic optimization problems, one that relies on an approximate probability measure (to be refined as the algorithm progresses) to obtain its iterates, allowing for the possibility of a quasi-gradient at each step without losing some of the inherent adaptive possibilities of the quasi-gradient algorithm.

PART II

Numerical Procedures

CHAPTER 2

APPROXIMATION TECHNIQUES IN STOCHASTIC PROGRAMMING

P. Kall, A. Ruszczyński, and K. Frauendorfer

2.1 Introduction

We start this section with a brief discussion of basic difficulties encountered in stochastic programming and overview main approaches for overcoming them. Next we describe fundamental ideas of approximation techniques, which we analyze in more detail in the next sections of this chapter.

2.1.1 The need to approximate stochastic programming problems

The basic feature that differs stochastic programming problems from other optimization problems is the way in which the objective function or constraint functions are defined. In stochastic programming problems values of some of these functions are numerical characteristics of random phenomena dependent on the decision variables. In particular, these can be

(i) *mathematical expectations* of functions dependent on our decision variables and some random parameters, or

(ii) *probabilities* of some random events which are controlled by the decision variables.

This feature gives rise to the main difficulty encountered in stochastic programming problems: the difficulty of calculating values and gradients (or subgradients) of the functions defining the problem.

To discuss this matter in more detail, let us suppose that the objective function $F(x)$ in a stochastic programming problem is defined as a mathematical expectation of a function $f(x, \xi)$, where $x \in R^n$ is the vector of decision variables and ξ is an m-dimensional vector of random parameters. Formally, the objective function can be expressed as follows:

$$F(x) = Ef(x, \xi) = \int_\Omega f(x, \xi(\omega)) P(d\omega), \qquad (2.1)$$

where Ω denotes an abstract probability space and P is the corresponding probability measure. In a special case, if ξ is a *discrete random vector* attaining only a finite number of values $\xi^1, \xi^2, \ldots, \xi^L$ with probabilities $p_1 > 0, p_2 > 0, \ldots, p_L > 0, \sum_{\ell=1}^{L} p_\ell = 1$, we can rewrite (2.1) as

$$F(x) = \sum_{\ell=1}^{L} p_\ell f(x, \xi^\ell). \qquad (2.1a)$$

But in another special case, if the random vector $\xi = (\xi_1, \xi_2, \ldots, \xi_m)$ has a *probability density function* $\varphi(\xi_1, \xi_2, \ldots, \xi_m)$ the general formula (2.1) takes on the form of a Riemann integral

$$F(x) = \int \int_{R^m} \cdots \int f(x, \xi) \varphi(\xi) d\xi_1 d\xi_2, \ldots, d\xi_m. \qquad (2.1b)$$

We see that to evaluate the objective function F at a given point x it is necessary to calculate a multiple integral with respect to the measure describing the distribution of ξ. If it is not possible to perform the integration analytically, we have to use numerical methods, which usually require much computational effort, which increases rapidly with the dimension of ξ and with the required accuracy.

Straightforward application of common nonlinear programming methods (see, e.g., [2], [16], [21]) to stochastic programming problems would require calculation of integrals of the form (2.1) at each point x^k, $k = 0, 1, 2, \ldots$, generated by the optimization algorithm. Difficulties increase if the optimization technique needs also gradients $\nabla F(x^k)$, $k = 0, 1, 2, \ldots$, which in our case turn out to be even more difficult to evaluate than the objective. Indeed, if the function $f(x, \xi)$ in (2.1) is continuously differentiable with respect to x for all ξ, then, under reasonable additional conditions (cf., e.g. [31]) $F(x)$ is continuously differentiable and

$$\nabla F(x) = \int_\Omega \nabla_x f(x, \xi(\omega)) P(d\omega), \qquad (2.2)$$

where $\nabla_x f(x, \xi)$ denotes the gradient of f with respect to x. In the two special cases considered above we obtain

$$\nabla F(x) = \sum_{\ell=1}^{L} p_\ell \nabla f(x, \xi^\ell) \qquad (2.2a)$$

and

$$\nabla F(x) = \int \int_{R^m} \cdots \int \nabla_x f(x, \xi) \varphi(\xi) d\xi_1 d\xi_2, \ldots, d\xi_m, \qquad (2.2b)$$

respectively. Since nonlinear programming methods usually need many iterations to reach a neighborhood of the solution, the total computational effort required may be beyond the cost that can be afforded.

There are two main approaches which overcome the difficulties discussed above: *approximation techniques* and *stochastic quasigradient methods*.

In approximation techniques we replace the original problem with a simpler one by approximating the random vector ξ by another random vector $\tilde{\xi}$ for which integral, (2.1) are easy to handle. Typically, we choose $\tilde{\xi}$ to be a discrete random vector and deal only with sums of the form (2.1a).

Stochastic quasigradient methods avoid at all computation of integrals of the form (2.1). The main idea of these methods is to make random steps in directions calculated on the basis of some statistical information about the problem gained at each step. Contrary to approximation techniques, they do not tend to get a global image of the properties of $F(x)$, but use random values $f(x, \xi^k)$ and corresponding gradients $\nabla_x f(x, \xi^k)$ (or subgradients in a nondifferentiable case) calculated at some sampled realizations ξ^k of ξ, $k = 0, 1, 2, \ldots$. In such a way a kind of self-learning method is constructed, in which each particular step may be inefficient, but their large number exhibits general statistical properties that imply convergence with probability one to a solution.

Stochastic quasigradient methods are discussed later in this volume, and from now on we shall concentrate on the approximation schemes. It is also worth mentioning here that recently, in [19], an attempt has been made to combine these two approaches.

I.1.2 Fundamentals of approximation techniques

When constructing approximations to stochastic programming problems we have to analyze the following mutually related questions.

First we have to find out a proper way of replacing the original random vector ξ with a discrete one.

Secondly, we have to study the relations between the original problem and the approximate problem and estimate the accuracy of approximation.

Thirdly, we need a method of improving the accuracy, if it is not sufficient, by constructing a better approximation to ξ.

Before investigating these problems in detail, let us introduce some basic ideas and mathematical properties of this approach.

Let $\Xi \subset R^m$ be the *support* of the random vector ξ (i.e. the smallest closed set in R^m such that $P\{\xi \in \Xi\} = 1$) and let S^L be a finite collection of subsets Ξ_l, $\ell = 1, 2, \ldots, L$, of Ξ satisfying the following conditions:

$$\bigcup_{\ell=1}^{L} \Xi_\ell = \Xi, \tag{2.3}$$

$$\Xi_i \cap \Xi_j = \emptyset \text{ for } i \neq j; \quad i, j = 1, 2, \ldots, L. \tag{2.4}$$

We shall call S^L a *partition* of Ξ.

For any partition we can rewrite integral (2.1) as follows

$$F(x) = \int_\Xi f(x, \xi) P(d\xi) = \sum_{\ell=1}^{L} \int_{\Xi_\ell} f(x, \xi) P(d\xi), \tag{2.5}$$

where we perform integration over the support $\Xi \subset R^m$ and use the description of the distribution of ξ in the space of its values.

In the particular case (2.1b), which is of special interest for us, (2.5) reads

$$F(x) = \sum_{\ell=1}^{L} \int \int \cdots \int_{\Xi_\ell} f(x,\xi)\varphi(\xi)d\xi_1 d\xi_2,\ldots,d\xi_m. \qquad (2.5a)$$

Proceeding as in the simplest method for calculating integrals we can now approximate each integral over Ξ_ℓ as follows

$$\int_{\Xi_\ell} f(x,\xi)P(d\xi) \sim f(x,\xi^\ell) \int_{\Xi_\ell} P(d\xi) = f(x,\xi^\ell)P\{\xi \in \Xi_\ell\}. \qquad (2.6)$$

where ξ^ℓ is a selected representative of the subset Ξ_ℓ. In other words, we approximate the function $f(x,\xi)$ by a step function in ξ, which is constant in each set Ξ_ℓ, $\ell = 1, 2, \ldots, L$. In this way we arrive to the following approximation of $F(x)$:

$$F^L(x) = \sum_{\ell=1}^{L} p_\ell f(x,\xi^\ell), \qquad (2.7)$$

with

$$p_\ell = P\{\xi \in \Xi_\ell\}.$$

Since by (2.3) and (2.4) we also have $\sum_{\ell=1}^{L} p_\ell = 1$, our approximation can be equivalently interpreted as an approximation of ξ by a discrete random vector $\tilde{\xi}$ attaining values ξ^ℓ with probabilities p_ℓ, $\ell = 1, 2, \ldots, L$, and our approximating formula (2.7) is exactly of the form (2.1a).

Generally, if the support Ξ is bounded and if $\max_{1 \le \ell \le L} P\{\xi \in \Xi_\ell\} \to 0$ as $L \to \infty$, then for each x, under reasonable assumptions of $f(x,\xi)$ we get a *pointwise convergence* of function values: $F^L(x) \to F(x)$ as $L \to \infty$. This fundamental and highly desirable property, however, is not sufficient for us, because we are rather interested in the convergence of the sequence of solutions $\hat{x}_L$ of approximate problems, or at least of its convergent subsequences to a solution of the original optimization problem. Some additional conditions, e.g. compactness of the feasible set for x together with the uniform convergence of F^L to F and continuity of F, are needed to ensure such a kind of convergence. We shall not go further into the analysis of these theoretical problems; a thorough discussion of them and various generalizations can be found in [1], [15], [30], [34]. Still, in many practical problems such conditions are satisfied. It is also often the case, that in practice a point $\tilde{x}$ is satisfactory, for which the objective value lies within a certain tolerance range with respect to the minimum value, and this is possible to achieve for a far broader class of problems.

Nevertheless, it is still very difficult to determine in advance how fine the partition should be to ensure the accuracy of approximation. Division of Ξ into many small pieces Ξ_ℓ, $\ell = 1, 2, \ldots, L$, without any strategy may dramatically increase the computational complexity of the approximate problem. To illustrate the difficulties that may arise, let us suppose that there are 10 independent scalar random variables in our original problem, so that $\xi = (\xi_1, \xi_2, \ldots, \xi_{10})$. If the support of each ξ_j, $j = 1, 2, \ldots, 10$, is divided into 10 subintervals, we get 10^{10} subsets Ξ_ℓ of the support Ξ of ξ, a number which is clearly beyond any computational capabilities.

To avoid such excessive numbers of subsets Ξ_ℓ we have to use nonuniform partitions which are suited to properties of $f(x, \xi)$ as a function of ξ. The problem of constructing such partitions is closely related to the way of choosing points $\xi^\ell \in \Xi_\ell$. Considering only convergence, these can be arbitrary points; however, if we choose them more carefully, namely as conditional expectations

$$\xi^\ell = E\{\xi(\omega)/\xi(\omega) \in \Xi_\ell\} \tag{2.8}$$

with probabilities

$$p_\ell = P\{\xi(\omega) \in \Xi_\ell\} \tag{2.9}$$

then we shall not only improve the accuracy of approximation in many cases, but also gain information that will help us to properly refine the partitioning if the accuracy shall not be sufficient.

Indeed, if the function $f(x, \xi)$ is linear with respect to ξ in the set Ξ_ℓ, then with ξ^ℓ defined by (2.8) we obtain strict equality in (2.6),

$$\int_{\Xi_\ell} f(x, \xi) P(d\xi) = f(x, \xi^\ell) P\{\xi \in \Xi_\ell\}. \tag{2.10}$$

This implies that further division of the subset Ξ_ℓ is useless for improving the accuracy of approximation at a given x. On the other hand, if $f(x, \cdot)$ is highly nonlinear in Ξ_ℓ, the approximation in Ξ_ℓ can be rather rough and a finer partition of Ξ_ℓ is desirable. Hence, the density of partitioning in various subregions of the support Ξ should be related to the nonlinearity of $f(x, \cdot)$.

Generally, we do not know in advance such detailed properties of the function $f(x, \xi)$, some information can be gained only in the course of solving a definite approximation problem. Furthermore, the properties of the function $f(x, \cdot)$ change when x changes, and we are interested in having a good partition for x close to the solution of our problem.

Thus we arrive at an idea of a *sequential approximation method* in which constructing a partition of Ξ and approximating a solution to the original problem are mutually related:

(1) Choose an initial partitioning Ξ_ℓ, $\ell = 1, 2, \ldots, L$, which satisfies (2.3) and (2.4).
(2) Choose points $\xi^\ell \in \Xi$ and probabilities p_ℓ, $\ell = 1, 2, \ldots, L$, according to (2.8) and (2.9).
(3) Solve the approximate problem.
(4) At the solution $\tilde{x}_L$ analyze the accuracy of approximation by investigating properties of the function $f(\tilde{x}_L, \xi)$ in each of the subsets Ξ_ℓ, $\ell = 1, 2, \ldots, L$,

choose those of them that should be further divided, if the accuracy is not sufficient, and repeat step 2.

Detailed realization of this procedure depends upon properties of the class of problems to which it is applied. In the next section we shall describe in more detail its application to a certain important class of stochastic programming problems.

2.2 Approximation Schemes for Linear Two-stage Problems of Stochastic Programming

In this section we consider a special class of stochastic programming problems, so-called two-stage problems, and we describe the realization of the sequential approximation method in this case. In 2.2.1 we formulate the problem and review its basic properties and in Section 2.2.2 we consider the special case with a discretely distributed random vector. Section 2.2.3 is devoted to estimates of the accuracy of approximation, which are followed in 2.2.4 by the analysis of refining strategies. The special case of so-called simple recourse is discussed separately in 2.2.5.

2.2.1   Basic properties of linear two-stage problems.

The *linear two-stage problem of stochastic programming* is defined as follows:

$$\text{minimize} \quad [\psi(x) = c^T x + \int_\Omega Q(x, \xi(\omega)) P(d\omega)]$$

$$\text{subject to} \quad Ax = b, \tag{2.11}$$

$$x \geq 0,$$

where $c \in R^{n_1}$, $b \in R^{m_1}$ and A of dimension $m_1 \times n_1$ are defined as in a common linear programming problem. The function $Q(x, \xi(\omega))$ that appears in the additional part of the objective in (2.11) is defined as the optimal value of another linear programming problem which has x as a parameter and involves random coefficients $\xi(\omega) = (q(\omega), h(\omega), T(\omega))$:

$$\text{minimize} \quad q^T(\omega) x$$

$$\text{subject to} \quad Wy = h(\omega) - T(\omega)x, \tag{2.12}$$

$$y \geq 0.$$

The linear programming problem (2.12) is called the *second stage problem*, or the *recourse problem*; it consists in finding the best *recourse decision* $y \in R_+^{n_2}$, when the *first stage decision* $x \in R_+^{n_1}$ and random realization of the parameters $q(\omega) \in R^{n_2}$, $h(\omega) \in R^{m_2}$ and $T(\omega)$ of dimension $m_2 \times n_1$ are already established. The $m_2 \times n_2$ matrix W is deterministic.

Since the expected value of the minimum *recourse cost* $Q(x, \xi(\omega))$ modifies the objective of the first-stage problem (2.11), the whole model (2.11)–(2.12) has a certain internal dynamical structure: when looking for an optimal first stage decision x we have to take into account not only the direct first stage cost $c^T x$ but also the expected value of the future recourse cost. If there is no feasible solution to (2.12) we assume $Q(x, \xi(\omega)) = +\infty$, and this should also be considered at the first stage.

We are especially interested in stochastic programming problems with recourse because of their wide application to modeling decision problems which involve random data. If some constraints, e.g. $Tx = h$, in a linear programming problem include random coefficients in T or h and we have to take the decision before knowing the realizations $T(\omega)$ and $h(\omega)$ of T and h, it is generally impossible to require that the equality

$$T(\omega)x = h(\omega) \tag{2.13}$$

be satisfied for each realization of the stochastic constraint parameters. The problem with recourse is a way of overcoming these modeling difficulties; the recourse decision y may be interpreted as a correction in (2.13), and the recourse cost $Q(x, \xi(\omega))$—as a penalty for discrepancy in (2.13).

In a more general model the matrix W in (2.12) could be random too, but for the ease of exposition we assume that it is deterministic; such a model is called the problem with *fixed recourse*. Most of the theory and computational methods have been developed for this class of linear two-stage problems.

Let us review briefly basic properties of the problem (2.11)–(2.12). The feasible set of (2.11) is the intersection of the set given by the first stage constraints

$$K_1 = \{x \in R^{n_1} : Ax = b, x \geq 0\} \tag{2.14}$$

and of the *induced feasible set*

$$K_2 = \{x \in R^{n_1} : Q(x, \xi(\omega)) < \infty \text{ with probability } 1\}. \tag{2.15}$$

While K_1 is described explicitly and easy to handle, the induced set K_2 is defined implicitly and hard to express analytically. However, if the matrix W in (2.12) is such that $\{Wy : y \geq 0\} = R^{m_2}$ (i.e. the corrections Wy in (2.12) can cancel any error), we have $K_2 = R^{n_1}$. Problems with such a property are called problems with *complete recourse*. In the special case of $W = [I, -I]$ we speak about *simple recourse*. Although generally the induced feasible set K_2 need not contain K_1 we still have the following property.

(a) *The sets K_1, K_2 and $K = K_1 \cap K_2$ are convex and closed.*

As far as the recourse cost $Q(x, \xi(\omega))$ is concerned, many interesting theoretical results are available. First, by the theory of duality in linear programming we know that $Q(x, \xi(\omega)) > -\infty$ (i.e. the second stage problem is bounded from below) if and only if one can find $u \in R^{m_2}$ such that $W^T u \leq q(\omega)$. Since the case of unboundedness is of no interest for us, we shall from now on assume that the above condition is satisfied for each realization of the random vector $q(\omega)$. Under this assumption the recourse function possesses the following properties.

(b) For any fixed $x \in K$ and any q the function $(h, T) \to Q(x, \xi = (q, h, T))$ is piecewise linear and convex.

(c) For any fixed $x \in K$ and any h and T the function $q \to Q(x, \xi = (q, h, T))$ is piecewise linear and concave.

(d) For any fixed $\xi = (q, h, T)$ the function $x \to Q(x, \xi)$ is a convex piecewise linear function on K.

Under the additional condition that the random variable $\xi(\omega) = (q(\omega), h(\omega), T(\omega))$ has finite second moments we finally obtain the following result.

(e) The function $Q(x) = \int_\Omega Q(x, \xi(\omega)) P(d\omega)$ is finite and convex in K.

A detailed discussion of properties of linear two-stage stochastic programming problems can be found in [12] and [35].

Properties (a)–(e) are of fundamental importance for the concepts and methods discussed in this chapter and will be frequently used in subsequent sections. We also assume that we deal with the case of complete recourse (no induced constraints). Motivation for the later assumption is rather obvious: with $K_2 \neq R^{n1}$ it would be extremely difficult to ensure that solutions to approximate problems are in the induced feasible set of the original problem.

2.2.2 The two-stage problem with a discrete random vector

Let us consider in more detail properties of stochastic programming problem with recourse in case of a discretely distributed random vector $\tilde{\xi}$ attaining values:

$$\xi^1 = (q^1, h^1, T^1) \text{ with probability } p_1 > 0,$$
$$\xi^2 = (q^2, h^2, T^2) \text{ with probability } p_2 > 0,$$
$$\dots \tag{2.16}$$
$$\xi^L = (q^L, h^L, T^L) \text{ with probability } p_L > 0,$$

where

$$\sum_{\ell=1}^{L} p_l = 1. \tag{2.17}$$

In this case the two-stage problem (2.11)–(2.12) takes on the form

$$\text{minimize} \quad [\tilde{\psi}(x) = c^T x + \sum_{\ell=1}^{L} p_\ell Q(x, \xi^\ell)]$$
$$\text{subject to} \quad Ax = b \tag{2.18}$$
$$x \geq 0,$$

where $Q(x, \xi^\ell)$ is the minimum objective value in the recourse problem

$$\text{minimize} \quad (q^\ell)^T y$$
$$\text{subject to} \quad W y = h^\ell - T^\ell x, \tag{2.19}$$
$$y \geq 0,$$

$\ell = 1, 2, \ldots, L$. If we denote by $\hat{y}^\ell(x)$, $\ell = 1, 2, \ldots, L$, the solutions to problems (2.19) at a given x, we can express the first stage objective as

$$\tilde{\psi}(x) = c^T x + \sum_{\ell=1}^{L} p_\ell (q^\ell)^T \hat{y}^\ell(x). \tag{2.20}$$

Of course, the solutions $\hat{y}^\ell(x)$ depend on x in a rather involved way, so that the products $(q^\ell)^T \hat{y}^\ell(x)$ are piecewise linear (cf. property (d) in 2.2.1). However, instead of considering (2.18)–(2.19) as a two-level problem, we can put together the first stage problem (2.18) and all realizations of the second stage problem (2.19) into a large linear programming model:

minimize $\quad c^T x + p_\ell (q^1)^T y^1 + p_2 (q^2)^T y^2 + \ldots p_L (q^L)^T y^L$

subject to

$$
\begin{array}{llll}
Ax & & = & b \\
T^x + Wy^1 & & = & h^1 \\
T^2 x & +Wy^2 & = & h^2 \\
\ldots & & \ddots & \ldots \\
T^L x & +Wy^L & = & h^L \\
x \geq 0, & y^1 \geq 0, y^2 \geq 0 & \ldots \quad y^L \geq 0.
\end{array}
\tag{2.21}
$$

Problems (2.18)–(2.19) and (2.21) are equivalent in the sense that they have the same set of solutions, as the first stage decision vector x is concerned, and the optimal values of $y^1, y^2, \ldots, y^L$ in (2.21) are solutions to the realizations of the second stage problem (2.19) at the optimal x.

Summing up, a two stage problem with a discretely distributed random vector $\tilde{\xi}$ turns out to be equivalent to a large-scale linear programming problem, which can be solved by powerful linear programming techniques, which take account of its special *dual block angular* structure. These techniques are discussed in detail in chapter 5 of this volume (see also [13], [28] and [33]).

2.2.3 Error estimates

Let us now investigate relations between a two-stage problem with an arbitrary distribution of the random parameter ξ and its approximation resulting from the discretization of ξ. Recall that, according to the ideas sketched in Section 2.1.2, the discretely distributed approximation $\tilde{\xi}$ to ξ is constructed for a given partition $S^L = (\Xi_1, \Xi_2, \ldots, \Xi_L)$ of the support Ξ of ξ as follows:

$$P\{\tilde{\xi} = \xi^\ell\} = p_\ell, \qquad \ell = 1, 2, \ldots, L, \tag{2.22}$$

where $\xi^1, \xi^2, \ldots, \xi^l$ are conditional expectations of ξ in Ξ_ℓ,

$$\xi^\ell = E\{\xi/\xi \in \Xi_\ell\}, \qquad \ell = 1, 2, \ldots, L \tag{2.23}$$

and

$$p_\ell = P\{\xi \in \Xi_\ell\}, \qquad \ell = 1, 2, \ldots, L, \tag{2.24}$$

$$\sum_{\ell=1}^{L} p_\ell = 1. \tag{2.25}$$

We expect (2.23) to be a good choice, since the conditional expectations minimizes $E\|\xi - \tilde{\xi}\|^2$ with respect to all discrete distributions corresponding to our partition [14].

After replacing ξ in (2.11)–(2.12) by the discrete variable $\tilde{\xi}$ we obtain an approximating problem of the form (2.21). Obviously, this problem is much easier to solve than the original one, but now we need estimates of errors caused by the approximation. Such estimates can be derived from general properties (a)–(d) of two-stage problems, discussed in Section 2.2.1.

Lower Bounds

Let us assume that all the subsets Ξ_ℓ, $\ell = 1, 2, \ldots, L$, are convex and the function $Q(x, \xi)$ in (2.11) is convex in ξ for each x. By property (b), the latter condition is satisfied if q in (2.12) is deterministic, and only $T(\omega)$ and $h(\omega)$ vary randomly.

Under this assumption, with ξ^ℓ and p_ℓ representing conditional expectations and probabilities defined by (2.23)–(2.24), for each block Ξ_ℓ from *Jensen's inequality* (see [14]) we obtain

$$\int_{\Xi_\ell} Q(x, \xi(\omega)) P(d\omega) \geq p_\ell Q(x, \xi^\ell), \qquad \ell = 1, 2, \ldots, L. \tag{2.26}$$

Thus for any x we have

$$\psi(x) = c^T x + \int_{\Xi} Q(x, \xi(\omega)) P(d\omega) \geq$$
$$\geq c^T x + \sum_{\ell=1}^{L} p_\ell Q(x, \xi^\ell) = \tilde{\psi}(x). \tag{2.27}$$

Hence, the objective value in the approximate problem (2.18) is a lower bound for the true objective value at a given x. Furthermore, in (2.18), or its extended LP form (2.21), we minimize $\tilde{\psi}(x)$ and therefore the minimal value $\tilde{\psi}(\tilde{x})$, where $\tilde{x}$ solves (2.21), is a lower bound for the least value of the true objective:

$$\psi(x) \geq \tilde{\psi}(\tilde{x}) \text{ for all feasible } x. \tag{2.28}$$

Another important feature of the Jensen's lower bound is its *monotonicity*: if S^{L+1} is a refinement of the partition S^L (i.e. results from S^L by division of some of its members), then a lower bound obtained for S^{L+1} is at least as good as the previous one (see [8]).

A more thorough discussion of applications of Jensen's inequality in stochastic programming can be found in [3], [8], [9], and [14].

One can also exploit the convexity of the function $Q(x, \cdot)$ by approximating it from below by a piecewise linear function ($Q(x, \cdot)$ is piecewise linear itself, but may contain a very large number of pieces).

By the duality theory in linear programming

$$Q(x, \xi(\omega)) = \min\{q^T y | W y = h(\omega) - T(\omega)x, y \geq 0\}$$
$$= \max\{(h(\omega) - T(\omega)x)^T u | W^T u \leq q\}, \tag{2.29}$$

where $u \in R^{m_2}$ is the vector of multipliers in (2.12). If u_ℓ, $\ell = 1, 2, \ldots, L$, are some feasible solutions to the dual program, then

$$Q(x, \xi(\omega)) \geq \max_{1 \leq \ell \leq L} (h(\omega) - T(\omega)x)^T u_\ell = \tilde{Q}(x, \xi(\omega)). \tag{2.30}$$

For a deterministic q the feasible set $W^T u \leq q$ in the dual problem (2.29) does not depend on ω, hence we can substitute for u_ℓ dual solutions to the second stage problem of any x and with any $\xi(\omega) = (T(\omega), h(\omega))$. In particular, if we choose u_ℓ to be optimal multiplier vectors at ξ^ℓ for a fixed x, then the graph of the linear function $\overline{Q}_\ell(\xi(\omega)) = (h(\omega) - T(\omega)x)^T u_\ell$ will support the graph of $Q(x, \cdot)$ at ξ^ℓ. Finally, taking the expectation of both sides of (2.30) we obtain a lower bound for $\psi(x)$:

$$\psi(x) \geq c^T x + E\{\max_{1 \leq \ell \leq L} (h(\omega) - T(\omega)x)^T u_\ell\} = \tilde{\psi}(x). \tag{2.31}$$

The two methods for calculating lower bounds are illustrated in Figure 2.1 and Figure 2.2. We see from these figures that the lower bound (2.27) results from approximating the function $Q(x, \cdot)$ by a step function $\tilde{Q}(x, \cdot)$ attaining in Ξ_ℓ the values $Q(x, \xi^\ell)$, $\ell = 1, 2, \ldots, L$, while the lower bound (2.31) results from approximating $Q(x, \cdot)$ by a convex piecewise linear function $\tilde{Q}(x, \cdot)$ defined by supporting hyperplanes at ξ^ℓ. The second approximation can be more accurate and the resulting bound sharper at a given x, but the evaluation of (2.31) requires an additional integration of the approximating piecewise linear function

$\tilde{Q}(x, \xi(\omega))$. Another difference is that Jensen's inequality is in some way consistent with the approximating problem (2.21) and provides a lower bound (2.28) for the minimum objective values, which in general is not true for (2.31) (to get a global lower bound one would have to minimize the right-hand side of (2.31) instead of solving (2.21)). An extensive discussion of the above techniques for constructing lower bounds can be found in [3], [8], and [14].

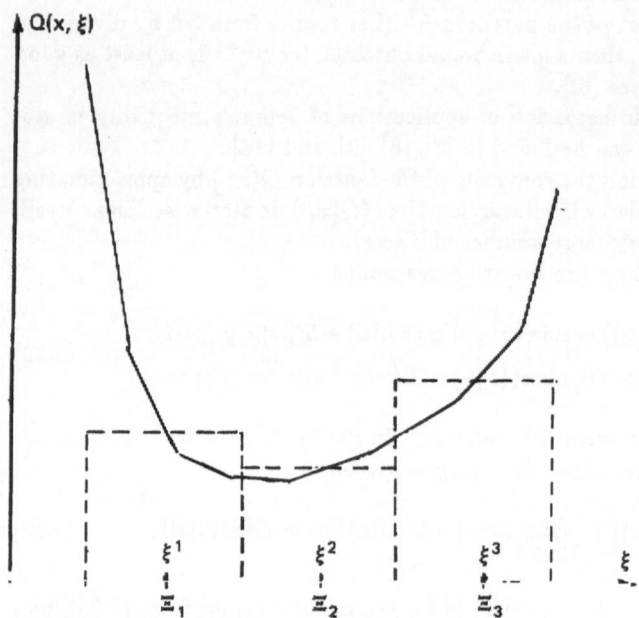

Figure 2.1   Lower bound by Jensen's inequality

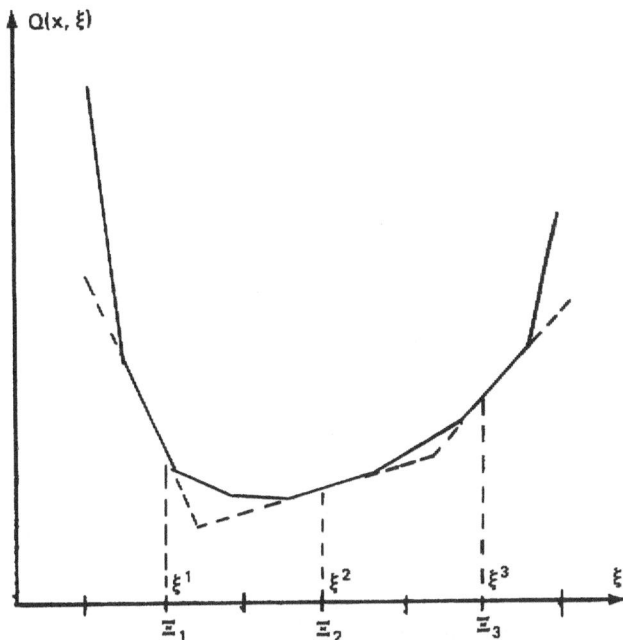

Figure 2.2 Lower bound by piecewise linear approximation

Upper bounds

Since in general we are not able to evaluate $\psi(x)$ exactly, we need also upper bounds on the objective value to compare them with our estimates of the minimum. Such bounds can be obtained from the *Edmundson-Madansky inequality* for expectations of convex functions.

To explain the main idea of constructing an upper bound, let us assume that ξ is a one-dimensional random variable with a support $\Xi = [a, b]$. Define now $\hat{\xi}$ to be a discrete random variable attaining values:

$$
\begin{aligned}
&a \text{ with probability } p_1 = \frac{b - \xi^0}{b - a}, \\
&b \text{ with probability } p_2 = \frac{\xi^0 - a}{b - a},
\end{aligned}
\tag{2.32}
$$

where $\xi^0 = E\xi = \int_a^b \xi P(d\xi)$. The Edmundson-Madansky inequality, when applied to our problem, says that

$$
EQ(x, \xi) \le EQ(x, \hat{\xi}),
\tag{2.33}
$$

provided that $Q(x, \cdot)$ is convex. Indeed, the convexity of $Q(x, \cdot)$ implies that

$(b-a)Q(x,\xi) \leq (b-\xi)Q(x,a) + (\xi-a)Q(x,b)$ for each $\xi \in [a,b]$, hence

$$
\begin{aligned}
EQ(x,\xi) &= \int_a^b Q(x,\xi)P(d\xi) \\
&\leq \int_a^b \left[\frac{b-\xi}{b-a}Q(x,a) + \frac{\xi-a}{b-a}Q(x,b) \right] P(d\xi) \\
&= \frac{b-\xi^0}{b-a}Q(x,a) + \frac{\xi^0-a}{b-a}Q(x,b) = EQ(x,\hat\xi).
\end{aligned}
\tag{2.34}
$$

If ξ is an m-dimensional random variable with independent components distributed in intervals $[a_j, b_j]$ with expectations ξ_j^0, $j = 1, 2, \ldots, m$, inequality (2.33) holds with a variable $\hat\xi$ having independent components $\hat\xi_j$ distributed in points a_j and b_j according to (2.32) (see [8], [9], [32]). The variable $\hat\xi$ constructed in this way is a discrete random variable attaining values only at vertices of the rectangle $\Xi = \mathsf{X}_{j=1}^m [a_j, b_j]$.

The distribution of $\hat\xi$ may be viewed as an extremal distribution in the following sense: among all distributions with support Ξ and the same expectation $\xi^0 = (\xi_1^0, \xi_2^0, \ldots, \xi_m^0)$, for any convex function $\varphi : \Xi \to R^1$ the distribution of $\hat\xi$ provides the maximum of the expected value of φ (cf. [6], [36], [32]). This property explains the essence of the upper bound (2.33) and can also be used for constructing *worst-case approximations* to stochastic programming problems (see Section 2.4).

Let us now consider the partition of Ξ into rectangles

$$
\Xi_\ell = \mathop{\mathsf{X}}_{j=1}^m [a_j^\ell, b_j^\ell) \quad \ell = 1, 2, \ldots, L.
\tag{2.35}
$$

Obviously,

$$
EQ(x,\xi) = \sum_{\ell=1}^L \int_{\Xi_\ell} Q(x,\xi)P(d\xi).
\tag{2.36}
$$

Each of the integrals in (2.36) can be estimated from above according to the Emundson-Madansky inequality, with the expected value of ξ replaced by the conditional expectation ξ^ℓ of ξ in Ξ_ℓ. This yields the upper bound

$$
EQ(x,\xi) \leq \sum_{\ell=1}^L p_\ell EQ(x,\hat\xi^\ell)
\tag{2.37}
$$

where each $\hat\xi^\ell$ is defined for the corresponding subset Ξ_ℓ according to (2.32) with ξ^0 replaced by the conditional expectations $\xi_j^\ell = E\{\xi_j / \xi_j \in [a_j^\ell, b_j^\ell)\}$, $j = 1, 2, \ldots, m$. Two equivalent interpretations of this procedure in a one-dimensional case are illustrated in Figure 2.3 and Figure 2.4, while in Figure 2.5 we show how an upper bound is constructed for a given Ξ^ℓ in a two-dimensional case.

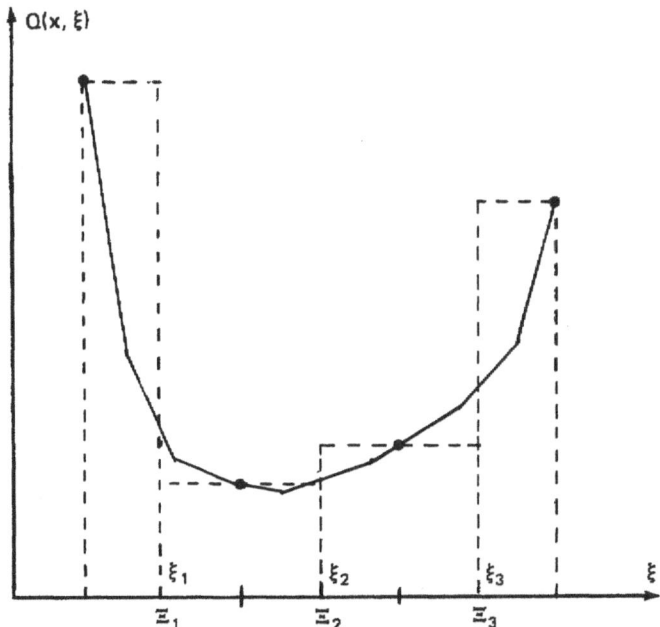

Figure 2.3 Upper bound by Edmundson-Madanski inequality

One can use inequality (2.37) in two ways. First, directly from (2.37) we obtain an upper bound for the value of the objective at any given point x

$$\psi(x) = c^T x + EQ(x,\xi) \leq c^T x + \sum_{\ell=1}^{L} p_\ell EQ(x,\hat{\xi}^\ell) = \hat{\psi}(x). \qquad (2.38)$$

We usually calculate this upper bound at the solution $\tilde{x}$ of (2.21), by solving the second stage problem at $\tilde{x}$ and at each vertex $\hat{\xi}^{\ell\nu}$, $\ell = 1,2,\ldots,L$, $\nu = 1,2,\ldots,2^m$ of our partition (note that most of vertices are common for many subsets).

Secondly, we can estimate from above the minimum value of $\psi(x)$ by finding a point $\hat{x}$ which solves the problem

$$\text{minimize } [\hat{\psi}(x) = c^T x + \sum_{\ell=1}^{L} p_\ell EQ(x,\hat{\xi}^\ell)]$$

$$\text{subject to } Ax = b, \qquad (2.39)$$

$$x \geq 0.$$

From (2.38) we get

$$\min \psi(x) \leq \hat{\psi}(\hat{x}) \leq \hat{\psi}(\tilde{x}) \qquad (2.40)$$

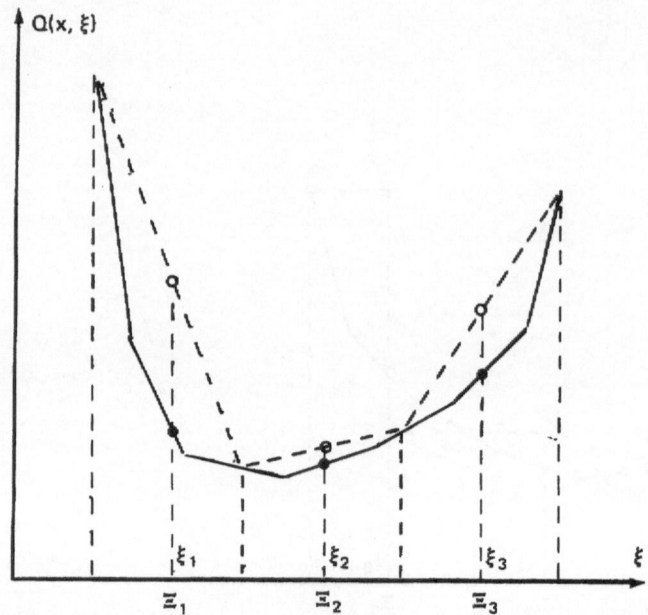

Figure 2.4 Equivalent interpretation of the upper bound in one-dimensional case

Problem (2.39) can be equivalently formulated as a large-scale linear programming problem of the same structure as (2.21). Indeed,

$$EQ(x, \hat{\xi}^\ell) = \sum_{\nu=1}^{2^m} p^{\ell\nu} Q(x, \hat{\xi}^{\ell\nu}), \qquad (2.41)$$

where $\hat{\xi}^{\ell\nu}$ are vertices of the subsets Ξ_ℓ, and the probabilities $p^{\ell\nu}$ are defined as follows

$$p^{\ell\nu} = P\{\hat{\xi}^\ell = \hat{\xi}^{\ell\nu}\} = \prod_{j=1}^{m} P\{\hat{\xi}_j^\ell = \hat{\xi}_j^{\ell\nu}\}. \qquad (2.42)$$

Each of the factors $p_j^{\ell\nu} = P\{\hat{\xi}_j^\ell = \hat{\xi}_j^{\ell\nu}\}$ is defined as in (2.32) with a, b and ξ^0 replaced by a_j^ℓ, b_j^ℓ and the conditional expectation ξ_j^ℓ of ξ_j in $[a_j^\ell, b_j^\ell)$. The number of blocks in the resulting linear programming problem will be equal to the number of vertices of our partition.

Consequently, on the one hand $\hat{\psi}(\hat{x})$ is a better upper bound than $\hat{\psi}(\tilde{x})$, but on the other hand its calculation requires solution of an additional large scale linear program.

Analogously to the Jensen's inequality, the upper bounds (2.40) possess the property of *monotonicity*: if we refine the partition (i.e. subdivide some of

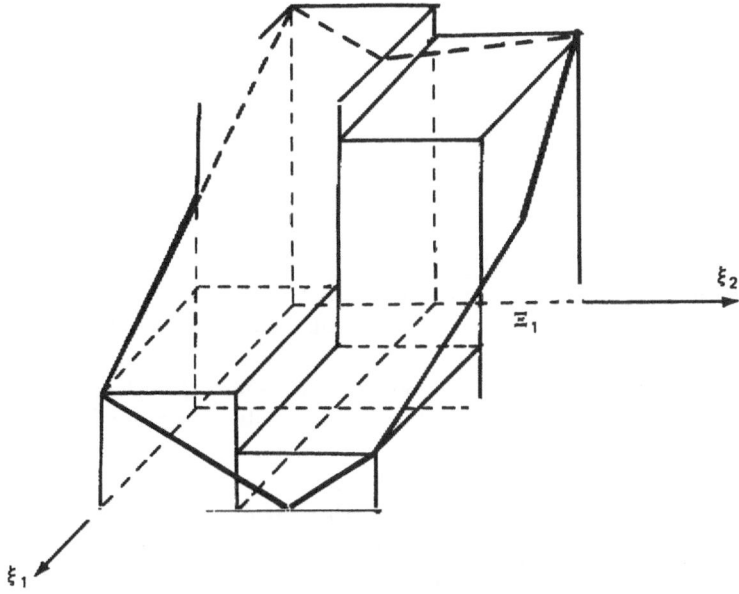

Figure 2.5 Upper bound in multi-dimensional case

its members), the new bounds will be at least as good as the previous ones (see [14], [8]).

We end this section by noting that in the absence of convexity of $Q(x, \cdot)$, which was crucial for our previous considerations, one can still derive some error bounds for linear two-stage problems (see [11]); the results are rather of theoretical than computational nature and substantiate convergence of approximation schemes for problems with recourse.

2.2.4   Refining Strategies

In the previous section we discussed methods for estimating errors that result from approximating the random variable ξ by a discrete one defined by a partition $\Xi_., \Xi_2, \ldots, \Xi_\ell$ of the support Ξ of ξ. Let us now consider the question how this partition should be refined so as to improve the accuracy of approximation.

The simplest and most obvious technique of refining is to cut each subset $\Xi_\ell, \ell = 1, 2, \ldots, L$, by hyperplanes orthogonal to coordinate axes in R^m. If the subsets Ξ_ℓ are hypercubes in R^m, this strategy divides each Ξ_ℓ into 2^m smaller cubes, hence after k steps we shall get $L = 2^{mk}$ subsets. Consequently, the size of the approximating linear programming problem (2.21) will increase so fast that after a small number of refining steps we shall no longer be able to solve it.

However, a more careful analysis of our problem shows that the computational effort can be considerably reduced by dividing only some properly chosen subsets along appropriate directions.

Let us at first discuss the question of selecting subsets (and the corresponding blocks in (2.21)) that should be further divided. To this end let us recall our results concerning error bounds and formulate them for subsets Ξ_ℓ, $\ell = 1, 2, \ldots, L$. The lower bound $\check{\psi}_\ell(\tilde{x})$ for $\int_{\Xi_\ell} Q(\tilde{x}, \xi) P(d\xi)$ we get from (2.26)

$$\check{\psi}_\ell(\tilde{x}) = p_\ell Q(\tilde{x}, \xi^\ell), \tag{2.43}$$

while the upper bound is given by (2.41):

$$\hat{\psi}_\ell(\tilde{x}) = p_\ell \sum_{y=1}^{2^m} p^{\ell y} Q(\tilde{x}, \hat{\xi}^{\ell y}). \tag{2.44}$$

It is now obvious that we need to divide only such blocks, for which differences between upper and lower bounds exceed the assumed tolerance. These differences depend on properties of the function $Q(\tilde{x}, \cdot)$ in Ξ_ℓ ; as mentioned in Section 2.1.2, if $Q(\tilde{x}, \cdot)$ is linear in Ξ_ℓ then there is no approximation error in this subset, $\check{\psi}_\ell(\tilde{x}) = \hat{\psi}_\ell(\tilde{x})$, and further division of Ξ_ℓ will not improve the accuracy of approximation at $\tilde{x}$. On the other hand, nonlinearity of $Q(\tilde{x}, \cdot)$ in Ξ_ℓ leads to differences between $\check{\psi}_\ell(\tilde{x})$ and $\hat{\psi}_\ell(\tilde{x})$ that indicate the necessity of dividing Ξ_ℓ.

Let us now discuss the choice of the direction along which a subset Ξ_ℓ should be split. Again, the efficiency of cuts in different directions is related to the linearity of the function $Q(\tilde{x}, \xi)$ with respect to coordinates $\xi_1, \xi_2, \ldots, \xi_m$ of ξ. As we see from the example in Figure 2.6, no improvement can be gained by splitting Ξ_ℓ with a cutting plane orthogonal to the coordinate ξ_1 in which Q is linear. On the other hand, if we cut Ξ_ℓ by a plane orthogonal to ξ_2 we may obtain two subregions in which $Q(\tilde{x}, \xi)$ will be linear in ξ, and our next upper and lower bounds in these subsets might become exact.

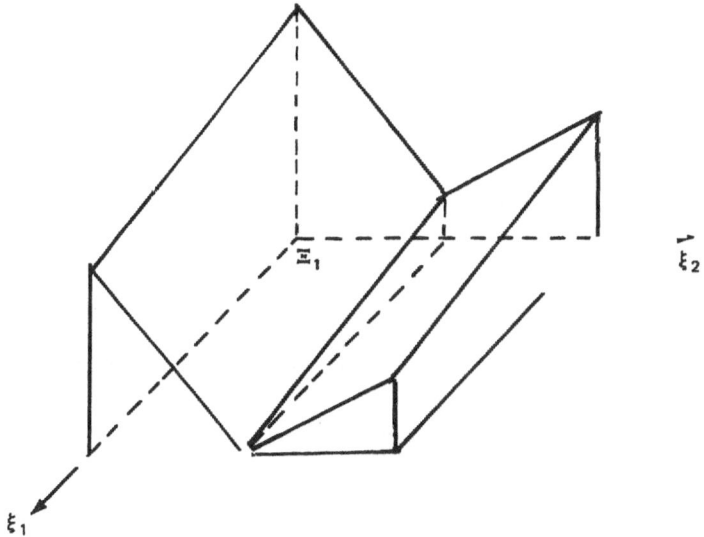

Figure 2.6   Illustration of the strategy of partitioning

Generally, it is very difficult to divide sets Ξ_ℓ into subregions in which $Q(\tilde{x}, \cdot)$ is linear. Moreover, it is convenient to use cutting planes orthogonal to coordinate axes in R^m, since independence of the components $\xi_1, \xi_2, \ldots, \xi_m$ in rectangular subregions is useful for calculating upper bounds. Still, for each Ξ_ℓ selected to be further divided we can choose the coordinate along which $Q(\tilde{x}, \cdot)$ is "mostly nonlinear".

How can we estimate the extent of the nonlinearity of $Q(\tilde{x}, \cdot)$ with respect to $\xi_1, \xi_2, \ldots, \xi_m$ in the subset Ξ_ℓ? Let us observe that for calculating the upper bound (2.44) we solve the second stage problem

$$\begin{aligned}
\text{minimize} \quad & q^T y \\
\text{subject to} \quad & Wy = h^{\ell\nu} - T^{\ell\nu}\tilde{x} \\
& y \geq 0,
\end{aligned} \tag{2.45}$$

at each vertex $\tilde{\xi}^{\ell\nu}$, $\nu = 1, 2, \ldots, 2^m$ of the rectangle Ξ_ℓ. From the theory of duality in linear programming we know that the vector of multipliers (prices) $\pi^{\ell\nu}$ corresponding to the constraints in (2.45) is a measure of sensitivity of Q with respect to the right-hand side $h - T\tilde{x}$ at $\tilde{\xi}^{\ell\nu} = (h^{\ell\nu}, T^{\ell\nu})$. If the multipliers are the same at each vertex then $Q(\tilde{x}, \cdot)$ is linear in Ξ_ℓ; otherwise we can select a direction in which the multipliers change most rapidly. One can use here various methods for comparing differences between multiplier vectors, giving rise to many particular strategies of refining, but the basic idea will always be to avoid inefficient cuts.

After dividing some of the subsets Ξ_ℓ we shall have to solve the approximate problem (2.21) again, with a larger number of blocks. This will give us a new point $\tilde{x}$, at which we shall have to repeat our analysis of upper and lower bounds and again select subsets to be divided and directions of cuts.

2.2.5 The case of simple recourse with random right-hand sides

In this section we improve and simplify our previous results concerning error bounds and refining strategies in a special case of linear two-stage problems of stochastic programming, so called problems with *simple recourse*. The main feature of these problems is that the matrix W in (2.12) is of the form

$$W = [I, -I]$$

where I denotes the *identity matrix* in R^{m_2}. We shall also assume that the cost vector q and the matrix T in (2.12) are deterministic, and only the right hand side $h(\omega)$ is random.

After substituting $Tx = \chi$ and $y = [y^+, y^-]$, $q = [q^+, q^-]$ we can rewrite the second stage problem (2.12) as follows

$$
\begin{aligned}
\text{minimize} \quad & (q^+)^T y^+ + (q^-)^T y^- \\
\text{subject to} \quad & y^+ - y^- = h(\omega) - \chi, \\
& y^+ \geq 0, y^- \geq 0.
\end{aligned}
\qquad (2.46)
$$

We shall denote the optimal value of this problem by $Q(\chi, h)$.

Owing to the special form of constraints in (2.46), we can now split it into m_2 independent linear problems with only one constraint:

$$
\begin{aligned}
\text{minimize} \quad & q_j^+ y_j^+ + q_j^- y_j^- \\
\text{subject to} \quad & y_j^+ - y_j^- = h_j(\omega) - \chi_{j'} \\
& y_j^+ \geq 0, y_j^- \geq 0,
\end{aligned}
\qquad (2.47)
$$

$j = 1, 2, \ldots, m_2$. If we denote the optimal objective values in subproblems (2.47) by $Q_j(\chi_j, h_j)$, $j = 1, 2, \ldots, m_2$, we may write

$$Q(\chi, h) = \sum_{j=1}^{m_2} Q_j(\chi_j, h_j). \qquad (2.48)$$

It is the above separable structure of the two stage problem (2.46) that substantially simplifies error bounds and refining strategies.

Before we pass on to this matter, let us briefly discuss conditions of solvability of the second stage problem. Observe that if $q_j^+ + q_j^- < 0$ for some j, then (2.47) has an unbounded solution: $y_j^- = t$, $y_j^+ = t + h_j(\omega) - \chi_j$, $t \to \infty$, for which $Q_j = t(q_j^+ + q_j^-) + q_j^+(h_j(\omega) - \chi_j) \to -\infty$, as $t \to +\infty$. Conversely, for $q_j^+ + q_j^- \geq 0$ problem (2.47) has an optimal solution defined as follows:

$$
\begin{aligned}
&\text{if } h_j(\omega) - \chi_j \geq 0 \text{ then } y_j^+ = h_j(\omega) - \chi_j, y_j^- = 0; \\
&\text{if } h_j(\omega) - \chi_j < 0 \text{ then } y_j^+ = 0, y_j^- = -h_j(\omega) + \chi_j.
\end{aligned}
\qquad (2.49)
$$

Therefore the condition $q^+ + q^- \geq 0$ is necessary and sufficient for solvability of the second stage problem (2.46) at any $h(\omega)$ and any $\chi = Tx$ (cf. [12]). From now on we shall assume that this condition is satisfied.

The first important observation concerning our problem is that the expected value $EQ(\chi, h(\omega))$ of the recourse function can be calculated exactly at any $\chi = Tx$. Indeed, by (2.48)

$$Eq(\chi, h(\omega)) = \sum_{j=1}^{m_2} EQ_j(\chi_j, h_j(\omega)), \tag{2.50}$$

where, according to (2.49), each Q_j is of the form

$$Q_j(\chi_j, h_j(\omega)) = \begin{cases} q_j^+ (h_j(\omega) - \chi_j) & \text{if } h_j(\omega) \geq \chi_j, \\ q_j^- (\chi_j - h_j(\omega)) & \text{if } h_j(\omega) < \chi_j, \end{cases}$$

The dependence of $Q_j(\chi_j, h_j)$ on h_j is illustrated in Figure 2.7. By the linearity of this function in the regions $\{h_j(\omega) \geq \chi_j\}$ and $\{h_j(\omega) < \chi_j\}$ we obtain

$$\begin{aligned} EQ_j(\chi_j, h_j(\omega)) &= q_j^+ (h_j^+(\chi_j) - \chi_j) p_j^+(\chi_j) \\ &+ q_j^- (\chi_j - h_j^-(\chi_j)) p_j^-(\chi_j), \end{aligned} \tag{2.51}$$

where

$$h_j^+(\chi_j) = E\{h_j(\omega)/h_j(\omega) \geq \chi_j\},$$
$$h_j^-(\chi_j) = E\{h_j(\omega)/h_j(\omega) < \chi_j\}$$

are conditional expectations of h_j in the areas of linearity of $Q_j(\chi_j, \cdot)$, and

$$p_j^+(\chi_j) = P\{h_j(\omega) \geq \chi_j\},$$
$$p_j^-(\chi_j) = P\{h_j(\omega) < \chi_j\}$$

are the corresponding probabilities. The function $EQ_j(\chi_j, h_j(\omega))$ is illustrated in Figure 2.8, where $[a_j, b_j]$ denotes the support of $h_j(\omega)$. We see that if $\chi_j < a_j$, then $p_j^+(\chi_j) = 1$, $p_j^-(\chi_j) = 0$ and the function is linear in χ_j with the slope $-q_j^+$. An analogous situation occurs for $\chi_j > b_j$ and the slope is equal to q_j^-. Within the support of $h(\omega)$ the function is convex and its minimum depends on q_j^+, q_j^- and of course on the distribution of $h_j(\omega)$.

From (2.50) and (2.51) we finally get

$$\begin{aligned} EQ(\chi, h(\omega) &= \sum_{j=1}^{m_2} [q_j^+ (h_j^+(\chi_j) - \chi_j) p_j^+(\chi_j) \\ &+ q_j^- (\chi_j - h_j^-(\chi_j)) p_j^-(\chi_j)]. \end{aligned} \tag{2.52}$$

Practical application of formula (2.52) is relatively easy, since it requires only one-dimensional integration for calculating the quantities p_j^+, p_j^-, h_j^+ and h_j^- at

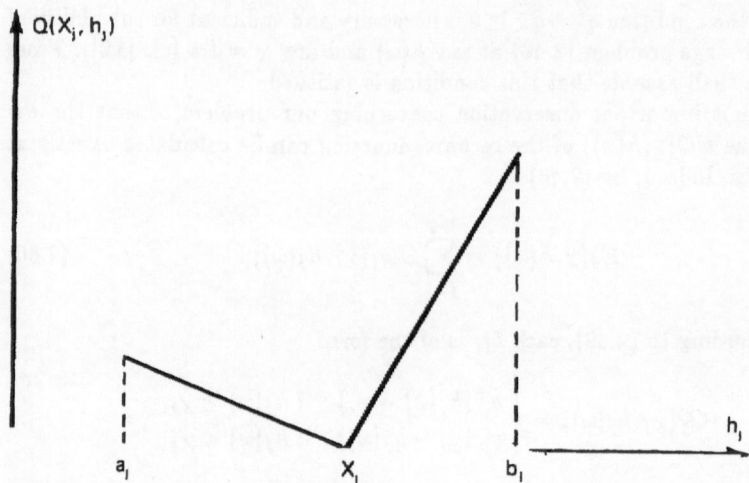

Figure 2.7   The j-th component of the recourse cost in a problem with simple recourse

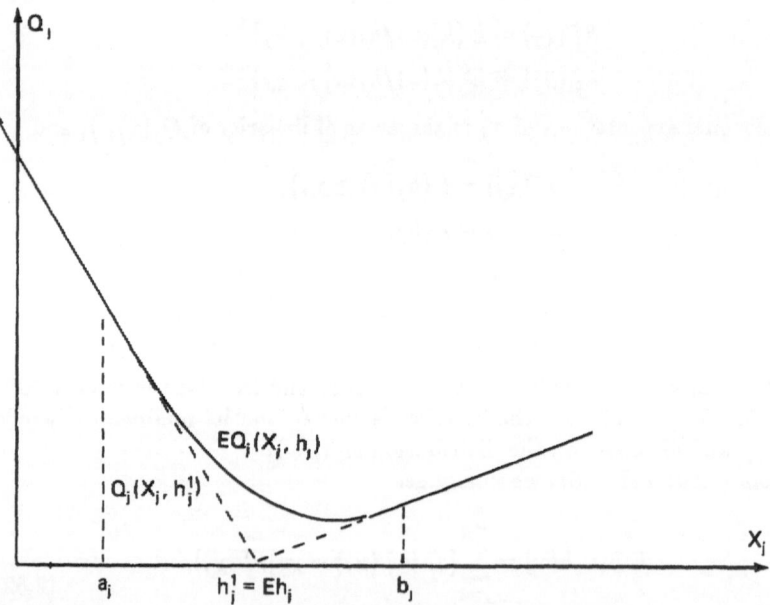

Figure 2.8   The j-th component of the expected recourse cost and its approximation

a given χ_j, contrary to the general two stage problem, where multidimensional integrals would have to be evaluated.

Since we can exactly evaluate the objective $\psi(x) = c^T x + EQ(Tx, h(\omega))$ at any x, we no longer need upper and lower bounds for this value. One may ask here, whether we need approximation methods at all, if the objective values can be computed easily. There is no general answer to this question, but approximation schemes may still prove useful, since the approximating problems are linear, while the original one is nonlinear in general, as we see from Figure 2.8. But if we use approximation methods we shall still need lower bounds for the *minimum* value of the objective $\psi(x)$ and appropriate refining strategies.

Similarly to the way of evaluating the objective, both these operations—calculation of lower bounds and refining of the partition—can be carried out separately for each coordinate of $h(\omega)$. Let the coordinates $h_j(\omega)$ be distributed in intervals $[a_j, b_j]$, $j = 1, 2, \ldots, m_2$, so that the support of $h(\omega)$ is contained in the hyper-rectangle $\Xi = X_{j=1}^{m_2}[a_j, b_j]$. In an analogous way to the case of complete recourse, we solve at first the approximating linear programming problem (2.21) with only one block $\Xi^1 = \Xi$:

$$
\begin{aligned}
\text{minimize} \quad & c^T x + (q^+)^T y^+ + (q^-)^T y^- \\
\text{subject to} \quad & Ax = b, \\
& Tx + Iy^+ - Iy^- = h^1, \\
& x \geq 0, y^+ \geq 0, y^- \geq 0,
\end{aligned}
\tag{2.53}
$$

where $h^1 = Eh(\omega)$. Let $(\tilde{x}, \tilde{y}^+, \tilde{y}^-)$ be a solution to this problem. Then obviously each pair $(\tilde{y}_j^+, \tilde{y}_j^-)$, $j = 1, 2, \ldots, m_2$, is a solution to the j-th piece of the second stage problem:

$$
\begin{aligned}
\text{minimize} \quad & q_j^+ y_j^+ + q_j^- y_j^- \\
\text{subject to} \quad & y_j^+ - y_j^- = h_j^1 - \tilde{\chi}_j, \\
& y_j^+ \geq 0, y_j^- \geq 0,
\end{aligned}
\tag{2.54}
$$

with $\tilde{\chi}_j$ being the j-th coordinate of $T\tilde{x}$ (cf. (2.47)). Since $h_j^1 = Eh_j(\omega)$ and the function $Q_j(\tilde{\chi}_j, \cdot)$ is convex (see Figure 2.7), from *Jensen's inequality* (cf. Section 2.2.3) we obtain

$$
Q_j(\tilde{\chi}_j, h_j^1) \leq EQ_j(\tilde{\chi}_j, h_j(\omega)), \quad j = 1, 2, \ldots, m_2,
\tag{2.55}
$$

and

$$
\begin{aligned}
c^T \tilde{x} + \sum_{j=1}^{m_2} Q_j(\tilde{\chi}_j, h_j^1) &\leq \min_{Ax=b, x \geq 0}[c^T x + EQ(Tx, h(\omega))] \\
&\leq c^T \tilde{x} + \sum_{j=1}^{m_2} EQ_j(\tilde{\chi}_j, h_j(\omega)).
\end{aligned}
\tag{2.56}
$$

The left side of (2.56) we obtain from the approximating problem (2.53), while the right side represents the objective value at $\tilde{x}$ and can be calculated by (2.52).

If we had equalities in (2.55) for all j, the point $\tilde{x}$ would minimize our original objective function, as follows from (2.56). If this is not the case, the differences between both sides of (2.55) are measures of accuracy of our approximation with respect to each coordinate $h_j(\omega)$ of $h(\omega)$, $j = 1, 2, \ldots, m_2$. Hence, we can select the coordinates for which the accuracy is not sufficient and split the corresponding intervals $[a_j, b_j]$. It follows from Figure 2.7 that it is most efficient to divide them at $\tilde{\chi}_j$, since the function $Q_j(\tilde{\chi}_j, \cdot)$ will be linear in the resulting subintervals $[a_j, \chi_j]$ and $[\chi_j, b_j]$. Obviously, $\chi_j \in [a_j, b_j]$ for the selected coordinates, since otherwise we would have either $h_j^+(\chi_j) = h_j^1$ or $h_j^-(\chi_j) = h_j^1$ and an equality in (2.55) (see Figure 2.8).

The partition of the intervals $[a_j, b_j]$ defines a new partition of the rectangle Ξ into subregions $\Xi_1, \Xi_2, \ldots, \Xi_L$. With this partition we solve (2.21) again and obtain a new point $\tilde{x}$ for which the analysis of accuracy can be also carried out component-wise. Indeed, in each subinterval $[a_j^k, b_j^k]$ of $[a_j, b_j]$ we have Jensen's inequality similar to (2.55),

$$Q_j(\tilde{\chi}_j, h_j^k) \leq E\{Q_j(\tilde{\chi}_j, h_j(\omega))/h_j(\omega) \in [a_j^k, b_j^k]\}, \qquad (2.57)$$

where h_j^k is the conditional expectation in $[a_j^k, b_j^k]$

$$h_j^k = E\{h_j(\omega)/h_j(\omega) \in [a_j^k, b_j^k]\}. \qquad (2.58)$$

In a similar way to (2.52) we can also calculate the exact value of the conditional expectation $E\{Q_j(\tilde{\chi}_j, h_j(\omega))/h_j(\omega) \in [a_j^k, b_j^k]\}$. Again, if $\tilde{\chi} \notin [a_j^k, b_j^k]$ then $Q_j(\tilde{\chi}_j, \cdot)$ is linear in $[a_j^k, b_j^k]$ and (2.57) becomes an equality. Therefore we divide only those subintervals, for which $\tilde{\chi}_j \in [a_j^k, b_j^k]$ and the corresponding accuracy in (2.57) is not sufficient (i.e. at most one subinterval for each component). This strategy of refining is illustrated in Figure 2.9.

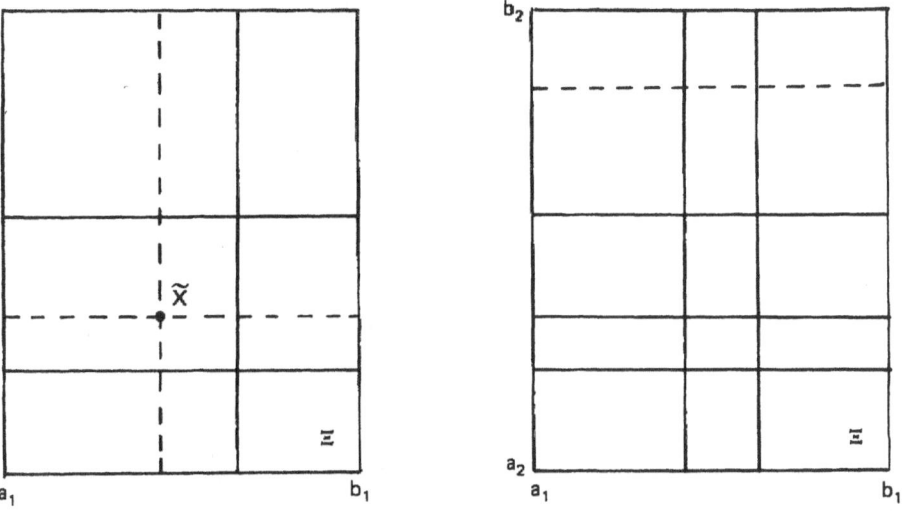

Figure 2.9 The strategy of partitioning in problems with simple recourse

2.3 Chance Constrained Programming

Another way of formulating optimization models for problems which involve random parameters is the use of *chance constraints*. If in our linear model with objective $c^T x$ and constraints $Tx \geq h$, $x \geq 0$ some entries of the matrix T or the right-hand side h are random, we can formulate the corresponding optimization problem as follows

$$
\begin{aligned}
\text{minimize} \quad & c^T x \\
\text{subject to} \quad & P\{T(\omega)x \geq h(\omega)\} \geq \alpha, \\
& x \geq 0,
\end{aligned}
\tag{2.59}
$$

where $0 \leq \alpha \leq 1$ is a prescribed reliability level. Problem (2.59) is called the stochastic programming problem with *joint chance constraints*. Another possibility of formulating such constraints is to impose reliability levels for each row of the relation $T(\omega)x \geq h(\omega)$ (so called *disjoint chance constraints*), which yields the problem

$$
\begin{aligned}
\text{minimize} \quad & c^T x \\
\text{subject to} \quad & P\{T_j(\omega)x \geq h_j(\omega)\} \geq \alpha_j, \quad j = 1, 2, \ldots, m, \\
& x \geq 0,
\end{aligned}
\tag{2.60}
$$

with $0 \leq \alpha_j \leq 1$, and $T_j(\omega)$ indicating the j-th row of $T(\omega)$. Problems (2.59) and (2.60) can be regarded as natural generalizations of common linear programming problems to the case of random constraint coefficients.

These problems, however, are no longer linear, since the constraint functions

$$g(x) = P\{T(\omega)x \geq h(\omega)\} \tag{2.61}$$

and

$$g_j(x) = P\{T_j(\omega)x \geq h_j(\omega)\}, \quad j = 1, 2, \ldots, m \tag{2.62}$$

are in general nonlinear. Moreover, it may turn out that these functions are not concave and the feasible sets

$$X_1(\alpha) = \{x \geq 0 : g(x) \geq \alpha\} \tag{2.63}$$

or

$$X_2(\alpha_1, \alpha_2, \ldots, \alpha_m) = \bigcap_{j=1}^{m} X_{2j}(\alpha_j), \tag{2.64}$$

$$X_{2j}(\alpha_j) = \{x \geq 0 : g_j(x) \geq \alpha_j\},$$

may be nonconvex and even disconnected. An extensive discussion of convexity properties of chance-constrained programs can be found in [12], [24] and [35]. Below we summarize only the simplest results.

If only the right-hand side $h(\omega)$ is random, then the sets $X_{2j}(\alpha_j)$ are convex for all $0 \leq \alpha_j \leq 1$, $j = 1, 2, \ldots, m$. Convexity properties of the set $X_1(\alpha)$ depend, however, on the distribution of $h(\omega)$. From the general theory of so-called *logarithmic concave* and *quasi-concave* probability measures (cf. [22], [24], [4] and [27]) it follows that for a normal distribution of $h(\omega)$ the set $X_1(\alpha)$ is convex and closed for each $0 \leq \alpha \leq 1$.

When also the technology matrix T is random, up to now no general convexity statements are available. We know that $X_{2j}(\alpha_j)$ are convex for normally distributed $T_j(\omega)$ and $h_j(\omega)$, under the condition that $\frac{1}{2} \leq \alpha_j \leq 1$. Special conditions have also been found for some other particular distributions (see [17], [23]).

Let us now discuss possible approaches to solving chance-constrained problems of the type (2.59) and (2.60). The most straightforward one is to use nonlinear programming techniques for constrained optimization. These techniques, however, require calculation of constraint functions (2.61) or (2.62) and their gradients (if they exist) at successively generated points, which in general is a rather difficult task involving multidimensional integration. Still, with only the right-hand side $h(\omega)$ random and for some special classes of distributions, application of fast simulation techniques (cf. [5]) makes this approach effective, as practical examples of [25] and [26] show.

We may also try to approximate (2.59) or (2.60) by another optimization problem which would be easier to solve.

One approach is to approximate the random variable $\xi(\omega) = (h(\omega), T(\omega))$ by a discretely distributed one. If $\tilde{\xi}$ is such an approximation, with

$$\begin{aligned}
P\{\tilde{\xi} = (h^1, T^1)\} &= p_1 > 0, \\
P\{\tilde{\xi} = (h^2, T^2)\} &= p_2 > 0, \\
&\cdots \\
P\{\tilde{\xi} = (h^L, T^L)\} &= p_L > 0,
\end{aligned} \tag{2.65}$$

$$\sum_{\ell=1}^{L} p_\ell = 1$$

then the problem that approximates (2.59) takes on the form

$$\text{minimize} \quad c^T x$$

$$\text{subject to} \quad \tilde{g}(x) = \sum_{\ell=1}^{L} p_\ell \gamma_\ell(x) \geq \alpha \tag{2.66}$$

$$x \geq 0,$$

where for $\ell = 1, 2, \ldots, L$

$$\gamma_\ell(x) = \begin{cases} 1 & \text{if } T^\ell x \geq h^\ell, \\ 0 & \text{otherwise.} \end{cases} \tag{2.67}$$

Interesting results concerning the convergence of the feasible set of (2.66) to the feasible set of (2.59) as the accuracy of discretization increases have been obtained in [29]. So far we do not know much about the practical efficiency of this approach. It may be, however, limited by the fact that the functions (2.67) are discontinuous and the feasible set of (2.66) may be nonconvex and disconnected, even in the feasible region of (2.59) is convex.

Another possibility is to replace (2.59) by a two-stage problem.

$$\underset{x \geq 0}{\text{minimize}}\, c^T x + EQ(x, \xi(\omega)) \tag{2.68}$$

where $Q(x, \xi(\omega))$ is the minimum objective value in the second stage problem

$$\text{minimize} \quad q^T y$$

$$\text{subject to} \quad Wy \geq h(\omega) - T(\omega)x, \tag{2.69}$$

$$y \geq 0$$

with some $q \in R^m$, $q \geq 0$, and a certain recourse matrix W. The simplest choice of these parameters would be $q_j = M$, $j = 1, 2, \ldots, m$, with some large $M > 0$ and $W = I$ (simple recourse). The idea of this approximation is to introduce the penalty $q^T y$ for not satisfying the constraints $T(\omega)x \geq h(\omega)$. We can see it directly in the simple recourse case: the solution $y(x, \omega)$ to (2.69) is given by

$$y_j(x, \omega) = \max\{0, h_j(\omega) - T_j(\omega)x\}, \quad j = 1, 2, \ldots, m, \tag{2.70}$$

and $E\{q^T y(x, \omega)\}$ is an average cost of violating the constraint $T(\omega)x \geq h(\omega)$.

Problems (2.68)-(2.69) and (2.59) are not equivalent, but under reasonable assumptions one can prove that the probability of satisfying $T(\omega)\hat{x} \geq h(\omega)$ at the solution $\hat{x}$ to (2.68)-(2.69) tends to 1, as $q_j \to +\infty$, $j = 1, 2, \ldots, m$. Of course, in practice we shall have to experiment with values of q so as to achieve the required level of probability of satisfying chance constraints with reasonable values of $c^T x$.

2.4 Game-theoretic Models and Worst-case Approximations

So far we analyzed stochastic programming models in which distributions of random parameters were known, and our main concern was to find efficient solution techniques. However, in practice we often encounter stochastic problems in which statistical properties of some parameters are known only to a certain extent, e.g. only their supports and expected values are available. In such situations a fundamental question arises, whether it is possible to properly define a concept of a solution and to develop methods for finding such solutions. We shall show that a special approximation of the original problem, so-called *worst-case approximation*, may help us to answer these questions.

To formulate the problem under consideration more precisely, let us assume that the objective function of our optimization problem is defined as a mathematical expectation

$$F(x) = Ef(x,\xi) = \int_{\Xi_*} f(x,\xi) P_*(d\xi), \tag{4.71}$$

where $x \in R^n$ denotes the decision vector, $\Xi_* \subset R^m$ is the support of the vector of random parameters ξ, P_* is the probability measure on Ξ_* describing the distribution of ξ, and $f : R^n \times R^m \to R^1$. Next, suppose that the distribution of ξ is not known exactly; we know only a certain outer approximation $\Xi \subset R^m$ of the support Ξ_*,

$$\Xi_* \subset \Xi, \tag{2.72}$$

and expectations of some functions $g_1, g_2, \ldots, g_k$,

$$Eg_i(\xi) = \int_{\Xi_*} g_i(\xi) P_*(d\xi) = \mu_i, \quad i = 1, 2, \ldots, k. \tag{2.73}$$

In particular, equations (2.73) may represent our knowledge about the moments of ξ: setting, for instance, $k = m$ and $g_i(\xi) = \xi_i$, $i = 1, 2, \ldots, m$, we obtain from (2.73) conditions on the expected value of ξ, $E\xi_i = \mu_i$, $i = 1, 2, \ldots, m$.

Since the distribution of ξ is not known, we are not able to calculate or approximate with a reasonable accuracy the value of the objective $F(x)$, and thus looking for a vector x that minimizes (2.71) is out of question. We have to reformulate the problem in such a way that the new formulation will involve only information that is really available. A *game-theoretic approach* initiated in the area of stochastic programming in [10], [36] provides a way to overcome this difficulty.

Let P be the class of probability measures P on R^m satisfying the following conditions:

$$P(\Xi) = 1, \tag{2.74}$$

$$\int_{\Xi} g_i(\xi) P(d\xi) = \mu_i, \quad i = 1, 2, \ldots, k. \tag{2.75}$$

It follows from (2.72) and (2.73) that the "true" measure P_* belongs to P; on the other hand all measures $P \in P$ cannot be distinguished on the basis of the information available to us. Therefore it seems reasonable to assume the worst case and consider the function

$$\overline{F}(x) = \sup_{P \in \mathcal{P}} \int_{\Xi} f(x, \xi) P(d\xi). \tag{2.76}$$

Obviously, for each x we have $\overline{F}(x) \geq F(x)$, hence after minimizing $\overline{F}$ with respect to x the value of the "real" objective will be at least as good as the value of $\overline{F}$.

The definition of $\overline{F}$ involves only information that is available, but it requires the operation of maximization with respect to probability distributions, which in general is extremely difficult and unsuitable for practical calculations. Still, it turns out that in many important cases we are able to carry out this operation analytically, and the distribution at which the maximum of the integral in (2.76) is attained does not depend on x and has a special and easy to handle form.

Let us assume that Ξ is a convex, closed and bounded polyhedron, and let ξ^ν, $\nu = 1, 2, \ldots, N$ denote its vertices. Furthermore, let the functions g_i, $i = 1, 2, \ldots, k$, in (2.75) be linear and the function $f(x, \xi)$ be convex in ξ for each x. Then one can prove that the supremum in (2.76) is attained at a measure $\hat{P}$ (generally dependent on x) concentrated at vertices ξ^ν:

$$\hat{P}(\{\xi^\nu\}) = p_\nu, \quad \nu = 1, 2, \ldots, N, \tag{2.77}$$

$$p_\nu \geq 0, \quad \nu = 1, 2, \ldots, N, \tag{2.78}$$

$$\sum_{\nu=1}^{N} p_\nu = 1. \tag{2.79}$$

Values of probabilities p_ν associated with the vertices of Ξ should satisfy, besides (2.78) and (2.79), the following equations that result from (2.75):

$$\sum_{\nu=1}^{N} p_\nu g_i(\xi^\nu) = \mu_i, \quad i = 1, 2, \ldots, k. \tag{2.80}$$

Hence, $\overline{F}(x)$ is the optimal value of the linear programming problem

$$\underset{\{p_\nu\}}{\text{maximize}} \quad \sum_{\nu=1}^{N} p_\nu f(x, \xi^\nu) \tag{2.81}$$

subject to $(2.78), (2.79)$ and (2.80)

Obviously, problem (2.81) is much easier to sove that (2.76) and makes the concept of worst-case approximations implementable.

In some cases the task of calculating the upper bound $\overline{F}(x)$ may be simplified even further, because it may turn out that the feasible set of (2.81), defined by (2.78)–(2.80), contains exactly one point. To illustrate this possibility, suppose that ξ is a scalar random variable, $\Xi = [a, b]$, and additional conditions (2.73) comprise only one equation regarding the expected value: $E\xi = \mu$. Then (2.78) -(2.80) uniquely determine the probabilities describing the extremal distribution $\hat{P}$:

$$\hat{P}(\{a\}) = p_1 = \frac{b - \mu}{b - a},$$

$$\hat{P}(\{b\}) = p_2 = \frac{\mu - a}{b - a}. \tag{2.82}$$

These probabilities do not depend on x, hence our worst-case approximation resolves itself to replacing the random variable ξ in (2.71) by the random variable $\hat{\xi}$, attaining values a and b with probabilities p_1 and p_2.

It is interesting to observe that the distribution of $\hat{\xi}$ defined in this way is identical with that used for the Edmundson-Madansky inequality (cf. (2.42), (2.43)), which is quite natural, because essentially we consider the same problem of finding an upper bound for the integral (2.71).

The above observations can be easily extended to the multidimensional case, provided that Ξ is a hyper-rectangle $X_{j=1}^m [a_j, b_j]$, $f(x, \xi)$ is *separable* with respect to the coordinates ξ_j, i.e.

$$f(x, \xi) = \sum_{j=1}^m f_j(x, \xi_j), \tag{2.83}$$

and conditions (2.73) are of the form

$$E\xi_j = \mu_j, \quad j = 1, 2, \ldots, m. \tag{2.84}$$

Under these assumptions, the worst-case approximation to (2.71) can be obtained by replacing ξ with a discrete random vector $\hat{\xi}$ having coordinates $\hat{\xi}_j, j = 1, 2, \ldots, m$, distributed similarly to (2.82):

$$P\{\hat{\xi}_j = a_j\} = \frac{b_j - \mu_i}{b_j - a_j},$$

$$P\{\hat{\xi}_j = b_j\} = \frac{\mu_j - a_j}{b_j - a_j} \tag{2.85}$$

The above result can be directly applied to two-stage problems with simple recourse (cf. Section 2.2.5), since objective functions of these problems possess the required property of separability with respect to the coordinates of the random vector h, see (2.48). One can further extend this result to some problems with a nonseparable objective $f(x, \xi)$. Namely, assuming that we know in advance that the coordinates $\xi_j, j = 1, 2, \ldots, m$, are *independent* random variables, we can restrict the class of measures considered to such probability measures on R^{in} that satisfy (2.74), (2.75) and can be expressed as products of measures with respect to the coordinates. Under this assumption, for a hyper-rectangle Ξ the worst-case distribution does not depend on x and is defined again by (2.85).

Interesting extensions and generalizations of the idea of worst-case approximations in stochastic programming can be found in [6] and [7].

References

[1] H. Attouch and R.J-B. Wets, "Approximation and convergence in nonlinear optimization", in *Nonlinear Programming* 4, eds. O. Mangasarian, R. Meyer and S. Robinson, Academic Press, New York, 1981, pp. 367–394.

[2] M.S. Bazaraa and C.M. Shetty, *Nonlinear programming: Theory and algorithms*, John Wiley, New York, 1979.

[3] J. Birge and R.J-B. Wets, "Designing approximation schemes for stochastic optimization problems, in particular for stochastic programs with recourse", Working Paper WP-83-111, IIASA, Laxenburg, 1983.

[4] C. Borell, "Convex set functions in d-spaces", *Period. Math. Hungar* 6(1975), 111–136.

[5] I. Deák, "Three digit accurate multiple normal probabilities", *Numerische Mathematik* 35(1980), 369–380.

[6] J. Dupačová, "Minimax stochastic programs with nonseparable penalties", Proceedings of the 9th IFIP Conference, Warsaw 1979. Part 1, *Lecture Notes in Control and Information Sciences* 22, Springer-Verlag, Berlin, 1980, pp. 157–163.

[7] Yu Ermoliev, A. Gaivoronski and C. Nedeva, "Stochastic optimization problems with incomplete information on distribution functions", Working Paper WP-83-113, IIASA, Laxenburg, 1983.

[8] P.B. Hausch and W.T. Ziemba, "Bounds on the value of information in uncertain decision problems", *Stochastics* 10(1983), 181–217.

[9] C. Huang, W.T. Ziemba and A. Ben-Tal, "Bounds on the expectation of a convex function with a random variable with applications to stochastic programming", *Operations Research* 25(1977), 315–325.

[10] M. Iosifescu and R. Theodorescu, "Sur la programmation lineaire", *C.R. Acad. Sci. Paris* 256(1963), 4831–4833.

[11] P. Kall, "Approximations to stochastic programs with complete fixed recourse", *Numerische Mathematik* 22(1974), 333–339.

[12] P. Kall, *Stochastic programming*, Springer-Verlag, Berlin, 1976.

[13] P. Kall, "Computational methods for solving two-stage stochastic linear programming problems", *Z. Angew. Math. Phys.* 30(1979), 261–271.

[14] P. Kall and D. Stoyan, "Computational methods for solving two-stage stochastic linear programming problems with recourse including error bounds", *Math. Operationsforsch. Statist., Ser. Opt.*, 13(1982), 431–447.

[15] P. Kannapian and S.M.A. Sastry, "Uniform convergence of convex optimization problems", *J. Math. Anal. Appl.* 96(1983), 1–12.

[16] D.G. Luenberger, *Introduction to linear and nonlinear programming*, Addison-Wesley, Reading, 1973.

[17] K. Marti, "Konvexitätsaussagen zum linearen Stochastischen Optimierungsproblem", *Z. Wahrscheinlichkeitstheorie u. verw. Geb.* 18(1971), 159–166.

[18] K. Marti, *Approximationen Stochastischer Optimierungsproblemen*, Verlag Anton Hain, Königstein, 1979.

[19] K. Marti, "Computations of descent directions in stochastic optimization problems having a discrete distribution", *ZAMM* 64(1984), T336–T338.

[20] C. van de Panne and W. Popp, "Minimum cost cattle feed under probabilistic problem constraints", *Management Sci.* 9(1963), 405–430.

[21] E. Polak, *Computational methods in optimization*, Academic Press, New York, 1971.

[22] A. Prékopa, "Logarithmic concave measures with applications to stochastic programming", *Acta Sci. Meth.* 32(1971), 301–316.

[23] A. Prékopa, "Programming under probabilistic constraints with a random technology matrix", *Math. Operationsforsch. Statist.* 5(1974), 109–116.

[24] A. Prékopa, "Logarithmic concave measures and related topics", in *Stochastic Programming*, ed. M. Dempster, Academic Press, London, 1980, pp. 63–82.

[25] A. Prékopa and P. Kelle, "Reliability type inventory models based on stochastic programming", *Mathematical Programming Study* 9(1983), 43–58.

[26] A. Prékopa and T. Szantai, "Flood control reservoir system design using stochastic programming, *Mathematical Programming Study* 9(1983), 138–151.

[27] Y. Rinott, "On convexity of measures", *Annals of Probability* 4(1976), 1020–1026.

[28] J.B. Rosen, "Primal partition programming for block diagonal matrices", *Numerische Mathematik* 6(1964), 250–260.

[29] G. Salinetti, "Approximations for chance-constrained programming problems", *Stochastics* 10(1983), 157–169.

[30] G. Salinetti and R.J-B. Wets, "On the relations between two types of convergence for convex functions", *J. Math. Anal. Appl.* 60(1977), 211–226.

[31] L. Schwartz, *Analyse mathématique*, Hermann, Paris, 1967.

[32] D. Stoyan, *Qualitative Eigenschaften und Abschatzungen stochastischer Modelle*, Oldenbourg Verlag, München-Wien, 1977.

[33] B. Strazicky, "On an algorithm for solution of the two-stage stochastic programming problem", *Methods of Operations Research* XIX(1974), 142–156.

[34] R. Wets, "Convergence of convex functions, variational inequalities and convex optimization problems", in *Variational Inequalities and Complementarity Problems*, eds. R. Cottle, F. Gianessi and J.-L. Lions, John Wiley, New York, 1980, pp. 375–403.

[35] R. Wets, "Stochastic programming: Solution techniques and approximation schemes", in *Mathematical programming: the state of the art*, eds. A. Bachem, M. Gröschel and B. Korte, Springer-Verlag, Berlin, 1983, pp. 566–603.

[36] J. Zácková, "On mimimax solutions of stochastic linear programming", *Casopis pro pestováni matematiky* 91(1966), 430–433.

CHAPTER 3

LARGE SCALE LINEAR PROGRAMMING TECHNIQUES

R.J-B. Wets

We study the use of large scale linear programming techniques for solving (linear) recourse problems[1] whose random elements have discrete distributions (with finite support) more precisely for problems of the type:

$$
\begin{aligned}
&\text{find } \ x \in R_+^{n_1} \\
&\text{such that } \ Ax = b \\
&\text{and } \ z = cx + \mathcal{Q}(x) \text{ is minimized}
\end{aligned}
\tag{3.1}
$$

where

$$
\mathcal{Q}(x) = \sum_{\ell=1}^{L} p_\ell Q(x, \xi^\ell) = E\{Q(x, \xi(\omega))\}
\tag{3.2}
$$

and for each $\ell = 1, \ldots, L$, the *recourse cost* $Q(x, \xi^\ell)$ is obtained by solving the *recourse problem*:

$$
Q(x, \xi^\ell) = \inf\{q^\ell y | W y = h^\ell - T^\ell x, y \in R_+^{n_2}\}
\tag{3.3}
$$

where

$$
\xi^\ell = (q^\ell, h^\ell, T^\ell) = (q_1^\ell, \ldots, q_{n_2}^\ell; h_1^\ell, \ldots, h_{m_2}^\ell; t_{11}^\ell, \ldots, t_{1n_1}^\ell, \ldots, t_{m_2 n_1}^\ell)
$$

i.e.

$$
\varsigma^\ell \in R^N \text{ with } N = n_2 + m_2 + m_2 \cdot n_1
$$

and

$$
p_\ell = \text{Prob } [\xi(\omega) = \xi^\ell].
$$

---

[1] The potential use of large scale programming techniques for solving stochastic programs with chance-constraints appears to be less promising and has not yet been investigated. The approximation scheme for chance-constraints proposed by Salinetti, 1983, would, if implemented require a detailed analysis of the structural properties of the resulting (large-scale) linear programs. Much of the analysis laid out in this Section would also be applicable to that case but it appears that further properties—namely the connections between the upper and lower bounding problems—should be exploited.

The sizes of the matrices are consistent with $x \in R^{n_1}, y \in R^{n_2}, b \in R^{m_1}$ and for all $\ell, h_\ell \in R^{m_2}$; for a more detailed description of the recourse model consult Part I of this Volume. Because W is nonstochastic we refer to this problem as a model with *fixed recourse*. The ensuing development is aimed at dealing with problems that exhibit no further structural properties. Problems with simple recourse for example, i.e. when $W = (I, -I)$, are best dealt with in a nonlinear programming framework, cf. Chapter 4.

Before we embark on the description of solution strategies for the problem at hand, it is useful to review some of the ways in which a problem of this type might arise in practice. First, the problem is indeed a linear recourse model whose random elements follow a known discrete distribution function. In that case either q or h or T is random, usually not all three matrices at once, but the number of independent random variables is liable to be relatively large and even if each one takes on only a moderate number of possible values, the total number L of possible vectors ξ^ℓ could be truly huge, for example a problem with 10 independent random variables each taking on 10 possible values leads us to consider 10 billion $(= L)$ 10-dimensional vectors ξ^ℓ. Certainly not the type of data we want, or can, keep in fast access memory.

Second, the original problem is again a stochastic optimization problem of the recourse type but (3.1) is the result of an approximation scheme, either a discretization of an absolutely continuous probability measure or a coarser discretization of a problem whose "finite" number of possible realizations is too large to contemplate; for more about approximation schemes consult Chapter 2. In this case L, the number of possible values taken on by $\xi(\cdot)$, could be relatively small, say a few hundreds, in particular if (3.1) is part of a sequential approximation scheme, details can be found in Chapter 2, see also Birge and Wets [2], for example.

Third, the original problem is a stochastic optimization problem but we have only very limited statistical information about the distribution of the random elements, and $\xi^1, \ldots, \xi^L$ represents all the statistical data available. Problem (3.1) will be solved using the empirical distribution, the idea being of submitting its solution to statistical analysis such as suggested by the work of Dupačová and Wets [7]. In this case L is usually quite small, we are thinking in terms of L less than 20 or 30.

Fourth, problem (3.1) resulted from an attempt at modeling uncertainty, with no accompanying statistical basis that allows for accurate descriptions of the phenomena by stochastic variables. As indicated in Chapter I, this mostly comes from situations when there is data uncertainty about some parameters (of a deterministic problem) or we want to analyse decision making or policy setting and the future is modeled in terms of scenarios (projections with tolerances for errors). In this case the number L of possible variants of a key scenario that we want to consider is liable to be quite small, say 5 to 20, and the ξ^ℓ can often be expressed as a sum:

$$\xi^\ell = \varsigma^0 + \eta_{1\ell}\,\varsigma^1 + \ldots + \eta_{K\ell}\,\varsigma^k$$

where for $k = 1, \ldots, K$, the $\varsigma^k \in R^N$ are fixed vectors and $(\eta_1(\cdot), \ldots, \eta_K(\cdot))$ are scalar random variables with possible values $\eta_{1\ell}, \ldots, \eta_{K\ell}$ for $\ell = 1, \ldots, L$. We think of K as being 2 or 3. The typical case being when we have a base projection: $\varsigma^\nu + \varsigma^1$, but we want to consider the possibility that certain factors may vary by as much as 25% (plus or minus). In such a case the model assigns to the (only) random variable $\eta_1(\cdot)$ some discrete distribution on the interval [.75,1.25].

With this as background to our study it is natural to search solution procedures for recourse problems with discrete distributions when there is either only a moderate number of vectors ξ^ℓ to consider (scenarios, limited statistical information, approximation) or there is a relatively large number of possible vectors ξ^ℓ that result from combinations of the values taken on by independent random variables. The techniques discussed further on, apply to both classes of problems, but the tendency is to think of software development that would be appropriate for problems with relatively small L, say from 5 to 1,000. Not just because this class of problems appears more manageable but also because when L is actually very large, although finite, the overall solution strategy would still rely on the solution of approximate problems with relatively small L.

3.1 Recourse Models as Large Scale Linear Programs

Substituting in (3.1) the expressions for $\mathcal{Q}$ and Q, we see that we can obtain the solution by solving the linear program:

$$\text{find} \quad x \in R_+^{n_1} \text{ and for } \ell = 1, \ldots, L, y^\ell \in R_+^{n_2}$$
$$\text{such that} \quad Ax = b,$$
$$T^\ell x + W y^\ell = h^\ell, \quad \ell = 1, \ldots, L \qquad (3.4)$$
$$\text{and} \quad z = cx + \sum_{\ell=1}^{L} p_\ell q^\ell y^\ell \text{ is minimized.}$$

To each recourse decision to be chosen if $\xi(\cdot)$ takes on the value $\xi^\ell = (q^\ell, h^\ell, T^\ell)$ corresponds the vector of variables y^ℓ. This is a linear program with

$$m_1 + m_2 \cdot L \quad \text{constraints,}$$

and

$$n_1 + n_2 \cdot L \quad \text{variables.}$$

The possibility of solving this problem using standard linear programming software depends very much on L, but even if it were possible to do so, in order to avoid making the solving of (3.4) prohibitively expensive—in terms of time and required computer memory—it is necessary to exploit the properties of this highly structured large scale linear program. The structure of the tableau of detached coefficients takes on the form:

$$c \quad p_1 q^1 \quad p_2 q^2 \quad \cdots \quad p_L q^L$$

$$A \qquad\qquad\qquad\qquad = b$$

$$T^1 \quad W \qquad\qquad\qquad = h^1$$

$$T^2 \qquad\quad W \qquad\qquad = h^2$$

$$\vdots \qquad\qquad\qquad \ddots \qquad\quad \vdots$$

$$T^L \qquad\qquad\qquad\qquad W = h^L$$

Figure 3.1   Structure of discrete stochastic program

We have here a so-called *dual block angular structure* with the important additional feature that all the matrices, except for A, along the block diagonal are the same. It is this feature that will lead us to the algorithms that are analysed in Section 3.3 and which up to now have provided us with the best computational results. It is also this feature which led Dantzig and Madansky [5], to suggest a solution procedure for (3.4) by way of the dual. Indeed, the following problem is a dual of (3.4):

$$\text{find} \quad \sigma \in R^{m_1}, \text{ and for } \ell = 1, \ldots, L, \pi^\ell \in R^{m_2}$$

$$\text{such that} \quad \sigma A + \sum_{\ell=1}^{L} p_\ell \pi^\ell T^\ell \le c,$$

$$\pi^\ell W \le q^\ell, \quad \ell = 1, \ldots, L \tag{3.5}$$

$$\text{and} \quad w = \sigma b + \sum_{\ell=1}^{L} p_\ell \pi^\ell h^\ell \text{ is maximized.}$$

Problem (3.5) is not quite the usual (formal) dual of (3.4). To obtain the classical linear program dual, set

$$\hat{\pi}^\ell = p_\ell \pi^\ell$$

and substitute in (3.5). This problem has *block angular structure*, the block diagonal consisting again of identical matrices W. The tableau with detached coefficients takes on the form:

$$b \quad p_1 h^1 \quad p_2 h^2 \quad \cdots \quad p_L h^L$$

$$A' \quad p_1 T_1' \quad p_2 T_2' \quad \cdots \quad p_L T_L' \quad \leq \quad c'$$

$$W' \qquad\qquad\qquad\qquad \leq \quad q^1$$

$$W' \qquad\qquad\qquad\qquad \leq \quad q^2$$

$$\ddots \qquad\qquad\qquad\qquad \vdots$$

$$W' \quad \leq \quad q^L$$

Figure 3.2 Structure of dual problem.

Transposition is denoted by $'$, e.g. W' is the transposed matrix of W. Observe that we have now fewer (unconstrained) variables but a larger number of constraints, assuming that $n_2 \geq m_2$, as is usual when the recourse problem (3.3) is given its canonical linear programming formulation. In Section 3.2 we review briefly the methods that rely on the structure of this dual problem for solving recourse models.

At least when the technology matrix T is nonstochastic, i.e. when $T^\ell = T$, a substitution of variables, mentioned in Wets [26], leads to a linear programming structure that has received a lot of attention in the literature devoted to large scale dynamical systems. Using the constraints of (3.4), it follows that for all $\ell = 1, \ldots, L - 1$,

$$Tx = h^\ell - W y^\ell$$

and substituting in the $(\ell + 1)$-th system, we obtain

$$-W y^\ell + W y^{\ell+1} = h^{\ell+1} - h^\ell.$$

Problem (3.4) is thus equivalent to

$$\begin{aligned} \text{find} \quad & x \in R_+^{n_1} \text{ and for } \ell = 1, \ldots, L, y^\ell \in R_+^{n_2} \\ \text{such that} \quad & Ax = b \\ & Tx + W y^1 = h^1 \\ & -W y^\ell - 1 + W y^\ell = h^\ell - h^\ell - 1, \quad \ell = 2, \ldots, L \end{aligned} \tag{3.6}$$

$$\text{and} \quad z = cx + \sum_{\ell=1}^{L} p_\ell q^\ell y^\ell \text{ is minimized.}$$

With $h^0 = 0$ and for $\ell = 1, \ldots, L$,

$$\hat{h}^\ell = h^\ell - h^{\ell-1},$$

the tableau of detached coefficients exhibits a *staircase structure:*

$$
\begin{array}{ccccc}
c & p_1 q^1 & p_2 q^2 & \cdots & p_L q^L
\end{array}
$$

$$
\begin{array}{ccccc}
A & & & & = b \\[2ex]
T & W & & & = \hat{h}^1 \\[2ex]
& -W & W & & = \hat{h}^2 \\[2ex]
& & \ddots & \ddots & \vdots \\[2ex]
& & & -W \quad W & = \hat{h}^L
\end{array}
$$

Figure 3.3   Equivalent staircase structure.

   We bring this to the fore in order to stress at the same time the close relationship and the basic difference between the problem at hand and those encountered in the context of dynamical systems, i.e. discrete version of continuous linear programs or linear control problems. Superficially, the problems are structurally similar, and indeed the matrix of a linear dynamical system may very well have precisely the structure of the matrix that appears in Figure 3.3. Hence, one may conclude that the results and the computational work for staircase dynamical systems, cf. in particular Perold and Dantzig [16], Fourer [8], and Saunders [19], is in some way transferrable to the stochastic programming case. Clearly some of the ideas and artifices that have proved their usefulness in the setting of linear (discrete time) dynamical systems should be explored, adapted and tried in the stochastic programming context. But one should at all times remain aware of the fact that dynamical systems have coefficients (data) that are 1-parameter dependent (time) whereas we can view the coefficients of stochastic problems as being multi-parameter dependent. In some sense, *the gap between Figure 3.2 and staircase structured linear programs that arise from dynamical systems is the same as that between ordinary differential equations and partial differential equations.* We are not dealing here with a phenomenon that goes forward (in time) but one which can spread all over R^N (which is only partially ordered)! Thus, it is not so surprising that from a computational viewpoint almost no effort has been made to exploit the structure Figure 3.3 to solve stochastic programs with recourse. However, the potential is there and should not remain unexplored.

3.2 Methods that Exploit the Dual Structure

Dantzig and Madansky [5], pointed out that the dual problem (3.5) with matrix structure Figure 3.2 is ripe for the application of the decomposition principle. It was also the properties of Figure 3.2 that led Strazicky [21], to suggest and implement a basis factorization scheme, further analysed and modified by Kall [11], Wets [29], and Birge in Chapter 12. We give a brief description of both methods and study the connections between these two procedures. We begin with the second one, giving a modified compact version of the original proposal.

We assume that W is of full row rank, if not the recourse problem (3.3) defining Q would be infeasible for some of the values of h^ℓ and T^ℓ unless all belong to the appropriate subspace of R^N in which case a row transformation would allow us to delete the redundant constraints. We also assume that A is of full row rank, (possibly 0 when there are no constraints of that type). Thus with the columns of A' and W' linearly independent (recall that the variables σ and π are unrestricted), and after introducing the slack variables ($s^0 \in R_+^{n_1}$ and $s^\ell \in R_+^{n_2}$ for $\ell = 1, \dots, L$), we see that each basic feasible solution will include at least n_2 variables of each subsystem

$$\pi^\ell W + s^\ell I = q^\ell, s^\ell \geq 0, \quad \ell = 1, \dots, L, \tag{3.7}$$

the (unrestricted) m_2 variables π^ℓ and a choice of at least $(n_2 - m_2)$ slack variables $(s_j^\ell, j = 1, \dots, n_2)$. Thus the portion of the basic columns that appear in the l-th subsystem can be subdivided into two parts

$$[B'_\ell, I'_{\ell 2}] = [(W', I'_{\ell 1}), I'_{\ell 2}]$$

where $(W', I'_{\ell 1})$ is an $(n_2 \times n_2)$ invertible matrix and the extra columns, if any, are relegated to $I_{\ell 2}$. Thus, schematically and up to a rearrangement of columns, a feasible basis $\hat{B}$ has the structure:

$$\hat{B}' = \begin{bmatrix} C', & D' \\ B', & N' \end{bmatrix},$$

and in a detached coefficient form:

$$\begin{array}{cccccccc} C'_1 & C'_2 & \cdots & C'_L & D' & & & \\ B'_1 & & & & I'_{12} & & & \\ & B'_2 & & & & I'_{22} & & \\ & & \ddots & & & & \ddots & \\ & & & B'_L & & & & I'_{L2} \end{array}$$

Figure 3.4 Basis structure of dual.

The matrix D' corresponding to the columns of (A', I'_{n_1}) that belong to this basis and for $\ell = 1, \ldots, L$, C'_ℓ is the $n_1 \times m_2$ matrix:

$$C'_\ell = [p_\ell T'_\ell, 0]$$

(recall that T'_ℓ is of dimension $n_1 \times m_2$). Each B'_ℓ, after possible rearrangement of row and columns, is of the following type:

$$B'_\ell = \begin{bmatrix} W'_{(\ell)} & 0 \\ & \begin{matrix} 1 & & \\ & \ddots & \\ & & 1 \end{matrix} \\ W'_{(c\ell)} & \end{bmatrix} = [W', I_{\ell 1}]$$

Figure 3.5   Structure of B'_ℓ.

where $W'_{(\ell)}$ is a $m_2 \times m_2$ invertible submatrix of W', and $W'_{(c\ell)}$ are the remaining rows of W' that correspond to the rows of the identity that have been included in B'_ℓ (through $I'_{\ell 1}$). The simplex multipliers associated with this basis $\hat{B}$, of dimension $n_1 + n_2 \cdot L$, are denoted by

$$\begin{pmatrix} x \\ y \end{pmatrix} = \begin{pmatrix} x \\ y^1 \\ \vdots \\ y^L \end{pmatrix}$$

and are given by the relations

$$\hat{B} \begin{pmatrix} x \\ y \end{pmatrix} = \begin{pmatrix} C & B \\ D & N \end{pmatrix} \begin{pmatrix} x \\ y \end{pmatrix} = \begin{pmatrix} \gamma \\ \beta \end{pmatrix} \tag{3.8}$$

where $[\gamma', \beta']$ is the appropriate rearrangement of the subvector of coefficients of the objective of Figure 3.2 that corresponds to the columns of $\hat{B}'$, with β' being the subvector of $[b', 0]$ whose components correspond to the columns of D'. This (dual feasible) basis is optimal if the vectors

$$(x, y^\ell, \ell = 1, \ldots, L)$$

defined through (3.8) are primal feasible, i.e. satisfy the constraints of (3.4). To obtain x and y we see that (3.8) yields

$$y = B^{-1}(\gamma - Cx)$$
$$x = (D - NB^{-1}C)^{-1}(\beta - NB^{-1}\gamma).$$

For every $\ell = 1, \ldots, L$,

$$y_\ell = B_\ell^{-1}(\gamma^\ell - C_\ell x) \tag{3.9}$$

where γ^ℓ is the subvector of $[p_\ell h^\ell, 0]$ that corresponds to the columns in B'_ℓ. We have used the fact that B is a block diagonal with invertible matrices $(B'_\ell, \ell = 1, \ldots, L)$ on the diagonal. Going one step further and using the properties of N and C, we get the system for x:

$$\left(D - \sum_{\ell=1}^{L} I_{\ell 2} B_\ell^{-1} C_\ell \right) x = \beta - \sum_{\ell=1}^{L} I_{\ell 2} B_\ell^{-1} \gamma_\ell \tag{3.10}$$

The system (3.10) involves n_1 equations in n_1 variables and the L systems (3.9) are of order n_2. Thus instead of calculating the inverse of $\hat{B}$—a square matrix of order $(n_1 + n_2 \cdot L)$—all that is needed is the inverse of L matrices of order n_2 and a square matrix of order n_1.

Similarly to calculate the values to assign to the basic variables associated to this basis, the same inverses is all that is really required, as can easily be verified. In order to implement this method one would need to work out the updating procedures to show that the simplex method can be performed in this compact form, i.e. that the updating procedures involve only the restricted inverses. But there are other features of which one should take advantage before one proceeds with implementation.

Recall that

$$B_\ell = \begin{pmatrix} W_{(\ell)} & W_{(c\ell)} \\ I & 0 \end{pmatrix} \tag{3.11}$$

where $W_{(\ell)}$ is an invertible matrix of size $m_2 \times m_2$. Then

$$B_\ell^{-1} = \begin{bmatrix} W_{(\ell)}^{-1} & -W_{(\ell)}^{-1} W_{(c\ell)} \\ 0, & I \end{bmatrix} \tag{3.12}$$

Thus it really suffices to know the inverse of $W_{(\ell)}$, and rather than keeping and updating the $n_2 \times n_2$–matrix B_ℓ^{-1}, all the information that is really needed can be handled by updating an $m_2 \times m_2$–matrix, relying on sparse updates whenever possible. This should result in substantial savings. The algorithm could even be more efficient by taking advantage of the repetition of similar (sub)bases $W_{(\ell)}$. We shall not pursue this any further at this time because all of these computational shortcuts are best handled in the framework of methods based on the decomposition principle that we describe next.

The decomposition principle, as used to solve the linear program (3.5), generates the master problem from the equations

$$\sigma A + \sum_{\ell=1}^{L} \pi^\ell (p_\ell T^\ell) \leq c,$$

by generating extreme points or directions of recession (directions of unboundedness) from the polyhedral regions determined by the L subproblems,

$$\pi^\ell W \leq q^\ell.$$

In order to simplify the comparison with the factorization method described earlier, let us assume that

$$\{\pi | \pi W \leq 0\} = \{0\},$$

i.e. there are no directions of recession other than 0, which means that for all ℓ, the polyhedra $\{\pi^\ell W \leq q^\ell\}$ are bounded; feasibility of (3.5) implying that they are nonempty. For $k = 1, \ldots, \nu$, let

$$\eta^k = (\eta^{1k}, \ldots, \eta^{\ell k}, \ldots, \eta^{Lk})$$

the extreme point generated by the k-th iteration of the decomposition method, i.e.

$$\eta^{\ell k} \in \operatorname{argmin}(p_\ell \pi^\ell (h^\ell - T^\ell x^k) | \pi^\ell W \leq q^\ell) \tag{3.13}$$

where $x^k = (x_j^k, j = 1, \ldots, n_1)$ are the multipliers associated to the first n_1 linear inequalities of the master problem :

$$\text{find} \quad \sigma \in R^{m_1}, \lambda_k \in R_+, k = 1, \ldots, \nu$$

$$\text{such that} \quad \sigma A + \sum_{k=1}^{\nu} \lambda_k \Big(\sum_{\ell=1}^{L} p_\ell \eta^{\ell k} T^\ell \Big) \leq c$$

$$\sum_{k=1}^{\nu} \lambda_k = 1 \tag{3.14}$$

$$\text{and} \quad w = \sigma b + \sum_{k=1}^{\nu} \lambda_k \Big(\sum_{\ell=1}^{L} p_\ell \eta^{\ell k} h^\ell \Big) \text{ is maximized.}$$

The basis associated to the master problem is $(n_1 \times n_1)$, whereas the basis for each subproblem is exactly of order n_2. In the process of solving the subproblems the iterations of the simplex method bring us from one basis of type (3.11) to another one of this type (all transposed, naturally) with inverses given by (3.12). Here again, the implementation should take advantage of this structural property, and updates should be in terms of the $m_2 \times m_2$ submatrices $W_{(l)}$. But we should also take advantage of the fact that all these subproblems are identical except for the right-hand sides and/or the cost coefficients, and this, in turn, would lead us to the use of bunching and sifting procedures of Section 3.4.

It is remarkable and important to observe that the basis factorization method *with the modifications* alluded to earlier and the decomposition method applied to the dual, as proposed by Dantzig and Madansky [5], require the same computational effort; J. Birge gives a detailed analysis in Chapter 12, independently B. Strazicky arrived at similar results. In view of all of this it is appropriate to view the method relying on basis factorization as a very close parent of the decomposition method as applied to the dual problem (3.5), but it does not give us the organizational flexibility provided by this latter algorithm. On conceptual ground, as well as in terms of computational efficiency, it is the decomposition based algorithm that should be retained for potential software implementation. In fact, this is essentially what has occurred, but it is a "primal" version of this decomposition algorithm, which in this class of (essentially) equivalent methods appears best suited for solving linear stochastic programs with recourse. It is a primal method—which means that we always have a feasible $x \in R_+^{n_1}$ at our disposal—and it allows us to take advantage in the most straightforward manner of some of the properties of recourse models to speed up computations.

3.3 Methods that are Primal Oriented

The great difference between the methods that we consider next and those of Section 3.2 is that finding x that solves the stochastic program (3.1) is now viewed as our major, if not exclusive, concern. Obtaining the corresponding recourse decisions $(y^\ell, \ell = 1, \ldots, L)$ or associated dual multipliers $(\pi^\ell, \ell = 1, \ldots, L)$ is of no real interest, and we only perform some of these calculations because the search for an optimal solution x requires knowing some of these quantities, at least in an amalgamated form. On the other hand, in the methods of Section 3.2 all the variables $(\sigma, \pi^1, \ldots, \pi^L)$ are treated as equals; to have the optimality criterion fail for some variable in subsystem ℓ (even when p_ℓ is relatively small) is handled with the same concern as having the optimality criteria fail for some of the $(\sigma_i, i = 1, \ldots, m_1)$ variables.

Another important property of these methods is their natural extension to stochastic programs with arbitrary distribution functions. In fact, they are particularly well-suited for use in a sequential scheme for solving stochastic programs by successive refinement of the discretization of the probability measure, each step involving the solution of a problem of type (3.1), cf. Chapter 2.

We stress these conceptual differences, because they may lead to different, more flexible, solution strategies; although we are very much aware of the fact that if at each stage of the algorithm all operations are carried out (to optimality), it is possible to find their exact counterpart in the algorithms described in Section 3.2; for the relationship between the L-shaped algorithm described here and the decomposition method applied to the dual, see Van Slyke and Wets [20]; between the above and the basis factorization method see Chapter 13; consult also Ho [10], for the relationship between various schemes for piecewise linear functions which are widely utilized for solving certain classes of stochastic programming problems, and Chapter 4.

The L-shaped algorithm, which takes its name from the matrix layout of the problem to be solved, was proposed by Van Slyke and Wets [20]; in Chapter 12, Birge describes his implementation of this method. It can be viewed as a cutting hyperplane algorithm (outer linearization) but to stay in the framework of our earlier development, it is best to interpret it here as a *partial* decomposition method. We begin with a description of a very crude version of the algorithm, only later do we elaborate the modifications that are vital to make the method really efficient. To describe the method it is useful to consider the problem in its original form (3.1) which we repeat here for easy reference:

$$\text{find} \quad x \in R_+^{n1}$$
$$\text{such that} \quad Ax = b, \tag{3.15}$$
$$\text{and} \quad z = cx + (x) \text{ is minimized.}$$

We assume that the problem is feasible and bounded, implementation of the algorithm would require an appropriate coding of the initialization step rep lying on the criteria for feasibility and boundedness such as found in Wets [27]. The method consists of three steps that can be interpreted as followes In Step 1, we solve an approximate of (3.15) obtained by replacing Q by are outer-linearization, this brings us to the solving of a linear programming whose constraints are $Ax = b, x \geq 0$ and the additional constraints (3.16) and (3.17) that come from:

(i) induced feasibility cuts generated by the fact that the choice of x must be restricted to those for which $Q(x)$ is finite, or equivalently for which $Q(x, \xi^\ell < +\infty$ for all $\ell = 1, \ldots L$ or still for which there exists $y^\ell \in R_+^{n2}$ such that $W y^\ell = h^\ell - T^\ell x$ for all $\ell = 1, \ldots, L$.

(ii) linear approximations to Q on its domain of finiteness.

These constraints are generated systematically through Steps 2 and 3 of the algorithm, when a proposed solution x^v of the linear program in Step 1 fails to satisfy the induced constraints, i.e. $Q(x^v) = \infty$ (Step 2) or if the approximating problem does not yet match the function Q at x^v (Step 3). The row-vector generated in Step 3 is actually a subgradient of Q at x^v . The convergence of the algorithm under the appropriate nondegeneracy assumptions, to an optimal solution of (3.15), is based on the fact that there are only a finite number of constraints of type (3.16) and (3.17) that can be generated by Steps 2 and 3 since each one corresponds to some basis of W and a pair (h^ℓ, T^ℓ) or to a basis of W and to one of a finite number of weighted averages of the $(q^\ell, \ell = 1, \ldots, L)$ and $((h^\ell, T^\ell), \ell = 1, \ldots L)$.

Step 0. Set $\nu = r = s = 0$.

Step 1. Set $\nu = \nu + 1$ and solve the linear program

$$\text{find} \quad x \in R_+^{n1}, \theta \in R$$
$$\text{such that} \quad Ax = b$$
$$D_k x \geq d_k, \quad k = 1, \ldots, r, \tag{3.16}$$
$$E_k x + \theta \geq e_k, \quad k = 1, \ldots, s, \tag{3.17}$$
$$\text{and} \quad cx + \theta = z \text{ is minimized.}$$

Let (x^v, θ^v) be an optimal solution. If there are no constraints of type (3.17), the variable θ is ignored in the computation of the optimal x^v, the value of θ^v is then fixed at $-\infty$.

Step 2. For $\ell = 1, \ldots, L$ solve the linear programs

$$\text{find} \quad y \in R_+^{n_2}, v^+ \in R_+^{m_2}, v^- \in R_+^{m_2}$$
$$\text{such that} \quad Wy + Iv^+ - Iv^- = h^\ell - T^\ell x^\nu \tag{3.18}$$
$$\text{and} \quad ev^+ + ev^- = v^\ell \text{ is minimized}$$

(here e denotes the row vector $(1,1,\ldots,1)$), until for some ℓ the optimal value $v^\ell > 0$. Let σ^ν be the associated simplex multipliers and define

$$D_{r+1} = \sigma^\nu T^\ell$$

and

$$d_{r+1} = \sigma^\nu h^\ell$$

to generate an induced feasibility cut. Return to Step 1 adding this new constraint of type (3.16) and set $r = r + 1$. If for all ℓ, the optimal value of the linear program (3.18) $v^\ell = 0$, go to Step 3.

Step 3. For every $\ell = 1, \ldots, L$, solve the linear program

$$\text{find} \quad y \in R_+^{n_2}$$
$$\text{such that} \quad Wy = h^\ell - T^\ell x^\nu, \tag{3.19}$$
$$\text{and} \quad q^\ell y = w^\ell \text{ is minimized.}$$

Let $\pi^{\ell\nu}$ be the multipliers associated with the optimal solution of problem ℓ. Set $t = t + 1$ and define

$$E_t = \sum_{\ell=1}^{L} p_\ell \pi^{\ell\nu} T^\ell,$$

$$e_t = \sum_{\ell=1}^{L} p_\ell \pi^{\ell\nu} h^\ell,$$

and

$$w^\nu = \sum_{\ell=1}^{L} p_\ell \pi^{\ell\nu} (h^\ell - T^\ell x^\nu) = e_t - E_t x^\nu.$$

If $\theta^\nu \geq w^\nu$, we stop; x^ν is the optimal solution. Otherwise, we return to Step 1 with a new constraint of type (3.17).

An efficient implementation of this algorithm, whose steps can be identified with those of the decomposition method applied to the dual problem (see Section 3.2), depends very much on the acceleration of Steps 2 and 3. This is made possible by relying on the specific properties of the problem at hand (3.15), and it is in order to exploit these properties that we have separated Steps 2 and 3 which are the counterparts of Phase I and Phase II of the simplex method as applied to the recourse problem (3.3). In practice one certainly does not start from scratch when solving the L linear programs in Step 3; Section 3.4

is devoted to the analysis of Step 3, i.e. how to take advantage of the fact that the L linear programs that need to be solved have the same technology matrix W as well as from the fact that the $\xi^\ell = (q^\ell, h^\ell T^\ell)$ are the realizations of a random vector. Here we concern ourselves with the improvements that could be made to speed up Step 2, and we see that in many instances, dramatic gains could be realized.

First and for all, Step 2 can be skipped altogether if the stochastic program is with *complete recourse*, i.e. when

$$\text{pos } W := \{t \,|\, t = Wy, y \geq 0\} = R^{m_2}, \qquad (3.20)$$

a quite common occurrence in practice. This means naturally that no induced feasibility constraints (3.16) need to be generated. This will also be the case if we have a problem with *relatively complete recourse* i.e. when for every x satisfying $Ax = b, x \geq 0$, and for every $\ell = 1, \ldots, L$, the linear system

$$Wy = h^\ell - T^\ell x, y \geq 0,$$

is feasible. This weaker condition is much more difficult to recognize, and to verify it would precisely require the procedure given in Step 2.

Even in the general case, it may be possible to substitute for Step 2: for some (h^ν, T^ν)

Step 2'. Solve the linear program

$$
\begin{aligned}
&\text{find} \quad y \in R_+^{n_2}, v^+ \in R_+^{m_2}, v^- \in R_+^{m_2} \\
&\text{such that} \quad Wy + Iv^+ - Iv^- = (h^\nu - T^\nu x^\nu) \qquad (3.21) \\
&\text{and} \quad ev^+ + ev^- = v^\nu \text{ is minimized.}
\end{aligned}
$$

Let σ^ν be the associated simplex multipliers and if the optimal value of $v^\nu > 0$, define

$$D_{r+1} = \sigma^\nu T^\nu,$$

and

$$d_{r+1} = \sigma^\nu h^\nu$$

to generate an induced feasibility cut of type (3.16). Return to Step 1 with $r = r + 1$. If the optimal value of $v^\nu = 0$, go to Step 3.

This means that we have replaced solving L linear programs by just solving 1 of them. In some other cases it may be necessary to solve a few problems of type (3.21) but the effort would in no way be commensurate with that of solving all L linear programs of Step 2. In Section 3.5 of Wets [28], one can find a detailed analysis of the cases when such a substitution is possible, as well as some procedures for the choice or construction of the quantities h^ν and T^ν that appear in the formulation of (3.21). Here we simply suggest the reasons why this simplification is possible and pay particular attention to the case when the matrix T is nonstochastic.

Let $<$ be the partial ordering induced by the closed convex polyhedral cone pos W, see (3.20), i.e. $a^1 < a^2$ if $a^2 - a^1 \in$ pos W. Then for given $x \in R^{n_1}$ and for every $\ell = 1, \ldots, L$, the linear system

$$Wy = h^\ell - T^\ell x^\nu, y \geq 0 \qquad (3.22)$$

is feasible, if there exists $a^\nu \in R^{m_2}$ such that for all $\ell = 1, \ldots, L$,

$$a^\nu < h^\ell - T^\ell x^\nu, \qquad (3.23)$$

and the linear system

$$Wy = a^\nu, y \geq 0 \qquad (3.24)$$

is feasible—or equivalently $a^\nu \in$ pos W. There always exists a^ν that satisfies (3.23), recall L is finite. If in addition, a^ν can be chosen so that

$$a^\nu = h^\nu - T^\nu x \qquad (3.25)$$

for $\nu \in \{1, \ldots, L\}$, then (3.22) is feasible for all ℓ *if and only if* (3.24) is feasible with a^ν as defined by (3.25). Although in general such an a^ν does not exist, in practice, at most a few extreme points of the set

$$S^\nu = \{a | a = h^\ell - T^\ell x^\nu, \ell = 1, \ldots, L\},$$

need to be considered in order to verify the feasibility of *all* the linear systems (3.22). Computing lower bounds of S^ν with respect to $<$ may require more work than we bargained for, but it really suffices, cf. Theorem 4.17 of Wets [28], to construct lower bounds of S^ν with respect to any closed cone contained in pos W, and this could be, and usually is taken to be, an orthant. In such a case obtaining a^ν is effortless.

Let us consider the case when T is nonstochastic and assume that pos W contains the positive orthant, if it contains another orthant simply multiply some rows by -1 making the corresponding adjustments in the vectors $(h^\ell, \ell = 1, \ldots, L)$. This certainly would be the case if slack variables are part of the y-vector, for example.

For $i = 1, \ldots, m_2,$

let $a_i = \min_\ell h_i^\ell$

If $a = h^\nu$ for some $\nu \in \{1, \ldots, L\}$, which would always be the case if the $(h_i(\cdot), i = 1, \ldots, m_2)$ are independent random variables, then it follows from the above that for $\ell = 1, \ldots, L$, the linear system

$$Wy = h^\ell - Tx^\nu, y \geq 0$$

is feasible if and only if

$$Wy = a - Tx^\nu, y \geq 0.$$

is feasible. Note that in this case the lower bound

$$a^\nu = a - Tx^\nu$$

is a simple function of x^ν.

In our description of the L-shaped algorithm the connections to large scale linear programming may have been somewhat lost, if anything it is how to deal with the "nonlinearity" of Q which has played center stage. To regain maybe a more *linear* programming perspective it may be useful to view the algorithm in the following light. Let us return to the dual block angular structure Figure 3.1 from which it is obvious that if we can adjust the simplex method so that it operates separately on the x-variables and the (y_ℓ-variables, $\ell = 1,\ldots,L$), it will be possible to take advantage of the block diagonal structure of the problem with respect to the (y^ℓ-variables, $\ell = 1,\ldots,L$). Given that some x^ν is known which satisfies the constraint $x \geq 0, Ax = b$, then finding the optimal solution of Figure 3.1, with the additional constraint $x = x^\nu$ leads to solving a linear program, whose tableau of detached coefficients has the structure:

$$
\begin{array}{ccccl}
p_1q^1 & p_2q^2 & \cdots & p_Lq^L & \\[2mm]
W & & & & = h^{1\nu} \\[2mm]
& W & & & = h^{2\nu} \\[2mm]
& & \ddots & & \vdots \\[2mm]
& & & W & = h^{L\nu}
\end{array}
$$

Figure 3.6   Structure of the y-problem.

where for $\ell = 1,\ldots,L, h^{\ell\nu} = h^\ell - T^\ell x^\nu$. Clearly, when confronted with such a problem we want to take advantage of its separability properties and this is precisely what is done in Steps 2 and 3 of the L-shaped algorithm.

The structure of Figure 3.6, with the same matrix W on the block diagonal, suggests that of a distributed system. A continuous version would take the form:

$$
\begin{aligned}
&\text{find} \quad y : \Omega \to R^{n_2} \\
&\text{such that} \quad \forall w \in \Omega \\
&\qquad y(w) \in \operatorname{argmin}[q(w)y | Wy = h^\nu(w), y \in R_+^{n_2}].
\end{aligned}
\tag{3.26}
$$

Because of the linearity of the objective function, the trajectory $w \mapsto y(w)$ will be linear with respect to h^ν if the same basis of W remains optimal. The main task in solving (3.26) would be to decompose Ω in regions of linearity of $y(\cdot)$. Once this decomposition is known the remainder is rather straightforward.

Finding this decomposition is essentially the subject of Section 3.4, which concerns itself with the organization of the computational work so as to bring the effort involve 1 to an acceptable level. Problem (3.26) again brings to the fore the connections between this work and that on dynamical systems (continuous linear programming). With not too much difficulty it should be possible to formulate a bang-bang principle for systems with distributed parameters space (here R^{m_2}) that would correspond to our scheme for decomposing Ω.

To conclude our discussion of the L-shaped algorithm, let us record a further modification suggested by L. Nazareth. When the matrix T is nonstochastic, say $T^\ell = T$ for all ℓ, then with $\chi = Tx$, $\Psi(\chi) = \Psi(T\tau) = (x)$, the linear program in Step 1 may be reformulated as

$$
\begin{aligned}
\text{find} \quad & x \in R_+^{n_1}, \chi \in R^{m_2}, \theta \in R \\
\text{such that} \quad & Ax = b \\
& Tx - \chi = 0 \\
& F_k\chi \geq f_k, \quad k = 1, \ldots, r \\
& G_k\chi + \theta \geq g_k, \quad k = 1, \ldots, s, \\
\text{and} \quad & cx + \theta = z \text{ is minimized.}
\end{aligned} \tag{3.27}
$$

The induced feasibility constraints are generated as earlier in Step 2 with

$$
F_{r+1} = \sigma^\nu, f_{r+1} = \sigma^\nu h^\ell
$$

The optimality cuts (approximation cuts) are generated in Step 3 with

$$
G_t = \sum_{\ell=1}^{L} p_\ell \pi^{\ell\nu},
$$

$$
g_t = \sum_{\ell=1}^{L} p_\ell \pi^{\ell\nu} h^\ell.
$$

The linear program that generates the σ^ν and $\pi^{\ell\nu}$ as (optimal) simplex multipliers of Phases I and II respectively, is given by

$$
\begin{aligned}
\text{find} \quad & y \in R_+^{n_2} \\
\text{such that} \quad & Wy = h^\ell - \chi^\nu, \\
\text{and} \quad & q^\ell y = w^\ell \text{ is minimized.}
\end{aligned}
$$

Note that now the "nonlinearity" is handled in a space of dimension m_2 which is liable to be much smaller than n_1, and we should reap all the advantages that usually come from a reduction in the number of nonlinear variables.

All of these simplifications come from the fact that when T is nonstochastic we can interpret the search for an optimal solution, as the search for an optimal

χ^*, "the certainty equivalent". It is easy to see that knowing χ^* would allow us to solve the original problem by simply solving

$$\text{find} \quad x \in R^n_+$$
$$\text{such that} \quad Ax = b, Tx = \chi^*, \tag{3.28}$$
$$\text{and} \quad z = cx \text{ is minimized.}$$

The sequence $\{\chi^\nu, \nu = 1, \ldots\}$ generated by the preceding algorithm can be viewed as a sequence of tenders (to be "bet" against the uncertainty represented by h). This then suggests other methods based on finding χ^* by considering the best possible convex combination of the tenders generated so far; these algorithms are based on generalized linear programming, see Nazareth and Wets [15], and Chapter 4 of this Volume. In the context of the general class of linear stochastic programming problems considered here, we have up to now very limited experience with this method. The algorithm would proceed as follows:

Step 0. Find a feasible $x^0 \in R^{n_1}_+$ such that $Ax^0 = b$

Set $\chi^0 = x^0$

Choose $\chi^1, \ldots, \chi^\nu$, potential tenders, $\nu \geq 0$.

Step 1. Find $(\sigma^\nu, \pi^\nu, \theta_\nu)$ the (optimal) simplex multipliers associated with the solution of the linear program:

$$\text{minimize} \quad cx + \sum_{\ell=0}^{\nu} \lambda_\ell \Psi(\chi^\ell)$$
$$Ax = b : \sigma^\nu$$
$$Ix - \sum_{\ell=0}^{\nu} \lambda_\ell \chi^\ell = 0 : \pi^\nu$$
$$\sum_{\ell=0}^{\nu} \lambda_\ell = 1 : \theta_\nu$$
$$x \geq 0, \lambda_\ell \geq 0$$

Step 2. Stop unless there exists $\chi^{\nu+1}$ such that

$$\Psi(\chi^{\nu+1}) + \pi^\nu \chi^{\nu+1} < \theta_\nu \tag{3.29}$$

in which case return to Step 1 with $\nu = \nu + 1$.

The attractiveness of this approach rests on the fact that the algorithm allows for the choice of a number of tenders (trial solutions) which would provide an excellent initial approximate solution to the problem as a whole just after 1 passage through Step 1, assuming of course that the tenders $\chi^1, \ldots, \chi^\nu$ are chosen by an informed problem solver. Note, however, that for each tender $\chi \in R^{n_1}$ we need to find the value of $\Psi(\chi) = \sum_{\ell=1}^{L} p_\ell \psi(\chi, \xi^\ell)$, i.e. solve the L linear programs

$$\text{find} \quad y \in R^{n_2}_+$$
$$\text{such that} \quad Wy = h^\ell - T^\ell \chi,$$
$$\text{and} \quad \psi(\chi, \xi^\ell) = q^\ell y \text{ is minimized.}$$

Of course in order to do so, we can take advantage of the techniques described in the next section.

As suggested by Nazareth [14], Step 2 should not be carried out to optimality, by which one means: find $\chi^{\nu+1}$ that minimizes $\Psi(\chi) + \pi^\nu \chi$. All what is really necessary is to find a tender that satisfies the condition (3.29) given in Step 2. Nazareth points out that if this strategy is followed, the complete set of calls to Step 2 will be of similar computational effort as that of solving problem (3.1), whereas carrying out Step 2 to optimality would require at each iteration essentially the same amount of work as solving (3.1). In fact Nazareth [14], suggests that Step 2 should be done with a nonsmooth optimizer (using the bunching techniques to be discussed in Section 3.4). This is also the direction of the algorithmic research recently undertaken by Kiwiel [12].

3.4 Sifting, Bunching and Bases Updates

In the final analysis, Step 3 of the L-shaped algorithm boils down to the calculation of the value of Q and of its gradient at x^ν. What it involves is solving a large number of similar linear programs, or if you prefer one linear program with matrix structure as in Figure 3.6. The same type of operations would be required for the actual carrying out of Step 2 of the algorithm based on the generation of tenders. The extent to which we are able to speed up these computations will determine the level of "stochasticity" that we are able to handle. This Section raises the question of how to organize the work so as to minimize the computational effort involved. We consider only the case of multiple right-hand sides, resulting, as the case may be, from h and/or T random; by duality, the analysis also applies to the case when only q is random (and h and T are nonstochastic). When both the cost coefficients and the right-hand sides of the recourse problem (3.3) include random variables a further refinement of the methods suggested here would be required. We shall not be concerned with special cases such as simple recourse $W = (I, -I)$, or network-structured problems when specific computational shortcuts are possible, e.g. Midler and Wollmer [13], Wallace [23], and Qi [17].

In its simplest form, the problem that we are concerned with is finding an efficient procedure for solving L linear programs with variable right-hand sides: for $\ell = 1, \ldots, L$,

$$\text{find} \quad y \in R_+^{n_2}$$
$$\text{such that} \quad W y = t^\ell, \tag{3.30}$$
$$qy = w^\ell \text{ is minimized.}$$

The cost coefficients are constant, we simply write q for $q^1 = q^2 = \ldots = q^L$. In terms of Step 3 of the L-shaped algorithm, the vectors $\tau = \{t^\ell, \ell = 1, \ldots, L\}$ come from $t^\ell = h^\ell - T^\ell x^\nu$ for some fixed x^ν.
For all l, (3.30) is feasible, i.e.

$$t^\ell \in \text{pos} \, W = \{t | t = W y, y \geq 0\}, \tag{3.31}$$

(this comes from the fact that x^ν or χ^ν satisfies the induced feasibility constraints). Moreover, by assumption we have that (3.30) is bounded, and hence for all ℓ, (3.30) is solvable. We shall denote the optimal solution by y^ℓ, and the associated simplex multipliers by π^ℓ. We have that

$$\pi^\ell W \leq q,$$

and

$$qy^\ell = \pi^\ell t^\ell.$$

The methods that we study can be divided into *sifting* (discrete parametric analysis) and *bunching* (basis by basis analysis) procedures. We begin with a description of a very crude bunching procedure, which nonetheless would be much more efficient than solving separately all L linear programs (3.30). This technique is easily modified to also take care of the case of random cost coefficients, cf. Wets [29], p.587.

Let B be an $m_2 \times m_2$ invertible submatrix of W with $\gamma B^{-1} W \leq q$ where γ is the subvector of q that corresponds to the columns of W in B; recall that W is assumed to be of full row rank. Then from the optimality conditions for linear programming, it follows that this basis B is optimal for any vector $t \in R^{m_2}$ such that

$$B^{-1}t \geq 0 \qquad\qquad (3.32)$$

and then the optimal simplex multipliers are given by

$$\pi = \gamma B^{-1}.$$

This means that pos W is decomposable into a number of simplicial cones of the type pos $= \{t | B^{-1}t \geq 0\}$, such that whenever $t \in \text{pos} B$ then B is an optimal basis for the linear program: find $y \in R_+^{n_2}$ such that $Wy = t$ and $w = qy$ is minimized. Moreover, on pos B, the (optimal) simplex multipliers remain constant. All of these observations can be rendered very precise and are summarized in the Basis Decomposition Theorem, Walkup and Wets [22]. The figure below illustrates such a decomposition.

Now suppose that we solve the linear program (3.30) for some ℓ, and $B_{(1)}$ is the corresponding optimal basis. Since $B_{(1)}^{-1}$ is readily available, finding the bunch of vectors t^ℓ for which $B_{(1)}$ is the optimal basis is relatively easy since all we need to do is to verify if

$$B_{(1)}^{-1}t^\ell \geq 0. \qquad\qquad (3.33)$$

Let B_1 be the family of all such vectors, $\pi_{(1)}$ be the corresponding simplex multipliers and the probability mass associated with B_1 given by

$$p_{(1)} = \sum_{t^\ell \in B_1} p_\ell.$$

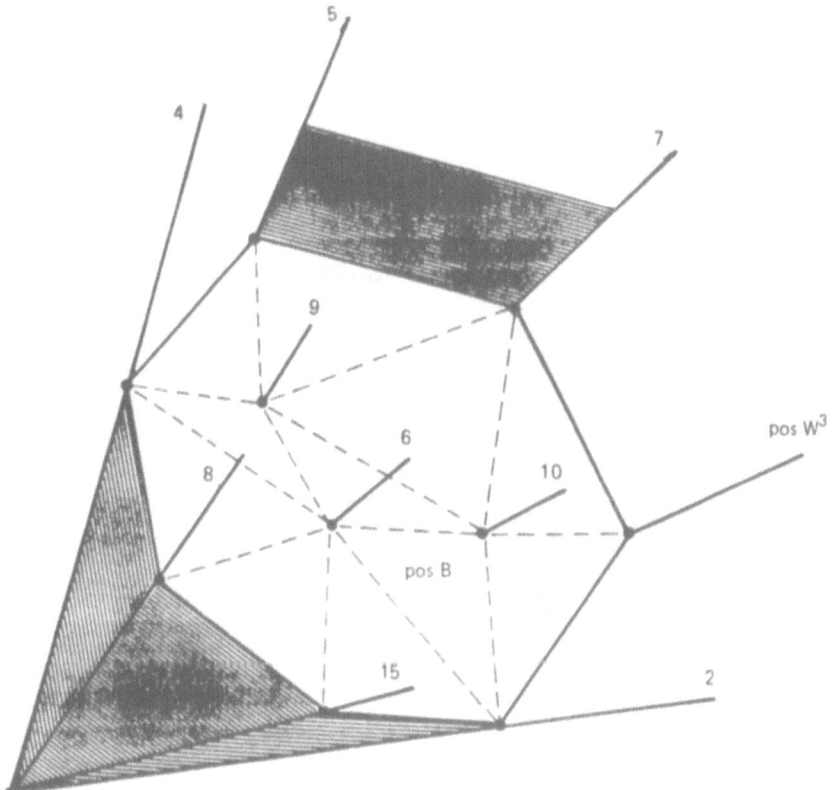

Figure 3.7 Decomposition of pos W.

All vectors t^ℓ that have failed the nonnegativity test (3.33) are in $\tau_1 = \tau \backslash B_1$. We are now in the same situation as at the outset. Picking a vector in τ_1, we obtain a new basis $B_{(2)}$, the corresponding vector $\pi_{(2)}$ the bunch B_2 and associated probability mass $p_{(2)}$. This process is continued until all $t^\ell \in \tau$ have been bunched. The expected value of these linear programs—the quantity that would correspond to (x^ν) or $\Psi(\chi^\nu)$—is given by:

$$\sum_k \pi_{(k)} \sum_{t^\ell \in B_k} p_\ell t^\ell.$$

The expected simplex multiplier—a quantity used in the construction of feasibility and optimality cuts—is given by :

$$\sum_k p_{(k)} \pi_{(k)}.$$

A number of computational shortcuts come immediately to mind as suggested by the decomposition of pos W. First, note that τ or even co τ (the convex hull of τ), is a subset of pos W that meets some—and usually only a few—of the simplicial cones that are part of this decomposition. Moreover, most of the vectors in τ will be found in adjacent cells, thus instead of just picking any vector t^ℓ that failed the (nonnegativity) test (3.33), we could choose a vector $\hat{t}$ in τ_1 such that *that* belongs to a neighboring cell, which necessarily means that $B_{(1)}^{-1}\hat{t}$ has exactly 1 negative entry; note that $B_{(1)}^{-1}t$ having exactly one negative entry does not automatically imply that t belongs to an adjacent cell of pos $B_{(1)}$. Passing from pos $B_{(1)}$ to a neighboring cell requires just one (dual) pivot step. It is clear that substantial computational savings could be realized by a systematic organization of the work.

One way is to proceed as suggested in Wets [29]: pick a vector $t \in \tau$, say t^1, and solve the linear program (3.30) with $\ell = 1$. Let $B_{(1)}$ be the optimal basis. Multiply each vector t in τ by $B_{(1)}^{-1}$. The bunch B_1 is the collection of all vectors t such that

$$\bar{t}_{(1)}^\ell = B_{(1)}^{-1} t^\ell \geq 0.$$

For each vector $\bar{t}_{(1)}^\ell \in \tau_1 = \tau \backslash B_1$, with necessarily at least 1 negative element, we record the actual number of negative entries as well as m_ℓ the magnitude of the most negative element. Now choose a vector t in τ_1 with a minimal number of negative entries and among them one with m_ℓ as small as possible. Pivot, relying on the criteria provided by the dual simplex method, to obtain the next (optimal) basis $B_{(2)}$, the associated multipliers $\pi_{(2)}$ and construct τ_2; and then continue in a similar manner.

What all of this comes down to is that we build a partitioning of that portion of pos W that covers τ (or co τ). What we need is the sublattice structure of the cells that contain τ. In certain cases it may be possible to work out the complete decomposition of pos W and then use it whenever we enter Step 3 of the L-shaped algorithm. Each subbasis of W that generates a cell of the decomposition is recorded with labels that point to the neighboring cells. The lattice generated by the decomposition in Figure 3.7, would take the graph structure given in Figure 3.8. The labeling of the nodes could be the indices of the columns in the basis.

The pointers would correspond to the pivot step required to pass from one basis to a neighboring one. Here this is a planar graph but that would not necessarily be the case if $m_2 > 3$. In general, working out the complete decomposition of pos W may be a serious undertaking, the number of cells could increase exponentially as a function of m_2 (for n_2 sufficiently large). Even for problems whose recourse matrix W have a network structure, the number of

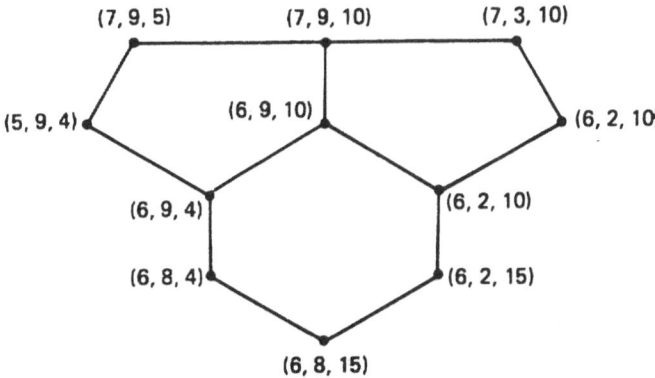

Figure 3.8 Lattice of the decomposition of pos W.

components in a complete decomposition of pos W may become unmanageable even for relatively "small" problems, see Wallace [23].

Short of first working out a complete decomposition and then finding a good path through the lattice, so as to minimize the number of operations, what could be done? What appears the most efficient approach to date is to bunch the elements of τ by a *trickling down procedure* that we describe next. Unless there are some good reasons for proceeding otherwise—for example the inverse of a "good" subbasis of W is available—we would start by finding the cell associated with $\bar{t}$, where

$$\bar{t} = \sum_{t^\ell \in \tau} p_\ell t^\ell$$

is the mean of the vectors in τ, geometrically: the centroid of τ. We have to solve the linear program:

$$\text{find} \quad y \in R_+^{n_2}$$
$$\text{such that} \quad W y = \bar{t},$$
$$\text{and} \quad qy \text{ is minimized.}$$

This yields an optimal basis $B_{(1)}$, its inverse $B_{(1)}^{-1}$ and associated multiplier $\pi_{(1)}$. We assume that $B_{(1)}^{-1}$ is stored as an explicit dense matrix. Now consider t^1 and sequentially perform the multiplications

$$[B_{(1)}^{-1}]_i \, t^1 = \hat{t}_i^1.$$

If $\hat{t}_i^1 \geq 0$ for all i, place t^1 in bunch 1, otherwise stop as soon as for some index $i, \hat{t}_i^1 < 0$. Perform *one* dual simplex step, with pivot in row i. In doing so we create a new basis $B_{(2)}$ with

$$[B_{(2)}^{-1}]_i \, t^1 \geq 0$$

(preserving dual feasibility). The branching from $B_{(1)}$ occurred on i. Repeat the same procedure with $B_{(2)}$ instead of $B_{(1)}$, branching if necessary (recording the branching index), otherwise assigning t^1 to bunch 2. If branching did occur, then continue until a basis $B_{(k)}$ is found such that $B_{(k)}^{-1} t^1 \geq 0$. This will necessarily take place since $t^1 \in \tau \subset \mathrm{pos}\, W$ by assumption, and the pivot path is a simplex path for the dual problem with the pivot choice determined by the first negative entry; degeneracy could be resolved by a random selection rule or Bland's rule. This procedure creates a tree, rooted at $B_{(1)}$, whose nodes correspond to the bases (associated with the cells of the decomposition of $\mathrm{pos}\, W$), the branches being determined by the first negative entry encountered when multiplying t by $B_{(1)}^{-1}$. Figure 3.9 gives part of such a tree for the decomposition of Figure 3.7 assuming that τ covers $\mathrm{pos}\, W$, and that

$$\bar{t} \in \mathrm{pos}(W^9, W^6, W^{10}).$$

The number on the branches indicating branching on the i-th entry that leads to the subsequent basis.

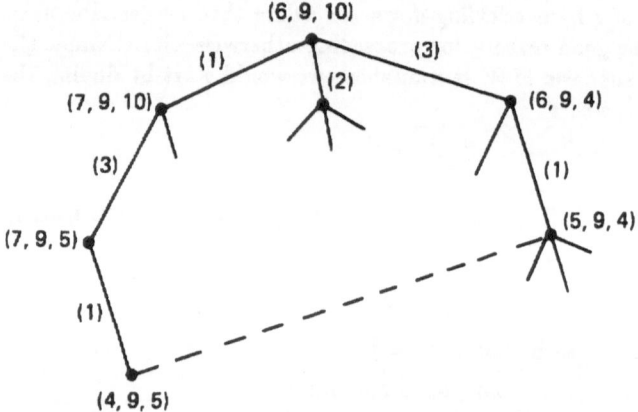

Figure 3.9   Tree generated by trickling down procedure

Note that the same cell may be discovered on different branches of the tree. No effort would be made to recognize that this is taking place, since too much computational effort would be involved in trying to identify such a situation, and only marginal gains could be reaped as will be clear from the subsequent development that concerns updates, i.e. the information necessary to pass from one node of the tree to the next.

It is clear that a great amount of calculations are bypassed by the trickling down procedure, by comparison to the "rough" version of the bunching procedure described at the beginning of this Section. However, it may appear that the storage of all inverse bases (corresponding to the nodes of the tree) as well as keeping track of pointers may negate all the advantages that may be gained from this bunching technique. This, however, can be overcome by relying on Schur-complement updates for the bases $B_{(k)}$. Updates of this type in the context of linear programming were first suggested by Bisschop and Meeraus [3], [4]. Suppose $B_{(k)}$ is obtained from $B_{(0)}$ by adding k columns—without loss of generality assume they are $W_{(k)} = [W^{j_1}, \ldots, W^{j_k}]$—and by pivoting out k columns. The equation

$$B_{(k)} y' = t$$

where $y' \in R^{m_2}$ can also be rewritten as

$$\begin{pmatrix} B_{(0)} & W_{(k)} \\ I_{(k)} & 0 \end{pmatrix} \begin{pmatrix} y' \\ z \end{pmatrix} = \begin{pmatrix} t \\ 0 \end{pmatrix}$$

where $I_{(k)}$ is part of an identity matrix with rows having their entry 1 corresponding to the columns that have to leave the basis when passing from $B_{(0)}$ to $B_{(k)}$. This matrix of coefficients can be written as a block LU product

$$\begin{pmatrix} B_{(0)} & W_{(k)} \\ I_{(k)} & 0 \end{pmatrix} = \begin{pmatrix} B_{(0)} & 0 \\ I_{(k)} & C_{(k)} \end{pmatrix} \begin{pmatrix} I & Y_{(k)} \\ 0 & I \end{pmatrix}$$

where the I''s in the last matrix are $m_2 \times m_2$ and $k \times k$ indentity matrices. We have that

$$Y_{(k)} = B_{(0)}^{-1} W_{(k)},$$
$$C_{(k)} = -I_{(k)} Y_{(k)},$$

and thus

$$C_{(k)} = -I_{(k)} B_{(0)}^{-1} W_{(k)}.$$

This matrix is $k \times k$ and is the only information that is needed to reconstruct all that is needed at the node associated with $B_{(k)}$, in addition to $B_{(0)}^{-1}$ which is supposed to be available (in an LU form, for example). This means that at depth 1 in the tree, only 1×1 updates are necessary; at depth 2, 2×2 updates. Since we reasonably expect to find the largest number of points of τ in the immediate neighborhood of $\bar{t}$ we do not expect to have to construct very long (deep) trees, and the updating information should be of manageable size.

Bunching by the trickling down procedure appears to minimize the amount of operations needed to assign a given $t \in \tau$ to its bunch, and by relying on Schur-complement updates the amount of information required at each node is kept very low. When k—the number of bunches—gets to be too large it may be necessary to start a tree with a new root. This approach to bunching can even be used effectively in specially structured problems such as worked out in Wallace [24], in the case of networks.

The *sifting procedure*, a sort of discrete parametric analysis, has been proposed by Garstka and Rutenberg [9]. It is designed for handling the case when the points in τ are the possible realizations of m_2 independent random variables, for example when T is nonstochastic and the $h_i(\cdot)$, $i = 1, \ldots, m$, are independent random variables. We assume that the vectors in $\tau \subset \operatorname{pos} W$ are obtained by setting for every $i = 1, \ldots, m_2$,

$$t_i = \tau_{i\ell}$$

for some $\ell \in \{1, \ldots, k_i\}$ where we have ordered the τ_{il} i.e.,

$$\tau_{i1} < \tau_{i2} < \cdots < \tau_{ik_j}.$$

We have thus a doubly indexed array:

$$
\begin{array}{cccc}
\tau_{11} & < & \tau_{12} & < \cdots < & \tau_{1k_1}, \\
\tau_{21} & < & \tau_{22} & < \cdots < & \tau_{2k_2}, \\
\cdots & & & & \\
\tau_{m_2,1} & < & \tau_{m_2,2} & < \cdots < & \tau_{m_2,k_m}.
\end{array}
\tag{3.34}
$$

We sift through this array in the following manner: let

$$t^1 = (\tau_{11}, \tau_{21}, \ldots, \tau_{m_2,1}),$$

and solve the linear program

$$\text{find} \quad y \in R_+^{n_2}$$
$$\text{such that} \quad Wy = t^1,$$
$$\text{and} \quad qy \text{ is minimized.}$$

Suppose that $B_{(1)}$ is the associated optimal basis, with

$$[B_{(1)}^{-1}] = [\beta^1, \beta^2, \ldots, \beta^{m_2}].$$

Recall that $t \in \operatorname{pos} B_{(1)}$ as long as $[B_{(1)}^{-1}]t \geq 0$. Hence to find out which subset of vectors belong to $\operatorname{pos} B_{(1)}$, for $\ell = m_2, \ldots, 1$ we study systematically the range of values of τ that satisfy:

$$\left(\sum_{\substack{j=1 \\ j \neq \ell}}^{m_2} \beta^j \tau_{j,e_j} \right) + \beta^\ell \tau \geq 0$$

for some fixed $\tau_{j,e_j} \in (\tau_{j1}, \ldots, \tau_{j,k_j})$ and record those values of $\tau_{\ell,q}$ that belong to that range; the corresponding t-vectors are then in $\operatorname{pos} B_{(1)}$. More specifically, identify first the largest index k such that

$$\left(\sum_{j=1}^{m_2-1} \beta^j \tau_{j1} \right) + \beta^{m_2} \tau_{m_2,k} \geq 0.$$

All vectors $(\tau_{11}, \tau_{21}, \ldots, \tau_{m_2-1,1}, \tau_{m_2,s})$ with $s = 1, \ldots, k$ are recorded as being in pos $B_{(1)}$. We then "move" $\tau_{m_2-1,1}$ to $\tau_{m_2-1,2}$ and repeat the same analysis on the last coordinate of t. If

$$\left\{ \tau \Big| \left(\sum_{j=1}^{m_2-2} \beta^j \tau_{j1} \right) + \beta^{m_2-1} \tau_{m_2-1,2} + \beta^{m_2} \tau \geq 0 \right\} \bigcap [\tau_{m_2,1}, \tau_{m_2,k_2}] = \emptyset,$$

we return the $(m_2 - 1)$-th coordinate of t to $\tau_{m_2-1,1}$ and increase the preceding element of t to its next higher value, otherwise it is the $(m_2 - 1)^{th}$ coordinate which is increased (discretely) to its next higher value, if possible; if not it is again the $(m_2 - 2)$-th coordinate which is pushed to its next value. This is continued, systematically, until the search with $B_{(1)}$ is exhausted. We now restart the procedure with the "lowest" vector

$$(\tau_{1j_1}, \tau_{2j_2}, \ldots, \tau_{m_2,jm})$$

which has not been included in the first bunch, i.e. for every $i = 1, \ldots, m_2$, the index j_i is as small as possible. The procedure is repeated until all possible vectors generated by the array have been assigned to a given bunch. Further details can be found in Garstka and Rutenber [9], who also report computational experience which would favor this approach with respect to the coarse bunching procedure described at the beginning of this section. However, to rely on this procedure we must be in this specific situation, i.e. when the vectors in τ can be given the array representation (3.34) and this is not always the case, we often deal with dependent random variables and if (3.1) is the result of an approximation scheme then the chosen discretization will usually not be of this type.

3.5 Conclusion

At this stage of algorithmic development for (linear) stochastic programs with recourse, decomposition-type methods aided by a number of shortcuts made possible by the structural properties of the problem, appear as the clear cut favorites. Of course, this is mostly due to the fact that they allow us to exploit to the fullest these structural properties, see Section 3.4, but there may be some other justification for using decomposition-type methods. Experiments, cf. Beer [1], have shown that with the decomposition method, a value near the optimum—Beer speaks of an error of no more than 3% —is reached at an early stage of the computation. Given on one hand the stability of the solution to stochastic programs—see Dupačová [6], Wang [25]—and on the other hand our limitations in the (precise) description of stochastic phenomena or other sources of uncertainties, as mentioned in Section 3.1, a rapid convergence to an approximate solution is all that is expected and required. If solving the discrete stochastic program (3.1) is part of a sequential scheme for solving a stochastic program with continuous probability distribution or with a discrete distribution involving many more points than L , then it would *not* be necessary to solve up to optimality before a further refinement is introduced. Again decomposition-type methods that exhibit rapid convergence to nearly optimal solutions would be ideally suited in such a scheme.

References

[1] K. Beer, "Solving linear programming problems by resource allocation methods", in *Large-scale Linear Programming* eds. G. Dantzig, M. Demp-ster, and M. Kallio, IIASA Collaborative Proceedings Series, Laxenburg, (1981) 409–424.

[2] J. Birge and R. Wets, "Designing approximation schemes for stochastic optimization problems, in particular stochastic programs with recourse", *Mathematical Programming Study* (1984).

[3] J. Bisschop and A. Meeraus, "Matrix augmentation and partitioning in the updating of the basis inverse", *Mathematical Programming* 13(1977), 241–254

[4] J. Bisschop and A. Meeraus, "Matrix augmentation and structure preserva-tion in linearly constrained control problems", *Mathematical Programming* 18(1980), 7–15.

[5] G. Dantzig and A. Madansky, "On the solution of two-stage linear pro-grams under uncertainty", *Proc. Fourth Berkeley Symposium on Mathe-matical Statistics and Probability, Vol.1*, Univ. California Press, Berkeley, 1961. 165–176

[6] J. Dupačová, "Stability in stochastic programming with recourse-estimated parameters", *Mathematical Programming* 28(1984), 72–83.

[7] J. Dupačová and R. Wets, "On the asymptotic behaviour of constrained es-timates and optimal decision". Manuscript, IIASA, Laxenburg (1985)(forth coming).

[8] R. Fourer, "Staircase matrices and systems", *SIAM Review* 26(1984), 1–70.

[9] S. Garstka and D. Rutenberg, "Computation in discrete stochastic pro-grams with recourse", *Operations Research* 21(1973), 112–122.

[10] J. Ho, "Equivalent piecewise linear formulations of separable convex pro-grams", Manuscript, Univ. Tennessee, Knoxville (1983).

[11] P. Kall, "Computational methods for solving two-stage stochastic linear programming problems", *Z. Angew. Math. Phys.* 30 (1979), 261–271.

[12] C. Kiwiel, "A descent algorithm for large-scale linearly constrained convex nonsmooth minimization", IIASA CP-84-15, Laxenburg, Austria (1984).

[13] J. Midler and R. Wollmer, "Stochastic programming models for scheduling airlift operations, *Naval Res. Logist. Quat.* 16(1969), 315–330.

[14] L. Nazareth, "Algorithms based upon generalized linear programming for stochastic programs with recourse", IIASA WP-84-81, Laxenburg, Austria (1984).

[15] L. Nazareth and R. Wets, "Algorithms for stochastic programs: the case of nonstochastic tenders", *Mathematical Programming Study*.

[16] A. Perold and G. Dantzig, "A basis factorization method for block triangular linear programs", in *Sparse Matrix Proceedings, 1978*, eds I. Duff and G. Stewart, SIAM Publications, Philadelphia, (1979) 283-312.

[17] L. Qi, "Forest iteration method for stochastic transportation problem", *Mathematical Programming Study* (1984).

[18] G. Salinetti, "Approximations for chance-constrained programming problems", *Stochastics* (1983).

[19] M. Saunders, Private Communication (1983).

[20] R. Van Slyke and R. Wets, "*L*-shaped linear programs with applications to optimal control and stochastic programming", *SIAM J. Appl.Math.* **17**(1969), 638-663.

[21] B. Strazicky, "Some results concerning an algorithm for the discrete recourse problem", in *Stochastic Programming* ed. M. Dempster, Academic Press, London (1980), 263-274.

[22] D. Walkup and R. Wets, "Lifting projections of convex polyhedra", *Pacific J. Mathem.* **28**(1969), 465-475.

[23] S. Wallace, "Decomposing the requirement space of a transportation problem into polyhedral cones", *Mathematical Programming Study* (1984).

[24] S. Wallace, "On network structured stochastic optimization problem", *Report no.* 842555-8, Chr. Michelsen Institute, Bergen Norway (1985).

[25] J.Wang, "Distribution sensitivity analysis for stochastic programs with recourse", *Mathematical Programming* (1984).

[26] R. Wets, "Programming under uncertainty: the equivalent convex program", *SIAM J. Appl. Math.* **14**(1966), 89-105.

[27] R. Wets, "Characterization theorems for stochastic programs", *Mathematical Programming* **2**(1972), 166-175.

[28] R. Wets, "Stochastic programs with fixed recourse: the equivalent deterministic program", *SIAM Review* **16**(1974), 309-339.

[29] R. Wets, "Stochastic programming: solution techniques and approximation schemes", in *Mathematical Programming: The State of the Art, 1982*, eds. A. Bachem, M. Grötschel and B. Korte, Springer Verlag (1983), 566-603.

CHAPTER 4

NONLINEAR PROGRAMMING TECHNIQUES APPLIED
TO STOCHASTIC PROGRAMS WITH RECOURSE

L. Nazareth and *R. J-B Wets*

Abstract

Stochastic convex programs with recourse can equivalently be formulated as nonlinear convex programming problems. These possess some rather marked characteristics. Firstly, the proportion of linear to nonlinear variables is often large and leads to a natural partition of the constraints and objective. Secondly, the objective function corresponding to the nonlinear variables can vary over a wide range of possibilities; under appropriate assumptions about the underlying stochastic program it could be, for example, a smooth function, a separable polyhedral function or a nonsmooth function whose values and gradients are very expensive to compute. Thirdly, the problems are often large-scale and linearly constrained with special structure in the constraints.

This paper is a comprehensive study of solution methods for stochastic programs with recourse viewed from the above standpoint. We describe a number of promising algorithmic approaches that are derived from methods of nonlinear programming. The discussion is a fairly general one, but the solution of two classes of stochastic programs with recourse are of particular interest. The first corresponds to stochastic linear programs with simple recourse and stochastic right-hand-side elements with given discrete probability distribution. The second corresponds to stochastic linear programs with complete recourse and stochastic right-hand-side vectors defined by a limited number of scenarios, each with given probability. A repeated theme is the use of the MINOS code of Murtagh and Saunders as a basis for developing suitable implementations.

4.1   Introduction

We consider stochastic linear programs of the type

$$\text{find} \quad x \in R^{n_1}$$
$$\text{such that} \quad Ax = b, x \geq 0 \tag{4.1}$$
$$\text{and} \quad z = E_w[c(\omega)x + Q(x,\omega)] \text{ is minimized}$$

where Q is calculated by finding for given decision x and event ω, an optimal recourse $y \in R^{n_2}$, viz.

$$Q(x,w) = \inf_{y \in C}[q(y,\omega)|Wy = h(\omega) - Tx]. \tag{4.2}$$

Here $A(m_1 \times n_1)$, $T(m_2 \times n_1)$, $W(m_2 \times n_2)$ and $b(m_1)$ are given (fixed) matrices, $c(\cdot)(n_1)$ and $h(\cdot)(m_2)$ are random vectors, $y \to q(y, \cdot) : R^{n_2} \to R$ is a random finite-valued convex function and C is a convex polyhedral subset of R^{n_2}, usually $C = R_+^{n_2}$. E denotes expectation.

With $c = E_\omega[c(\omega)]$, an equivalent form to (4.1) is

$$\text{minimize} \quad cx + Q(x)$$
$$\text{subject to} \quad Ax = b \tag{4.3}$$
$$x \geq 0$$

where $Q(x) = E_\omega[Q(x,\omega)]$. Usually $q(y,w)$ will also be a linear nonstochastic function qy. (For convenience, we shall, throughout this paper, write cx and qy instead of $c^T x$ and $q^T y$.)

Two instances of the above problem are of particular interest:

(C1) Problems with simple recourse i.e. with $W = [I, -I]$, stochastic right-hand-side elements with given discrete probability distribution and penalty vectors q^+ and q^- associated with shortage and surplus in the recourse stage (4.2).

(C2) Problems with complete recourse and stochastic right-hand-side vectors defined by a limited number of scenarios, each with given probability.

Henceforth, for convenience, we shall refer to these as C1 and C2 problems respectively. They can be regarded as a natural extension of linear and nonlinear programming models into the domain of stochastic programming. More general stochastic programs with recourse can sometimes be solved by an iterative procedure involving definition (for example, using approximation or sampling) of a sequence of C1 or C2 problems.

Within each of several categories of nonlinear programming methods, we summarize briefly the main underlying approach for smooth problems, give where appropriate extensions to solve nonsmooth problems and then discuss how these lead to methods for solving stochastic programs with recourse. Thus, in each case, we begin with a rather broadly based statement of the solution strategy, and then narrow down the discussion to focus on methods and computational considerations for stochastic programs with recourse, where the special structure of the problem is now always in the background. (During the course of the discussion we occasionally consider other related formulations, in particular the model with probabilistic constraints. However it is our intention to concentrate upon the recourse model. (We do not discuss questions concerning approximation of distribution functions, except very briefly at one or two points in the text). This paper is *not* intended to provide a complete survey. Rather, our aim is to establish some framework of discussion within the theme set by the title of this paper and within it to concentrate on a number of promising lines of algorithmic development. We try to strike a balance between the specific (what is practicable using current techniques, in particular, for C1 and C2 problems) and the speculative (what should be possible by extending current techniques). *An important theme will be the use of MINOS* (the Mathematical Programming System of Murtagh and Saunders [49],[50]) *as a basis for implementation*. Finally we seek to set the stage for the description of an optimization system based upon MINOS for solving C1 problems, see Nazareth [55].

We shall assume that the reader is acquainted with the main families of optimization methods, in particular,

(a) univariate minimization,

(b) Newton, quasi-Newton and Lagrangian methods for nonlinear minimization,

(c) subgradient (nonmonotonic) minimization of nonsmooth functions, possibly using space dilation (variable metric), and the main descent methods of nonsmooth minimization,

(d) stochastic quasi-gradient methods,

(e) the simplex method of linear programming and its reduced-gradient extensions.

Good references for background material are Fletcher [20], Gill et al. [23], Bertsekas [4], Lemarechal [42], Shor [66], Ermoliev [16], Dantzig [11], Murtagh & Saunders [49].

We shall concentrate upon methods of nonlinear programming which seem to us to be of particular relevance to stochastic programming with recourse and discuss them under the following main headings:

1. Problem Redefinition
2. Linearization Methods
3. Variable Reduction (Partitioning) Methods
4. Lagrange Multiplier Methods

A nonlinear programming algorithm will often draw upon more than one of these groups and there is, in fact, significant overlap between them. However, for purposes of discussion, the above categorization is useful.

4.2   Problem Redefinition

By problem redefinition we mean a restructuring of a nonlinear programming problem to obtain a new problem which is then addressed in place of the original one. This redefinition may be achieved by introducing new variables, exploiting separability, dualizing the original problem and so on. For example, consider the minimization of a polyhedral function given by

$$\min_{x \in R^n} \max_{j=1,\ldots,m} [(a^j)^T x + b^j] \tag{4.4a}$$

This can be accomplished by transforming the problem into a linear program

$$\begin{aligned} \text{minimize} \quad & v \\ \text{such that} \quad & v \geq (a^j)^T x + b^j, \quad j = 1,\ldots,m \end{aligned} \tag{4.4b}$$

which can then be solved by the simplex method.

Problem redefinition often precedes the application of other solution methods discussed in later sections of this paper.

4.2.1   Application to Recourse Problems

The following two transformations of recourse problems will prove useful:

(a) When the technology matrix is fixed, new variables χ, termed *tenders*, can be introduced into (4.3). This gives an equivalent form as follows:

$$\begin{aligned} \text{minimize} \quad & cx + \Psi(\chi) \\ \text{subject to} \quad & Ax = b \\ & Tx - \chi = 0 \\ & x \geq 0. \end{aligned} \tag{4.5}$$

(4.5) is useful because it is a nonlinear program in which the number of variables occurring nonlinearly is m_2 instead of n_1 and usually $m_2 \ll n_1$. For a more detailed discussion of the use of tenders in algorithms for solving stochastic linear programs with recourse, see Nazareth and Wets [56].

(b) Another useful transformation involves introducing second stage activities into the first stage. It is shown in Nazareth [51] that that an alternative form equivalent to (4.5) is

$$\begin{aligned} \text{minimize} \quad & cx + qy + \Psi(\chi) \\ \text{subject to} \quad & Ax = b \\ & Tx + Wy - \chi = 0 \\ & x \geq 0, y \geq 0. \end{aligned} \tag{4.6}$$

This transformation also has significant advantages from a computational standpoint, as we shall see below. These stem, in part, from the fact that dual feasible variables, say (ρ, π) satisfy $W^T \pi \leq q$.

For C1 problems, $\Psi(\chi)$ in (4.5) is separable, i.e. $\Psi(\chi) = \sum_{i=1}^{m_2} \Psi_i(\chi_i)$. In such problems, each component of $h(\cdot)$ is assumed to be discretely distributed, say with $h_i(\cdot)$ given by levels $h_{i1}, \ldots, h_{ik_i}$ and associated probabilities $p_{i1}, \ldots, p_{ik_i}$; also $q(y)$ in (4.2) is two-piece linear and can be replaced by $q^+ y^+ + q^- y^-$, $y^+ \geq 0, y^- \geq 0$ in (4.6). This implies that each $\Psi_i(\chi_i)$ is piecewise linear with slopes, say, $s_{i\ell}, \ell = 0, \ldots, k_i$. By introducing new bounded variables $z_{i\ell}, \ell = 0, \ldots, k_i$ we can reexpress χ_i as

$$\chi_i = h_{i0} + \sum_{\ell=0}^{k_i} z_{i\ell}$$

where h_{i0} is the i-th component of h_0 the *base* tender. Then (4.5) takes the form:

$$\text{minimize} \quad cx + \sum_{i=1}^{m_2} \sum_{\ell=0}^{k_i} s_{i\ell} z_{i\ell}$$

$$\text{subject to} \quad Ax = b$$

$$T^i x - \sum_{\ell=0}^{k_i} z_{i\ell} = h_{i0}, \quad i = 1, \ldots, m_2 \tag{4.7}$$

$$x \geq 0, 0 \leq z_{i\ell} \leq d_{i\ell}, \quad \ell = 0, \ldots, k_i$$

$$\text{with } d_{i\ell} = h_{i,\ell+1} - h_{i\ell}.$$

T^i denotes the i-th row of T. Optionally we can use the transformation (4.6) to introduce $W = [I, -I]$ into the first stage. Details of an algorithm based upon (4.7) can be found in Wets [72] and an alternative simpler version of this algorithm can be found in Nazareth & Wets [56]. The latter algorithm is implemented in the optimization system described [55], where further discussion and computational considerations may be found.

4.2.2 Extensions

The device of introducing new bounded variables, which was used to obtain (4.7), can be applied to a wider class of recourse problems. The assumptions of discrete distribution of $h(\cdot)$ and of two (or more) piece linearity of recourse objective are not central, although one must still retain the assumptions of simple recourse and separable recourse objective. Suppose, for example, the distribution function of $h_i(\cdot)$ which need not be continuous, is piecewise linear with knots $h_{i1}, \ldots, h_{ik_i}$, and $q = (q^+, q^-)$. Then $\Psi_i(\chi_i)$ is piecewise quadratic. In general, if the distribution is defined in terms of splines of order s at knots $h_{i1}, \ldots, h_{ik_i}$ and $q(y)$ is separable, say, $\sum_{i=1}^{n_2} q_i(y_i)$ with each $q_i(y_i)$ convex, then $\Psi_i(\chi_i)$ can be shown to be convex and piecewise smooth. Suppose it is given by pieces $\Psi_{i\ell}(\chi_i)$ over intervals $(h_{i\ell}, h_{i,\ell+1})$. Then, analogously to (4.7) we can transform the problem (4.5) into the structured and *smooth nonlinear program*

$$\text{minimize} \quad cx + \sum_{i=1}^{m_2} \sum_{\ell=0}^{k_i} \Psi_{i\ell}(h_{i\ell} + z_{i\ell}) - \Psi_{i\ell}(h_{i\ell})$$

subject to $Ax = b$

$$T^i x - \sum_{\ell=0}^{k_i} z_{i\ell} = h_{i0}, \quad i = 1, \ldots, m_2 \tag{4.8}$$

$$x \geq 0, 0 \leq z_{i\ell} \leq d_{i\ell}, \quad \ell = 0, \ldots, k_i$$

with $d_{i\ell} = h_{i,\ell+1} - h_{i\ell}.$

(Here again we could use the transformation (4.6) to introduce $W = [I, -I]$ into the first stage). Note that (4.7) is a special case of (4.8). The optimal solution of (4.8) has an important property which is easy to prove. This result makes the nonlinear program (4.8) very amenable to solution by MINOS-like techniques and it is given by the following proposition:

Proposition. *In the optimal solution of (4.8), say $(x^*, z_{i\ell}^*)$, if for some $t, z_{it}^* < d_{it}^*$ then $z_{i\ell}^* = d_{i\ell}$ for all $\ell < t$.*

Outline of Proof: Regard each $\Psi_{i\ell}(\chi_i)$ as the limit of a piecewise linear function, and then appeal to the standard argument used in the piecewise-linear case.

The above proposition tells us that there are, at most, m_2 superbasic variables (see Section 4.4 for terminology) in the optimal solution of (4.8). This would be to the advantage of a routine like MINOS, which thrives on keeping the number of superbasics low. These remarks will become clearer after looking at Section 4.4. Note also that Wets [70] discusses a special case of (4.8) when $\Psi_{i\ell}(\chi_i)$ are piecewise quadratic. A well-structured code for solving (4.7), which uses only the LP facilities of MINOS, would be capable of a natural extension to solve nonlinear problems of the form (4.8). MINOS was really designed to solve problem of this type.

The above approach remains limited in scope, because of the need to assume that recourse is simple and that the recourse objective is separable. Therefore we wou'd not expect it to be useful for C2 problems.

The transformations given by (4.5) and (4.6) are very useful prior to the application of other techniques discussed in the following sections of this chapter. Let us consider some possibilities.

1. When T' is nonstochastic, use of the transformation (4.5) in the methods described by Kall [30] or the L-shaped algorithm of Van Slyke and Wets [68] (see also Birge [6]) would lead to *fewer nonzero elements* in the representation of the associated large-scale linear programs.

2. When T is not a fixed matrix, typically only a few columns (activities), say $T_2(\omega)$, would be stochastic. Say these correspond to variables $\hat{x}$, with $x = (\tilde{x}, \hat{x})$. We could then introduce a redefinition of the problem in which a tender is associated with the nonstochastic columns, say T_1 of T; then the degree of nonlinearity of the equivalent deterministic nonlinear programming problem would be m_2+ dimension ($\hat{x}$) instead of n_1. For example, for simple recourse with $q = (q^+, q^-)$ we would have

$$\psi(\chi, \hat{x}, w) = \min_{y^+, y^- \geq 0} [q^+ y^+ + q^- y^- \,|\, y^+ - y^- = h(\omega) - \chi - T_2(\omega)\hat{x}]$$
$$\Psi(\chi, \hat{x}) = E_w[\psi(\chi, \hat{x}, \omega)]$$

Note that $\Psi(\chi, \hat{x})$ continues to be separable in χ, i.e. $\Psi(\chi, \hat{x}) = \sum_{i=1}^{m_2} \Psi_i(\chi_i, \hat{x})$. These observations and the further developments that they imply would be useful in a practical implementation.

3. Another interesting example of the use of the transformations involving tenders is given in Nazareth [52] where they are used in the solution of deterministic staircase-structured linear programs.

4.3 Linearization Methods

A prominent feature of methods in this group is that they solve sequences of linear programs. One can distinguish single-point and multi-point linearization. In both approaches convexity of functions is normally assumed.

4.3.1   Single-Point Linearization Methods

We discuss this case very briefly.

Consider the problem minimize$_{x \in K} f(x)$, where K is polyhedral and $f(x)$ is smooth. The approach consists of solving a sequence of problems of the form:

$$\operatorname*{minimize}_{x \in \overline{K}} \nabla f(x_k)^{\mathrm{T}} (x - x_k) \qquad (4.9)$$

where $\overline{K}$ is the original polyhedral set K, possibly augmented by some additional constraints. This leads to a variety of methods. When $K = \overline{K}$ we obtain the Frank–Wolfe [21] method, in which the solution, say x_k^*, defines a search direction $d_k = x_k^* - x_k$. The method has the virtue that the solution is found in one step if the original problem is linear. If K is augmented by the constraints $\|x - x_k\|_\infty \leq \delta$ for some small positive constant δ we obtain the Griffith & Stewart [26] method of approximate programming (MAP); for minimax applications see Madsen & Schjaer-Jacobsen [46] and for extensions to the domain of general nonsmooth optimization see the monograph of Demyanov & Vasiliev [12].

4.3.1.1   Applications to Recourse Problems

For simple recourse when the equivalent (deterministic) nonlinear program is smooth, algorithms are given, for example, by Ziemba [77]. Kallberg and Ziemba [33] use the Frank–Wolfe method in a setting where only estimates of functions and gradients can be obtained. The approach has been widely studied within the context of the general expectation model, see Ermoliev [16] and models with probabilistic constraints, see Komaroni [37] and references cited there. In this latter context, however, one needs to rely on a variant of the standard Frank–Wolfe method to take into account nondifferentiability (infinite slope) of the objective at the boundary of the feasible region. Given a stochastic program with probabilistic constraints of the type

$$
\begin{aligned}
\text{minimize} \quad & cx \\
\text{subject to} \quad & Ax = b \\
& \text{Prob}\,[w|Tx > h(w)] \geq \alpha \\
& x \geq 0
\end{aligned}
$$

we see that it is equivalent to

$$
\begin{aligned}
\text{minimize} \quad & cx \\
\text{subject to} \quad & Ax = b \\
& Tx - \chi \geq 0 \\
& g(\chi) \geq 0 \\
& x \geq 0 \\
\text{where} \quad & g(\chi) = \ln(\text{Prob}\,[w|\chi > h(w)] - \alpha).
\end{aligned}
$$

Assuming that the probability measure is log-concave, it follows that g is concave and thus we are dealing with a convex optimization problem with one nonlinear constraint. Its dual is

$$\text{maximize} \quad ub + \rho(v)$$
$$\text{subject to} \quad uA + vT \le c$$
$$v \ge 0$$
$$\rho(v) = \inf[v\chi | g(\chi) \ge 0].$$

The function ρ is a sublinear (concave and positively homogeneous) finite-valued (only) on the positive orthant. If the probability measure is strictly log-concave, the function ρ is differentiable on the interior of the positive orthant and thus we could use the Frank–Wolfe procedure to solve this dual problem as long as the iterates (u^s, v^s) are such that $v^s \in$ interior $R_+^{m_2}$; when v^s is on the boundary of $R_+^{m_2}$, the standard procedure must be modified to handle the 'infinite' slope case, see Komaroni [37].

4.3.2 Multi-Point Linearization Methods

Consider the problem

$$\text{minimize } f(x) \text{ where } g_i(x) \le 0, \qquad i = 1, \dots, m, x \in X \qquad (4.10)$$

where all functions are convex, but not necessarily differentiable, and X is a compact set. We shall concentrate in this section on the generalized linear programming method (GLP) of Wolfe (see Dantzig [11], Shapiro [65]) which solves a sequence of problems obtained by *inner (or grid) linearization* of (4.10) over the convex hull of a set of points $x^1, \dots, x^K$, to give the following master program:

$$\text{minimize} \quad \sum_{i=1}^{K} \lambda_i f(x^i)$$

$$\text{subject to} \quad u_i^{(K)} : \sum_{i=1}^{K} \lambda_i g_i(x^i) \le 0, \qquad i = 1, \dots, m \qquad (4.11a)$$

$$w^{(K)} : \sum_{i=1}^{K} \lambda_i = 1, \lambda_i \ge 0$$

where $u^{(K)}$ and $w^{(K)}$ represent the dual variables associated with the optimal solution of (4.11a). The dual of (4.11a) is

$$\text{maximize} \quad w$$
$$\text{subject to} \quad w \le f(x^{(k)}) + u g(x^{(k)}), \quad k = 1, \dots, K \qquad (4.11b)$$
$$u \ge 0.$$

The next grid point x^{K+1} is obtained by solving the Lagrangian subproblem

$$\underset{x \in X}{\text{minimize}}[f(x) + u^{(K)}g(x)] \tag{4.12}$$

where $u^{(K)}$ is also the optimal solution of (4.11b). Convergence is obtained when

$$f(x^{(K+1)}) + u^{(K)}g(x^{(K+1)}) \geq w^{(K)}.$$

Since the dual of (4.10) is

$$\underset{u \geq 0}{\text{maximize}}\, h(u), \text{ where } h(u) = \underset{x \in X}{\text{min}}(f(x) + ug(x)) \tag{4.13}$$

and $h(u)$ is readily shown to be concave, an alternative viewpoint is to regard the GLP method as a *dual* cutting plane (or outer linearization) method on (4.13) yielding (4.11b); new grid points obtained from (4.12) yield a supporting hyperplane to $h(u)$ at $u^{(K)}$.

It is worth emphasizing again that an important advantage of the inner-linearization approach is that it can be directly applied to the solution of non-smooth convex problems without extensions.

Outer linearization could be applied directly to the functions in (4.10) to give a *primal* cutting plane method which also solves sequences of linear programs. For details, see Kelley [34], Zangwill [76] and Eaves & Zangwill [13].

4.3.2.1   Applications to Recourse Problems

For recourse problems, particularly with the form (4.5) using tenders, the GLP approach looks very promising.

Using GLP to solve *simple recourse* problems has an early history. It was first suggested by Williams [74], in the context of computation of error bounds and also used at an early date by Beale [2]. Parikh [57] describes many algorithmic details. The method has also been implemented for specialized applications (e.g. see Ziemba [78], for an application to portfolio selection). However, as a general computational technique in particular, for *nonsimple recourse* it has apparently not been studied until recently, see Nazareth and Wets [56] and Nazareth [51].

The GLP method applied to (4.5) yields the following master program:

$$\text{minimize}\quad cx + \sum_{i=1}^{K} \lambda_k \Psi(\chi^{(k)})$$

$$\text{subject to}\quad \rho^{(K)} : Ax = b$$

$$\pi^{(K)} : Tx - \sum_{k=1}^{K} \lambda_k \chi^k = 0 \tag{4.14a}$$

$$\theta^{(K)} : \sum_{k=1}^{K} \lambda_k = 1$$

$$x \geq 0, \lambda_k \geq 0.$$

The associated subproblem is

$$\underset{\chi \in X}{\text{minimize}} \; \Psi(\chi) + \pi^{(K)}\chi \qquad (4.14b)$$

In order to complete the description of the algorithm it is necessary to specify X, and if this is not a compact set, to extend the master program (4.14a) by introducing directions of recession (whose associated variables do not appear in the convexity row). In addition, a suitable set of starting tenders which span R_2^m should be specified. As discussed in Nazareth [51] these considerations can be largely circumvented by using the equivalent form (4.6) and solving master programs of the form:

$$\text{minimize} \quad cx + qy + \sum_{k=1}^{K} \lambda_k \Psi(\chi^{(k)})$$

$$\text{subject to} \quad Ax = b$$

$$Tx + Wy - \sum_{k=1}^{K} \lambda_k \chi^{(k)} = 0 \qquad (4.15)$$

$$\sum_{k=1}^{K} \lambda_k = 1$$

$$x \geq 0, y \geq 0, \lambda_k \geq 0.$$

As discussed in more detail in Nazareth & Wets [56], we expect the above algorithm to perform well because normally only a few tenders will have nonzero coefficients in the optimal solution and because one can expect to obtain a good set of starting tenders from the underlying recourse program.

Still at issue is how readily one can compute $\Psi(\chi^{(k)})$ and its subgradients at a given point χ^k. This in turn determines the ease with which one can solve the subproblem (4.14b) and obtain coefficients in the objective row of the master.

For C1 problems $\Psi(\chi)$ is separable and easy to specify explicitly (see Wets [72]). Algorithms have been given by Parikh [57] and Nazareth [51]. A practical implementation is given in Nazareth [55] where further details may be found.

For C2 problems (i.e. with complete recourse and a relatively small set of scenarios say, $h^\ell, \ell = 1, \ldots, L$ with known probabilities $f_\ell, \ell = 1, \ldots, L$) one can solve the subproblem (4.14b) and compute $\Psi(\chi^{(k)})$ in one of two ways, as discussed in Nazareth [51]:

(i) Formulate (4.14b) as a linear program which can be efficiently solved by Schur-Complement techniques, see Bisschop & Meeraus [9], and Gill et al. [24]. The values $\Psi(\chi^{(k)})$ are a part of the solution of this linear program.

(ii) Use unconstrained nonsmooth optimization techniques, see Lemarechal [40],[41], Kiwiel [35] and Shor [66]. Information needed by such methods is $\Psi(\chi^{(k)})$ and its subgradient $g(\chi^{(k)})$ and this can be computed by solving a set of linear programs of the form:

$$\psi(\chi^{(k)}, h^\ell) = \min_{y \geq 0}[qy|Wy = h^\ell - \chi^{(k)}].$$

Suppose π^ℓ are the optimal dual multipliers of the above problem. Then

$$\Psi(\chi^{(k)}) = \sum_{\ell=1}^{L} f_\ell \psi(\chi^{(k)}, h^\ell)$$

$$g(\chi^{(k)}) = \sum_{\ell=1}^{L} f_\ell \pi^\ell.$$

This can be carried out very efficiently using the dual simplex method coupled with techniques discussed by Wets in [73].

The method based upon outer linearization mentioned at the end of the previous section has been widely used to solve stochastic programs with recourse (see Van Slyke & Wets [68], Wets [73] and Birge [6]). This is a particular form of Benders' decomposition [3] and it is well known that approaches based upon Benders' decomposition can solve a wider class of *nonlinear* convex programs than approaches based upon the Dantzig–Wolfe decomposition, see, for example, Lasdon [39]). We shall not however discuss this approach in any detail here because it is already studied, in depth, in the references just cited.

4.3.2.2 Extensions

When $\Psi(\chi^k)$ and its subgradients are difficult to compute, the GLP approach continues to appear very promising but many open questions remain that center on convergence.

Two broad approaches can be distinguished:

(i) *Sampling*: Stochastic estimates of $\Psi(\chi)$ and its subgradient can be obtained by sampling the distribution. An approach that uses samples of fixed size and carries out the minimization of the Lagrangian subproblem (4.14b) using smoothing techniques is described by Nazareth [51]. Methods for minimizing noisy functions suggested recently by Atkinson et al. [1] would also be useful in this context. With a fixed level of noise, convergence proofs can rely upon the results of Poljak [58].

Another variant is to use samples of progressively increasing size tied to the progress of the algorithm and to solve the Lagrangian subproblem using stochastic quasi-gradient methods, see Ermoliev & Gaivoronski [18]. A particular algorithm (suggested jointly with A. Gaivoronski) is to replace $\Psi(\chi^{(k)})$ in (3.6) by some estimate $\Psi_s(\chi^{(k)})$ which is based upon a suitable sample size N. When no further progress is made, then this sample size is incremented by ΔN and the approximation refined for all $\chi^{(k)}$ in the current basis. There are, of course, many further details that must be specified, but under appropriate assumptions convergence can be established.

(ii) *Approximate Distribution and Compute Bounds*: At issue here is how to simultaneously combine approximation and optimization. For example, Birge [7] assumes that converging approximations $\Psi_K^U(\chi)$ and $\Psi_K^L(\chi)$ are available for $K = 1, 2, \ldots$ and replaces $\Psi(\chi^{(k)})$ in (4.14a) by the upper bound $\Psi_k^U(\chi(k)), k = 1, \ldots, K$. In the subproblem (4.14b), if

$$\Psi_{K+1}^U(\chi^{(K+1)}) + \pi^{(K)}\chi^{(K+1)} \geq \theta^{(K)}$$

then $\Psi_{K+1}^L(\chi^{(K+1)})$ is computed. If, further, the above inequality is satisfied using this lower bound in place of the upper bound, then $\chi^{(k+1)}$ is optimal. Otherwise the approximation is refined and the process continued. Approximation schemes for obtaining bounds rely on the properties of recourse problems, instead of purely on the distance between the given probability distribution and the approximating ones; this allows for sequential schemes that involve much fewer points as discussed by Kall & Stoyan [31] and Birge & Wets [8].

The interpretation of the optimal solution in Nazareth [51], suggests the possibility of an alternative approach to approximation by an increasingly large number of points. It is shown that if $\lambda_{k_j}^*$ and $\chi^{(k_j)}, j = 1, \ldots, (m_2 + 1)$ give the optimal solution of (4.14a), then the problem (4.5) is equivalent to the associated discretized problem obtained by replacing the distribution of $h(w)$ by the distribution whose values are $\chi^{(k_j)}, j = 1, \ldots, (m_2 + 1)$ with associated probabilities $\lambda_{k_j}^*$. Note that $\sum_{i=1}^{m_2+1} \lambda_{k_j}^* = 1, \lambda_{k_j}^* \geq 0$, so that these quantities do indeed define a probability distribution.

Let us conclude this section with a discussion of some other possibilities.

1. When the technology matrix is nonlinear, i.e. when T is replaced by a smooth nonlinear function, we have the possibility of a generalized programming algorithm where the *master program itself is nonlinear*. The question of convergence is open. Here an implementation based upon MINOS would be able to immediately draw upon the ability of this routine to solve programs with nonlinear constraints.

2. When some columns of T are stochastic, the transformation discussed at the end of Section 4.2 can also be used within the context of the GLP algorithm to keep the degree of nonlinearity low. This time inner approximation of $\Psi(\chi, \hat{x})$ would be carried out over the convex hull of $(\chi^{(k)}, \hat{x}^{(k)}), k = 1, \ldots, K$.

3. Generalized programming techniques appear to be useful for solving programs with probabilistic constraints, for example, of the form:

$$\begin{aligned}
\text{minimize} \quad & cx \\
\text{subject to} \quad & Ax = b \\
& \text{Prob}\,[\omega|Tx > h(\omega)] \geq \alpha \\
& x \geq 0.
\end{aligned}$$

With the usual definition of tenders $T x = \chi$ and under the appropriate assumptions on the distribution of $h(\cdot)$, we can express the above problem as:

$$\begin{aligned}
\text{minimize} \quad & cx \\
\text{subject to} \quad & Ax = b \\
& Tx - \chi = 0 \\
& g(\chi) \le 0 \\
& x \ge 0
\end{aligned}$$

where $g(\chi) = \alpha - \operatorname{Prob}\,[\omega | h(\omega) < \chi]$ is a nonlinear function which is log-concave for a wide variety of distribution functions, in which case the set $[\chi | g(\chi) \le 0]$ is convex. In such a situation we can reformulate the constraint

$$g(\chi) \le 0,$$

as

$$\chi \in D = \{y | g(y) \le 0\}$$

where

$$g(y) = \alpha - \int_{\varsigma < \chi} p(\varsigma)\,d\varsigma.$$

Here $p(\cdot)$ denotes the density function of the random vector $h(\cdot)$. Assuming that we have already generated $\chi^1, \ldots, \chi^K$ in D such that

$$\{\chi = Tx | Ax = b, x \ge 0\} \cap \operatorname{co}\{\chi^1, \ldots, \chi^K\} \ne \emptyset,$$

we would be confronted at step K with the master problem:

$$\begin{aligned}
\text{minimize} \quad & cx \\
\text{subject to} \quad & \sigma^K : Ax = b \\
& \pi^K : Tx - \sum_{i=1}^{K} \lambda_i \chi^i = 0 \\
& \theta^K : \sum \lambda_i = 1 \\
& x \ge 0, \lambda_i \ge 0, \qquad i = 1, \ldots, K
\end{aligned}$$

where $(\sigma^K, \pi^K, \theta^K)$ represent the dual variables associated with the optimal solution (χ^K, λ^K) of this master problem. The next tender χ^{K+1} is obtained by solving the Lagrangian subproblem, involving only χ:

$$\text{minimize } [\pi^K \chi | \chi \in D]$$

and this χ^{K+1} is introduced in the master problem unless

$$\pi^K \chi \ge \theta^K,$$

in which case x^K is an optimal solution of the master problem. To find

$$\chi^{K+1} \in \text{argmin}[\pi^K \chi | \int_{\varsigma < \chi} p(\varsigma) d\varsigma \leq \alpha]$$

we consider the Lagrangian function

$$\ell(\chi, \beta) = \pi^K \chi + \beta(\int_{\varsigma < \chi} p(\varsigma) d\varsigma - \alpha), \beta \geq 0$$

and the dual problem

$$\text{maximize } h(\beta), \beta \geq 0,$$

where

$$h(\beta) = \inf[\ell(\chi, \beta) | \chi].$$

The function h is an ℓ-dimensional concave function, its (generalized) derivative is a monotone increasing function, and, moreover, under strict log-concavity of the probability measure, its maximum is attained at a unique point. To search for the optimal β^K we can use a secant method for finding the zero of a monotone function. We have that for fixed β,

$$\chi(\beta) = \text{argmin } \ell(\chi, \beta)$$

is obtained by solving the following system of equations:

$$-\pi_i^K / \beta = \int_{\{\varsigma_k < \chi_k | k \neq i\}} p(\varsigma_1, \ldots, \varsigma_{i-1}, \ldots, \chi_i, \varsigma_{i+1}, \ldots, \varsigma_{m_2}) d\varsigma, i = 1, \ldots, m_2.$$

If p is simple enough, or if it does not depend on too many variables then this system can be solved by a quasi-Newton procedure that avoids multidimensional integration.

This application to chance-constrained stochastic linear programming is an open area and certainly deserves further investigation.

4. It is also worth pointing out that generalized programming methods have been recently applied to the study of problems with partially known distribution functions (incomplete information), see Ermoliev et al. [17] and Gaivoronski [22].

4.4   Variable Reduction (Partitioning) Methods

Methods in this group seek to restrict the search region to one defined by a sub-set of the variables and carry out one or more iterations of a gradient (or subgradient) based search procedure. The search region is then revised and the process continued. We can make a distinction between 'homogeneous' and 'global' methods (using the terminology of Lemarechal [42]). Homogeneous or active set methods, in the linearly constrained case, restrict the region of search to an *affine subspace* within which unconstrained minimization techniques can be used. We shall concentrate on the reduced gradient formulation of Murtagh & Saunders [49],[50] as implemented in MINOS and seek extensions of an approach which has proved effective for large smooth problems. However the fact that extension is necessary, in contrast to the methods of the previous section, and the fact that there are theoretical issues of convergence that remain to be settled mean that such methods are still very much in the development stage.

Global methods treat all constraints *simultaneously* and define direction finding subproblems which usually involve minimization subject to inequality constraints (often just simple bound constraints). Convergence issues are more easily settled here. We shall consider some methods of this type.

We also include here approaches where the partition of variables is more directly determined by the problem structure, in particular the grouping into linear and nonlinear variables.

Consider first the problem defined by

$$
\begin{aligned}
\text{minimize} \quad & f(x) \\
\text{subject to} \quad & Ax = b \\
& x \geq 0
\end{aligned}
\tag{4.16}
$$

where, initially, $f(x)$ is assumed to be smooth.

The variables at each cycle of the Murtagh and Saunders [49] reduced gradient method are partitioned into three groups, (x_B, x_S, x_N) representing m basic variables, s superbasic variables, and $nb = n - m - s$ nonbasic variables respectively. Non-basics are at their bound. A is partitioned as $[B|S|N]$ where B is an $m \times m$ nonsingular matrix, S is an $m \times s$ matrix, and N is an $m \times nb$ matrix. Let $g = \nabla f(x)$ be similarly partitioned as (g_B, g_S, g_N).

Each cycle of the method can be viewed as being roughly equivalent to:

RG1 one or more iterations of a quasi-Newton method on an unconstrained optimization problem of dimension s determined by the *active* set $Ax = b, x_N = 0$. Here a reduced gradient is computed as

$$
\mu = g_S - (g_B^T B^{-1})S = [-(B^{-1}S)^T|I_{s \times s}|0]g = Z_S^T g. \tag{4.17}
$$

The columns of Z_S span the space in which the quasi-Newton search direction lies, and this is given by $p = -Z_S H Z_S^T g$ where H is an inverse Hessian approximation obtained by quasi-Newton update methods and defines the variable metric, e.g. $H = I$ gives the usual projected gradient direction. Along p a line search is usually performed. (Note that in actual computation H would not be computed. Instead we would work with approximations to the Hessian and solve systems of linear equations to compute the search direction p.)

RG2 an iteration of the revised simplex method on a linear program of dimension $m \times nb$. Here components of the reduced gradient (Lagrange multipliers) corresponding to the nonbasic components are computed by

$$\lambda = g_N - (g_B^T B^{-1})N \tag{4.18}$$

$$\lambda = [-(B^{-1}N)^T|0|I_{nb \times nb}]g = Z_N^T g. \tag{4.19}$$

This is completely analogous to the computation of μ in (4.17) above. The difference is in the way that λ is used, namely to revise the active set. In each case above prices π can be computed by $\pi = g_B^T B^{-1}$ and μ and λ computed as

$$\mu = g_S - \pi^T S, \lambda = g_N - \pi^T N \tag{4.20}$$

(It is worth noting that the *convex simplex method* is a special case of the above where (RG1) is omitted and (RG2) is replaced by a coordinate line search along a single coordinate direction in the reduced space given by $(Z_N)_k$, say, for which $\lambda_k < 0$. When there are nonlinear constraints present the above method can also be suitably generalized.)

In the *nonsmooth* case we can proceed along three main directions:

1. Compute μ and λ in place of the above by

$$\mu = Z_S^T \{\operatorname{argmin}[g^T (Z_S Z_S^T)g | g \in \partial f(x)]\}$$
$$\lambda = Z_N^T \{\operatorname{argmin}[g^T (Z_N Z_N^T)g | g \in \partial f(x)]\}$$

where $\partial f(x)$ is the subdifferential of $f(x)$ at x. In effect we are computing steepest descent directions in the appropriate subspaces. Note that it is, in general, not correct to first compute a steepest descent direction $\bar{g}$ from

$$\bar{g} = \operatorname{argmin}[g^T g | g \in \partial f(x)]$$

and then reduce $\bar{g}$ to give

$$\mu = Z_S^T \bar{g}$$
$$\lambda = Z_N^T \bar{g}. \tag{4.22}$$

The reason for this is that the operations of minimization and projection are not interchangeable. However this approach does make it possible to restore use of the π vector and therefore yields useful heuristic methods, as we shall see in the next section. In order to ensure convergence, it is necessary to replace $\partial f(x)$ by $\partial_\epsilon f(x)$—the ϵ-subdifferential (except in special circumstances e.g. when $f(x)$ is polyhedral and line searches are exact). This is useful from a theoretical standpoint. However, from the point of view of computation it is usually impractical to use the subdifferential, let alone the ϵ-subdifferential (except again in rather special circumstances). One such instance is when the subdifferential is defined by a small set of vectors, say, $g_1, \ldots, g_N$. Then (4.21) leads to the problem:

$$
\begin{aligned}
\text{minimize} \quad & g^T Z_S H Z_S^T g \\
\text{subject to} \quad & g = \sum_{i=1}^{N} \lambda_i (Z_S^T g_i) \\
& \sum_{i=1}^{N} \lambda_i = 1 \\
& \lambda_i \geq 0.
\end{aligned}
\tag{4.23}
$$

If g^* is its solution, then $\mu = Z_S^T g^*$, with a similar computation for Z_N^T. We also have $p = -Z_S H Z_S^T g^*$.

2. Utilize bundle methods in which the subdifferential is replaced by an approximation composed from subgradients obtained at a number of prior iterations. For the unconstrained case algorithms are given by Lemarechal [40],[41] and an implementable version is given by Kiwiel [35]. An extension of [40] to handle linear constraints in the reduced gradient setting is given by Lemarechal et al. [45]. However, as the authors point out theoretical issues of convergence remain to be settled in the latter case.

3. Utilize nonmonotonic methods (see, for example, Shor [66]) which require only a single subgradient at each iteration. In effect nonmonotonic iterations will be carried out in subspaces (see RG1 and RG2 above) determined by Z_S and Z_N, using reduced subgradients $Z_S^T g$ and $Z_N^T g$. Again convergence issues remain open.

Line searches suitable for use in the above cases (1) and (2) are given by Mifflin [48] and Lemarechal [43].

The reduced gradient method as formulated above benefits from additional structure in objective and constraints, in particular the partition between variables that occur linearly and variables that occur nonlinearly. We shall see instances of this in the discussion of recourse problems. In particular, it is easy to show that when $f(x)$ is replaced by $cx + \Psi(\chi)$, an optimal solution exists for which the number of superbasics does not exceed the number of nonlinear variables χ.

Instead of obtaining an active set from $x_N = 0$, another approach which gives a 'global' method is to reduce the gradient or subgradient only through the equality constraints $Ax = b$ (these are always active) and define reduced problems to find the search direction involving bound constraints on the x_N variables. This is discussed in Bihain [5]. (See also Strodiot et al. [67].)

Reduced gradient methods, as discussed above, benefit from the partition of the problem into linear and nonlinear variables, but they do not *explicitly* utilize it. It is however possible to take more immediate advantage of this partition. Possible approaches are given, for example, by Rosen [62] and by Ermoliev [15]. Consider the problem

$$\text{minimize} \quad cx + F(y)$$
$$\text{subject to} \quad Ax + By = b$$
$$x \geq 0, y \geq 0$$

If the nonlinear variables y are fixed at certain values we obtain a simpler problem, in this case a linear program (which may have further structure, for example, when A is block-diagonal). The optimal dual multipliers π^* of this linear program (assumed feasible), can then be used to define a reduced sub-problem, for example, $F(y) - (\pi^*)^T By, y \geq 0$. This is then solved to revise the current values of y, for example, by computing a reduced subgradient by $g - \pi^* B, g \in F(y)$ and carrying out (nonmonotonic) iterations in the positive orthant of the y variables (see Ermoliev [15]). An alternative approach is given by Rosen [62].

4.4.1 Applications to Recourse Problems

Since the number of nonlinear variables χ in (4.5) is usually small relative to the number of linear variables, the reduced gradient approach outlined above is a natural choice. When $\Psi(\chi)$ is smooth (and the gradient is computable) the reduced gradient method can be used directly. In the form of the convex simplex method, which is a special case of the reduced gradient method, it has been suggested for the simple recourse problem by Wets [69] and Ziemba [77]. Wets [71] extends the convex simplex method to solve problems with simple recourse when the objective is nonsmooth.

For C1 problems $\partial \Psi_i(\chi_i) = [v_i^-, v_i^+]$ (see Nazareth & Wets [56]). The computation of μ and λ in (4.21) thus requires that we solve *bound constrained quadratic programs*. We can utilize structure in the basis matrix in defining these quadratic programs. Since the χ variables are unrestricted, they can be assumed to be always in the basis. A basis matrix will thus have the form

$$B = \left(\begin{array}{c|c} D & 0 \\ \hline E & I \end{array} \right) \tag{4.24a}$$

and its inverse (never, of course, computed directly) will therefore be given by

$$B^{-1} = \left(\begin{array}{c|c} D^{-1} & 0 \\ \hline -ED^{-1} & I \end{array} \right). \tag{4.24b}$$

Let $g_B = (c_B, g_\chi)$ where c_B are coefficients of the objective row corresponding to the x variables in the basis and g_χ is a subgradient of $\Psi(\chi)$ at the current value of χ. Also, since superbasics and nonbasics are always drawn from c, we shall use c_S and c_N in place of g_S and g_N. Thus we define $g = (c_B, g_\chi, c_S, c_N)$. The quadratic programs (4.21) then takes the form

$$
\begin{aligned}
\text{minimize} \quad & g^T Z_S Z_S^T g \\
\text{subject to} \quad & v_i^- \le (g_\chi)_i \le v_i^+, \quad i = 1, \ldots, m_2
\end{aligned}
\tag{4.25}
$$

where g is defined above, $Z_S^T = (-(B^{-1}S)^T | I_{s \times s} | 0)$ with B defined by (4.24a). Note that usually g_χ will have relatively few components. A similar bound constrained quadratic program can be defined for Z_N^T. Both can be solved very efficiently using a routine like QPSOL, see [25]. The above approach also requires a line search and an efficient one based upon a specialized version of generalized upper bounding, is given in Nazareth [54]. An implementation could thus be based upon MINOS, QPSOL and this line search.

It is possible to avoid the use of quadratic programming by using a heuristic technique in which a steepest-descent direction is first computed as the solution of the expression preceding (4.22). This is given by:

$$
\begin{aligned}
\text{minimize} \quad & g_\chi^T g_\chi \\
\text{subject to} \quad & v_i^- \le (g_\chi)_i \le v_i^+, \quad i = 1, \ldots, m_2
\end{aligned}
\tag{4.26}
$$

The solution $\bar{g}_\chi$ is given explicitly by:

$$
(\bar{g}_\chi)_i = \begin{cases} v_i^- & \text{if } v_i^- > 0 \\ 0 & \text{if } 0 \in [v_i^-, v_i^+] \\ v_i^+ & \text{if } v_i^+ < 0 \end{cases}.
\tag{4.27}
$$

Projected quantities $Z_S^T \bar{g}$ and $Z_N^T \bar{g}$ can then be computed with $\bar{g}$ defined analogously to g (just before expression (4.27)). This and use of the line search in Nazareth [54] suggests a very convenient *heuristic* extension of MINOS. Even the construction of a specialized line search can be avoided by utilizing line search methods designed for smooth problems (again heuristic in this context) as discussed by Lemarechal [44].

For C2 problems, computing μ and λ by (4.21) again requires that we solve the following special structured quadratic program (Nazareth & Wets [56]):

$$
\text{find } g \in R_2^m \text{ such that } \|g\|_M^2 \text{ is minimized}
$$

such that

$$
g_\chi = \sum \ell = 1 L f_{\ell\pi}^\ell \text{ and } \pi^\ell W \le q, \pi^\ell (h^\ell - \chi) \ge \Psi(\chi, h^\ell), \ell = 1, \ldots, L
$$

where h^ℓ and f^ℓ define the probability distribution of the scenarios, as in Section 4.3.2.1. M defines the metric and for different choices, the objective takes the form $g^T g$ (or equivalently, in this case, $g_\chi^T g_\chi$), $g^T Z_S Z_S^T g$ or $g^T Z_N Z_N^T g$. Again

special purpose techniques can be devised to solve such problems. It is however often impractical to consider use of the above steepest descent approach
because only $\Psi(\chi)$ and a subgradient are available. In this case an algorithm
would have to be designed around bundle techniques or nonmonotonic optimization as discussed in Section 4.4, items (2) and (3) (after expression (4.23)),
using reduced subgradients given by $Z_S^T g$ and $Z_N^T g$, with g and other quantities
defined as in the paragraph preceding (4.25) In this case an implementation
could be based upon a routine for minimizing nonsmooth functions, see Bihain
[5].

In the above methods the χ variables would normally always be in the
basis, since they have no bounds on their value. This means that there are
always some variables in the basis which correspond to the nonsmooth part of
the objective function. An alternative approach is to try and restore a more
simple pricing strategy by keeping the χ variables always *superbasic* and define
a basis only in the x variables. The alternating method of Qi [61] is an attempt
in this direction although it is not implementable in the form given in [61].
Other methods along these lines are given by Birge [6]. However, the numerical
results given by Birge [6] show that the approach may not be as promising
as the method based upon outer linearization (the so-called L-shaped method)
mentioned at the end of Section 4.3.2.1.

4.4.2 Extensions

As with generalized linear programming, we think that much can be done by
extending the above approach, when $\Psi(\chi)$ and its subgradient are hard to
compute, but there are many open questions. As in Section 4.3.2.2, two broad
approaches can be followed:

(i) *Sampling*: Potentially the most valuable approach seems to be an *alternating* method in which one would carry out iterations in the χ space and
combine them in some suitable way with subgradient (or stochastic quasigradient) iterations in the x space (along the lines suggested by Ermoliev
[15]). It is also possible to consider 'homogeneous' or active set methods
which extend the reduced gradient approach and interleave iterations involving two projection operators into the space defined by superbasic and
nonbasic variables respectively.

(ii) *Approximate Distribution and Compute Bounds*: For a discussion of this
approach see Birge & Wets [8].

4.5   Lagrange Multiplier Methods

We conclude this chapter with very brief mention of methods which have recently achieved much popularity for smooth and nonsmooth optimization and are thus likely to lead to useful methods for solving recourse problems. Bertsekas [4] and Powell [59] give comprehensive reviews in the smooth case. Lemarechal [42] explains connections with minimax optimization and other methods of nonsmooth optimization.

A distinguishing feature of methods in this category is that they combine cutting plane techniques with use of a quadratic penalty term in the computation of search directions and that they often treat the constraints 'globally', again in the sense of Lemarechal [42]. For an example of the use of a (parameterized) quadratic penalty term in unconstrained minimization see the proximal point method of Rockafellar [63]; in smooth nonlinear programming, see Wilson [75] and in nonsmooth optimization, see Pschenichnyi & Danilin [60].

Consider the problem

$$\begin{aligned} \text{minimize} \quad & f(x) \\ \text{subject to} \quad & Ax = b \\ & x \geq 0. \end{aligned} \tag{4.28}$$

The search direction finding problem then takes the form:

$$\begin{aligned} \text{minimize} \quad & v + (1/2)d^T B d \\ \text{subject to} \quad & v \geq -\alpha_i + g_i^T d, i \in I \\ & Ax = b \\ & x \geq 0 \end{aligned} \tag{4.29}$$

where I denotes an index set and g_i, $i \in I$ a set of subgradients of $f(x)$. α_i is a scalar. If $B = 0$, I has only one element and $f(x)$ is smooth (so that g_i corresponds to a gradient), note the connection with the method of Frank & Wolfe [21] (see also Section 4.3.1). When $B = I$, the identity matrix, we have the method suggested by Pschenichnyi & Danilin, see [60].

By dualizing (4.29) it is easy to establish ties with steepest descent methods determined by bundles of subgradients in the appropriate reduced space together with the appropriate definition of a metric (see (4.23) and also Han [27],[28], Lemarechal [42], Kiwiel [35] and Demyanov & Vasiliev [12]). Recently Kiwiel [36] has suggested a method which further exploits the structure in (4.23) and has also considered extensions of methods under consideration in this section when there is uncertainty in the value of the function.

Finally, for application of ideas underlying Lagrange multiplier methods to stochastic programs with recourse, see Rockafellar & Wets [64], Merkovsky, Dempster & Gunn [47].

References

[1] E.N. Atkinson, B.W. Brown, J.E. Dennis, J.R. Thompson, "Minimization of noisy functions", presented at SIAM Conference on Numerical Optimization. Boulder, Colorado, June 1984.

[2] E.M.L. Beale, Private communication.

[3] J.F. Benders, "Partitioning procedures for solving mixed variables programming problems", *Numerische Mathematik* 4(1962), 238–252.

[4] D.P. Bertsekas, "Constrained Optimization and Lagrange Multiplier Methods", Academic Press, New York (1982).

[5] A. Bihain, "Numerical and algorithmic contributions to the constrained optimization of some classes of nondifferentiable functions", Ph.D. dissertation, Facultes Universitaires Notre-Dame De La Paix, Belgium (1984).

[6] J. Birge, "Decomposition and partitioning methods for multistage stochastic linear programs", Tech. Report 82–6, Dept. of IE & OR, University of Michigan (1982).

[7] J. Birge, "Using sequential approximations in the L-shaped and generalized programming algorithms for stochastic linear programs", Tech. Report 83–12, Dept. of IE & OR, University of Michigan (1983).

[8] J. Birge and R. J-B. Wets, "Designing approximation schemes for stochastic optimization problems, in particular for stochastic programs with recourse", WP-83-111, IIASA, Laxenburg, Austria (1983) and also in A. Prékopa and R.J-B. Wets eds., *Mathematical Programming Study*, to appear.

[9] J. Bisschop and A. Meeraus, "Matrix augmentation and partitioning in the updating of the basis inverse", *Mathematical Programming* 19(1977), 7–15.

[10] H. Cleef, "A solution procedure for the two-stage stochastic program with simple recourse", *Z. Operations Research* 35(1981), 1–13.

[11] G.B. Dantzig, *Linear Programming and Extensions*, Princeton University Press (1980).

[12] V.F. Demyanov and L.V. Vasiliev, "Nondifferentiable Optimization", (in Russian), Nauka (1981).

[13] B.C. Eaves and W. Zangwill, "Generalized cutting plane algorithms", *SIAM J. Control* 9(1971), 529–542.

[14] J. Edwards, J. Birge, A. King and L. Nazareth, "A standard input format for computer codes which solve stochastic programs with recourse and a library of utilities to simplify its use", WP-85-03, IIASA, Laxenburg, Austria (1984). (Also this volume, see [19].)

[15] Yu. Ermoliev, "Methods of nondifferentiable and stochastic optimization and their applications", in: E. Nurminski, ed., Progress in Nondifferentiable Optimization, CP-82-S8, IIASA, Laxenburg, Austria (1982).

[16] Yu. Ermoliev, "Stochastic quasi-gradient methods and their application in systems optimization", *Stochastics* 8(1983).

[17] Yu. Ermoliev, A. Gaivoronski and C. Nedeva, "Stochastic optimization

problems with incomplete information on distribution functions", WP-83-113, IIASA, Laxenburg, Austria (1983).

[18] Yu. Ermoliev and A. Gaivoronski, "Stochastic quasigradient methods and their implementation", WP-84-55, IIASA, Laxenburg, Austria (1984).

[19] Yu. Ermoliev and R. J-B. Wets, eds., *Numerical Techniques for Stochastic Optimization Problems*, Springer-Verlag, (to appear).

[20] R. Fletcher, *Practical Methods of Optimization, Vol. 2: Constrained Optimization*, J. Wiley, Chichester, 1981.

[21] M. Frank and P. Wolfe, "An algorithm for quadratic programming", *Naval Research Logistics Quarterly*, III(1956), 95–110.

[22] A. Gaivoronski, "Optimization of functionals which depend on distribution functions: 1. Nonlinear functional and linear constraints", WP-83-114, IIASA, Laxenburg, Austria (1983).

[23] P.E. Gill, W. Murray and M.H. Wright, *Practical Optimization*, Academic Press, London and New York, 1981.

[24] P.E. Gill, W. Murray, M.A. Saunders, and M.H. Wright, "Sparse matrix methods in optimization", Tech. Report SOL-82-17, Systems Optimization Lab., Dept. of Operations Research, Stanford University (1983).

[25] P.E. Gill, W. Murray, M.A. Saunders and M.H. Wright, "User's guide for SOL/QPSOL: A FORTRAN package for quadratic programming", SOL 83-7, Systems Optimization Laboratory, Stanford University (1983).

[26] R.E. Griffith and R.A.Stewart, "A nonlinear programming technique for the optimization of continuous processing systems", *Management Science* 7(1961), 379–392.

[27] S.P. Han, "Superlinearly convergent variable metric algorithms for general nonlinear programming problems", *Mathematical Programming* 11(1976), 263–282.

[28] S.P. Han, "Variable metric method for minimizing a class of nondifferentiable functions", *Mathematical Programming* 20(1981), 1–12.

[29] IBM Document H20-0476-2, "Mathematical Programming System/360 Version 2, Linear and Separable Programming – User's Manual", IBM Corporation, New York.

[30] P. Kall, "Computational methods for solving two-stage stochastic linear programs", *Z. Agnew.-Math. Phys.* 30(1979), 261–271.

[31] P. Kall and D. Stoyan, "Solving stochastic programming problems with recourse including error bounds", *Math. Operations Forsch. Statist. Ser. Optimization* 13(1982), 431–447.

[32] J. Kallberg and M. Kusy, "Code Instruction for S.L.P.R., a stochastic linear program with simple recourse", Tech. Report, University of British Columbia (1976).

[33] J. Kallberg and W. Ziemba, "An extended Frank–Wolfe algorithm with application to portfolio selection", in: P. Kall and A. Prékopa, eds., Springer-Verlag, Berlin (1981).

[34] J.E. Kelley, "The cutting plane method for solving convex programs", *J. Soc. Ind. Appl. Math.* 8(1960), 703–712.

[35 K.C. Kiwiel, "An aggregate subgradient method for nonsmooth convex minimization", *Mathematical Programming* **27**(1983) 320–341.

[36] K.C. Kiviel, "A descent algorithm for large-scale linearly constrained convex nonsmooth minimization", CP-84-15, IIASA, Laxenburg, Austria (1984).

[37] E. Komaroni, "A dual method for probabilistic constrained problems", A. Prékopa and R.J-B. Wets, eds., *Mathematical Programming Study*, to appear.

[38] M. Kusy and W. Ziemba, "A bank asset and liability management model", CP-83-59, IIASA, Laxenburg, Austria (1983).

[39] L.S. Lasdon, "Optimization Theory for Large Scale Systems", Macmillan, London, 1970.

[40] C. Lemarechal, "Nonsmooth Optimization and Descent Methods", RR-78-4, IIASA, Laxenburg, Austria (1977).

[41] C. Lemarechal, "Bundle methods in nonsmooth optimization", in: C. Lemarechal and R. Mifflin, eds., Nonsmooth Optimization, IIASA Proceedings Series, Vol. 3, 79–102. Laxenburg, Austria (1977).

[42] C. Lemarechal, "Nonlinear programming and nonsmooth minimization: A unification", Report No.332, INRIA, France (1978).

[43] C. Lemarechal, "A view of line searches", in: A. Auslender, W. Otelli, and J. Stoer, eds., *Optimization and Optimal Control.* Lecture Notes in Control and Information Science 30, Sprinter Verlag, Berlin, 1981.

[44] C. Lemarechal, "Numerical experiments in nonsmooth optimization", in: E.A. Nurminski, ed., *Progress in Nondifferentiable Optimization*, CP-82-S8, 61–84, IIASA, Laxenburg, Austria (1982).

[45] C. Lemarechal, J.J Strodiot and A. Bihain, "On a bundle algorithm for nonsmooth optimization", in: O.L. Mangasarian, R.R. Meyer and S.M. Robinson, eds., *Nonlinear Programming*, Academic Press, New York (1981).

[46] K. Madsen and H. Schjaer-Jacobsen, "Linearly constrained minimax optimization", *Mathematical Programming* **14**(1978), 208–223.

[47] R. Merkovsky, M.A.H. Dempster and E.A. Gunn, "Some Lagrangian approaches to stochastic programming with recourse", Presented at 12-th International Symposium on Mathematical Programming, Boston, Mass., August 1985.

[48] R. Mifflin, "Stationarity and superlinear convergence of an algorithm for univariate locally Lipschitz constrained minimization", *Mathematical Programming* **28**(1984), 50–71.

[49] B. Murtagh and M. Saunders, "Large-Scale linearly constrained optimization", *Mathematical Programming* **14**(1978), 41–72.

[50] B.A. Murtagh and M.A. Saunders, "MINOS 5.0 User's Guide", Report No. SOL 83–20, Systems Optimization Laboratory, Stanford University (1983).

[51] J.L. Nazareth, "Algorithms based upon generalized linear programming for stochastic programs with recourse", in: F. Archetti, ed., *Proceedings of IFIP International Workshop on Stochastic Programming: Algorithms and Applications.* Springer-Verlag, (to appear).

[52] J.L. Nazareth, "Variants on Dantzig–Wolfe decomposition with applications to multistage problems", WP-83-61, IIASA, Laxenburg, Austria (1983).

[53] J.L. Nazareth, "Hierarchical implementation of optimization methods", in: P. Boggs, R. Byrd and B. Schnabel, eds., *Numerical Optimization, 1984*, SIAM, Philadelphia, 199–210 (1985).

[54] J.L. Nazareth, "An efficient algorithm for minimizing a multivariate polyhedral function along a line", in R.W. Cottle, ed., *Mathematical Programming Study*, (to appear).

[55] J.L. Nazareth, "Design and implementation of a stochastic programming optimizer with recourse and tenders", IIASA WP-85-63. (Also this volume, see [19].)

[56] J.L. Nazareth and R.J-B. Wets, "Algorithms for stochastic programs: the case of non-stochastic tenders", A. Prékopa and R.J-B. Wets, eds., *Mathematical Programming Study*, (to appear).

[57] S.C. Parikh, Lecture notes on stochastic programming, unpublished, University of California, Berkeley (1968).

[58] B. Poljak, "Nonlinear programming methods in the presence of noise", *Mathematical Programming* 14(1978), 87–97.

[59] M.J.D. Powell, "Algorithms for nonlinear constraints that use Lagrangian functions", *Mathematical Programming* 14(1978), 224–248.

[60] B.N. Pshenichnyi and Yu. Danilin, "Methodes Numeriques dans les Problemes d'extremum", Editions de Moscou, Paris (1977).

[61] L. Qi, "An alternating method to solve stochastic programming with simple recourse", Tech. Report No.515, Computer Science Department, University of Wisconsin (1983).

[62] J.B. Rosen, "Convex partition programming", in: R.L. Graves and P. Wolfe, eds., *Recent Advances in Mathematical Programming*, 159–176, McGraw-Hill, New York, 1963.

[63] R.T. Rockafellar, "Monotone operators and the proximal point algorithm", *SIAM Journal on Control and Optimization* 14(1976) 877–898.

[64] R.T. Rockafellar and R.J-B. Wets, "A lagrangian finite generation technique for solving linear–quadratic problems in stochastic programming", WP-84-25, Laxenburg, Austria (1984).

[65] J.F. Shapiro, "Mathematical programming: structures and algorithms", John Wiley, New York, 1979.

[66] N.Z. Shor, "Generalized gradient methods of nondifferentiable optimization employing space dilation operators", in: A. Bachem, M. Groetschel and B. Korte, eds., *Mathematical Programming: The State of the Art*, 501–529. Springer Verlag, Berlin, 1983.

[67] J.J. Strodiot, V.H. Nguyen and N. Heukmes, "Eps-optimal solutions in nondifferentiable convex programming and some related questions", *Mathematical Programming* 25(1983), 307–328.

[68] R. Van Slyke and R.J-B. Wets, "*L*-shaped linear program with applications to optimal control and stochastic linear programs", *SIAM Journal on*

Applied Mathematics **17**(1969), 638–663.

[69] R. Wets, "Programming under uncertainty: the complete problem", *Z. Wahrsch. verw. Gebiete* **4**(1966), 316–339.

[70] R. Wets, "Solving stochastic programs with simple recourse II", Proceedings of Johns Hopkins Symposium on Systems and Information, 1–6 (1975).

[71] R. Wets, "A statistical approach to the solution of stochastic programs with (convex) simple recourse", in: A. Wierzbicki, ed., *Generalized Lagrangians in Systems and Economic Theory*, IIASA Proceedings Series, Pergamon Press, Oxford, 1983.

[72] R. Wets, "Solving stochastic programs with simple recourse", *Stochastics* **10**(1983), 219–242.

[73] R. Wets, "Stochastic programming: solution techniques and approximation schemes", in: A. Bachem, M. Groetschel and B. Korte, eds., *Mathematical Programming: The State-of-the-Art*, 566–603, Springer-Verlag, Berlin, 1983.

[74] A.C. Williams, "Approximation formulas for stochastic linear programming", *SIAM Journal on Applied Mathematics* **14**(1966), 668–677.

[75] R.B. Wilson, "A simplicial algorithm for concave programming", Ph.D. Thesis, Graduate School of Business Administration, Harvard University (1963).

[76] W.I. Zangwill, *Nonlinear Programming: A Unified Approach*, Prentice-Hall (1963).

[77] W.T. Ziemba, "Computational algorithms for convex stochastic programs with simple recourse", *Operations Research* **18**(1970), 415–431.

[78] W.T. Ziemba, "Solving nonlinear programming problems with stochastic objective functions", *Journal of Financial and Quantitative Analysis* VII (1972), 1809–1827.

[79] W.T. Ziemba, "Stochastic programs with simple recourse", in: P. Hammer and G. Zoutendijk, eds., *Mathematical Programming in Theory and Practice*, 213–274, North-Holland, Amsterdam, 1974.

CHAPTER 5

NUMERICAL SOLUTION OF PROBABILISTIC CONSTRAINED PROGRAMMING PROBLEMS

A. Prékopa

5.1 Introduction

In this paper we present solution techniques to problems of the following kind

$$\text{minimize} \quad h(x)$$
$$\text{subject to} \quad h_0(x) = P(g_1(x,\xi) \geq 0, \ldots, g_r(x,\xi) \geq 0) \geq p, \qquad (5.1)$$
$$h_1(x) \geq p_1, \ldots, h_m(x) \geq p_m,$$

where for the sake of simplicity we assume that the functions $h, h_1, \ldots, h_m$ are defined on the whole n-dimensional space. Similarly, the functions $g_1(x,y), \ldots, g_r(x,y)$ are supposed to be defined on the whole $n+q$-dimensional space, $x \in R^n$, $y \in R^q$. For the probability p the notation p_0 will also be used.

Various engineering and economic problems can be cast into this form. Now we do not intend to survey the applicational models belonging to this category. We only refer to a few papers [7]–[11], where the interested reader may find model formulations and references to applications.

The most important special case of Problem (5.1) is obtained by specializing the functions $g_i(x,y)$, $i = 1, \ldots, r$ so that

$$g_i(x,y) = T_i x - y_i, \qquad i = 1, \ldots, r$$

where $T_1, \ldots, T_r$ are rows of an $r \times n$ matrix T. In this case the probabilistic constraint in Problem (5.1) takes the form

$$P(Tx \geq \xi) \geq p. \qquad (5.2)$$

Introducing the notation $F(z)$ for the joint probability distribution function of the components of the random vector ξ, i.e., $F(z) = P(\xi \leq z)$, the constraint (5.2) can be written in the following manner

$$F(Tx) \geq p. \qquad (5.3)$$

Before proceeding to describe the numerical solution techniques to Problem (5.1) we mention the following theorem that serves as a basis of the convergence theory in many special cases. For the proof of the theorem we refer to the summarizing paper [12] and the references there.

Theorem 5.1. *If $g_1(x,y), \ldots, g_r(x,y)$ are concave functions in R^{n+q} and ξ has a continuous probability distribution with logarithmically concave probability density function f, i.e., for every $x_1, x_2 \in R^n$ and $0 < \lambda < 1$ we have*

$$f(\lambda x_1 + (1-\lambda)x_2) \geq f[(x_1)]^{\lambda}[f(x_2)]^{1-\lambda},$$

then the function h_0 is also logarithmically concave in R^n.

This theorem implies that if ξ has the required property then the function standing on the left hand side of (5.2) is logarithmically concave.

Maximization of Probability and the Method of Two Phases

Together with problem (5.1) we also formulate the problem

$$\begin{aligned} &\text{maximize } h_0(x) = P(g_1(x,\xi) \geq 0, \ldots g_r(x,\xi) \geq 0)\\ &\text{subject to} \quad h_1(x) \geq p_1, \ldots, h_m(x) \geq p_m. \end{aligned} \tag{5.4}$$

This problem has practical importance too. Many reliability problems belong to this category. For one practical application we refer to the paper [13] where a sequential decision process consists of a sequence of problems of the type (5.4).

Another importance of problem (5.4) is that when solving problem (5.1) a two-phase method can be applied where in the first phase we seek a feasible solution and in the second phase we solve the original problem. Assuming that we possess a method to find a feasible solution to the system of inequalities $h_1(x) \geq p_1, \ldots, h_m(x) \geq p_m$, a feasible solution to problem (5.1) can be found in such a way that we start to solve problem (5.4) and stop the procedure when we reach an x satisfying $h_0(x) \geq p$. This x is a feasible solution to problem (5.1).

For the solution of problem (5.1) we propose the application of suitable nonlinear programming methods supplied by Monte Carlo simulation procedures to find function values and gradients of the function h_0. There exist other proposals too to solve stochastic programming problems among which the stochastic quasi gradient method of Yu. Ermolev and his collaborators should be mentioned. There is, however, little experience regarding how this method works in case of problem (5.1) and (5.4). On the other hand the application of the already well developed theory and techniques of nonlinear programming seems to be advantageous to apply. In this case, among others, we are able to present optimality criterion which helps us to check the termination of the applied optimization procedure.

A nonlinear programming problem which is proved to be effective in case of deterministic nonlinear programming problems is not necessarily effective in case of the solution of problems (5.1) and (5.4). The reason is that in problems (5.1) and (5.4) each value of the function h_0 is the probability of a set in R^q and these values furthermore the values of the gradient of h_0 are calculated by Monte Carlo simulation. This letter gives us a satisfactorily accurate value provided the sample size is chosen large enough. However, we are able to do so only in the case if the effect of the Monte Carlo simulation can be well controlled, i.e., the effect of this kind of randomness can clearly be seen throughout the procedure and the numerically unstable steps can be avoided or at least controlled.

5.2 The SUMT Method with Logarithmic Penalty Function

We introduce the following assumptions:

- $0 < p < 1, p_1 > 0, \ldots, p_m > 0$, h is convex in R^n,
- $h_1, \ldots, h_m$ are continuous logconcave functions in R^n,
- $g_1, \ldots, g_r$ are concave functions in R^{n+q},
- the set of feasible solutions is compact,
- there exists an x satisfying $h_i(x) > p_i$, $i = 0, \ldots, m$,
- ξ has a continuous probability distribution with logarithmically concave density.

The Sequential Unconstrained Minimization Technique [2] applied to our problem works in the following manner [10]. We define the penalty function

$$T(x, s) = h(x) - s \sum_{i=0}^{m} \ln \frac{h_i(x) - p_i}{M_i} \tag{5.5}$$

for every x satisfying $h_i(x) > p_i$, $i = 0, \ldots, m$ and for every fixed $s > 0$ where M_i is the maximum of $h_i(x) - p_i$ on the set of feasible solutions. Take a positive sequence $s^1 > s^2 \ldots$ with the property that $\lim_{k \to \infty} s^k = 0$ and minimize the function $T(x, s^k)$ for every fixed s^k. As the set of feasible solutions is compact then the minimum of $T(x, s^k)$ exists. Let x^k be an optimal solution to this problem. Then we have the relation

$$\lim_{k \to \infty} T(x^k, s^k) = \lim_{k \to \infty} h(x^k) = \min_{x \in D} h(x) \tag{5.6}$$

where D denotes the set of feasible solutions. It is remarkable that under the mentioned assumptions the function $T(x, s)$ is a convex function for every fixed s thus various unconstrained optimization techniques work effectively. To compute the values and the gradients of h_o remain difficult problems to which we return later. The sequence $s^1, s^2, \ldots$ in practice is chosen as a geometric sequence and the procedure frequently stops after a few number of steps.

Below we prove two theorems which help to check properties generally required when solving optimization problems by the SUMT method.

Theorem 5.2.1. *If a function h is logconcave on the convex set given by the relation*

$$H = \{x \mid h(x) \geq p\},$$

where p is a fixed probability satisfying the inequality $0 < p < 1$ then the function $h(x) - p$ is also logconcave on the set H.

Proof. Let $x, y \in H$, $x \neq y$ and $0 < \lambda < 1$. Then since h is logconcave on H we have the inequality

$$h(\lambda x + (1 - \lambda)y) - p \geq [h(x)]^\lambda [h(y)]^{1-\lambda} - p.$$

Setting $h(x) = a$, $h(y) = b$, it will be enough to prove the inequality

$$a^\lambda b^{1-\lambda} - p \geq (a-p)^\lambda (b-p)^{1-\lambda}.$$

Dividing by $a^\lambda b^{1-\lambda}$ on both sides we obtain

$$1 - \left(\frac{p}{a}\right)^\lambda \left(\frac{p}{b}\right)^{1-\lambda} \geq \left(\frac{a-p}{a}\right)^\lambda \left(\frac{b-p}{b}\right)^{1-\lambda}.$$

Now, using the arithmetic mean-geometric mean inequality, we derive

$$\left(\frac{p}{a}\right)^\lambda \left(\frac{p}{b}\right)^{1-\lambda} + \left(\frac{a-p}{a}\right)^\lambda \left(\frac{b-p}{b}\right)^{1-\lambda}$$

$$\leq \lambda \frac{p}{a} + (1-\lambda)\frac{p}{b} + \lambda \frac{a-p}{a} + (1-\lambda)\frac{b-p}{b} = 1.$$

This proves the theorem.

Theorem 5.2.1 shows that under the conditions introduced in the beginning of this section the function $T(x, s)$ is convex for every fixed $s > 0$ on the set of x vectors satisfying the inequalities $h_i(x) > p_i$, $i = 0, 1, \ldots, m$.

Theorem 5.2.2. *Suppose that in problem (5.1) the assumptions introduced in Section 5.2 hold and let z be a nonboundary point of the set of feasible solutions. Then we have*

$$h_i(z) > p_i, i = 0, 1, \ldots, m.$$

Proof. By the assumptions introduced in the beginning of this section there exists an x satisfying the inequalities

$$h_i(x) > p_i, i = 0, 1, \ldots, m.$$

We may assume that $z \neq x$. For some $\mu > 1$ the point

$$y = x + \mu(z - x)$$

is a boundary point of the set of feasible solutions. Using the notation $\lambda = 1/\mu$ we obtain

$$z = \lambda y + (1 - \lambda)x.$$

By the logconcavity of the constraining functions and taking into account the inequalities $p_i > 0$, $i = 0, 1, \ldots, m$, we obtain

$$h_i(z) = h_i(\lambda y + (1-\lambda)x) \geq [h_i(y)]^\lambda [h_i(x)]^{1-\lambda}$$

$$\geq p_i^\lambda [h_i(x)]^{1-\lambda} > p_i^\lambda p_i^{1-\lambda} = p_i,$$

$$i = 0, 1, \ldots, m.$$

This proves the theorem.

Theorem 5.2.2 states that on every nonboundary feasible solution of problem (5.1) the penalty function (5.5) is defined and this makes possible the proof of the limit relation (5.6) also in the case if the optimal solution is on the boundary of the set of feasible solutions.

Finally we remark that the application of the SUMT method is particularly advantageous in cases when the calculation of the gradients of h_0 (and eventually also of h_i, $i = 1, \ldots, m$) would be sophisticated not so much because of the probabilistic nature of h_0 but because of the special structure of the functions $g_1, \ldots, g_m$. In such cases gradient-free techniques may be applied to minimize $T(x, s)$.

5.3 Solution by the Method of Feasible Directions

The following assumptions are introduced:

- The probabilistic constraint has the form (5.3),
- h is convex and has continuous gradient in R^n,
- $h_1, \ldots, h_m$ are quasi-concave and have continuous gradients in R^n,
- The constraints in which the constraining functions are linear determine a bounded set,
- there exists an x satisfying $h_i(x) > p_i$, $i = 0, \ldots, m$,
- ξ has a continuous probability distribution with logarithmically concave density.

The method uses subsequent linearization of the constraints and the objective function. We start from an arbitrary feasible vector x^1 and if $x^1, \ldots, x^k$ are already fixed then first we solve the following direction finding problem:

$$\text{minimize } y$$
$$\begin{aligned} \text{subject to} \quad & \nabla h(x^k)(x - x^k) \leq y \\ & h_i(x^k) + \nabla h_i(x^k)(x - x^k) \geq p_i, \\ & \nabla h_i(x^k)(x - x^k) + \theta_i y \geq 0, \text{ if } h_i(x^k) = p_i, \\ \text{and} \quad & h_i \text{ is a nonlinear function}, i = 0, 1, \ldots, m, \end{aligned} \tag{5.7}$$

where the θ_i are fixed positive numbers not depending on the individual problems (5.7). If x_k^* is an optimal solution of problem (5.7) then we solve the following step length finding problem:

$$\min_{\lambda} h(x^k + \lambda(x_k^* - x^k)), \tag{5.8}$$

where the minimization is extended over such λ values for which $x^k + \lambda(x_k^* - x^k)$ is feasible. If λ^k is an optimal solution of problem (5.8) then we define

$$x^{k+1} = x^k + \lambda^k(x_k^* - x^k).$$

Under the assumptions introduced in the beginning of this section the following limit relation holds

$$\lim_{k \to \infty} h(x^k) = \min_{x \in D} h(x).\tag{5.9}$$

The above procedure was published by Zoutendijk [16]. The convergence proof under the mentioned conditions is presented in [5]. Of particular interest is the case where all constraining functions but h_0 are linear. Writing $h_i(x) = a_i'x$, $i = 1, \ldots, m$ and $h(x) = c'x$, the problem is to

$$\text{minimize } c'x$$
$$\text{subject to } \quad P(Tx \geq \xi) \geq p\tag{5.10}$$
$$a_i'x \geq p_i, \qquad i = 1, \ldots, m.$$

The first phase problem is to find a feasible solution to (5.10) is the following

$$\text{maximize } P(Tx \geq \xi)$$
$$\text{subject to } \quad a_i'x \geq p_i, \qquad i = 1, \ldots, m.\tag{5.11}$$

When maximizing the objective function in problem (5.11) we can stop the procedure whenever we reach an x satisfying

$$P(Tx \geq \xi) \geq p.\tag{5.12}$$

On the other hand if we perform it as long as the inequality (5.12) holds strictly we have numerical evidence that the regularity condition (the second to the last condition) holds true.

If the probability $P(Tx \geq \xi)$ is positive in the set of feasible solutions then we take its negative logarithm and minimize this rather than maximize the original probability. Thus the new problem, equivalent to problem (5.11), is the following

$$\text{minimize } -\log P(Tx \geq \xi)$$
$$\text{subject to } \quad a_i'x \geq p_i, \qquad i = 1, \ldots, m.\tag{5.13}$$

The gradient of the objective function in problem (5.13) can be computed on the bases of the equality

$$\nabla \log P(Tx \geq \xi) \geq \frac{1}{P(Tx \geq \xi)} \nabla P(Tx \geq \xi).$$

The method of feasible directions is considered today a slow method to solve nonlinear programming problems. Taking into account aspects that arise concerning probabilistic constrained programming problems we cannot be as dissatisfied with its performance. Problems (5.7) and (5.8) clearly show how accurately we have to compute the function values and the gradient values in order to obtain good approximations.

5.4 Solution by the Supporting Hyperplane Method

We introduce the following assumptions:

- there exists a bounded convex polyhedron K^1 such that the set of feasible solutions is contained in K^1,
- the functions $-h, h_1, \ldots, h_m$ are quasi-concave and have continuous gradients on K^1,
- there exists an x such that $h_i(x) > p_i, i = 0, \ldots, m$,
- ξ has continuous probability distribution and logconcave density in R^n furthermore h_0 has continuous gradient in R^n. We assume that we have an initial feasible vector x^1. Then we perform subsequent iterations where the kth iteration in this method consists of two subsequent steps.

Step 1. Solve the problem

$$\text{minimize} \quad h(x)$$
$$\text{subject to} \quad x \in K^k,$$

where K^k is a convex polyhedron. Let x^k be an optimal solution to this problem. If $h_i(x^k) \geq p_i$, $i = 0, \ldots, m$ then x^k is an optimal solution to problem (5.7). Otherwise go to Step 2.

Step 2. Let λ^k be the largest $\lambda(0 \leq \lambda \leq 1)$ for which the following inequality holds

$$h_i(x^1 + \lambda(x^k - x^1)) \geq p_i, \qquad i = 0, \ldots, m.$$

Various one-dimensional methods can be applied to solve this problem. Let

$$y^k = x^1 + \lambda^k(x^k - x^1).$$

If $h(y^k) - h(x^k) \leq \varepsilon$ where ε is a previously chosen small positive number then we stop and accept y^k as an approximate solution to the optimization problem. Otherwise choose a subscript i_k for which $h_{i_k}(y^k) = 0$ and define

$$K^{k+1} = \{x | x \in K^k, \nabla h_{i_k}(y^k)(x - y^k) \geq 0\}$$

and go to Step 1 using $k + 1$ instead of k. Under the mentioned assumptions the procedure is convergent in the sense that

$$\lim_{k \to \infty} h(x^k) = \min_{x \in D} h(x).$$

This method was published in [14] and applied to solve probabilistic constrained programming problems in [9].

5.5   Solution by a Variant of the General Reduced Gradient Method

A variant of the GRG method [1] suitably adapted to problem (5.1) where the stochastic constraint reduces to (5.2) and the other constraints are linear has been reported in [4]. It differs from the GRG method primarily in the formulation of the direction finding problem. Here we generate always feasible solutions and thus we avoid the application of intermediate methods to return to the feasible set which is very important because our function values are noisy.

The problem to be solved is now formulated in the following form:

$$\begin{aligned}
\text{minimize} \quad & h(x) \\
\text{subject to} \quad & h_0(x) = P(Tx \geq \xi) \geq p \\
& Ax = b \\
& x \geq 0.
\end{aligned} \tag{5.14}$$

Concerning this problem the following assumptions are introduced:

– the random variable ξ has a continuous probability distribution with log-concave density function,
– $\nabla h_0(x)$ is Lipschitz-continuous and bounded in R^n,
– there exists a feasible x such that $h_0(x) > p$,
– the $m \times n$ matrix A has rank equal to m and for every feasible x there exists a basis B such that $x_i > 0$ for $i \in I_B$ and I_B is the set of subscripts of the basis vectors.

We start from a feasible solution x to problem (5.14) and assume that a basis B of the columns of A can be found which, for the sake of simplicity is assumed to consist of the first m columns of A, with the property that when applying the partition $A = (B, C)$ and the corresponding partition of x is $x' = (w', z')$ then all components of w are strictly positive. We will have a direction finding problem and a setp length determination problem.

Direction finding problem.   First we formulate the following problem

$$\begin{aligned}
\text{minimize} \quad & y \\
\text{subject to} \quad & \nabla_w h(x)u + \nabla_z h(x)v \leq y \\
& \nabla_w h_0(x)u + \nabla_z h_0(x)v + \theta y \geq 0, \text{ if } h_0(x) = p, \\
& Bu + Cv = 0, \\
& v_i \geq 0, \text{ if } z_i = 0, \quad i = 1, \ldots, n - m, \|v\| \leq 1.
\end{aligned} \tag{5.15}$$

Here $\theta > 0$ is a fixed number and the partition $t' = (u', v')$ corresponds to the partition of $x' = (w', z')$. Introducing the row vectors

$$r = \nabla_z h(x) - \nabla_w h(x)B^{-1}C,$$
$$s = \nabla_z h_0(x) - \nabla_w h_0(x)B^{-1}C,$$

which are called reduced gradients, problem (5.15) can be rewritten in the following manner

$$\min y$$

$$rv \leq y$$
$$sv + \theta y \geq 0, \text{ if } h_o(x) = p,$$
$$v_i \geq 0, \text{ if } z_i = 0, \quad i = 1, \ldots, n - m, \tag{5.16}$$
$$\|v\| \leq 1.$$

It can easily be proved that the optimum value of (5.16) is equal to zero if and only if x is a Kuhn-Tucker point. If this is not the case then the optimal value of problem (5.16) is negative and if v^*, y^* is an optimal solution of this problem furthermore $u^* = -B^{-1}Cv^*$ then

$$t^* = \begin{pmatrix} u^* \\ v^* \end{pmatrix}$$

is a feasible direction such that along this the function h is strictly locally decreasing.

If the norm $\|v\|$ is chosen in the following manner $\|v\| = \max_i |v_i|$ then problem (5.16) becomes a two row LP with individual lower resp. upper bounds which can easily be handled. Here we are able to take into account the inaccuracy in the evaluation of ∇h_o. The accuracy can be increased by taking a larger sample in the Monte Carlo evaluation. We remark that when updating the reduced gradients standard LP technique can be used.

Step length determination. Starting from the interval allowed by the nonnegativity restrictions we apply a linear search technique to find a point for which the nonlinear restriction holds with equality. Then we minimize the objective function on the line segment between x and this point. In this one dimensional optimization we optimize with respect to λ i.e. we solve the problem

$$\min_{\lambda} h(x + \lambda t^*).$$

If its optimal solution is λ^* then the new feasible solution will be

$$x^{(1)} = \begin{pmatrix} w^{(1)} \\ z^{(1)} \end{pmatrix} = \begin{pmatrix} w \\ z \end{pmatrix} + \lambda^* \begin{pmatrix} w^* \\ z^* \end{pmatrix}$$

provided all components of w are strictly positive. Otherwise by applying subsequent pivoting we find a basis $B^{(1)}$ with the property that the corresponding components of $x^{(1)}$ are already strictly positive.

For the sake of simplicity, we did not include into the algorithm all technicalities ensuring the convergence. The paper (4) already referred to gives a full description of these.

5.6   Solution by a Primal-dual Type Algorithm

The problem to be solved has the following form:

$$\text{minimize} \quad c'x$$
$$\text{subject to} \quad F(y) \geq p \tag{5.17}$$
$$Tx \geq y, Bx \geq d,$$

where $x \in R^n$ and $y \in R^r$. We assume that the multivariate probability distribution function F is strictly logarithmically concave and has continuous gradient in R^n. We will shortly describe the method proposed in (3).

To this problem we assign a problem that we will call dual problem although it is not dual in the classical sense. This dual is the following:

$$T'u + B'v = c$$
$$u \geq 0, v \geq 0, \tag{5.18}$$
$$\max\left[\min_{F(y) \geq p} u'y + v'd \right].$$

The procedure works in the following manner. First we assume that a pair of vectors (u^1, v^1) is available for which

$$(u^1, v^1) \in V = \{u, v \,|\, T'u + B'v = c, v \geq 0\}.$$

Suppose that (u^k, v^k) has already been chosen, where $u^k \geq 0$. Then the following steps have to be performed.

Step 1. Solve the problem

$$\text{minimize} \quad y^{k'}y$$
$$\text{subject to} \quad F(y) \geq p.$$

Let $y(u^k)$ denote the optimal solution to this problem. Then we solve the following direction finding problem

$$\text{maximize} \quad [u'y(u^k) + d'v]$$
$$\text{subject to} \quad (u, v) \in V.$$

Let (u_k^*, v_k^*) be an optimal solution to this problem. If $u_k^* = \rho u^k$ then (u_k^*, v_k^*) is an optimal solution of the dual problem and the pair $\hat{x}, y(u^k)$ is an optimal solution of the primal problem where $\hat{x}$ is an optimal solution of the linear programming problem:

$$\text{minimize} \quad c'x$$
$$\text{subject to} \quad Tx \geq y(u^k), Bx \geq d.$$

Otherwise go to

Step 2. Find $\lambda^k (0 < \lambda^k < 1)$ satisfying

$$u_k^{*'} y \left(\frac{\lambda^k}{1 - \lambda^k} u^k + u_k^* \right) > u^{k'} y(u^k) + v^{k'} d.$$

Then we define

$$u^{k+1} = \lambda^k u^k + (1 - \lambda^k) u_k^*,$$
$$v^{k+1} = \lambda^k v^k + (1 - \lambda^k) v_k^*.$$

If the procedure is infinite then it can be proved that the sequence (u^k, v^k) converges and the limiting pair has the same property as (u_k^*, v_k^*) in Step 1.

5.7 The Polynomial Distribution

A special multivariate probability distribution has been introduced by the author to approximate the distribution of ξ. This is defined on the unit square of the n-dimensional space by its probability distribution function as follows:

$$F(z_1 \ldots, z_n) = \frac{1}{\sum_{i=1}^{N} c_i z_i^{\alpha_{i1}} \ldots z_n^{\alpha_{in}}},$$

if

$$0 < z_i \leq 1, i = 1, \ldots, N, \tag{5.19}$$

$F(z_1, \ldots, z_n)$ is suitable defined otherwise. Here $\alpha_{i1} \leq 0, \ldots, \alpha_{in} \leq 0$, $\alpha_{i1} + \ldots + \alpha_{in} < 0$, $i = 1, \ldots, N$ and $c_1 > 0, \ldots, c_N > 0$; furthermore these are constants.

If a mathematical programming problem has the form of a geometric programming problem and in addition a probabilistic constraint of the type $F(z) \geq p$ is included where $F(z)$ is of the above type then the new problem is again a geometric programming problem for which methods of solution are available.

We will not consider the algorithmic solution of problems of this type in detail. Our purpose here is to show that under certain conditions the function (5.19) will in fact be a probability distribution function. To illustrate the situation we restrict ourselves to the case of $n = 2$.

Theorem 5.7.1. *If the following conditions holds:*

$$\alpha_{11} \leq \alpha_{12} \leq \ldots \leq \alpha_{1n},$$
$$\alpha_{21} \geq \alpha_{22} \geq \ldots \geq \alpha_{2n},$$

then the function (5.19) is a probability distribution function in the unit square $0 < z_1, z_2 < 1$.

Proof. The only property that we need to show is that

$$\frac{\partial^2 F(z_1, z_2)}{\partial z_1 \partial z_2} \geq 0, \text{ if } 0 < z_1, z_2 < 1. \tag{5.21}$$

The other properties of a two-dimensional probability distribution are satisfied. Introducing the notation:

$$\sum = \sum_{i=1}^{N} c_i z_1^{\alpha_{i1}} z_2^{\alpha_{i2}},$$

the function F can be written as $F = 1/\sum$. By differentiating we obtain

$$\frac{\partial^2 F(z_1, z_2)}{\partial z_1 \partial z_2} = \frac{2}{\sum^3} \frac{\partial \sum}{\partial z_1} \frac{\partial \sum}{\partial z_2} - \frac{1}{\sum^2} \frac{\partial^2 \sum}{\partial z_1 \partial z_2}.$$

The requirement that this be non-negative is equivalent to the following inequality:

$$\frac{2 \partial \sum}{\partial z_1} \frac{\partial \sum}{\partial z_2} - \sum \frac{\partial^2 \sum}{\partial z_1 \partial z_2} \geq 0,$$

or in a more detailed form:

$$2 \sum_{i=1}^{N} c_i \alpha_{i1} z_1^{\alpha_{i1}-1} z_2^{\alpha_{i2}} \sum_{j=1}^{N} c_j \alpha_{j2} z_1^{\alpha_{j1}} z_2^{\alpha_{j2}-1} \geq$$

$$\geq \sum_{i=1}^{N} c_i z_1^{\alpha_{i1}} z_2^{\alpha_{i2}} \sum_{j=1}^{N} c_j \alpha_{j1} \alpha_{j2} z_1^{\alpha_{j1}-1} z_2^{\alpha_{j2}-1}. \tag{5.22}$$

Multiplying by $z_1 z_2$ on both sides in (5.22) we get the equivalent inequality

$$2 \sum i = 1^N c_i \alpha_{i1} z_1^{\alpha_{i1}} z_2^{\alpha_{i2}} \sum_{j=1}^{N} c_j \alpha_{j2} z_1^{\alpha_{j1}} z_2^{\alpha_{j2}} \geq$$

$$\geq \sum_{i=1}^{N} c_i z_1^{\alpha_{i1}} z_2^{\alpha_{i2}} \sum_{j=1}^{N} c_j \alpha_{j1} \alpha_{j2} z_1^{\alpha_{j1}} z_2^{\alpha_{j2}}. \tag{5.23}$$

Let us introduce the notation:

$$\lambda_i = \frac{c_i z_1^{\alpha_{i1}} z_2^{\alpha_{i2}}}{\sum}, \qquad i = 1, \ldots, N.$$

Then (5.23) is equivalent to

$$2 \sum_{i=1}^{N} \alpha_{i1} \lambda_i \sum j = 1^N \alpha_{j2} \lambda_j \geq \sum_{i=1}^{N} \alpha_{i1} \alpha_{i2} \lambda_i. \tag{5.24}$$

Since

$$\lambda_i > 0, \quad i = 1, \ldots N, \quad \lambda_1 + \cdots + \lambda_N = 1,$$

$$\sum_{i=1}^{N} \alpha_{i1} \alpha_{i2} \lambda_i - \sum_{i=1}^{N} \alpha_{i1} \lambda_i \sum j = 1^N \alpha_{j2} \lambda_j \tag{5.25}$$

is the covariance of the two sequences

$$\alpha_{11}, \alpha_{12}, \ldots, \alpha_{1N}$$

$$\alpha_{21}, \alpha_{22}, \ldots, \alpha_{2N}$$

where to the corresponding pairs we assign the probabilities $\lambda_1, \lambda_2, \ldots, \lambda_N$, respectively. Assumption (5.20) implies that the covariance (5.25) is nonpositive (as it can be seen very easily). Hence (5.24) holds true which is the same as (5.20) and the theorem is proved.

The following theorem is useful when considering probabilistic constraint of the form

$$F(z_1, \ldots, z_n) \geq p, \quad 0 < z_i \leq 1, \quad i = 1, \ldots, n, \tag{5.26}$$

where $0 < p < 1$ is a fixed probability.

Theorem 5.7.2. *The function $F(z_1, \ldots, z_n)$ is logconcave in the unit cube* $0 < z_1, \ldots, z_n \leq 1$.

Proof. A well-known theorem due to Artin states that the sum of logconvex functions defined on the same convex set is a logconvex function on the same set.

Since $\alpha_{i1} \leq 0, \ldots, \alpha_{iN} \leq 0$, $i = 1, \ldots, N$, it follows that each term

$$c_i z_1^{\alpha_{i1}} \ldots z_n^{\alpha_{in}}$$

is a logconvex function in the unit cube, hence the same holds for their sum which is equal to $\sum$. Now $F = 1/\sum$ and this implies that F is a logconcave function in the n-dimensional unit cube. This proves the theorem.

Theorem 5.7.2 shows that the set of n-tuples $z_1, \ldots, z_n$ determined by the inequality (5.26) is a convex set for every fixed probability p.

5.8 Calculation of Function Values and Gradients

In this section we consider the problem how to compute the gradient of the function $F(Tx)$. It turns out that many special probability distributions allow the computation of the gradient of $F(Tx)$ as we illustrate it in two special cases which are: the multivariate normal distribution and a special type of multivariate gamma distribution.

Under suitable differentiability assumptions the following equality holds true in all cases:

$$\frac{\partial F(z)}{\partial z_i} = F(z_j, j = 1, \ldots, r, j \neq i | z_i) f_i(z_i), \quad i = 1, \ldots, r, \tag{5.27}$$

where f_i is the probability density function of the random variable ξ_i.

Let us first consider the case of the multivariate normal distribution. It will be convenient to assume that the joint distribution of the variables $\xi_i, \ldots, \xi_r$ is nondegenerated, furthermore $E(\xi_i) = 0, E(\xi_i^2) = 1, i = 1, \ldots, r$. Then the joint probability distribution function is $\Phi(z; R)$ where R is the correlation matrix. It is well-known that

$$\frac{\partial \Phi(z; R)}{\partial z_i} = \Phi\left(\frac{z_j - r_{ji} z_i}{\sqrt{1 - r_{ji}^2}}, \quad j = 1, \ldots, r j \neq i; R_i\right) \varphi(z_i) \qquad (5.28)$$

where R_i is the $(r-1) \times (r-1)$ correlation matrix consisting of the correlations

$$s_{jk} = \frac{r_{jk} - r_{ji} r_{ki}}{\sqrt{1 - r_{ji}^2} \sqrt{1 - r_{ki}^2}}, \quad j, k = 1, \ldots, r j \neq i, k \neq i; \qquad (5.29)$$

and φ is the one-dimensional standard normal probability density function. It turns out that the gradient of $\Phi(z; R)$ can be computed in a similar way as the function value $\Phi(r; R)$. The same subroutine can be used in the $r - 1$ and r-dimensional cases, respectively.

The second example is the multivariate gamma distribution introduced in (8). Suppose that the random vector ξ has the form

$$\xi = A\eta \qquad (5.30)$$

where A is an $r \times (2^r - 1)$ matrix the columns of which are the different nonzero vectors having 0,1 components and η is a $2^r - 1$-dimensional random vector with independent, standard gamma distributed components (some of them may be equal to 0). Then the conditional probability distribution function in formula (5.27) can be written in the form

$$\begin{aligned}
&P(\xi_2 < z_2, \ldots, \xi_r < z_r | \xi_1 = z_1) = \\
&= P(\xi_2^{(1)} + \xi_2^{(2)} < z_2, \ldots, \xi_r^{(1)} + \xi_r^{(2)} < z_r | \xi_1 = z_1) = \\
&= P(z_1 \frac{\xi_2^{(1)}}{\xi_1} + \xi_2^{(2)} < z_2, \ldots, z_1 \frac{\xi_r^{(1)}}{\xi_1} + \xi_r^{(2)} < z_r | \xi_1 = z_1) = \\
&= P(z_1 \frac{\xi_2^{(1)}}{\xi_1} + \xi_2^{(2)} < z_2, \ldots, z_1 \frac{\xi_r^{(1)}}{\xi_1} + \xi_r^{(2)} < z_r | \xi_1 = z_1),
\end{aligned} \qquad (5.31)$$

where

$$\xi_2^{(2)} = \xi_2 - \xi_2^{(1)}, \ldots, \xi_r^{(2)} = \xi_r - \xi_r^{(1)}$$

and $\xi_2^{(1)}, \ldots, \xi_r^{(1)}$ are the sums of the joint η terms of $\xi_2, \ldots, \xi_r$ and ξ_1, respectively. Thus the conditional probability distribution function equals the unconditional probability distribution function of the sum $z_1 \beta + \gamma$, where γ has an $r - 1$-dimensional multigamma distribution of the same type that ξ has and β has similar structure but instead of partial sums of standard gamma variables now we use partial sums of components of a random vector having Dirichlet distribution. Moreover, β and γ are independent.

5.9 The case of Discrete Probability Distributions

The following problem will be considered

$$\text{minimize} \quad c'x$$
$$\text{subject to} \quad F(z) \geq p, \tag{5.32}$$
$$Tx \geq z, Bx \geq b,$$

where F is the probability distribution function of the random vector ξ. If ξ has possible values $z_1, \ldots, z_N$ such that all positive values of F are among $F(z_1), \ldots, F(z_N)$, then the above problem is equivalent to the following mixed variable problem

$$\text{minimize} \quad c'x$$
$$\text{subject to} \quad y_1 F(z_1) + \cdots + y_N F(z_N) \geq p,$$
$$y_1 + \ldots + y_N = 1 y_1, \ldots, y_N \geq 0, \uparrow \textit{integers}, \tag{5.33}$$
$$Tx \geq y_1 z_1 + \cdots + y_N z_N$$
$$Bx \geq b.$$

Taking a random vector uniformly distributed in the n-dimensional unit cube and discretizing it by a step length h which is chosen in such a way that

$$1 - nh = p \tag{5.34}$$

Vizvári [15] proves that the number of lattice points satisfying the probabilistic constraint is equal to

$$\binom{2n}{n}$$

which is a large number for a large n but small as compared to all lattice points (of distance h) in the unit cube, e.g. if $p = 0.95$ and $m = 5$ then $h = 0.01$. The total number of lattice points is 5^{101} whereas the number of those which satisfy the probabilistic constraint is only $\frac{(10}{5)=252}$.

Computational experiments show that handling problem (5.32) in the form of (5.33) provides us with satisfactory solution methodology if n is not very large.

Another mixed variable formulation will be illustrated in the case when ξ is a two-dimensional random vector the possible values of which are nonnegative lattice points with coordinates $\leq N, M$, respectively. The mixed variable reformulation of the problem is the following

$$\text{minimize} \quad c'x$$
$$\text{subject to} \quad p_{00}y_{00} + \cdots + p_{N0}y_{N0} + p_{01}y_{01} + \cdots +$$
$$p_{N1}y_{N1} + \cdots p_{0M}y_{0M} + \cdots + p_{NM}y_{NM} \geq p,$$
$$y_{00} + \cdots + y_{N0} = z_1,$$
$$y_{00} + y_{01} + \cdots + y_{0M} = z_2,$$

$$Tx \geq z,$$

$$Bx \geq b,$$

$$y_{ik} \leq y_{i-l,k}, \quad i = 1, \ldots, N; \quad k = 0, \ldots, M.$$

$$y_{ik} \leq y_{i,k-1}, \quad i = 0, \ldots, N; \quad k = 1, \ldots, M.$$

$$y_{ik} = 0 \text{ or } 1, \quad \text{for all } i, k \text{ and } z_1 \leq N_1 z_2 \leq M.$$

These models can be used in connection with continuously distributed random vector ξ too when approximating its distribution by a discrete distribution. In the higher dimensional case, however, the number of 0,1 variables becomes too large.

References

[1] J. Abadie, and J. Carpentier, "Géneralisation de la méthode du gradient réduit de Wolfe au cas des contraintes non-linéaires, in: Proc. IFORS Conf., eds. D.B. Hertz and J. Melese, Wiley, New York, (1966) 1041-1053.

[2] A.V. Fiacco and G.P. McCormick, *Nonlinear programming: sequential unconstrained minimization technique.* Wiley, New York (1968).

[3] E. Komáromi, "A dual approach to probabilistic constrained programming problem". Mathematical Programming Study. (forthcoming).

[4] J. Mayer, "A nonlinear programming method for the solution of a stochastic programming model of A. Prékopa". Survey of Mathematical Programming. North Holland Publishing Co., New York. Vol. 2,(1980)129-139.

[5] A. Prékopa, "Eine Erweiterung der sog", *Methode der Zulässigen Richtungen. Math. Operationsforschung und Statistik* 5 (1974), 281-293.

[6] A. Prékopa, "Contributions to the theory of stochastic programming", *Mathematical Programming* 4(1973), 202-221.

[7] A. Prékopa, I. Deák, J. Ganczer and K. Patyi, "The STABIL Stochastic programming model and its experimental application to the electrical energy sector of the Hungarian economy", in: *Stochastic Programming, Proceedings of the International Conference on Stochastic Programming*, Oxford, England, edited by M.A.H. Dempster, Academic Press, London (1980), 369-385.

[8] A. Prékopa and T. Szántai, "A new multivariate gamma distribution and its fitting to empirical streamflow data", *Water Resources Research* 14 (1978), 19-24.

[9] A. Prékopa and T. Szántai, "Flood control reservoir system design", *Mathematical Programming Study* 9, 138-151.

[10] A. Prékopa and P. Kelle, "Reliability type inventory models based on stochastic programming", *Mathematical Programming Study* 9(1978), 43-58.

[11] A. Prékopa, "Network planning using two-stage programming under uncertainty", in: Recent Results in Stochastic Programming (Proceedings of the International Conference on Stochastic Programming, Oberwolfach,

Germany, 1979). Lecture Notes in Economics and Mathematical Systems 179, Springer Verlag (1980), 215–237.

[12] A. Prékopa, "Logarithmic concave measures and related topics", in: Stochastic Programming. Ed. M.A.H. Dempster, Academic Press, London (1980), 63–82.

[13] A. Prékopa and T. Szántai, "On optimal regulation of a storage level with application to the water level regulation of a lake", *Survey of Mathematical Programming*, North Holland Publishing Co., New York (1981), pp. 183–210.

[14] A.F. Veinott, "The supporting hyperplane method for unimodal programming", *Operations Research* (1967), 147–152.

[15] B. Vizári, "On the discretization of probabilistic constrained programming problems", (manuscript in Hungarian).

[16] G. Zoutendijk, *Methods of feasible directions*, Elsevier Publishing Co., Amsterdam and New York (1960).

CHAPTER 6

STOCHASTIC QUASIGRADIENT METHODS

Yu. Ermoliev

As it follows from the brief discussion of the Chapter 1, the main purpose of the stochastic quasigradient (SQG) methods is the solution of optimization problems with a complex nature of objective functions and constraints. For the stochastic programming problems, SQG methods generalize the well-known stochastic approximation methods for unconstrained optimization of the expectation of random function (see for instance Wasan [45]) to problems involving general constraints and nondifferentiable functions. For deterministic nonlinear programming problems SQG methods can be regarded as methods of random search (see for instance [42], [67], [68]).

The purpose of this chapter is a discussion of the main direction of development of SQG procedures, their applications and an overview of ideas involved in the proofs. The contents of this chapter is close to that of the paper [69].

6.1 The General Idea

Consider the problem of minimization:

$$\text{minimize} \quad F^0(x) \tag{6.1}$$

$$\text{subject to} \quad F^i(x) \leq 0, i = 1 : m, \tag{6.2}$$

$$x \in X \subseteq R^n. \tag{6.3}$$

To start with, let us assume that the functions $F^\nu(x), \nu = 0 : m$ are convex. Then for every x we have the inequality

$$F^\nu(z) - F^\nu(x) \geq \langle F_x^\nu(x), z - x \rangle, \quad \forall z \in X,$$

where F_x^ν is a subgradient (generalized gradient). We denote as $\partial F^\nu(x)$ the whole set of subgradients at x—the subgradient set. In stochastic quasigradient methods the sequence of approximates x^s, $s = 0, 1, \ldots$ is constructed by using statistic estimates of the $F^\nu(x^s)$ and $F_x^\nu(x^s)$—random numbers $\eta_\nu(s)$ and vectors $\xi^\nu(s)$ which in average are close to the $F^\nu(x^s), F_x^\nu(x^s)$. These quantities are constructed by using information about the past history of the optimization process, generated by the path $(x^0, \ldots, x^s)$ and some other variables, for instance the Lagrangian multipliers. We denote this history as B_s and for the sake of simplicity we usually assume that it is the $(x^0, \ldots, x^s)$. Then for the $\eta_\nu(s)$, $\xi^\nu(s)$ we have the conditional mathematical expectation

$$E\{\eta_\nu(s)|x^0, \ldots, x^s\} = F^\nu(x^s) + a_\nu(s); \tag{6.4}$$

$$E\{\xi^\nu(s)|x^0,\ldots,x^s\} = F_x^\nu(x^s) + b^\nu(s), \qquad (6.5)$$

where the numbers $a_\nu(s)$ and the vectors $b^\nu(s)$ may depend on $(x^0,\ldots,x^s)$. For exact convergence to an optimal solution, the values $a_\nu(s)$, $\|b^\nu(s)\|$ must be small (in a certain sense) when $s \to \infty$. At some time we must have that

$$a_\nu(s) \to 0, \|b^\nu(s)\| \to 0 \qquad (6.6)$$

directly or in such a way that

$$F^\nu(x^*) - F^\nu(x^s) \ge \langle E\{\xi^\nu|x^0,\ldots,x^s\}, x^* - x^s\rangle + \gamma_\nu(s), \qquad (6.7)$$

where $\gamma_\nu(s) \to 0$ as $s \to \infty$ and x^* an optimal solution. The vector $\xi^s(s)$ is called a stochastic quasi-gradient when $b^\nu(s) \not\equiv 0$, or stochastic subgradient, stochastic generalized gradient (stochastic gradient for differentiable function $F^\nu(x)$) when $b^\nu(s) \equiv 0$.

It turns out that for many important classes of optimization problems with functions $F^\nu(x)$, $v = 0 : m$ of a complex structure it is much easier to generate statistic estimates $\eta_\nu(s)$, $\xi^\nu(s)$ then to calculate exact values $F^\nu(x^s)$ and its subgradients $F_x^\nu(x^s)$. For stochastic programming problems when

$$F^\nu(x) = Ef^\nu(x,\omega), \quad \nu = 0 : m \qquad (6.8)$$

typically one can take $\xi^\nu(s)$ equal to a subgradient (gradient in the differentiable case) of $f^\nu(\cdot,\omega)$ at x^s

$$\xi^\nu(s) = f_x^\nu(x^s,\omega^s) \qquad (6.9)$$

where ω^s is an observation of ω, since usually with an appropriate definition of the subgradient-set, we have

$$\partial F^\nu(x) = \int \partial f^\nu(x,\omega)P(d\omega).$$

More generally

$$\xi^\nu(s) = \frac{1}{N_s}\sum_{k=1}^{N_s} f_x^\nu(x^s,\omega^{sk})$$

with a collection of independent samples $\omega^s k$, $k = 1 : N_s$, $N_s \ge 0$. Similarly we can take

$$\eta_\nu(s) = f^\nu(x^s,\omega^s) \qquad (6.10)$$

or more generally

$$\eta_\nu(s) = \frac{1}{N_s}\sum_{k=1}^{N_s} f^\nu(x^s,\omega^{sk}),$$

since according to the definition of functions $F^\nu(x)$

$$F^\nu(x^s) = E\{f^\nu(x^s,\omega)|x^s\}.$$

We consider different special rules for computing $\xi_\nu(s)$, $\eta_\nu(s)$ in Sections 6.7–6.13

6.2 Methods for Convex Functions

6.2.1 The Projection Method

Suppose we have to minimize a convex continuous function $F^0(x)$ in $x \in X \subset R^n$, where X is a closed convex set such that a projection on X can easily be calculated: $\pi_X(y) = \operatorname{argmin}\{\|y - x\|^2 : x \in X\}$. For instance, if X is a hypercube $a \leq x \leq b$, then $\pi_X(y) = \max[a, \min\{x, b\}]$. Let X^* be a set of optimal solutions. The method is defined by the relations:

$$x^s + 1 = \pi_X[x^s - \rho_s \xi^0(s)], s = 0, 1, \dots \qquad (6.11)$$

$$F^0(x^s) - F^0(x^s) \geq \langle E\{\xi^0(s)|x^0, \dots, x^s\}, x^* - x^s\rangle + \gamma_0(s), \qquad (6.12)$$

where ρ_s is the step size, $\gamma_0(s)$ may depend on $(x^0, \dots, x^s), x^* \in X^*$. Let us notice, that if vector $\xi^\nu(s)$ satisfies (6.5), then

$$\gamma^0(s) = -\langle b^0(s), x^* - x^s\rangle. \qquad (6.13)$$

This method was proposed and studied in [1]–[3], [5]. If $\xi^0(s) = F_x^0(x^s)$, we obtain the generalized gradient method which was suggested by Shor [34] and was studied by Ermoliev [35] and Poljak [36]. If $X = R^n$,

$$F^0(x) = Ef^0(x, \omega),$$

$$\xi^0(s) = \sum_{j=1}^{n} \frac{f^0(x^s + \Delta_s e^j, \omega^{sj}) - f^0(x^s, \omega^s 0)}{\Delta_s} e^j,$$

then the method suggested by (6.11) corresponds to the well-known stochastic approximation methods which were developed by Robbins and Monro, Kiefer and Wolfowitz, Dvoretsky, Blum and others (see [45]).

It was shown that under natural assumptions, that are also those of interest in practice, the sequence $\{x^s\}$ defined by (6.11), converges to a set of minimum points of the original problem with probability 1. The proof of this fact is based on the notion of a stochastic quasi-Feyer sequence [3]. A sequence $\{z^s\}_{s=0}^{\infty}$ is a Feyer sequence for a set $Z \subset R^n$, if [66]

$$\|z - z^{s+1}\| < \|z - z^s\|, \quad \forall z \in Z.$$

A sequence of random vectors $\{z^s\}_{s=0}^{\infty}$ defined on a probability space (Θ, R, μ) is a stochastic quasi-Feyer sequence [3] for a set $Z \subset R^n$, if $E\|z^0\|^2 < \infty$, and for any $z \in Z$

$$E\{\|z - z^{s+1}\|^2 | z^0, \dots, z^s\} \leq \|z - z^s\|^2 + r_s, \quad s = 0, 1, \dots \qquad (6.14)$$

$$r_s \geq 0, \sum_{s=0}^{\infty} Er_s < \infty.$$

Theorem 6.1. [5, p.98]. *If $\{z^s\}$ is a stochastic quasi-Feyer sequence for a set Z, then:*

(a) *the sequence $\|z - z^{s+1}\|^2, s = 0,1$, converges with probability 1 for any $z \in Z, E\|z - z^s\|^2 < C < \infty$,*

(b) *the set of accumulation points of $\{z^s(\theta)\}$ is not empty for almost all θ,*

(c) *if $z'(\theta), z''(\theta)$ are a two distinct accumulation points of the sequence $\{z^s(\theta)\}$ which do not belong to the set Z then Z lies in the hyperplane equidistant from the point $z'(\theta), z''(\theta)$.*

In the simplest case when γ_s is independent of $(z^0, \ldots, z^s)$ the fact (a) would follow from convergence of super martingale

$$v_s = \|z - z^s\|^2 + \sum_{k=s}^{\infty} r_k, v_s \geq 0,$$

$$E\{v_{s+1}|v_s\} \leq v_s.$$

The (c) follows from the equality

$$\|z - z'\|^2 - \|z - z''\|^2 = 2(z, z'' - z') + \|z'\|^2 + \|z''\|^2 = 0.$$

Consider now a simpler version of the convergence theorem for the iterative procedure (6.11) to illustrate the techniques of proof.

Theorem 6.2. *Assume that*

(a) $F^0(x)$ *is a convex continuous function,*

(b) X *is a convex compact set,*

(c) *Parameters $\rho_s, \gamma_0(s)$ satisfy with probability 1 the conditions*

$$\rho_s \geq 0, \sum_{s=0}^{\infty} \rho_s = \infty, \sum_{s=0}^{\infty} E\{\rho_s|\gamma_0(s)| + \rho_s^2\|\xi^0(s)\|^2\} < \infty, \qquad (6.15)$$

Then $\lim x^s \in X^*$ with probability 1.

Consider function $F^0(x) = Ef^0(x, \omega)$ with uniformly bounded in X second derivatives. Then for

$$\xi^0(s) = \frac{3}{r_s} \sum_{k=1}^{r_s} \frac{f^0(x^s + \Delta_s h^{sk}, \omega^{sk}) - f^0(x^s, \omega^{s0})}{\Delta_s} h^s k$$

we have

$$E\{\xi^0|x^s\} = F_x^0(x^s) + O(\Delta_s),$$

where $\{\omega^{s0}, \ldots, \omega^{sr_s}\}$ is a collection of ω-observations independent of $(x^0, \ldots, x^s)$ and $\{h^{s1}, \ldots, h^{sr_s}\}$ is a collection of observations of vector $h = (h_1, \ldots, h_n)$ whose components are independently and uniformly distributed over $[-1, 1]$. In this case condition (6.15) signifies that numbers ρ_s, Δ_s, which may depend on

$(x^0, \ldots, x^s)$, must be subjected to the conditions (taking into account the (6.13) and boundness of X:

$$\rho_s \geq 0, \sum_{s=0}^{\infty} \rho_s = \infty, \sum_{s=0}^{\infty} E(\rho_s \Delta_s + \rho_s^2/\Delta_s^2) < \infty;$$

$\rho_s = 1/s$, $\Delta_s = s^{-1/(2+\epsilon)}$ for any $0 < \epsilon < \frac{1}{2}$ are such sequences.

Let us notice that if we take

$$\xi^0(s) = \frac{3}{r_s} \sum_{k=1}^{r_s} \frac{f^0(x^s + \Delta_s h^{sk}, \omega) - f^0(x^s, \omega)}{\Delta_s} h^{sk}$$

and $f(x, \omega)$ satisfy Lipshitz condition within respect to x uniformly over u then

$$E\|\xi^0(s)\|^2 < \text{const} < \infty$$

when random parameters have finite distribution and $x^s \in X$. In this case condition (6.15) leads to the following requirement on ρ_s, Δ_s:

$$\rho_s \geq 0, \sum_{s=0}^{\infty} \rho_s = \infty, \sum_{s=0}^{\infty} E(\rho_s \Delta_s + \rho_s^2) < \infty.$$

In what follows we often make the assumption that $E\|\xi^0(s)\|^2$ is bounded for simplicity of restrictions on ρ_s, Δ_s. Such an assumption is not too stringent for most applications. In practice it is the consequence of (b) and the fact that estimates of subgradients are often unbiased and distributions of random parameters are finite.

Proof of Theorem 6.2: The properties of the projection π_X yield for any $x^* \in X$

$$\begin{aligned}
E\{\|x^* - x^{s+1}\|^2 | x^0, \ldots, x^s\} &\leq \|x^* - x^s\|^2 \\
&+ 2\rho_s\langle E\{\xi^0(s)|x^0, \ldots, x^s\}, x^* - x^s\rangle \\
&+ \rho_s E\{\|\xi^0(s)\|^2 | x^0, \ldots, x^s\}.
\end{aligned}$$

By the assumption (c) and (6.12) (taking into account that $F(x^*) - F(x^s) \leq 0$)

$$E\{\|x^* - x^{s+1}\|^2 | x^0, \ldots, x^s\} \leq \|x^* - x^s\|^2 + C(\rho_s|\gamma_0(s)| + \rho_s^2\|\xi^0(s)\|^2),$$

where C is a constant.

In view of (6.15) and by the definition (6.14), it means that $\{x^s\}$ is indeed a stochastic quasi-Feyer sequence for the set X^*. Consequently, the sequence $\|x^* - x^s\|$, $s = 0, 1, \ldots$ converges with probability 1 for any $x^* \in X^*$, the set of accumulation points of $\{x^s\}$ is not empty. If we show that one of the accumulation points of $\{x^s(\theta)\}$ belongs to X^* for almost all θ, then from assertion (c) of Theorem 6.1 would follow the convergence of $\{x^s\}$ with probability 1 to a point of X^*.

Consider the inequality

$$E\|x^* - x^{s+1}\|^2 \le E\|x^* - x^0\|^2 + 2E\sum_{k=0}^{s}\rho_k\langle E\{\xi^0(k)|x^0,\ldots,x^k\}, x^* - x^k\rangle$$

$$+ C\sum_{k=0}^{s}E\rho_k^2\|\xi^0(k)\|^2.$$

Due to the inequality (6.12)

$$E\|x^* - x^{s+1}\|^2 \le E\|x^* - x^0\|^2 + 2E\sum_{k=0}^{s}\rho_k(F^0(x^*) - F^0(x^k))$$

$$+ C\sum_{k=0}^{s}E\{\rho_k|\gamma_0(k)| + \rho_k^2\|\xi^0(k)\|^2\}$$

from which we get

$$E\sum_{k=0}^{\infty}\rho_k(F^0(x^*) - F^0(x^k)) < \infty.$$

Since

$$\sum_{k=0}^{\infty}\rho_k = \infty \text{ and } F^0(x^*) - F^0(x^k) \le 0,$$

there exists a subsequence x^{k_s} such that $F^0(x^*) - F(x^{k_s}) \to 0$, and this completes the proof.

The methods which we shall consider below, converge under conditions approximately analogous to those mentioned above. Theorem 6.2 establishes the convergence of the iterative procedure (6.11) with probability 1. Such a convergence is important in many applications. If $\gamma_0(s) \equiv 0$ and if instead of (6.15) only the conditions

$$\rho_s \downarrow 0, \sum_{s=0}^{\infty}\rho_s = \infty$$

hold, then it can be shown [5], that

$$\inf_{x^*} E\|x^* - x^s\|^2 \to 0.$$

In [62] the following idea was proposed for estimating the efficiency vector

$$\bar{x}^s = \left(\sum_{k=0}^{s}\rho_k x^k\right)\left(\sum_{k=0}^{s}\rho_k\right)^{-1}.$$

From the inequality

$$E\|x^* - x^{s+1}\|^2 \leq E\|x^* - x^0\|^2 + 2E \sum_{k=0}^{s} \rho_k (F^0(x^*) - F^0(x^k))$$

$$+ C \sum_{k=0}^{s} E\{\rho_k |\gamma_0(k)| + \rho_k^2 \|\xi^0(k)\|^2\}$$

we have that

$$2E \sum_{k=0}^{s} \rho_k (F^0(x^k) - F^0(x^*)) \leq E\|x^* - x^0\|^2$$

$$+ C \sum_{k=0}^{s} E\{\rho_k |\gamma_0(k)| + \rho_k^2 \|\xi^0(k)\|^2\}$$

If the ρ_k are independent of $(x^0, \ldots, x^k)$, then

$$\left(\sum_{k=0}^{s} \rho_k\right)^{-1} E \sum_{k=0}^{s} \rho_k (F^0(x^k) - F^0(x^*)) \geq EF^0(\overline{x}^s) - F^0(x^*)$$

and we have such estimation

$$EF^0(\overline{x}^s) - F^0(x^*) \leq \left(2 \sum_{k=0}^{s} \rho_k\right)^{-1}$$

$$\left[E\|x^* - x^0\|^2 + C \sum_{k=0}^{s} (\rho_k |\gamma_0(k)| + \rho_k^2 \|\xi^0(k)\|^2)\right].$$

6.2.3 The Lagrange Multiplier Method

The method is characterized by the relations

$$x^{s+1} = \pi_X \left(x^s - \rho_s \left[\xi^0(s) + \sum_{i=1}^{m} u_i^s \xi^i(s)\right]\right),$$

$$u_i^{s+1} = \max\{0, u_i(s) + \delta_s \eta_i(s)\}$$

and when $X = R^n$, $\delta_s \equiv \rho_s \equiv \text{const}$, $\xi^\nu(s) = F_x^\nu(x^s)$, $\eta_i(s) = F^i(x^s)$, $i = 1 : m$, and the $f^\nu(x)$, $\nu = 0 : m$ are smooth it is a deterministic algorithm proposed in [52]. The stochastic version of this method was studied in [1], [5], where it was proved that the $\min_{k \leq s} F^0(x^k)$ converge to $\min F^0(x)$ with probability 1, provided that $F^0(x)$ is strictly convex and $\delta_s \equiv \rho_s$. The convergence for convex functions $F^0(x)$—not necessarily strictly convex—was studied in [21] with assumptions that $\rho_s/\delta_s \to 0$.

6.2.3   Penalty Function Methods. Averaging Operation

Constraints of type (6.2) of the general problem (6.1)–(6.3) can be taken into account by means of penalty functions and instead of the original problem, we can minimize a penalized function, for instance

$$\Psi(x,c) = F^0(x) + c \sum_{i=1}^{m} \min\{0, F^i(x)\}$$

on the set X, where c is a big enough number. A generalized gradient of $\Psi(x,c)$ at $x = x^s$ is

$$F_x^0(x^s) + c \sum_{i=1}^{m} \min\{0, F^i(x^s)\} F_x^i(x^s).$$

If the exact values of $F^i(x^s)$, $F^0(x^s)$, $F_x^i(x^s)$ are known, then a deterministic generalized gradient procedure can be used for minimizing $\psi(x,c)$. The penalty function methods for a problem with known values of the constraint functions $F^i(x^s)$ was considered in [46], [63]. In such cases the projection method (6.11) is applicable to minimizing $\psi(x,c)$, since the estimate of the subgradient $f_x(x^s,c)$ is vector

$$\xi^0(s) + c \sum_{i=1}^{m} \min\{0, F_i(x^s)\} \xi^i(s).$$

In general, if instead of the values $F^\nu(x^s)$, $F_x^n u$, $\nu = 0 : m$, only statistical estimations $\eta_\nu(s)$, $\xi^\nu(s)$ are available, it is impossible to actually find $\min\{0, F^i(x^s)\}$. How to handle this situation was studied in [4], [5].

Consider the following variant of the iterative scheme studied in the previous section.

$$x^{s+1} = \pi_X(x^s - \rho_s[\xi^0(s) + c \sum_{i=1}^{m} \min\{0, \overline{F}_i(s)\} \xi^i(s)]), \qquad (6.16)$$

$$\overline{F}_i(s+1) = \psi_s \eta_i(s+1) + (1 - \psi_s)\overline{F}_i(s), i = 1 : m, \qquad (6.17)$$

where ψ_s is the step-size, $0 \le \psi_s \le 1$, $\overline{F}_i(0) = \eta_i(0)$,

$$E\{\eta_i(s)|x^0, \ldots, x^s\} = F^i(x^s) + a_i(s),$$

$$F^\nu(x^*) - F^\nu(x^s) \ge \langle E\{\xi^\nu(s)|x^0, \ldots, x^s\}, x^* - x^s\rangle + \gamma_\nu(s).$$

For convergence with probability 1 of these kinds of procedures in addition to (6.15), we must demand that with probability 1

$$\psi_s \ge 0, \rho_s/\psi_s \to 0, \sum_{s=0}^{\infty} E\psi_s^2 < \infty,$$

$$\sum_{s=0}^{\infty} E\{\rho_s|\gamma_i(s)| + \psi_s|a_i(s)|\} < \infty, i = 1 : m$$

It is worthwhile to note that the above mentioned method may not converge when $\psi_s \equiv 1$. I.e., for $\overline{F}_i(s) \equiv \eta_i(s)$. If $\delta_s = 1/(s+1)$ then

$$\overline{F}_i(s) = \frac{1}{s} \sum_{k=0}^{s} \eta_i(k).$$

This is why the (6.17) was called the averaging operation. In the case when $F^i(x) = Ef^i(x,\omega)$,

$$\overline{F}_i(s+1) = \psi_s f^i(x^{s+1}, \omega^s + 1) + (1 - \psi_s)\overline{F}_i(s), i = 1 : m \qquad (6.18)$$

The averaging procedure proved to be very useful of stochastic and non-differentiable optimization, the following general fact is decisive concerning this operation. Consider the auxiliary procedure (6.17) itself for a given sequence $\{x^s\}_{s=0}^{\infty}$. The procedure (6.17) has the following general form

$$\beta(s+1) = \beta(s) - \psi_s[\beta(s) - \eta(s+1)], s = 0, 1, \dots \qquad (6.19)$$

where ψ_s is B_s-measurable function and $\eta(s)$ is a random observation of a vector $V(s)$:

$$E\{\eta(s)|B_s\} = V(s) + a(s), \qquad (6.20)$$

which in the case of method (6.16)(6.17) takes on the form $V(s) = F(x^s) = (F^1(x^s), \dots, F^m(x^s))$. Under rather general assumptions (see, for instance [10]], p.46) provided that with probability 1

$$\|V(s+1) - V(s)\|/\psi_s \to 0 \qquad (6.21)$$

$$\psi_s \geq 0, \sum_{s=0}^{\infty} E\psi_s^2 < \infty, \|a(s)\|/\psi_s \to 0 \qquad (6.22)$$

it can be shown that with probability 1

$$\|\beta(s) - V(s)\| \to 0 \text{ for } s \to \infty \qquad (6.23).$$

Therefore the $\beta(s)$ estimates vector $V(s)$ with increasing precision and we can "substitute" unknown $V(s)$ by $\beta(s)$. If $F_i(x)$, $i = 1 : m$ are Lipschitz continuous functions in X and points x^{s+1}, x^s are connected through the equation (6.16), $|\eta_i(s)| < $ const, $\|\xi^\nu(s)\| < $ const, $i = 1 : m$, $\nu = 0 : m$, then assumption (6.21) follows from the condition

$$\rho_s/\psi_s \to 0 \text{ for } s \to \infty \qquad (6.24)$$

with probability 1.

The assertion (6.23) has close connections to the general theorem 5 concerning the convergence of nonstationary optimization procedures, since the step direction

$$2[\beta(s) - \eta(s+1)]$$

of the (6.19) is the stochastic quasigradient of the time-depending function

$$\Phi^s(\beta) = E\|\beta - \eta(s+1)\|^2 \tag{6.25}$$

at $\beta = \beta(s)$.

The averaging operation enables us to elaborate many stochastic analogues of known deterministic methods. Gupal [8] has studied the following stochastic version of the deterministic procedure described in paper [36]:

$$x^{s+1} = \pi_X[x^s - \rho_s \varsigma^s], \tag{6.26}$$

$$\varsigma^s = \begin{cases} \xi^0(s), & \text{if } \overline{F}_{i_s}(s) = \max_{1 \leq i \leq m} \overline{F}_i(s) \leq 0, \\ \xi^{i_s}, & \text{if } \overline{F}_{i_s}(s) > 0. \end{cases}$$

The requirements for convergence of this method are similar to those for the method (6.16).

Consider now some other methods for which the averaging operation appeared to be crucial.

6.2.4 Mixed Stochastic Quasigradient Method

Bajenov and Gupal [25] were first to apply the averaging procedure to step directions. The method is defined by the relations

$$x^{s+1} = \pi_X[x^s - \rho_s d^s] \tag{6.27}$$

$$d^{s+1} = \delta_s \xi^0(s+1) + (1 - \delta_s) d^s = d^s + \delta_s[\xi^0(s+1) - d^s], \tag{6.28}$$

$$E\{\xi^0(s)|x^0, d^0, \ldots, x^s, d^s\} = F_x^0(x^s) + b^0(s), \tag{6.29}$$

$s = 0, 1 \ldots$ with initial $d^0 = \xi^0(0)$. Such types of methods have also been studied in [10], [70], [71], [73].

The sequence $\{x^s\}$ converges with probability 1 to an optimal solution provided that in addition to requirements a), b), c) of the theorem 2 the scalars ρ_s, δ_s are chosen so as to satisfy with probability 1

$$\rho_s \geq 0, \delta_s \geq 0, \rho_s/\delta_s \to 0, \|b^0(s)\| \to 0, \tag{6.30}$$

$$\sum_{s=0}^{\infty} \rho_s = \infty, \sum_{s=0}^{\infty} E(\rho_s^2 + \delta_s^2) < \infty.$$

The vector d^s defined the recurrent formula (6.28) is called the averaged, aggregated, or mixed stochastic quasigradient.

6.3 Nonconvex Nondifferentiable Functions - Finite Difference Approximations Schemes

The convergence of SQG methods for nonconvex objective functions and constraints functions have been studied by many authors (see [5], [10], [12]). In [12], Nurminski generalized method (6.11) for the case of nonconvex and nondifferentiable objective functions satisfying the inequality

$$F^0(z) - F^0(x) \geq \langle F_x^0(x), z - x \rangle + o(\|z - x\|)$$

when $z \to x$ for all x from a compact set. Such functions are called weakly convex. The class of weakly convex functions includes convex functions as well as nonconvex differentiable. Moreover, the maximum of a collection of weakly convex functions is also the weakly convex function. Significant results in elaborating SQG methods for the nonconvex and nondifferentiable functions were obtained in [9], [10], [31]. In these papers the following stochastic versions of the finite difference approximation schemes were proposed.

If values of the functions $F^\nu(x), \nu = 0 : m$ can be easily calculated and $F^\nu(x)$ are differentiable functions, then there exist methods using a finite difference approximations of the gradients $F_x^\nu(x^s)$ at current point x^s:

$$F_x^\nu(x^s) \sim \sum_{j=1}^{n} \frac{F^\nu(x^s + \Delta_s e^j) - F^\nu(x^s)}{\Delta_s} e^j, \tag{6.31}$$

$$F_x^\nu(x^s) \sim \sum_{j=1}^{n} \frac{F^\nu(x^s + \Delta_s e^j) - F^\nu(x^s - \Delta_s e^j)}{2\Delta_s} e^j, \tag{6.32}$$

where e^j is the unit vector on the j-th axis and $\Delta_s > 0$. Although the finite difference approximations exist for nondifferentiable functions, the use of them does not guarantee the convergence of optimization procedures. The proposed modification of finite-difference approximation schemes consists a slight randomization of them:

$$F_x^\nu(x^x) \sim \xi^\nu(s) = \sum_{j=1}^{n} \frac{F^\nu(\overline{x}^s + \Delta_s e^j) - F^\nu(\overline{x}^s)}{\Delta_s} e^j, \tag{6.33}$$

$$F_x^\nu(x^s) \sim \xi^\nu(s) = \sum_{j=1}^{n} \frac{F^\nu(\overline{x}^{sj} + \Delta_s e^j) - F^\nu(\overline{x}^{sj} - \Delta_s e^j)}{\Delta_s} e^j, \tag{6.34}$$

where $F_x^\nu(x)$ is a subgradient; $\overline{x}^s = (x_1^s + h_1^s, \ldots, x_j^s + h_j^s, \ldots, x_n^s + h_n^s), \overline{x}^{sj} = (x_1^s + h_1^s, \ldots, x_j - 1^s + h_{j-1}^s, x_j^s, x_{j+1}^s + h_{j+1}^s, \ldots, x_n^s + h_n^s), j = \overline{1, n}$ and h_j^s are independent random quantities uniformly distributed on interval $[-\frac{\Delta_s}{2}, \frac{\Delta_s}{2}]$.

The convergence of corresponding optimization procedures is based on the fact that with probability 1

$$\min_{F_x^\nu(x^s)} \left\| E\left\{ \sum_{j=1}^n \frac{F^\nu\left(\overline{x}^{sj} + \Delta_s e^j\right) - F^\nu\left(\overline{x}^{sj} - \Delta_s e^j\right)}{\Delta_s} e^j | x^s \right\} - F_x^\nu(x^s) \right\| \to 0$$

(6.35)

when $\Delta_s \to 0$ and $F^\nu(x)$ are local Lipschitz functions. Therefore vectors $\xi^\nu(s)$ defined by the (6.33), (6.34) are also statistical estimates of the subgradient $F_x^\nu(x^s)$, satisfying general requirements (6.4), (6.5), (6.6).

For stochastic programming problems when $F^\nu(x) = Ef^\nu(x, \omega)$, we have analogues of the (6.33), (6.34)

$$\xi^\nu(s) = \sum_{j=1}^n \frac{f^\nu\left(\overline{x}^s + \Delta_s e^j, \omega^{sj}\right) - f^\nu\left(\overline{x}^s, \omega^s 0\right)}{\Delta_s} e^j$$

(6.36)

$$\xi^\nu(s) = \sum_{j=1}^n \frac{f^\nu\left(\overline{x}^{sj} + \Delta_s e^j, \omega^{sj}\right) - f^\nu\left(\overline{x}^{sj} - \Delta_s e^j, \omega^s, n + j\right)}{\Delta_s} e^j$$

(6.37)

which also satisfy the relation (6.35).

Different generalizations of SEG methods to the case of local Lipschitz functions $F^\nu(x)$ making use of the (6.33), (6.34), (6.36), (6.37) type approximations have been studied in papers [10], [73], [74].

Let us discuss the general idea of such procedures with more details.

6.4 Simultaneous Optimization and Approximation Procedures, Nonstationary Optimization

Suppose we have to minimize a function $f^0(x)$ of a rather complex nature, for example, it does not have continuous derivatives. Consider the sequence of the "good" functions $\{F^0(x, s)\}$, for instance smooth, converging to $f^0(x)$ for $s \to \infty$. Now consider the procedure

$$x^{s+1} = x^s - \rho_s F_x^0(x, s), s = 0, 1, \ldots$$

(6.38)

Under rather general conditions $(\rho_s \downarrow 0, \Sigma \rho_s = \infty)$ it is possible to show (see [5], [14], [17] and Theorem 6.3) that $F^0(x^s, s) \to \min f^0(x)$.

Often approximate functions may have the form of mathematical expectations

$$F^0(x, s) = \int f^0(x + h) P_s(dh) = Ef^0(x + h(s)),$$

(6.39)

where the measure $P_s(d\omega)$ for $s \to \infty$ is centered at the point 0. Hence instead of the procedure given by (6.38) that requires the exact value of the gradient of the mathematical expectation, we can use the ideas of the stochastic quasigradient methods.

For example, see [9], let $h(s)$ be random vectors with independent components uniformly distributed on $[-\Delta_s/2, \Delta_s/2]$, $\Delta_s \to 0$ for $s \to \infty$, and suppose that $f^0(x, s)$ is continuous differentiable and $F^0(x, s) \to f^0(x)$ uniformly on any bounded domain. Consider the stochastic procedure with the (6.34) type approximation

$$x^{s+1} = x^s - \rho_s \xi^0(s), s = 0, 1 \ldots$$

$$\xi^0(s) = \sum_{j=1}^{n} \frac{f^0(\overline{x}^{sj} + \Delta_s e^j) - f^0(\overline{x}^{sj} - \Delta_s e^j)}{\Delta_s} e^j. \tag{6.40}$$

It can be shown that

$$E\{\xi^0(s)|x^s\} = F_x^0(x^s, s)$$

where $F^0(x, s)$ is defined by (6.39).

In other words the method (6.40) is a stochastic analogue of the method (6.38). Procedures (6.38), (6.40) are examples of simultaneous optimization and estimation procedures. The development of such procedures is connected with the following general problem of nonstationary optimization [15]–[20], [53], [75].

The objective function $F^0(x, s)$ and the feasible set X_s of the nonstationary problem depend on the iteration number $s = 0, 1, \ldots$. It is necessary to create a sequence of approximate solutions $\{x^s\}$, that tends, in some sense to follow the time path of the optimal solutions: for $s \to \infty$,

$$\lim[F^0(x^s, s) - \min\{F^0(x, s)|x \in X_s\}] = 0.$$

The case when there exist $\lim F^0(x, s)$ and $\lim X_s$ (in some sense) for $s \to \infty$ was called the limit extremal problems [14], [17], [5]. The optimization problems with time-varying functions and known trend of the optimal solutions is considered in [53], [54], [60].

To illustrate the ideas involved in the proof of convergence results, let us consider the following simple case:

Theorem 6.3. *Assume that:*
(a) $F^0(x, s), f^0(x)$ *are convex continuous functions,*
(b) X *is a convex compact set,*
(c) $F^0(x, s) \to f^0(x)$ *uniformly in* X,
(d) $\|F_x^0(x^s, s)\| \le$ const,

$$x^{s+1} = \pi_X[x^s - \rho_s F_x^0(x^s, s)] \tag{6.41}$$

and the parameters ρ_s *satisfy the conditions*

$$\rho_s \to 0, \sum_{s=0}^{\infty} \rho_s = \infty$$

Then $F^0(x^s, s) \to f^0(x^*) = \min\{f^0(x) | x \in X\}$.

    The principal difficulties associated with the convergence of procedure (6.41) are connected with the choice of the step-size ρ_s. There is no guarantee that the new approximate solution x^{s+1} will belong to the domain of the smaller values of functions $F^0(x,t)$ for $t \geq s+1$ (see Figure 6.1). Therefore even for $X = R^n$ and differentiable (continuously) functions $F^0(x,s)$, the (6.41) is essentially nonmonotonic optimization procedure. There is one more difficulty. In the general case without the assumption c), the aim of $\{x^s\}$ is to track the set of optimal solutions

$$X_s^* = \{x | F^0(x, s) = \min F^0(x, s), x \in X_s\}.$$

Unfortunately the Hausdorf distance between X_s^* and X_{s+1}^* may be large even for small distance between $F^0(x, s)$ and $F^0(x, s+1)$, as it shows in the Figure 6.2.

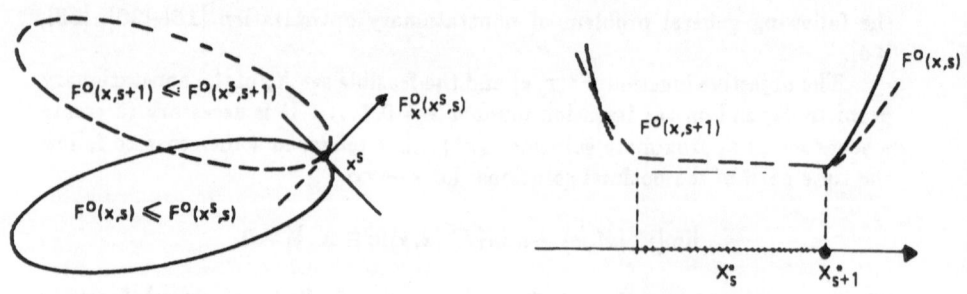

Figure 6.1.   Figure 6.2.

    The convergence study of the (6.38), (6.40), (6.41) type procedures in general case involves the sets of ε-solutions (see [18], [75], [76]).

    The essentially nonmonotonic solution procedures need an appropriate technique to prove their convergence. Often the necessary analysis can be based on the following result [5], [11].

Theorem 6.4. [5, p. 181] *Suppose that $X^* \subset R^n$ is closed and $\{x^s\}$ is a sequence of vectors in R^n*

(1) for all s $x^s \in K$, with K compact

(2) for any subsequence $\{x^{s_k}\}$ with $\lim x^{s_k} = x'$

    (a) if $x' \in X^*$, then $\|x^{s_k+1} - x^{s_k}\| \to 0$ as $k \to \infty$

    (b) if $x' \notin X^*$, then for ϵ sufficiently small and for any s_k

$$\tau_k = \min\{s | s \geq s_k, \|x^{s_k} - x^s\| < \epsilon\} > \infty.$$

(3) there exists a continuous function $V(x)$ attaining on X^* an at most countable set of values and

$$\lim_{k \to \infty} V(x^{\tau_k}) < \lim_{k \to \infty} V(x^{s_k}).$$

Then the sequence $\{V(x^s)\}$ converges and all accumulating points of the sequence $\{x^s\}$ belong to X^*.

The conditions of this theorem are similar to necessary and sufficient convergence conditions, proposed by Zangwill (see [65]). However, Zangwill's conditions are very difficult to verify for a nonmonotonic procedure.

Conditions (2) of Theorem 6.3 prevent all sequence $\{x^s\}$ converge to limit point x', which does not belong to the set X^*. However, condition (2) alone does not prevent "cycling", i.e., such a behavior of $\{x^s\}$ that it will be visiting any neighborhood of $x' \notin X^*$ infinitely many times. To exclude such a case the condition (3) is imposed, which guarantees that the sequence $\{x^s\}$ will be leaving a neighborhood of x' with decreasing values of some Lyapunov functions $V(x)$. Let us now illustrate the use of this theorem.

Proof of Theorem 6.3: The conditions 1,2(a) of Theorem 6.4 are fulfilled. It suffices to verify the conditions 2(b) and 3. Let $x^{s_k} \to x' \in X^*$, we need to show that $\tau_k < \infty$. We argue by contradiction, to suppose the contrary that $\tau_k = \infty$. For this purpose, we consider the function $V(x) = \min_{x^* \in X^*} \|x^* - x\|^2$. We have that

$$V(x^{s+1}) = \min_{x^* \in X^*} \|x^* - x^{s+1}\|^2 = \|x^*(s+1) - x^{s+1}\|^2 \leq \|x^*(s) - x^{s+1}\|^2$$

$$= V(x^s) + 2\rho_s \langle F_x^0(x^s, s), x^*(s) - x^s \rangle + \rho_s^2 \|F_x^0(x^s, s)\|^2.$$

Since $x^{s_k} \to x' \overline{\in} X^*$ and $\|x^s - x^{s_k}\| < \epsilon$ for sufficiently large s and any ϵ. Then there exists $\delta > 0$ such that

$$f^0(x^*) - f^0(x^s) < -\delta$$

and for $x^* \in X^*$ we have

$$\langle F_x^0(x^s, s), x^* - x^s \rangle \leq F^0(x^*, s) - F^0(x^s, s)$$

$$\leq F^0(x^*, s) - f^0(x^*) + f^0(x^s) - F^0(x^s, s)$$

$$< -\frac{\delta}{2}.$$

Therefore

$$V\left(x^{s+1}\right) \le V\left(x^s\right) - \delta\rho_s + c\rho_s^2 = V\left(x^s\right) - \rho_s(\delta - c\rho_s)$$

$$\le V\left(x^{s_k}\right) - \frac{\delta}{2}\sum_{\ell=s_k}^{s}\rho_\ell$$

and for a sufficiently large s, this contradicts the fact that $|V\left(x\right)| < \text{const}$ when $x \in X^*$. So, condition 2 is satisfied. Looking at condition 3, it is easy to realize that

$$V\left(x^{\tau_k}\right) \le V\left(x^{s_k}\right) - \frac{\delta}{2}\sum_{\ell=s_k}^{s}\rho_\ell.$$

Hence, in view of the properties of π_x,

$$\varepsilon < \|x^{\tau_k} - x^{s_k}\| \le \sum_{s=s_k}^{\tau_k-1}\|x^{s+1} - x^s\| \le C\sum_{s=s_k}^{\tau_k-1}\rho_s,$$

where C is a constant. Then

$$V\left(x^{\tau_k}\right) \le V\left(x^{s_k}\right) - \frac{\varepsilon\delta}{2C}$$

or equivalently

$$\overline{\lim} \, V\left(x^{\tau_k}\right) < \lim V\left(x^{s_k}\right)$$

and this completes the proof.

Consider now more general procedure

$$x^{s+1} = \pi_X[x^s - \rho_s\xi^0(s)], \quad s = 0, 1, \ldots, \tag{6.42}$$

$$E\{\xi^0(s)|x^0, \ldots, x^s\} = F_s(x^s, s) + a^0(s)$$

Theorem 6.5. [19] *Assume that*
(a) $F^0(x, s)$ *are convex continuous functions,*
(b) X *is a convex compact set,*
(c) $\max\limits_{x \in X}|F^{s+1}(x) - F^s(x)| \le \delta_s$, $E\|\xi^0(s)\| < \text{const}$,
(d) *with probability 1*

$$\delta_s/\rho_s \to 0, \|a^0(s)\| \to 0, \rho_s \ge 0, \sum_{s=0}^{\infty}\rho_s = \infty, \sum_{s=0}^{\infty}E\rho_s^2 < \infty.$$

Then with probability 1

$$|F^0\left(x^s, s\right) - \min\{F^0\left(x, s\right)|x \in X\}| \to 0 \text{ for } s \to \infty.$$

6.5 Feasibile Directions Methods

Consider the minimization of a continuously differentiable function $F^0(x)$ in a compact convex set X. If $F^0(x^s)$ and $F_x^0(x^s)$ are known, then the standard linearization method is defined by the relations

$$x^{s+1} = x^s + \rho_s(\overline{x}^s - x^s),$$

$$\langle F_x^0(x^s), \overline{x}^s \rangle = \min\{\langle F_x^0(x^s), x \rangle | x \in X\}$$

$$F^0(x^{s+1}) = \min_{0 \le \rho \le 1} F^0(x^s + \rho(\overline{x}^s - x^s)).$$

The stochastic variant of this method has been studied in [5], [6], [10], [30] and is defined by the relations

$$x^{s+1} = x^s + \rho_s(\overline{x}^s - x^s), \tag{6.43}$$

$$\langle d^s, \overline{x}^s \rangle = \min\{\langle d^s, x \rangle | x \in X\}$$

$$d^{s+1} = \delta_s \xi^0(s+1) + (1 - \delta_s)d^s = d^s + \delta_s[\xi^0(s+1) - d^s],$$

where ρ_s, δ_s satisfy conditions similar to those of Section 6.2. Notice that if instead of d^s the vectors $\xi^0(s)$ are used ($\delta_s \equiv 1$) then, some simple examples show that the method may not converge.

The linearization method usually is applied when X is defined by linear constraints. In such case this method requires at each iteration a solution of linear subproblem in contrast to the projection method (6.11), which requires the solution of quadratic subproblem. Let us notice that only small perturbations occur in the objective function of the subproblem at each $s > 0$, therefore for $s > 0$ only small adjustments of the preceding solution are needed in order to obtain a solution of the current subproblem.

Consider now the case when

$$X = \{x | F^i(x) \le 0, i = 1 : m\}.$$

Assume that $F^\nu(x), \nu = 0 : m$ are continuously differentiable functions, the set X is compact, and the gradient $F_x^0(\cdot)$ is Lipschitz continuous on X. Let sequences $\{x^s\}$ and $\{v^s\}$ be defined by the relations [10], [78]-[80]

$$x^{s+1} = x^s + \rho_s v^s \tag{6.44}$$

$$d^{s+1} = d^s + \delta_s(\xi^0(s+1) - d^s), d^0 = \xi^0(0),$$

$$E\{\xi^0(s)|B_s\} = F_x^0(x^s) + b^0(s),$$

where B_s is σ-field generated by points $\{(x^0, v^0, d^0), \ldots, (x^s, v^s, d^s)\}$ and v^s is a solution of the subproblem:

$$\max\{\tau | \langle d^s, v \rangle + \tau \le 0, \langle F_x^i(x^s), v \rangle + \delta \le 0, i \in I^s, -1 \le v_j \le 1, j = 1 : n\}, \tag{6.45}$$

$$I^s = \{i : -\varepsilon_s \le F_i(x^s) \le 0\}, \varepsilon_s \downarrow 0$$

Therefore it is assumed that we can calculate exact values $F_x^i(x), i = 1 : m$.
Consider

$$\rho_s' = \max\{\rho | x^s + \rho v^s \in X, \rho \ge 0\}$$

and let

$$\rho_s = \min\{\rho_s', \rho_s''\}, \rho_s'' \ge 0.$$

Theorem 6.6. (see [10], p.113) *If with probability 1*

$$\rho_s'' \ge 0, \sum_{s=0}^{\infty} \rho_s'' = \infty, \rho_s''/\delta_s \to 0, \sum_{s=0}^{\infty} E\delta_s^2 < \infty, \varepsilon_s \to 0,$$

$$E\|\xi^0(s)\|^2 \le C < \infty, E\|b^0(s)\|^2 \le C, E\{\|b^0(s)\| \|B_s\} \to 0$$

for some constant C, then the sequence $\{F^0(x^s)\}$ converges with probability 1 and all cluster points of the sequence $\{x^s\}$ satisfy the necessary optimality conditions of the problem.

Ruszczynski [80] modified the method (6.43) for nonconvex objective function with the following property: there exist $\delta \ge 0$ and $\mu \ge 0$ such that for all $x \in X$ all z satisfying $\|z - x\| \le \delta$

$$F^0(z) - F^0(x) \ge \langle F_x^0(x), z - x \rangle - \mu \|z - x\|^2,$$

where X is a compact set and $F_x^0(x)$ is a subgradient. This class of functions is identical with the family of functions, which in some open neighborhood of x have a representation [81]:

$$F^0(x) = \max_{u \in U} \varphi(x, u),$$

where U is a compact and $\varphi(\cdot, u)$ has second derivatives continuous in (x, u). In the method the following direction-finding subproblem is used instead of the subproblem (6.45):

$$\min_y \{\langle d^s, y - x^s \rangle + \tfrac{1}{2} \|y - x^s\|^2 | F^i(x^s) + \langle F_x^i(x^s), y - x^s \rangle \le 0, \tag{6.46}$$
$$i = 1 : m, y \in X\},$$

where $F^i(x), i = 1 : m$ are supposed to be convex and differentiable in X functions, X is a convex compact. If y^s is a solution of the subproblem then

$$v^s = y^s - x^s$$

is used in equation (6.44). The convergence theorem is similar to that of the method (6.44) provided in addition to the mentioned above alternations that with probability 1

$$b^s(s) \equiv 0, \delta_s = \alpha\rho_s, 0 < \rho_s \le \min(1, 1/\alpha)$$

$$\sum_{s=0}^{\infty} \rho_s = \infty, \sum_{s=0}^{\infty} E\rho_s^2 < \infty,$$

where scalar ρ_s may as usually depend on the past history generated by the $(x^0, d^0, \ldots, x^s, d^s)$.

The paper [80] contains a rather general requirement on the choice of direction v^s, which enables different modifications of the subproblem. In papers [10], [30], procedures (6.43), (6.44) were generalized to the minimization of local Lipschitz functions making use of approximations (6.33), (6.34), (6.36), (6.37).

6.6 Adaptive SQG Procedure

The success of the application of SQG methods depends on the particular rules for choosing their parameters—step sizes and step directions. The general convergence theorems provide a wide freedom in choosing them adaptively as a functions of the (random) history B_s, for instance $(x^0, \ldots, x^s)$. What is the best choice?

The behavior of SQG methods is unusual as compared with deterministic methods. The convergent with probability 1 sequence of approximate solutions $\{x^s\}$ defines the set of pathes (realizations) leading from the initial point x^0 to the set of optimal solutions (Figure 6.3).

In the case of unique solution the procedure may approach a neighborhood of the solution in different ways. The choice $\rho_s = 1/s$ serves all pathes in the same way, independently of the current situation and cannot be the best strategy. Of course the definition of the best strategy is the consequence of the performance function definition. If the performance function is defined on the whole set of pathes and if this function deals only with the asymptotic behavior, then the choice $\rho_s = 1/s$ with the appropriate constant a depending only on the unique solution might be the best opportunity (see pioneering papers [84], [85]). Unfortunately this conclusion about the "optimality" of the $\rho_s = 1/s$ mislead in the use of stochastic approximation type procedures. The asymptotic approach is really rather unsatisfactory for practical application, since it does not make any use of the valuable information which accumulates during solution, in particular, the starting point. The practical aim usually is to reach some neighborhood of the solution rather than to find the precise value of the solution itself. SQG methods are quite good enough for this purpose. They have been applied to various practical problems (see, for instance, [5], [7]) and there always have been used only adaptive principles for choosing their parameters (this is discussed in details in Chapter 15-17).

The adequate choice of the parameters at a nonmonotonic procedure is not trivial problem as it shows even the simplest deterministic analogue of the method (6.11)—so-called generalized gradient method (see [5], [38])

$$x^{s+1} = x^s - \rho_s \frac{F_x^0(x^s)}{\|F_x(x^s)\|}, \quad s = 0, 1, \ldots \quad (6.47)$$

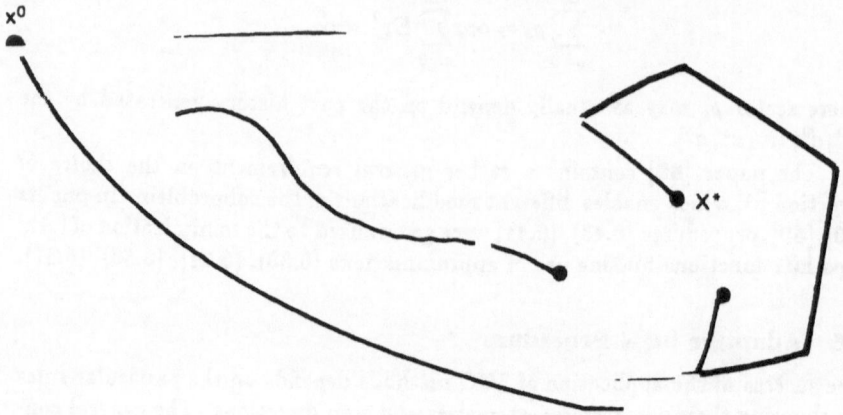

Figure 6.3.

where $F_x^0(x^s)$. Since there is no guarantee that the objective function is decreased in the direction $F_x^0(x^s)$ (see Figures 6.4 and 6.5), then for any choice of the ρ_s, satisfying the convergence conditions (see Theorem 6.3)

$$\rho_s \to 0, \sum_{s=0}^{\infty} \rho_s = \infty$$

the sequence $\{F^0(x^s)\}$ shows oscillatory behavior with tendency of decreasing in the "average". Stochastic version (6.11) is much more difficult since exact values of the objective function are not available.

A rather general way of changing the ρ_s would be to begin with a sufficiently large value for the first few iterations, and decrease ρ_s if additional tests show that the current point is in the vicinity of the optimum. The averaging procedure (see Sections 6.2.2, 6.2.3) appeared to be useful in tests of this types:

$$\overline{F}_x^\nu(s+1) = \overline{F}_x^\nu(s) + \delta_s[\xi^\nu(s+1) - \overline{F}_x^\nu(s)]$$
$$\overline{F}^\nu(s+1) = \overline{F}^\nu(s) + \psi_s[\eta_\nu(s+1) - \overline{F}^\nu(s)],$$

since $\min\{\|\overline{F}_x^\nu(s) - z\| \mid z \in \partial F_x^\nu(x^s)\}, |\overline{F}^\nu(s) - F^\nu(x^s)| \to 0$ under rather general conditions.

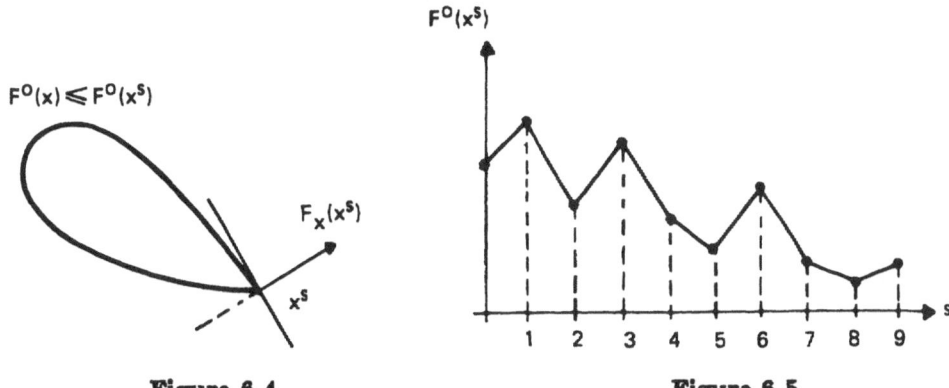

Figure 6.4.   **Figure 6.5.**

Therefore, the averaging results in the use increasingly precise estimates of the gradients (subgradients) and values of the functions without intensification of the observations. To avoid the influence of long tails of the past, it is sometimes more useful to adopt the averaging of the type

$$\overline{F}_x^{\nu}(s, \ell_s) = \frac{1}{s - \ell_s} \sum_{k=\ell_s}^{s} \xi^{\nu}(k),$$

$$\overline{F}^{\nu}(s, \ell_s) = \frac{1}{s - \ell_s} \sum_{k=\ell_s}^{s} \eta_{\nu}(k), 0 \le \ell_s \le s.$$

The decision as to whether to change the ρ_s or other parameters (steps of finite difference approximations, the smoothing parameters) may then be based on two modes:

- interactive mode
- automatic mode

By using the interactive mode it is assumed that the user can monitor the progress of the optimization process and can intervene to change the values of the step size and other parameters. These decisions should be based on the behavior of the averaged values $\overline{F}^\nu(s)$, $\overline{F}^\nu_x(s)$ and its different combinations and must partially be made by the user on the basic of the visually observed behavior of these quantities. For instance, in the case when observed behavior of $\overline{F}^0(s)$ shows a regular oscillations (see Figure 6.6 interval [a,b]).

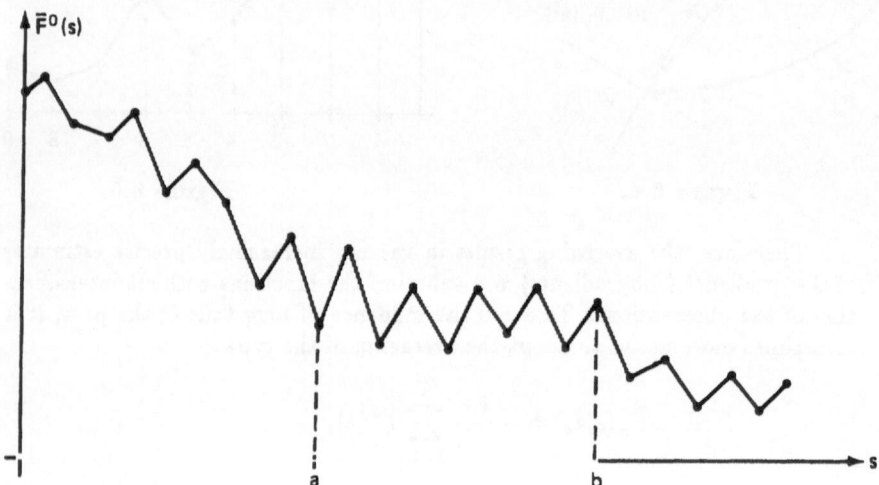

Figure 6.6.

In automatic mode the decisions about changing parameters is made automatically on the basic some tests which formalize the actions of "oscillatory behavior".

There is strong evidence that the interactive mode cannot be completely avoided in the stochastic optimization. There is only the question up to what extent to develop the automatic mode. The situation here is very much resembled to driving a car. Of course if road conditions are deterministic, it is possible to imagine an automat which drives the car. But since the road conditions are far away from the well formalized situation, the user himself drives the car using some minimal information about its construction.

Different concrete rules of choosing the parameters of SQG methods adaptively are discussed in Chapters 15–17.

6.7 Optimization of Stochastic Systems – General Standard Problem

In this and next sections we are going to discuss some applications of SQG methods to the stochastic programming problems when $F^\nu(x) = Ef^\nu(x,\omega), \nu = 0 : m$. From the discussion of the Chapter 1 it follows that taking into account the influence of uncertain random factors in optimization of systems leads to stochastic programming problems of the following standard form:

$$\text{minimize} \quad F^0(x) = Ef^0(x,\omega) \tag{6.48}$$

$$\text{subject to} \quad F^i(x) = Ef^i(x,\omega) \le 0, \quad i = 1 : m \tag{6.49}$$

$$x \in X \subset R^n, \tag{6.50}$$

where E is the operation of mathematical expectation with respect to some probability space (Ω, A, P).

The problem (6.48)–(6.50) is a model for stochastic systems optimization, when the decision (values to assign to the system parameters) x is chosen in advance, before the random factors ω is observed. A stochastic model tends to take into account all possible eventualities for stablizing the optimal solution with respect to perturbations of the data. There may also be a class of models, when the decision x is chosen only after an experiment over ω is realized and x is based on the actual knowledge of the outcomes of this experiment. Such situations occur in real-time control and short-term planning. In practice, these problems are usually rediced to problems of the type (6.48)–(6.50) via decision rules (see Chapter 1).

Consider some particular formulas for computing the estimates of values $F^\nu(x^s)$, $F_x^\nu(x^s)$. Suppose that it is possible to calculate the value of random functions $f^\nu(x^s, \omega)$. Then we can take

$$\eta_\nu(s) = \frac{1}{N_s} \sum_{k=1}^{N_s} f^\nu(x^s, \omega^{sk}), \nu = 0 : m \tag{6.51}$$

where the number $N_s \ge 1$ may depend on the past random history B_s of the stochastic procedure—the minimum σ-subfield that at least includes σ-algebra generated by the path $\{x^0, \ldots, x^s\}$ and may be some other random pathes associated with such quantities as Lagrange multipliers, averaged subgradient, etc. The collection $\{\omega^{s,1}, \ldots, \omega^{s,N_s}\}$ is result from samples of ω, which are mutually independent with respect to $s = 0, 1, \ldots$. By definition we have

$$E\{\eta_\nu(s)|B_s\} = \frac{1}{N_s} \sum_{k=1}^{N_s} E\{f^\nu(x^s, \omega)|x^s\} = F^\nu(x^s).$$

If the functions $F^\nu(x)$ have uniformly bounded second derivatives then for the random vectors

$$\xi^\nu(s) = \sum_{j=1}^{n} \frac{f^\nu(x^s + \Delta_s e^j, \omega^{sj}) - f^\nu(x^s, \omega^s 0)}{\Delta_s} e^j \tag{6.52}$$

we would have

$$E\{\xi^\nu(s)|x^s\} = F_x^\nu(x^s) + b^\nu(s), \|b^\nu(s)\| \le \text{const}.\Delta_s,$$

where e^j is the unit vector on the j-th axis $\Delta_s > 0$; $\{(\omega^{s0},\dots,\omega^{sn})\}_{s=0}^\infty$ are a result of independent $s = 0,1,\dots$, samples of ω (we could have $\omega^{s0} = \omega^{s1} = \dots = \omega^{sn}$). For the vector

$$\xi^\nu(s) = \frac{3}{r_s}\sum_{k=1}^{r_s} \frac{f^\nu(x^s + \Delta_s h^{sk}, \omega^{sk}) - f^\nu(x^s, \omega^{s0})}{\Delta_s} h^{sk}, \tag{6.53}$$

where $h^{s1},\dots,h^{sr_s}$ are independent of B_s observations of the random vector $h = (h_1,\dots,h_n)$ whose components are independently and uniformly distributed over $[-1,1]$; number $r_s \ge 1$ depends on B_s

$$E\{\xi_j^\nu(s)|B_s\} = \frac{3}{r_s}E\sum_{k=1}^{r_s} \frac{F^\nu(x^s + \Delta_s h^{sk}) - F^\nu(x^s)}{\Delta_s} h_j^{sk}$$

$$= \frac{3}{r_s}E\sum_{k=1}^{r_s} F_{x_j}^\nu(x^s)h_i^{sk}h_j^{sk} + \Delta_s\alpha_j(s) = F_{x_j}(x^s) + \Delta_s\alpha_j(s),$$

where $|\alpha_j(s)| < \text{const}$.

For nondifferentiable functions $F^\nu(x)$ typically one can take $\xi^\nu(s)$ equal to a subgradient of $f^\nu(x,\omega)$ at $x = x^s$:

$$\xi^\nu(x) = f_x^\nu(x^s, \omega^s),$$

where ω^s is a sample of ω independent of B_s; more generally similar to the (6.51):

$$\xi^\nu(s) = \frac{1}{N_s}\sum_{k=1}^{N_s} f_x^\nu(x^s, \omega^{sk}), \tag{6.54}$$

since under appropriate integrability conditions and the definition of the subgradient-set, we have

$$\partial F^\nu(x) = \int \partial f^v(x,\omega)P(d\omega).$$

For recourse and minimax problems referred to in Sections 6.8 and 6.9 such rules were firstly used in [2], [3], [5]. General framework provide results of papers [86], [87], [92]. If the functions $F^\nu(x)$ satisfy a local Lipschitz condition, then formulas (6.52), (6.53) can be modified respectively

$$\xi^\nu(x) = \sum_{j=1}^n \frac{f^\nu(\overline{x}^s + \Delta_s e^j, \omega^{sj}) - f^\nu(\overline{x}^s, \omega^{s0})}{\Delta_s} e^j \tag{6.55}$$

$$\xi^\nu(s) = \frac{3}{r_s}\sum_{k=1}^{r_s} \frac{f^\nu(\overline{x}^s + \Delta_s h^{sk}, \omega^{sk}) - f^\nu(x^s, \omega^{s0})}{\Delta_s} h^{sk}, \tag{6.56}$$

where $\overline{x}^s = (x_1^s + \tau_1^s, \ldots, x_n^s + \tau_n^s)$, random vector $\tau^s = (\tau_1^s, \ldots, \tau_n^s)$ independent of B_s with uniformly distributed on $[-\gamma_s/2, \gamma_s/2]$ components.

It is easy to see that in both cases (6.55) and (6.56)

$$E\{\xi^\nu(s)|x^s\} = F_x^\nu(x^s, s) + b^\nu(s),$$

where $\|b^\nu(s)\| \le \text{const} \frac{\Delta_s}{\gamma_s}$ and $F_x^\nu(x, s)$ is the gradient of the differentiable function

$$F^\nu(x, s) = Ef^\nu(x + \tau^s, \omega) = \frac{1}{(2\gamma_s)^n} E \int_{-\gamma_s}^{\gamma_s} \cdots \int_{-\gamma_s}^{\gamma_s} f^\nu(x + y, \omega) dy$$

with the property

$$\min\{\|F^\nu(x, s) - z\| \,|\, z \in \partial F^\nu(x)\} \to 0 \text{ for } \gamma_s \to 0.$$

The vector (see [9], [10])

$$\xi^\nu(x) = \sum_{j=1}^n \frac{f^\nu(\overline{x}^{sj} + \Delta_s e^j, \omega^{sj}) - f^\nu(\overline{x}^{sj} + \Delta_s e^j, \omega^{s,j-1})}{2\Delta_s} e^j \qquad (6.57)$$

is an unbiased estimate of the gradient $F_x^\nu(x, s)$

$$E\{\xi^\nu(s)|x^s\} = F_x^\nu(x^s, s).$$

Averaging operations (see Section 6.2) give us new opportunity to build wide range of the estimates $\overline{F}^\nu(s)$, $\overline{F}_x^\nu(s)$ from known, defined, for instance, through the (6.51)–(6.56):

$$\overline{F}^\nu(x + 1) = \overline{F}^\nu(s) + \psi_s[\eta_\nu(s + 1) - \overline{F}^\nu(s)],$$
$$\overline{F}_x^\nu(s + 1) = \overline{F}_x^\nu(s) + d_s[\xi^\nu(s + 1) - \overline{F}_x^\nu(s)].$$

Consider now more concrete classes of the estimates for some particular classes of problems.

6.8 The Stochastic Minimax Problems

The objective function of the simplest stochastic minimax problem (see [3], [5], [13], [32]) takes on the form

$$F^0(x) = E \max_{1 \le \ell \le m} \left[\sum_{j=1}^n a_{ij}(\omega) x_j + b_j(\omega) \right]. \qquad (6.58)$$

Many inventory models have such type of objective functions. Consider the simple example.

In a store of capacity r it is necessary to create a stock x of a homogeneous product whose demand is characterized by the random variable χ. The cost associated with the stock x on the condition that the demand is equal to χ is characterized by the function

$$f^0(x,\omega) = \begin{cases} \alpha(x-\chi), & x \geq \chi, \\ \beta(\chi-x), & x < \chi \end{cases}$$

or

$$f^0(x,\omega) = \max\{\alpha(x-\chi),\beta(\chi-x)\}$$

where α is the unit storage cost and β is the unit shortage cost. The decision about the stock-size x must be made before the information about the demand χ is available and the minimization of the expected cost leads to the minimization of the function

$$F^0(x) = E\max\{\alpha(x-\chi),\beta(\chi-x)\} \tag{6.59}$$

subject to $0 \leq x \leq r$.

For the function (6.59) and the more general

$$F^0(x) = Ef^0(x,\omega) = E\max_{y\in Y} g(x,y,\omega) = Eg(x,y(x,\omega),\omega)$$

a statistical estimate of the subgradient takes on the form (under reasonable assumptions)

$$\xi^0(s) = g_x(x^s,y,\omega^s)|_{y=y(x^s,\omega^s)}, \tag{6.60}$$

where g_x is a subgradient of $g(\cdot,y,\omega^s)$ at $x=x^s$.

To see that

$$E\{\xi^0(s)|x^s\} = F_x^0(x^s)$$

for a convex function $g(\cdot,y,\omega)$, we can write

$$g(x,y(x,\omega^s),\omega^s) - g(x^s,y(x^s,\omega^s),\omega^s) \geq g(x,y(x^s,\omega^s),\omega^s) -$$

$$-g(x^s,y(x^s,\omega^s),\omega^s) \geq \langle g_x(x^s,y(x^s,\omega^s),x-x^s\rangle = \langle \xi^0(s),x-x^s\rangle.$$

Taking conditional expectation on both side, we get

$$F^0(x) - F^0(x^s) \geq \langle E\{\xi^0(s)|x^s\},x-x^s\rangle,$$

from which the assertion follows.

Instead of $y(x^s,\omega^s)$ we can use also y^s such that $y^s \in Y$ and

$$g(x^s,y(x^s,\omega^s),\omega^s) - g(x^s,y^s,\omega^s) \leq \varepsilon_s,$$

where $\varepsilon_s \to 0$ as $s \to \infty$. It is easy to see that

$$\xi^0(s) = g_x(x^s,y,\omega^s)|_{y=y^s} \tag{6.61}$$

satisfy the condition (6.7). In (6.60), (6.61) we can apply also the approximations (6.52), (6.53), (6.55), (6.57) with $f^0(x,\omega) = \max_{y \in Y} g(x,y,\omega)$ for the gradient or subgradient g_x. According to the (6.60) for the objective function (6.58) we obtain the following expression for the $\xi^0(s) = (\xi^0_1(s), \ldots, \xi^0_n(s))$,

$$\xi^0_j(s) = a_{i_s j}(\omega^s), j = 1 : n \tag{6.62}$$

where

$$i_s = \operatorname*{argmax}_i \left[\sum_{j=1}^n a_{ij}(\omega^s) x^s_j + b_i(\omega^s) \right].$$

It means that for the stock problem (6.59) the scalar

$$\xi^0(s) = \begin{cases} \alpha, & \text{if } x^s \geq \chi^s \\ -\beta, & \text{if } x^s < \chi^s \end{cases} \tag{6.63}$$

and we have the following simple version of the method (6.11).

Let x^0 be an arbitrary initial approximation and x^s be the approximation obtained after the s-th iteration. A value χ^s is observed according to the distribution of the demand, for instance, through the Monte-Carlo simulation model. Since $X = [0, r]$, it is easy to perform the operation of projection onto X and get the new approximate solution

$$x^{s+1} = \max\{0, \min[r, x^s - \rho_s \xi^0(s)]\}, \quad s = 0, 1, \ldots \tag{6.64}$$

with the $\xi^0(s)$ computed according to the (6.63).

The usual approach to the solution of the problem (6.59) consists in the following. It is easily seen that

$$F^0(x) = \alpha \int_0^x (x - z) dH(z) + \beta \int_x^\infty (z - x) dH(z),$$

and if the $H(z)$ has the density (the distribution is absolutely continuous), then the function is found to be continuous differentiable. Then the solution is the nearest to the interval point satisfying the equation $F^0_x(x) = 0$, which is equivalent to the following

$$H(x) = \frac{\beta}{\alpha + \beta}.$$

If there exists an algorithm for calculating $H(x)$ then the solution of this equation presents no difficulties.

In applying the method (6.64) it is not required the differentiability of $F^0(x)$ the existence of the density. The distribution may also be given implicitly. it requires only observations $\chi^0, \chi^1, \ldots, \chi^s, \ldots$ and this feature makes the (6.64)-type methods applicable in cases when there is only the Monte-Carlo procedure available to simulate a possible demand. Consider the following problem which is discussed also in Chapter 21.

Suppose that we have to determine the amount x_i of materials, facilities, etc., required at points $i = 1 : n$ in order to meet a demand

$$\chi_i = \sum_{k=1}^{\ell} \varepsilon_k i,$$

where ε_{ki} is the random flow of users from the residence point $k = 1 : \ell$ to the demand point $i = 1 : n$. The users of residence point k are choosing the point i with given probability p_{ki}, $k = 1 : \ell, i = 1 : n$, and there are also relations

$$\sum_{i=1}^{n} \varepsilon_{ki} = b_k, k = 1 : \ell,$$

where b_k is the random quantity with known distribution function. The problem is to determine the size x_i in order to minimize the cost function

$$F^0(x^1, \ldots, x^n) = \sum_{i=1}^{n} E \max\{\alpha_i(x_i - \chi_i), \beta_i(\chi_i - x_i)\}$$

subject to $0 \le x_i \le r_i$, $i = 1 : n$.

The algorithm (6.11) with $\xi^0(s)$ as (6.62) takes the similar to the (6.64) form

$$x_i^{s+1} = \max\{0, \min[r_i, x_i^s - \rho_s \xi_i^0(s)]\}, \tag{6.65}$$

$$\xi_i^0(s) = \begin{cases} \alpha_i, & \text{if } x_i^s \ge \chi_i^s \\ -\beta_i, & \text{if } x_i^s < \chi_i^s, \end{cases}$$

$$\chi_i^s = \sum_{k=1}^{\ell} \varepsilon_{ki}^s, \sum_{i=1}^{n} \varepsilon_{ki}^s = b_k^s, k = \overline{1, \ell}, i = \overline{1, n},$$

where b_k^s, ε_{ki}^s, χ_i^s are observations of the amount of users at point k, the flow variables and the demand at point i respectively.

We note again that for the procedure (6.65) the distribution of the demand χ_i need not to be known: it is sufficient to have only a sequence of independent observations $\chi_i^0, \chi_i^1, \ldots, \chi_i^s, \ldots$ for each $i = 1 : n$. This circumstance allows us to solve by SQG methods fairly general inventory control problems (see [5], [7]). In the above discussed problem the distribution of the demands is hard to be found.

6.9 Recourse Problems

One of the simplest recourse problem (see chapter 1) may be formulated in the following way: to find a vector $x \geq 0$ minimizes the function

$$F^0(x) = \langle c, x \rangle + E \min_y \{ \langle q, y \rangle | Tx + Wy \leq h, y \geq 0 \}, \qquad (6.66)$$

where all elements of T, W, h, q, may be random variables. Here the decision x is made in advance, before observation of $\omega = (T, W, h, q)$, a corrective solution y is derived from the known ω and x.

Consider more general problem with the objective function

$$F^0(x) = E \min_y \{ g^0(x, y, \omega) | g^i(x, y, \omega) \leq 0, \quad i = 1 : m, y \in Y \}, \qquad (6.67)$$

where $g^\nu(\cdot, \omega), \nu = 0 : m$ are convex functions, Y a convex compact set.

Suppose that for each (x, ω) there is a feasible second stage solution y (we can always obtain it by introducing some additional variables) and a saddle point $(y(x, \omega), u(x, \omega))$ of the Lagrange function

$$g^0(x, \cdot, \omega) + \sum_{i=1}^m u_i g^i(x, \cdot, \omega),$$

where $y(x, \omega)$ is a second stage solution. Then an estimate of a subgradient of the function (6.67) takes on the form

$$\xi^0(s) = g_x^0(x^s, y, \omega^s) + \sum_{i=1}^m u_i(x^s, \omega^s) g_x^i(x^s, y, \omega^s)|_{y=y(x^s, \omega^s)} \qquad (6.68)$$

Let us show that (under reasonable assumptions of measurability and integrability) for the vector (6.68)

$$F^0(x) - F^0(x^s) \geq \langle E\{\xi^0(s)|x^s\}, x - x^s \rangle.$$

We have

$$g^0(x, y(x, \omega), \omega) = g^0(x, y(x, \omega), \omega) + \sum_{i=1}^m u_i(x, \omega) g^i(x, y(x, \omega), \omega)$$

for all (x, ω). Let us denote $q(x, \omega) = g^0(x, y(x, \omega), \omega)$. Then, taking into account the last relation, we have

$$q(x, \omega^s) - q(x^s, \omega^s) \geq g^0(x, y(x, \omega^s), \omega^s) - g^0(x^s, y(x^s, \omega^s), \omega^s)$$

$$+ \sum_{i=1}^m u_i(x^s, \omega^s)[g^i(x, y(x, \omega^s), \omega^s) - g^i(x^s, y(x^s, \omega^s), \omega^s)]$$

$$\geq \langle g_x^0(x^s, y(x^s, \omega^s), \omega^s) + \sum_{i=1}^m u_i(x^s, \omega^s) g_x^i(x^s, y(x^s, \omega^s), \omega^s), x - x^s \rangle$$

$$+ \langle g_y^0(x^s, y(x^s, \omega^s), \omega^s)$$

$$+ \sum_{i=1}^m u_i(x^s, \omega^s) g_y^i(x^s, y(x^s, \omega^s), \omega^s), y(x, \omega^s) - y(x^s, \omega^s) \rangle.$$

Since $y(x^s, \omega^s)$ minimizes the Lagrange function, then we get

$$q(x, \omega^s) - q(x^s, \omega^s) \geq \langle g_x^0(x^s, y(x^s, \omega^s), \omega^s)$$

$$+ \sum_{i=1}^m u_i(x^s, \omega^s) g^i(x^s, y(x^s, \omega^s), \omega^s), x - x^s \rangle.$$

The assertion now follows from taking conditional expectation on both side of this inequality.

From the formula (6.68) for the function (6.66) we get the estimate

$$\xi^0(s) = c + u(x^s, \omega^s) T(\omega^s) \tag{6.69}$$

where $\omega^0, \ldots, \omega^s, \ldots$ are mutually independent samples of ω and the $u(x^s, \omega^s)$ are a dual variables corresponding to a second-stage optimal plan $y(x^s, \omega^s)$. From formula (6.69) and the convergence of the procedure given by (6.11) we can obtain the following method for solving a recourse problem.

(i) For given x^s observe the random realization of h, q, T, W, which we note as $h(s), q(s), T(s), W(s)$;

(ii) solve the problem

$$\langle q^s, y \rangle = \min,$$

$$W(s)y \leq h(s) - T(s)x^s,$$

$$y \geq 0$$

and calculate the dual variables $u(x^s, \omega^s)$.

(iii) Get

$$\xi^0(x) = c + u(x^s, \omega^s) T(\omega^s)$$

and change x^s:

$$x^{s+1} = \max[0, x^s - \rho_s \xi^0(x)]. \tag{6.70}$$

It is worthwhile to note that this method can be regarded as a stochastic iterative procedure for the decomposition of large scale problems. For instance, if ω has a discrete distribution, i.e., $\omega \in \{1, 2, \ldots, N\}$ and $\omega = k$ with probability p_k, then the recourse problem (6.66) is equivalent to the problem:

$\langle c, x \rangle$	$+p_1 \langle q(1), y(1) \rangle$	$+p_2 \langle q(2), y(2) \rangle \ldots$	$+p_N \langle q(N), y(N) \rangle$	$= \min$
$T(1)x$	$+W(1)y(1)$			$\leq h(1)$
$T(2)x$		$+W(2)y(2)$		$\leq h(2)$
$\ldots$	$\ldots$	$\ldots$	$\ldots$	$\ldots$
$T(N)x$			$+W(N)y(N)$	$\leq h(N)$
$x \geq 0,$	$y(1) \geq 0,$	$y(2) \geq 0,$	$y(N) \geq 0,$	

were $y(k)$ is the correction of the plan x if $\omega = k$. The number N may be very large. If only the vector $h = (h_1, \ldots, h_m)$ is random and each of the components has two independent outcomes, then $N = 2^m$. Then the SQG procedure (6.70) allows us to solve extremely large-scale problems.

The formula (6.69) is also applicable for the dynamic version (see chapter 1) of the problem (6.66): find a sequence $x = (x(0), x(1), \ldots, x(T))$ minimizing the function

$$F^0(x) = \sum_{t=0}^{T} [< c(t), x(t) > + E \min_y (t) \{\{d(t), y(t)\}\} | \sum_{k=0}^{t} [A_t k x(k) \tag{6.71}$$
$$+ B_t y(t)] \leq h(t), y(t) \geq 0\}$$

subject to $\quad x(t) \geq 0, t = 0 : T$.

The estimate takes on the form:

(i) For given $x^s = (x^s(0), x^s(1), \ldots, x^s(T))$ observe a random realization of $d(t), h(t) \ A_t k, B_t$ for $k = 0 : t, t = 0 : T$, which we denote as $d^s(t), b^s(t), A_t k^s, B_t^s$;

(ii) Solve the problem

$\langle d^s(t), y(T) \rangle = \min,$

$\sum_{k=0}^{t} [A_{tk}^s x^s(k) + B_t^s y(t)] \leq h^s(t),$

$y(t) \geq 0,$

for $t = 0 : T$ and calculate the dual variables $u^s(t)$ to an optimal solution $y^s(t)$.

(iii) Calculate

$$\xi^{0k}(s) = c(k) + \sum_{t=k}^{T} u^s(t) A_{tk}^s, \quad k = 0 : T.$$

The vector $\xi^s(0) = (\xi^{00}(s), \ldots, \xi^{0T}(s))$ is an estimate of a subgradient $F_x^0(x^s)$ (according to the rule (6.69)).

Therefore the method (6.70) applying to the problem (6.71) takes on the form: in addition to (i)–(iii) change x^s according to the formula

$$x^{s+1}(t) = \max[0, x^s(t) - \rho_s \xi^{0t}(s)], t = 0 : T$$

and repeat (i)–(iii) with $x^{s+1} = (x^{s+1}(0), x^{s+1}(1), \ldots, x^{s+1}(T))$, etc.

The general formula (6.68) as well as (6.61) can also be modified according to all universal rules discussed in the Section 6.7.

6.10   Stochastic Problems with Composite Functions

Until now we have discussed solution procedures for the problem (6.48)–(6.50)
assuming that we know exact values of random function $f^\nu(x,\omega)$, $v = 0 : m$ for
fixed x, ω. Meanwhile there are important problems in which these values are
not known—problems with so-called composite (objective, constraints) function
F^ν of the following structure

$$
\begin{aligned}
F^\nu(x) &= Ef^\nu(x,\omega), \nu = 0 : m, \\
f^\nu(x,\omega) &= q^\nu(Eg^1(x,\omega),\dots,Eg^\ell(x,\omega),x,\omega),
\end{aligned}
\tag{6.72}
$$

where some of functions $g^1,\dots,g^\ell$ itself may have the same type of structure,
etc.

The penalty functions of the problem (6.48)–(6.59), for instance

$$
Ef^0(x,\omega) + C\sum_{i=1}^m E\min\{0, Ef^i(x,\omega)\}
$$

are examples of such objective function.

The moments

$$
E\prod_{k=1}^\ell [g^k(x,\omega) - Eg^k(x,\omega)]^{\tau_k}, \tau_k \geq 0
$$

are also such type of functions, where $g^1(x,\omega),\dots,g^\ell(x,\omega)$ are given random
functions.

For composite functions there are difficulties with computing the estimates
of values $F^\nu(x^s)$. The averaging operation allows us to overcome these difficulties in the similar way to the procedure (6.16), (6.17). Consider an illustrative
example (see [5]:pp 201–215 for more details).

Let us suppose that there are mathematical expectations of only two levels:
in functions q^ν and F^ν, therefore let us suppose that the values $g^1(x,\omega),\dots,g^\ell$
(x,ω) are calculated exactly for each (x,ω) and let $\{x^s\}_{s=0}^\infty$ be a bounded sequence of approximate solutions. Define the estimates $\bar{g}(s)$ by the formula

$$
\bar{g}(s+1) = \bar{g}(s) + \psi_s[g(x^{s+1},\omega^{s+1}) - \bar{g}(s)], s = 0,1,\dots
\tag{6.73}
$$

where $g(x^s,\omega^s) = (g^1(x^s,\omega^s),\dots,g^\ell(x^s,\omega^s))$. According to the general result
(6.23) under rather general requirements with probability 1

$$
\|\bar{g}(s) - E\{g(x^s,\omega^s)|x^s\}\| \to 0, \text{ for } s \to \infty
$$

Therefore, $\bar{g}(s)$ is an estimate of the $Eg(x^s,\omega)$ and $q^\nu(\bar{g}(s),x^s,\omega)$ can be used
as an estimate of the value $f^\nu(x^s,\omega)$.

Assume that

(i) $q^\nu(\cdot,\omega)$, $g^k(\cdot,\omega)$, $\nu = 0 : m$, $k = 1 : l$ are continuously differentiable functions, $\forall \omega \in \Omega$

(ii) values $q_x^\nu(z,x,\omega)$, $q_z^\nu(z,x,\omega)$, $g_x^k(x,\omega)$, $\nu = 0 : m$, $k = 1 : \ell$ are calculated exactly.

Then vectors

$$\xi^\nu(s) = q_x^\nu(\bar{g}(s), x^s, \omega) + \sum_{k=1}^{\ell} q_{z_k}(\bar{g}(s), x^s, \omega) g_x^k(x^s, \omega)$$

can be used as estimates of $F_x^\nu(x^s)$ in different type of solution procedures discussed in Sections 6.2-6.3. There might be also modifications with finite-difference approximations of gradients involved in the $\xi^\nu(s)$ and generalizations to the case of nondifferentiable functions.

6.11 Problems of Optimal Control

From the discussion of the Chapter 1 it follows that rather general problems of optimal control with discrete time can also be viewed as the (6.48)-(6.50) type of problem with implicitly given objective and constraints functions. We consider only discrete time problems here. Continuous time problems are very often only mathematical approximations of really discrete time problems obtained in the hope of simplifying analytical formulas. From the computational point of view they are required to be approximated by discrete time analogous again. Under natural assumptions an optimal control law of continuous time problems (without "pathologies") can be approximated (in terms of objective function) by optimal control law of the discrete time analogous (see [33], [29]).

Suppose we are interested only in time values $t = 0, 1, 2, \ldots, T$ and variables $\bar{x} = (x(0), x(1), \ldots, x(T))$, $\bar{z} = (z(0), z(1), \ldots, z(T))$ represent control actions and the state of the system over a given time-horizon, respectively. The problem is to find a control $x(t), t = 0 : T$ as a function of t which minimizes the objective function

$$F^0(\bar{x}) = E\gamma^0(z(0), \ldots, z(T), x(0), \ldots, x(T-1), \omega) \qquad (6.74)$$

subject to

$$F^i(\bar{x}) = E\gamma^i(z(0), \ldots, z(T), x(0), \ldots, x(T-1), \omega) \leq 0, \quad i = 1 : m, \qquad (6.75)$$

where variables x, z are connected by the system of stochastic equations

$$\begin{aligned} z(t+1) &= g(t, z(t), x(t), \omega), \\ z(0) &= z^0, t = 0, 1, \ldots, T-1. \end{aligned} \qquad (6.76)$$

with the vector function $g = (g^1, \ldots, g^r)$. Suppose that $g^k(t, y, x, \omega)$, $k = 1 : r$ are continuously differentiable functions with respect to (y, x) for $t = 1 : T-1$, $\omega \in \Omega$ and let $\gamma_{(z,x)}^\nu(\bar{z}, \bar{x}, \omega)$ be a subgradient of $\gamma^\nu(\bar{z}, \bar{x}, \omega)$. According to the equation (6.76) $\bar{z} = (z(0), \ldots, z(T))$ is implicit function of $\bar{x} = (x(0), \ldots, x(T-1))$, therefore,

$$\gamma^\nu(\bar{z}, \bar{x}, \omega) = f^\nu(\bar{x}, \omega).$$

If we denote components of vectors $\gamma_{(z,x)}^\nu$, $f_{\bar{x}}^\nu(\bar{x}, \omega)$ as the following

$$\gamma_{(z,x)}^\nu = \left(\gamma_{z(0)}^\nu, \ldots, \gamma_{z(T)}^\nu, \ldots, \gamma_{x(0)}^\nu, \ldots, \gamma_{x(T-1)}^\nu\right)$$

$$f_{\bar{x}}^\nu = \left(f_{x(0)}^\nu, \ldots, f_{x(T-1)}^\nu\right),$$

then in rather general cases (see [5], [29], [33])

$$f_{x(t)}^\nu = \gamma_{x(t)}^\nu(\bar{z}, \bar{x}, \omega) - \sum_{k=1}^r \lambda_k(t+1) g_x^k(t, z(t), x(t), \omega), \qquad (6.77)$$

where

$$\begin{cases} \lambda(t) & = \sum_{k=1}^r \lambda_k(t+1) g_z^k(t, z(t), x(t), \omega) - \gamma_z(t)^\nu(\bar{z}, \bar{x}, \omega), \\ \lambda(T) & = -\gamma_{z(T)}^\nu(\bar{z}, \bar{x}, \omega), \quad t = T-1, T-2, \ldots, 0. \end{cases} \qquad (6.78)$$

For instance, if, in addition, vectors $\gamma_{x(t)}^\nu(\bar{z}, \bar{x}, \omega)$, $\gamma_{z(t)}^\nu(\bar{z}, \bar{x}, \omega)$, $g(t, z, x)$, $g_x^k(z, x, \omega)$, $g_z^k(z, x, \omega)$, $\nu = 0 : m$, $k = 1 : r$ are bounded in any bounded set of $(\bar{z}, \bar{x})$; functions $\gamma^\nu(\bar{z}, \bar{x}, \omega)$ are weakly convex with respect to $(\bar{z}, \bar{x})$, then functions $F^\nu(\bar{x})$ are weakly convex and its subgradient

$$F_x^\nu(\bar{x}) = E f_x^\nu(\bar{x}, \omega)$$

Therefore, the estimate of the subgradient F_x^ν

$$\xi^\nu(s) = \left(f_{x(0)}^\nu(\bar{z}^s, \bar{x}^s, \omega^s), \ldots, f_{x(T-1)}^\nu(\bar{z}^s, \bar{x}^s, \omega^s)\right)$$

where $\bar{x}^s$ is the current approximate solution, ω^s is independent of B_s an observation of ω, $\bar{z}^s$ is the solution of the equation (6.76) for given $\omega = \bar{\omega}$, $x = \bar{x}^s$.

And again instead of exact gradient g_x^k, g_z^k and subgradients $\gamma_{x(t)}^\nu$, $\gamma_{z(t)}^\nu$ their finite difference approximations might be applied (see Section 6.7).

Consider more concrete example of problem (6.74)–(6.76). Suppose the system's equations are linear

$$g(t, z(t), x(t), \omega) = A(t, \omega) z(t) + B(t, \omega) x(t) + C(t, \omega), \qquad (6.79)$$

$$\gamma^0(\bar{z}, \bar{x}, \omega) = \max_{0 \le t \le T} \|z(t) - z^*(t)\|^2$$

where $z^*(t)$ is a given (observed or prescribed) trajectory.

The problem is to find such a vector $\bar{x}$ minimizes the expected deviation from the trajectory $z^*(t)$, $t = 0 : T$.

$$F^0(x) = E \max_{0 \le t \le T} \|z(t) - z^*(t)\|^2. \qquad (6.80)$$

The system (6.78) takes on the form

$$\begin{cases} \lambda(t) & = \lambda(t+1)A(t,\omega) + f_{z(t)}^0 \\ \lambda(T) & = -f_{z(t)}^0, t = T-1,\dots,0. \end{cases} \quad (6.81)$$

where

$$f_z(t)^0 = \begin{cases} 2(z(t) - z^*(t)), & \text{if } \|z(t) - z^*(t)\| = \max_{0 \le t \le T} \|z(t) - z^*(t)\|, \\ 0, & \text{in other case}, \end{cases}$$

for $t = 0 : T$. A stochastic subgradient $\xi^0(s) = (f_{z(0)}^0,\dots,f_{z(T-1)}^0)$

$$f_{z(t)}^0 = -\lambda(t+1)B(t,\omega).$$

Therefore, the method (6.11) applying to this problem is reduced to the following computation. Suppose that $x(t) \in X(t)$—a convex compact set.

(i) Let $\bar{x}^s$ be the current approximate solution. Make observation ω^s and find the solution $z^s(t)$, $t = 0 : T$ of the equations

$$z^s(t+1) = A(t,\omega^s)z^s(t) + B(t,\omega^s)x^s + C(t,\omega^s)$$

$$z(0) = z^0, t = 0 : T-1$$

(ii) Find the solution $\lambda^s(t)$, $t = T, T-1,\dots,0$ of the equations (6.81) with $z(t) = z^s(t)$, $t = 0 : T$, $\omega = \omega^s$.

(iii) Compute $\xi^0(s) = (f_{z(0)}^\nu,\dots,f_{z(T-1)}^\nu)$,

$$f_{z(t)}^0 = -\lambda^s(t+1)B(t,\omega^s)$$

and the new approximate solution $\bar{x}^{s+1} = (x^{s+1}(0),\dots,x^{s+1}(T-1))$, where

$$x^{s+1}(t) = \pi_{X(t)}[x^s(t) + \rho_s\lambda^s B(t,\omega^s)], \quad T = 0 : T-1, s = 0,1,\dots$$

6.12   Optimization Involving a Preference Structure

Many complex decision problems involve multiple conflicting objectives. Generally we cannot optimize several objective simultaneously; for instance, minimize cost and at the same time maximize reliability. If we can find some function (utility function) that combines all objectives into a scale index of preferability, then the problem of decision making can be put into the format of the standard optimization problem: to find $x \in X$ to optimize the utility function. The finding of a utility function may be a very difficult problem and often it is easy to have a preference ordering (preference structure) among feasible solutions $x \in X$ and deal with this structure directly to get the preferred solution. This ordering may be based on the decision maker's judgment or other rules. So let us assume that the decision maker has a preference structure at different points $x \in X$ and there exists a utility function (unknown) $U(x)$ such that

$$x' \sim x'' \Longleftrightarrow U(x) = U(x''), x' \succsim x'' \Longleftrightarrow U(x') > U(x'').$$

Consider the procedure

$$x^{s+1} = \pi_X\left(x^s + \rho_s \xi^0(s)\right),$$

$$\xi^0(s) = \begin{cases} h_s & \text{if } x^s + \Delta_s h^s \succsim x^s, \\ -h_s & \text{if } x^s + \Delta_s h^s \precsim x^s, \text{ for some } \Delta_s > 0. \end{cases}$$

where $h^0, h^1, \ldots, h^s, \ldots$ are the results of independent samples of the random vector $h = (h_1, \ldots, h_n)$ uniformly distributed over the unit sphere. It can be shown [7] that

$$\frac{E\{\xi^0(x)|x^s\} = \alpha U_x(x^s)}{\|U_x(x^s)\|},$$

for differentiable $U(x)$, where α is positive number. Therefore, the convergence of this procedure follows from the general conditions of the procedure given by (6.11) (with small corrections). A series of similar procedures for general constrained problems with nondifferentiable utility functions was investigated in [64].

6.13 Mathematical Statistics Problems and the Stochastic Optimization

Many problems of the mathematical statistics can be formulated as special cases of the stochastic optimization problems. Such interpretation allows us to bring ideas of mathematical statistics into stochastic optimization. Simultaneously, it gives opportunity to apply the developed optimization technique in the mathematical statistics. Consider some possible applications of SQG methods (see also [5], [28]). These methods allow us to construct iterative procedures which can be performed on line and can use *a priori* information concerning the structure of the system for improving estimates.

Many problems of statistical estimation deal with the problem of estimating the true value z^* of unknown parameters from the elements of a sample $\theta^0, \theta^1, \ldots, \theta^s, \ldots$ assumed to have been drawn from a distribution function $H(y, z^*) = P\{\theta \leq y\}$. There may be different formulations of optimization problems concerning such problems of estimation depending on our knowledge about $H(y, z^*)$.

(i) There is no information about $H(y, z^*)$ except the sample $\theta^0, \theta^1, \ldots, \theta^s, \ldots$ and $z^* = E\theta$. Therefore the problem is to estimate z^*, where

$$\theta^s = z^* + \varepsilon(s), E\varepsilon(s) = 0, \quad s = 0, 1, \ldots$$

The required parameter z^* minimizes the function

$$F^0(x) = E\|x - \theta\|^2, \tag{6.82}$$

because $x = E\theta$ satisfies the optimality conditions

$$F^0_{x_i} = 2(x_i - E\theta_i) = 0, \quad i = 1 : n.$$

If *a priori* knowledge about the unknown x is introduced as $x \in X$, then from (6.11) we could obtain the following iterative procedure for finding z^* (with $\xi^0(s) = 2(x^s - \theta^{s+1})$):

$$x^{s+1} = \pi_X(x^s - \rho_s(x^s - \theta^{s+1})), \quad s = 0, 1, \ldots. \tag{6.83}$$

If $X = R^n$, $\rho_s = \frac{1}{(s+1)}$, then

$$x^{s+1} = x^s - \frac{1}{s+1}(x^s - \theta^{s+1}) = \frac{1}{s+1}\sum_{k=1}^{s+1}\theta^k \tag{6.84}$$

The estimation (6.84) is the sample mean. The advantages of the estimation (6.83) when compared to (6.84) are

(a) if $X \neq R^n$, then from (6.83) it follows that $x^s \in X$ for all $s = 0, 1, \ldots$, whereas in (6.84) only $\lim x^s \in X$. Therefore the estimations from (6.83) must be better for small samples.

(b) possibilities of choosing ρ_s as a function of $(x^0, \ldots, x^s)$ in order to decrease the value of the objective function (6.82). It can be done by using adaptive ways of choosing ρ_s (interactive or automatic), as it is described in Section 6.6. It leads to different nonlinear estimations of z^* in contrast to the estimate (6.84) which is the linear function of observations.

Problems of estimation of the moments

$$E\theta^\ell, E|\theta|^\ell, E(\theta - E\theta)^\ell$$

may also be formulated as minimization problems

$$F_1^0(x) = \|x - \theta^\ell\|^2, F_2^0(x) = E\|x - |\theta|^\ell\|^2,$$
$$F_3^0(x) = E\|x - (\theta - E\theta)^\ell\|^2,$$

where for the sake of simplicity we denote

$$\theta^\ell = (\theta_1^\ell, \ldots, \theta_n^\ell), |\theta|^\ell = (|\theta_s|^\ell, \ldots, |\theta_n|^\ell), (\theta - E\theta)^\ell$$
$$= ((\theta_1 - E\theta_1)^\ell, \ldots, (\theta_n - E\theta_n)^\ell).$$

The stochastic gradients of these functions are:

$$\xi_1^0(s) = 2(x^s - (\theta^{s+1})^\ell), \xi_2^0 = 2(x^s - |\theta^{s+1}|^\ell),$$
$$\xi_3^0(s) = 2(x^s - \prod_{k=1}^{\ell} (\theta^{s+1} - \theta^s + 1 + k)).$$

(ii) Suppose now that we have the information

$$E\theta = V(z)|_{z=z^*},$$

where $V(z)$ is a given function and z^* is an unknown vector. Then z^* minimizes the function

$$E\|V(z) - \theta\|^2.$$

(iii) If we have information about the density $p(y, z^*)$ of $H(y, z^*)$ with a measure $\mu(dy)$, then it could be shown that z^* maximizes the function

$$E \ln p(x, \theta) = \int \ln p(x, y)p(y, z^*)\mu(dy).$$

These problems are reformulations of well-known principles for the least square, i.e., minimization of the function

$$\frac{1}{N} \sum_{k=1}^{N} \|V(z) - \theta^k\|^2$$

and maximum likelihood, i.e., maximization of the function

$$\frac{1}{N} \sum_{k=1}^{N} \ln p\left(x, \theta^k\right).$$

It gives us a good opportunity to apply SQG methods.

The above mentioned problems are the problems of pure estimation. Very often the main reasons for estimation and identification are control or optimization. In some cases, the task of optimization and estimation can be separated and optimization is performed after estimation. However, in the problems of adaptation it is usually necessary to optimize and estimate simultaneously. For instance, optimization cannot be separated from estimation if the observation of unknown parameters depends on the current value of the control variables.

Arising in such environment optimization task requires the development of a new optimization technique which have much in common with minimization of time-varying functions—the nonstationary optimization (see Section 6.4).

Consider an illustrative example—minimization of the differentiable function

$$F^0(x) = \psi(x, z^*), \quad x \in R^n$$

where z^* is a vector of unknown parameters. At each iteration $s = 0, 1, \ldots$, an observation θ^s is available which has the form of a direct observation of the parameter vector z^*:

$$E\theta^s = z^*, \quad s = 0, 1, \ldots \tag{6.85}$$

The problem is to create a sequence $\{x^s\}_{s=0}^{\infty}$ which converges to the set of optimal solutions. Note that $F^0(x)$ cannot be optimized directly because of the unknown parameters z^*. However, at iteration s we could obtain a statistical estimate z^s such that $z^s \to z^*$ with probability 1 and a sequence of functions $F^0(x, s) = \psi(x, z^s)$ such that

$$F^0(x, s) \to F^0(x)$$

with probability 1 for $s \to \infty$.

Let us notice that at iteration s only the function $F^0(x, s)$ is available. Therefore we led to the procedures of the nonstationary optimization

$$x^{s+1} = x^s - \rho_s F_x^0(x^s, s), \quad s = 0, 1, \ldots \tag{6.86}$$

$$F_x^0(x, s) = \psi_x(x, z^s).$$

In the case of stochastic programming problems z^* may correspond to the vector of unknown parameters of the probability measure $P(\cdot, z^*)$

$$\psi(x, z^*) = \int f^0(x, \omega) P(d\omega, z^*).$$

If $\{z^s\}$ is a sequence of estimates $z^s \to z^*$ with probability 1, then we led to the following type procedures

$$x^{s+1} = x^s - \rho_s \xi^0(s),$$

where $\xi^0(s)$ is an estimate of the $F_x^0(x, s)$ at $x = x^s$,

$$F^0(x, s) = \psi(x, z^s) = \int f^0(x, \omega) P(d\omega, z^s).$$

For instance, similar to the Section 6.7,

$$\xi^0(s) = f_x^0(x^s, \omega^s),$$

where ω^s is an independent of the B_s sample of the ω drawn from the non-stationary distribution $P(\cdot, z^s)$. We can also use more complicated estimates (similar to (6.52), (6.53)) More difficult problems arise when θ^s, $s = 0, 1, \ldots$ are not direct observations of the vector z^*. In other words, if, instead of the relationship (6.85), we have the following (see [20], [75], [76]).

$$E\{\theta^s | x^s\} = \varphi(x^s, z^*),$$

which may depend on the current approximate solution x^s. Since we do not know x^s in advance, then the (6.86) type procedure that directly solves an optimization problem and simultaneously estimates the z^* is needed again.

References

[1] Yu.M. Ermoliev and Z.V. Nekrylova, "The Method Stochastic Subgradients and its Applications". Notes, Seminar on the Theory of Optimal Solution. Academy of Sciences of the U.S.S.R., Kiev (1967).

[2] Yu.M. Ermoliev and N.Z. Shor, "Method of random walk for two-stage problem and its generalization", *Kibernetika*, 1(1968).

[3] Yu.M. Ermoliev, "On the stochastic quasi-gradient method and stochastic quasi-Feyer sequences", *Kibernetika*, 2(1969).

[4] Yu.M. Ermoliev, "General problem of stochastic programming", *Kibernetika*, 3(1971).

[5] Yu.M. Ermoliev, *Stochastic Programming Methods*. Moscow: Nauka (1976).

[6] A.M. Gupal and L.G. Bajenov, "Stochastic linearization", *Kibernetika*, 1(1972).

[7] Yu.M. Ermoliev and A.I. Jastremskiy, *Stochastic Models and Methods in Economic Planning*. Moscow: Nauka(1979).

[8] A.M. Gupal, "On the stochastic programming problem with constraints", *Kibernetika*, 6(1974).

[9] A.M. Gupal, "Methods of almost-differentiable function minimization", *Kibernetika*, 1(1977).

[10] A.M. Gupal, *Stochastic Methods of Nonsmooth Optimization*. Kiev: Naukova Dumka(1979).

[11] E.A. Nurminski, "Convergence conditions of algorithms of stochastic programming", *Kibernetika*, 3(1973).

[12] E.A. Nurminski, "Quasigradient method for solving problems on nonlinear programming", *Kibernetika*, 1(1973).

[13] E.A. Nurminski, *Numerical Methods for Solving Deterministic and Stochastic Minimax Problems*. Kiev: Naukova Dumka(1979).

[14] Yu.M. Ermoliev and E.A. Nurminski, "Limit extremal problems", *Kibernetika*, 1(1973).

[15] A.M. Gupal, "Optimization method for problems with time-varying functions", *Kibernetika*, 2(1974).

[16] E.A. Nurminski, "The problem of nonstationary optimization", *Kibernetika*, 2(1977).

[17] P.I. Vertchenko, "Limit Extremum Problems of Stochastic Optimization", Abstract of dissertation, Press of the Institute of Cybernetics, Kiev(1977).

[18] A.A. Gaivoronskiy, "Methods of Stochastic Nonstationary Optimization", Collection: Operations Research and Systems Reliability. Press of the Institute of Cybernetics, Kiev(1978).

[19] A.A. Gaivoronskiy, "Nonstationary stochastic programming problems", *Kibernetika*, 4(1978).

[20] A.A. Gaivoronskiy and Yu.M. Ermoliev, "Stochastic optimization and simultaneous parameter estimation", Izvestia Academii Nauk SSSR, Technischeskaj Kibernetika, 4(1979).

[21] E.A. Nurminski and P.I. Verchenko, "On a convergence of saddle-point algorithms", *Kibernetika*, 3(1977).

[22] A.N. Golodnikov, "Finding of Optimal Distribution Function in Stochastic Programming Problems", Abstract of dissertation, Institute of Cybernetics press, Kiev(1979).

[23] Yu.M. Ermoliev and Yu.M. Kaniovskiy, "Asymptotic behavior of stochastic programming methods with permanent step-size multiplier", *USSR, Computational Mathematics and mathematical Physics*, 2(1979).

[24] Yu.M. Kaniovskiy, P.S. Knopov, and Z.V. Nekrylova, *Limiting Theorems of Stochastic Programming Processes*. Kiev: Naukova Dumka(1980).

[25] L.G. Bajenov and A.M. Gupal, "Stochastic analog of conjugate gradients method", *Kibernetika*, 1(1972).

[26] A.M. Gupal, "Stochastic method of feasible directions of nondifferentiable optimization", *Kibernetika*, 2(1978).

[27] Yu.M. Ermoliev, "Stochastic models and methods of optimization", *Kibernetika*, 4(1975).

[28] Yu.M. Ermoliev and E.A. Nurminski, "Extremum Problems in Statistics and Numerical Methods of Stochastic Programming". Collection: Some Problems of Systems Control and Modeling. Press of the Institute of Mathematics, Ukrainian Academy of Sciences, Kiev(1973).

[29] Yu.M. Ermoliev, "The stochastic problem of optimal control", *Kibernetika*, 1(1972).

[30] sYu.M. Ermoliev and A.M. Gupal, "The linearization method in nondifferentiable optimization", *Kibernetika*, 1(1978).

[31] A.M. Gupal and V.P. Norkin, "Method of discontinuous optimization", *Kibernetika*, 2(1977).

[32] Yu.M. Ermoliev and E.A. Nurminski, "Stochastic quasigradient algorithms for minimax problems". Edited by M. Dempster. *Proceedings of the International Conference on Stochastic Programming*. London: Academic Press(1980).

[33] Yu.M. Ermoliev, V.P. Gulenko, and T.I. Tsarenko, *Finite-Difference Method in Optimal Control*. Kiev: Naukova Dumka(1978).

[34] N.Z. Shor, "Application of the Gradient Method for the Solution of Network Transportation Problems". Notes, Scientific seminar on theory and applications of cybernetics and operations research. Kiev: Academy of Sciences USSR(1962).

[35] Yu.M. Ermoliev, "Methods of solution of nonlinear extremal problems", *Kibernetika* 4(1966).

[36] B.T. Poljak, "A general method for solving extremal problems", *Soviet Mathematic Doklady*, 8(1967).

[37] P. Wolfe and M.L. Balinski (eds.), *Nondifferentiable Optimization*. Mathematical Programming Study 3, North-Holland Publishing Co(1975).

[38] N.Z. Shor, *The Methods of Nondifferentiable Optimization and their Applications*, Kiev: Naukova Dumka(1979).

[39] L.G. Bajenov, "Convergence conditions of almost-differentiable function minimization", *Kibernetika*, 4(1972).

[40] B.T. Poljak, "Convergence and rate of convergence of iterative stochastic algorithms", *Augomatic and Remote Control*, 12(1976).

[41] V.Ya. Katkovnik, *Linear Estimation and Stochastic Optimization Problems*, Moscow: Nauka(1976).

[42] L.A. Rastrigin, *Extremal Control Systems*. Moscow: Nauka(1974).

[43] V.V. Fjedorov, *Numerical Methods of Maxmin Problems*, Moscow: Nauka (1979).

[44] B.T. Poljak, "Nonlinear programming methods in the presence of noise", *Mathematical Programming*, 14(1978).

[45] M.T. Wasan, "Stochastic Approximations", *Cambridge Transactions in Math. and Math. Phys. 58*. Cambridge: Cambridge University Press (1969).

[46] V. Fabian, "Stochastic Approximation of Constrained Minima". In: *Transactions of the 4th Prague Conference on Information Theory*, Statistical Decision Functions and Random Processes, 1965, (Prague, 1967).

[47] H.J. Kushner, "Stochastic approximation algorithms for constrained optimization problems", *Annals of Statistics*, 2(4)(1974).

[48] H.J. Kushner and T. Gavin, "Stochastic approximation type methods for

constrained systems: algorithms and numerical results", *IEEE Transactions Automatic Control*, **19**(1974).

[49] H.J. Kushner and E. Sanvincente, "Stochastic approximation of constrained systems with system and constraint noise", *Automatic*, **11**(1975).

[50] K. Marti, "On approximate solutions of stochastic dominance and stochastic penalty methods", *Proceedings of the IX. Mathematical Programming Symposium*. Amsterdam: North-Holland(1976).

[51] J-B. Hiriart-Urruty, "Contributions à la Programmation Mathematique: Cas Deterministe et Stochastique", Thèse, D.Sc. Mathematiques, Université de Clermont-Ferrand 11(1977).

[52] K.J. Arrow, L. Hurwicz, and H. Uzawa, eds. *Studies in Linear and Non-Linear Programming*, Stanford, Calif.: Stanford University Press(1958).

[53] V. Dupač, "A dynamic stochastic approximation method", *Annals of Mathematical Statistics*, **6**(1965).

[54] V. Dupač, "The continuous dynamic Robbins-Monroe procedure", *Kybernetika*, **12**, N6 (1976).

[55] K. Uosaki, "Some generalizations of dynamic stochastic approximation procedures", *The Annals of Statistics*, **2** N5 (1976).

[56] S. Fujita and T. Fukao, "Convergence conditions of dynamic stochastic approximation method for non-linear stochastic discrete-time dynamic systems", *IEEE Transactions on Automatic Control*, AC–17, N5 (1972).

[57] Yu.M. Ermoliev and G. Leonardi, "Some proposals for stochastic facility location models", *Mathematical Modeling* **3**(1982)407–420.

[58] R.T. Rockafellar and R. Wets, "Stochastic convex programming: singular multipliers and extended duality", *Pacific J. Math.* **62**(1976).

[59] P.D. Roberts, "Multilevel approaches to the combined problem of systems optimization and parameter identification", *International Journal of Systems Science*, **8**(3) (1977).

[60] Ya.Z. Tsypkin, *Adaptation and Learning in Automatic Systems*. New York: Academic Press (1971).

[61] Yu.M. Ermoliev, G. Leonardi, and J. Vira, "Stochastic Quasi-Gradient Methods Applied to a Facility Location Problem", WP-81-14. Laxenburg, Austria: International Institute for Applied Systems Analysis (1980).

[62] A.S. Nemizovsky and D.B. Judin, *Complexity of Problems and Efficiency of Optimization Methods*. Moscow: Nauka (1979).

[63] H.J. Kushner and D.S. Clark, "Stochastic approximation methods for constrained and unconstrained systems", *Applied Mathematical Sciences*, **26** (1978).

[64] M.V. Michalevitch, "On optimization nondifferentiable utility functions". Collection: Methods nondifferentiable and stochastic optimization. Press of the Institute of Cybernetics Ukrainian Academy of Sciences, Kiev(1979).

[65] W.I. Zangill, "Convergence conditions for nonlinear programming", *Mang. Sci.* **16**(1) (1969).

[66] T.S. Motzhin and J.J. Schanberg, "The relaxation method for linear inequalities", *Canad.J.Math.*, **6**,3 (1954).

[67] D.C. Karnopp, "Random search for optimization problems", *Automatica*, **1**(1963) 111–121.

[68] K. Marti, "On accelerations of the convergence in random search methods", Forschungsschwerpunkt simulation und optimierung deterministischer und stochastischer dynamischer systeme, Hochschule der Bundeswehr, 8014 Neubiberg(1980).

[69] Yu.M. Ermoliev, "Stochastic quasigradient methods and their applications to systems optimization", *Stochastics*, **9**(1983) 1–36.

[70] A. Ruszczynski and W. Syski, "Stochastic approximation algorithm with gradient averaging for unconstrained problems", *IEEE Transactions on Automatic Control* AC–28 (1983), pp.1097–1105.

[71] K. Kiwiel, "An aggregate subgradient method for nonsmooth convex minimization", *Mathematical Programming* **27**(1983), 320–341.

[72] E.A. Nurminski, *Numerical Methods for Solving Deterministic and Stochastic Minimax Problems*, Kiev: Naukova Dumka(1979).

[73] N.D. Chepurnoi, "Methods of nondifferentiable optimization with averaged subgradients", Abstract of dissertation, Press of the Institute of Cybernetics, Kiev(1982).

[74] V.I. Norkin, "Generalized gradient method in nonconvex nondifferentiable optimization", Abstract of dissertation, Press of the Institute of cybernetics, Kiev(1983).

[75] Yu.M. Ermoliev and A. Gaivoronskiy, "Simultaneous nonstationary optimization, estimation, and approximation procedures", CP-82-16. Laxenburg, Austria: International Institute for Applied Systems Analysis (1982).

[76] A.A. Gaivoronskiy, "Methods of nonstationary optimization", Abstract of dissertation, Press of the Institute of Cybernetics, Kiev(1979).

[77] A.M. Gupal, "On feasible directions methods for problems of nondifferentiable optimization", *Kibernetika*, **2**.

[78] L.G. Bajenov and A.M. Gupal, "Stochastic analog of feasible directions methods", *Kibernetica*, **4**(1973).

[79] A. Ruszczynski, "Feasible direction methods for stochastic programming problems", *Mathematical Programming* **19**(1980),220–229.

[80] A. Ruszczynski, "A linearization method for nonsmooth stochastic programming problems", Institute für Operation Research der Universität Zürich (1984).

[81] R.T. Rockafellar, "Favorable classes of Lipschitz continuous functions in subgradient optimization", WP-81-1. Laxenburg, Austria: International Institute for Applied Systems Analysis(1981).

[82] R.T. Rockafellar and R.J-B. Wets, "On the interchange of subdifferentiation and conditional expectation for convex functions", *Stochastics*(1982).

[83] J-B. Hiriart-Urruty, "About properties of the mean value functional and the continuos inf-involution in stochastic convex analysis", in: *Optmization Techniques Modeling and Optimization in the Service of Man*. Ed., J. Cea, Springer-Verlag Lecture Notes in Computer Science, Berlin(1976). pp.763–789.

[84] V. Fabian, "Stochastic approximation of minima with improved asymptotic speed", *Ann. Math. Statist.* **38**(1967) 191–200.

[85] V. Fabian, "On asymptotic normality in stochastic approximation", *Ann. Math. Statist.* **39**(1968) 1327–1332.

[86] R.T. Rockafellar, "Lagrange multipliers and subderivatives of optimal value functions in nonlinear programming", *Mathematical Programming Study* **17**(1982) 28–66.

[87] J. Gauvin and F. Dubean, "Differential properties of the marginal function in mathematical programming", *Mathematical Programming Study* **19**(1982)101–119.

[88] G.Ch. Pflug, On the determination of the step-size in stochastic quasigradient methods. CP-83-25. Laxenburg, Austria: International Institute for Applied Systems Analysis (1983).

[89] S. Uriasiev, Adaptive methods step-size regulation in stochastic optimization methods. Abstract of dissertation, Press of the Institute of Cybernetics, Kiev (1982).

[90] A.P. Korostelev, *Stochastic Recoursive Procedures (Local Properties).* Moscow: Nauka (1984).

CHAPTER 7

MULTIDIMENSIONAL INTEGRATION AND STOCHASTIC PROGRAMMING

I. Deák

Abstract

A survey of well-known techniques and some recent results in multidimensional integration is presented together with a list of references. Methods are investigated with emphasis on their applications in stochastic programming. Also some results are reported on the Monte Carlo computation of the distribution function and probabilities of rectangles in case of multinormal distribution.

7.1 Introduction

In several problems of stochastic programming the evaluation of some kind of n-dimensional integrals is required. Of course, multidimensional integration is necessary in many other fields, too. Generally when one takes more aspects of the problem into account at the same time and wants to obtain a kind of general assessment of the problem one is faced with multidimensional integration.

There are some survey papers on multidimensional integration, e.g. Haber [24], Halton [25] and also there are the books Stroud [49], Ermakov [15] and that of Davis and Rabinowitz [5]. Especially this last one can be recommended for interested readers. Unfortunately no recent attempt to give a survey of the state of the art is known to the author (the survey paper of Niederreiter summarizes only quasi Monte Carlo methods). Since the subject of multidimensional integration is rapidly extending and no unique solution procedure can be judged at present to be the best, it is necessary to give at least an overview of the main streams at the moment.

First we describe some problems in stochastic programming where evaluation of multidimensional integrals is required. In Section 7.3 general methods of multidimensional integration are discussed with emphasis on those applicable in higher dimensions. Here we point out the advantages and drawbacks of the methods. In Section 7.4 some general difficulties encountered in multidimensional integration are considered.

In order to show the power of the Monte Carlo method we present some results in computing the multinormal distribution function and probabilities of rectangles in Section 7.5. Finally in the last section solution strategies are given a possible user how she or he should choose the method depending on the problem and the number of dimensions.

7.2   Integration problems in stochastic programming

Problems of evaluating multidimensional integrals generally can be written as

$$\int_D g(\underline{x})v(\underline{x})dF(\underline{x}) \tag{7.1}$$

where D is a d-dimensional set, $g,(\underline{x})$ is a function to be integrated, $v,(\underline{x})$ is a weight function (sometimes $g(\underline{x})$ or $v(\underline{x})$ or both of them equal to 1), $F(\underline{x})$ is a distribution function of a probability distribution.

In probability-constrained models presented by Prékopa [44], distribution function values and its gradients are needed. For example in the following STABIL model (see Prékopa et al [45])

$$\text{minimize} \quad \min \underline{c}'\underline{x}$$
$$\text{subject to} \quad A\underline{x} = \underline{b}$$
$$\underline{x} \geq \underline{0}$$
$$P\{T\underline{x} \geq \underline{\xi}\}p,$$

where $\underline{\xi}$ is a random vector with distribution function F we need the evaluation of the integral

$$F(\underline{t}) = \int_{-\infty}^{t_d} \ldots \int_{-\infty}^{t_1} dF(\underline{x}).$$

In some other cases, e.g. in the approximative solution strategy devised by Kall [30] for the two-stage stochastic programming model probabilities of rectangles are used (i.e. D is a rectangle, $g(\underline{x}) = v(\underline{x}) = 1$ in (7.1)).

The case when D is a simplicial cone is of interest, since fast computation of such integrals would make possible another solution procedure for the two-stage model as it was pointed out by Wets [55].

The two-stage stochastic programming problem is the following

$$\text{minimize} \quad \underline{c}'\underline{x} + \psi(\underline{x})$$
$$\text{subject to} \quad \psi(\underline{x}) = E(Q(\underline{x},\underline{\xi})),$$
$$Q(\underline{x},\underline{\xi}) = \inf_{y}\{\underline{q}'\underline{y}|W\underline{y} = \underline{h} - T\underline{x}\}$$

where all components of $\underline{h}, q, W$ and T might be random variables, $\underline{\xi}$ is the random vector comprising all the random terms on the right hand side, and its distribution function will be denoted by Φ. Thus we have

$$\psi(\underline{x}) = \int Q(\underline{x},\underline{y})d\Phi(\underline{y}).$$

The evaluation of this function ψ seems to be no simple problem (the integrand is a sophisticated function, to obtain only one value one has to solve a linear programming problem). It is highly probable that here direct integration cannot be applied, only reformulation of the problem or approximation schemes suggested by Strazicky [48], Kall [30], Kall and Stoyan [31], Wets [56] could overcome this stumbling block.

7.3 General Methods of Multidimensional Integration

There is a wide variety of methods applicable only in low dimensions $d = 2, 3, 4$ with good effect. We deny by no means the merits of these techniques but from the point of view of stochastic programming they seem to be of little value if they cannot be generalized or applied in greater dimensions. Just to give some examples of works in this category we mention some references. Donelly [12] expanded and integrated the two-dimensional normal density function with a very high degree of precision. Milton [38] and Dutt [14] suggested methods for computing normal integrals up to six and four dimensions respectively. Similar work for two-dimensional cases were published by Brown [3] and Drezner [13]. Integral formulas for the three-dimensional sphere were developed by Freeden [19] and Lebedev [36]. See also the papers of Terras [53] and Tsuda [54].

In multidimensional integration an important role is played by the change of the order of integration, approximations to the integrand, the many different kinds of integrand transforms and composite integration rules. Since most of these methods seem to be specific ad hoc methods, in what follows we will focus on the general methods of multidimensional integration, especially on those applicable in higher dimensions $(d \geq 5)$.

7.3.1 Product Rules

By an integration rule $R(\underline{x}_i, w_i)$ we mean a set of points $\underline{x}_i$, weights w_i, $i = 1, \ldots, M$ and the approximation of the integral:

$$\int_D f(\underline{x}) d\underline{x} \approx \sum_{i=1}^{M} w_i f(\underline{x}_i). \qquad (7.2)$$

By a product rule we mean a product of two lower dimensional rules. More precisely assume that $D = B \times C$ where x denotes the Cartesian product, $B \subset R^{d_1}, C \subset R^{d_2}$, $d_1 + d_2 = d$ furthermore we have an integration rule $R_1(\underline{x}_{1i}, w_{1i})$ with M_1 points in B and another rule $R_2(\underline{x}_{2i}, w_{2i})$ with M_2 points. The $R = R_1 \times R_2$ produce rule consists of the $(\underline{x}_{1i}, \underline{x}_{2j})$ points with weights (w_{1i}, w_{2j}) $i = 1, \ldots, M_1, j = 1, \ldots, M_2$ and gives the approximations

$$\sum_{i=1}^{M_1} \sum_{j=1}^{M_2} w_{1i} w_{2j} f(\underline{x}_{1i}, \underline{x}_{2j}) \qquad (7.3)$$

Application of product rules, especially if the same, say, one-dimensional rule is applied repeatedly, is easy. However there are cases where problems arise when D cannot be decomposed into a Cartesian product and also when the number of points in the product rule grows very big and thus the application of product rules are doomed to failure.

7.3.2 Rules Exact for Monomials

These rules are developed for exactly integrating monomials of type $\Pi_{i=1}^{d} x_i^{\alpha_i}$, up to a certain degree $k = \alpha_1 + \cdots + \alpha_d$. E.g. a rule exact with degree 2 can be determined by solving the following nonlinear system of equations for $w_i, \underline{x}_i; i = 1, \ldots, M$,

$$\text{Degree 0} \quad \sum_{i=1}^{M} w_i = \int_D d\underline{x},$$

$$\text{Degree 1} \quad \sum_{i=1}^{M} w_i x_{ij} = \int_D x_j d\underline{x}, \qquad j = 1, \ldots, d,$$

$$\text{Degree 2} \quad \sum_{i=1}^{M} w_i x_{ij} x_{i\ell} = \int_D x_j x_\ell d\underline{x}, \qquad \ell = 1, \ldots, d, j = 1, \ldots, d.$$

The number M of the points necessary for integrating monomials with degree k cannot be determined explicitly as a function of k and d, but according to well-known theorems the inequality

$$\binom{d + [k/2]}{[k/2]} \leq M \leq \binom{d + k}{k} \tag{7.4}$$

holds where [ ] means the integer part function. Generally the number of equations to be solved may be quite large, though some work has been done by Keast and Diaz (1983) in reducing the number of equations in a special case.

7.3.3 Quasi Monte Carlo Methods

These methods, contrary to what might be suggested by their name, use a carefully selected, deterministic sequence of points. Such sequences do not look like random sequences and nobody forces us to believe it. some papers call these methods also number theoretical methods.

Consider a sequence of points $S = \{\underline{x}_1, \ldots, \underline{x}_N\}$ in the unit cube $K = \{\underline{x} | \underline{0} \leq \underline{x} \leq \underline{1}\}$, then for the error of the approximating sum $\frac{1}{N} \sum_{i=1}^{N} f(\underline{x}_i)$ we have

$$\left| \int f(x) d\underline{x} - \frac{1}{N} \sum_{i+1}^{N} f(\underline{x}_i) \right| \leq D_N(S) V(f) \tag{7.5}$$

Here $D_N(s)$ denotes the discrepancy of the sequence S and is given by

$$D_N(S) = \sup_{\underline{0} \leq \underline{b} \leq \underline{1}} \frac{\#\{\underline{x}_i | \underline{x}_i \leq \underline{b}, \quad i = 1, \ldots, N\}}{N} - b_1 \ldots b_d$$

where $\#$ denotes the number of points in the set, $V(f)$ is the d-dimensional variation of the function f in the sense of Hardy and Krause (see Zaremba

[59] and Niederreiter [41]). Since we can do little about $V(f)$ we try to select sequences with small discrepancy. There can be found a sequence for which

$$D_N(S) \le c_1 \frac{(\log N)^{d-1}}{N}$$

with a constant c_1, but according to a result of Roth for *any* sequence S we have

$$D_N(S) \ge c_2 \frac{(\log N)^{(d-1)/2}}{N}$$

so this limits our hopes to find good sequences.

Research related to this field has been done by Zaremba [59], Sugihara and Murota [50], Cranley and Patterson [4]. For a comprehensive treatment the reader is referred to Niederreiter [41] which contains almost four hundred references.

Similar very closely connected research carried out by Korobov [34], Hlawka, Zaremba [61], Niederreiter [39], [42], [43] is called the theory of good lattice points (or optimal multipliers). This research consists of finding such a vector $\underline{a}$ for which the error of approximation given by

$$\left| \int_D f(\underline{x}) d\underline{x} - \frac{1}{M} \cdot \sum_{i=1}^{M} f(\{\frac{i}{M}\underline{a}\}) \right| \tag{7.6}$$

would be small, where $\{\ \}$ means the fractional part of the number.

The advantage of these methods is the fast convergence since the error is $O(\log N/N)$. There are several successful implementations in low dimensions (about $2 \le d \le 10$) but in higher dimensions the method is likely to run into difficulties.

For regions different from rectangles and for some simple function the theory has not been yet developed. E.g. consider the function $g(x,y) = O$ if $x < y$ otherwise $g(x,y) = 0$; this function has unbounded variation $V(g)$. In connection with the discrepancies some research has been made by Braaten [2] who defined a discrepancy measure invariant under reflections. Probably some similar results would be needed for the variation $V(f)$.

7.3.4   Monte Carlo Methods

This is a kind of integration where one uses—in theory—random points, the-oretical justifications hold for this case. In practice points produced by deter-ministic procedures are used, that look like random (sometimes they are called pseudorandom, more frequently random). The essence and the main types of Monte Carlo computations are elegantly described by Hammersley and Hand-scomb [26].

The integral is approximated by the estimator

$$\frac{1}{N} \sum_{i+1}^{N} f(\underline{\xi}_i) \tag{7.7}$$

where $\underline{\xi}_1, \ldots, \underline{\xi}_N$ are samples from the uniform distribution in D. The standard deviation of (7.7) is $D(f(\underline{\xi}))/\sqrt{N}$ this quantity is used as the error of the result in Monte Carlo computation. Generally this error is quite large and thus one is bound to use variance reduction techniques i.e. to construct estimators having less variance than the estimator (7.7).

Several such techniques have been devised, e.g. importance sampling, strat-ified sampling, the method of control variates and that of antithetic variates. Ermakov and Zolotukhin [16] proposed the expansion of the intergrand into a sum of orthogonal functions; this method was recently supplemented by details that make it computationally feasible by Bogues et al [1].

As an interesting approach we mention the work of Yakovitz et al [58] who gave estimator containing nonlinear combinations of the functions values and thus obtained convergence faster than $O(1/\sqrt{N})$ but only up to dimension $d = 4$.

The implementation of the Monte Carlo method is easy and can be done for almost every kind of function and integration domains (infinite ranges of integration have to be truncated). The deviation of the estimator (the error) can be computed with little additional effort and is sharp. Also note that integrals in very high dimensions can be computed by Monte Carlo method, e.g. in Deák [7] an example in $d = 50$ dimensions was presented. The trouble in Monte Carlo computations is with the accuracy which usually covers two or three digits only and with the very slow convergence.

7.4 Difficulties in Multidimensional Integration

Compared to the one-dimensional integration one encounters much more difficulties in integrating d-dimensional functions.

First of all the variety of domains of integration should be observed. In the one-dimensional case only the interval and the half-line have to be considered as possible domains, while even in two dimensions we have an infinite number of domains that cannot be transformed by affine transformations into each other. In practice simple regions, like cube, sphere, cone, simplex-torus, etc., are selected. Sometimes there is a possibility of transforming one of them into another (e.g. a sphere into a rectangle via polar transformation) but the resulting clumsy function in most cases deters the user from applying it. Also there is the possibility of subdividing a cube, say into simplices (see for example the paper of Good and Gaskins [22]) but in most cases the number of resulting subregions makes feasible this procedure in low dimensions only. Subdivision, if any may be done only in an adaptive or even in an interactive way, as can be seen from the papers of Friedman and Wright [20] and Kahaner and Wells [29].

Another problem is the so-called dimensional effect. In the application of product formulas we have to tackle with the following inconvenient phenomenon. If we need M points for integration in one dimension to achieve a given accuracy (in the sense that polynomials of a given order can be integrated exactly), then applying this rule repeatedly in d dimensions we require M^d points. In the case of nonproduct formulas $\binom{n+\lfloor k/2 \rfloor}{n}$ points are required at least for exactly integrating polynominals of degree k. It means that the necessary number of points (amount of work) grows much faster than the number of dimensions. One possibility to conquer the dimensional effect is to use Monte Carlo methods in great dimensions.

Generally the estimation of error is difficult; usually two rules are independently applied and the difference between the two results is used as the error. This way however, we are likely to overestimate the true error by orders of magnitude. One may always resort to Monte Carlo methods, nevertheless a better idea is proposed by Laurie [35], or the more general way of randomization of deterministic methods of Cranley and Patterson [4] can be recommended. This last one creates a family of rules by introducing a random parameter, and sampling from this family enables the construction of confidence intervals for the magnitude of error.

One should observe the use of the optimization and the mathematical programming in the field of the multi-dimensional integration as in Mantel and Rabinowitz [37], as well as Friedman and Wright [20]. Maybe this is the way to make adaptive subdivision really practical?

Finally we mention that the theory of orthogonal polynomials, so fruitful in one dimension, does not carry over to the d-dimensional case, only some part of the whole can be saved (see Davis and Rabinowitz [5]).

7.5   Computation of Multinormal Probabilities by Monte Carlo Methods

Here we summarize some results on computing the distribution function Φ of the multinormal distribution, that is the value

$$p = \Phi(\underline{h}) = \sum_{-\infty}^{h_d} \cdots \sum_{-\infty}^{h_1} \varphi(\underline{x}) d\underline{x}, \qquad (7.8)$$

where

$$\phi(\underline{x}) = \frac{1}{(2\pi)^d/2|R|^1/2} \exp\{-\frac{1}{2}x'R^{-1}\underline{x}\}$$

is the density function of the n-dimensional normal distribution with expectation $\underline{O}$ and correlation matrix R, furthermore the probability q of a rectangle $Q = \{\underline{x}|\underline{a} \le \underline{x} \le \underline{b}\}$ that is the value

$$q = \int_Q \varphi(\underline{x}) d\underline{x}. \qquad (7.9)$$

The main result on the evaluation of the value p has been described in Deák [**7**], while details on the computation of q can be found in Deák [**10**]. However the main idea will be demonstrated here. Denote by $g(\underline{x})$ the characteristic function of Q, that is $g(\underline{x}) = 1$ if $\underline{x} \in Q$, and $g(\underline{x}) = 0$ otherwise. Let $\underline{\xi}$ be a random vector with density φ, it can be written as $\underline{\xi} = \chi \underline{\eta}$ where χ is a χ-distributed random variable with d degrees of freedom (its distribution function is F_d), $\underline{\eta}$ is uniformly distributed on the surface of the hyperellipsoid $E_s = \{\underline{x}|\underline{x}'R^{-1}\underline{x} = 1\}$, its distribution function will be denoted by $V(\underline{y})$. Using these notations we can decompose (7.9) as

$$q = \int_Q \varphi(\underline{x}) d\underline{x} = \int_{R^d} g(\underline{x}) d\Phi(\underline{x}) = \int_{E_S} \int_O^\infty g(k\underline{y}) dF_d(k) dV(\underline{y}). \qquad (7.10)$$

Let r_1 and r_2 be the entry and exit constants of a vector $\underline{y}$ with respect to the domain Q, that is $r\underline{y} \in Q$ holds if $O \le r_1 \le r \le r_2$. Then define the function e as the probability content of the line $\underline{y}$ as follows:

$$e(\underline{y}) = \int_O^\infty g(k\underline{y}) dF_d(k) = F_d(r_2) - F_d(r_1)$$

Thus from (5.3) we have the following unbiased estimator of q.

$$\Theta_1 = \frac{1}{N} \sum_{i=1}^N e(\underline{y}_i)$$

where $\underline{y}_1 \ldots, \underline{y}_N$ are independent realizations of the random variable $\underline{\eta}$. An estimator with smaller variance can be obtained if we use a set of dependent vectors, an orthonormalised system of vectors instead of the independent vector $\underline{y}_j$. The estimator O_k is obtained if the sum of k vectors from this orthonormalised system of vectors is employed instead of the vector $\underline{y}_j$.

In the paper Deák [7] very fast machine coded random number generators were used to compute probabilities p. Recently we made an attempt to develop an easy-to-use subroutine system on an IBM 3031 computer. It was completely FORTRAN coded and only standard, very well-known techniques were used for random generation (a multiplicative congruential uniform generator in double precision and the polar method for generating normal samples). Some execution times are given in the following Table 7.1.

Table 7.1

Empty loop	3 μsec
Uniform generator	70 μsec
Polar method	186 μsec
Square root	56 μsec

We implemented only the estimator O_2 for computing distribution function values p. In order to obtain probabilities with error less than 0.01 (i.e. their standard deviation is less than 0.01).

Table 7.2

d	time(sec)
3	0.1
6	0.01
10	0.08
15	0.19
20	0.36

Times necessary to compute d-dimensional distribution functions values with two accurate digits we need less than 0.4 sec (up to 20 dimensions) see Table 7.2. Times necessary to compute d-dimensional probabilities q of rectangles with two accurate digits do not exceed 0.6 sec up to $d = 20$ dimensions, see Table 7.3. More details can be found in Deák [9] or in Deák [10].

Table 7.3

d	time(sec)
2	0.02
4	0.02
6	0.1
10	0.14
20	0.56

Times necessary to compute d-dimensional probabilities of rectangles with two accurate digits.

7.6 Solution strategies

In order to solve a stochastic programming problem, where multidimensional integrations are also involved, one has to experiment with several approaches. In this section we propose an order of priority of the different solution techniques.

First try

Solve the problem for specific practical cases explicitly; as for example Hansotia [27] solved the two-stage stochastic programming problem in case of normal distribution, or as Ewbank [17] gave a closed form expression for the distribution function of the maximum in a stochastic linear programming problem.

Second try

Consider an approximating discrete distribution and solve the resulting system Strazicky [48], Kall [30], Kall and Stoyan [31], Wets [56], Wets [57]. One must note here that sometimes the astronomical number of approximating problems or the size of the problem render the solution practically impossible.

Third try

Experiment with product forms or rules exact with a given degree in low $(d \leq 5)$ dimensions and with quasi-Monte Carlo methods in low and medium dimensions $(d \leq 10)$.

Fourth try

Use Monte Carlo methods in dimensions $(d \geq 5)$

Fifth try

Reduce the variance of the Monte Carlo estimator, developing special techniques for the given problem.

In the following Table 7.4 we summarize our preferences on the usage of the different multidimensional integration methods. The greater dimension we have the more random will be the method applied. This is not a coincidence since there is very strong evidence to do so.

The Sarma-Eberlein error estimations indicate that in very high dimensions Monte Carlo methods becomes best (see Stroud [49]), the work of Yakowitz and al. [58] demonstrates that the convergence rate of the nonlinear estimator decreases with the number of dimensions and finally in Deák [7] computer experiences showed the simpler estimator's performance to be better with the increase of the number of dimensions.

Table 7.4

References

[1] K. Bogues, C.R. Morrow, and T.N.L. Patterson, "An implementation of the method of Ermakov and Zolotukhin for multidimensional integration and interpolation", *Numer. Math.* **37**(1981), 49–60.

[2] E. Braaten, "An improved measure on the non-uniformity of the distribution of a sequence", *SIAM J. Numer Anal.* **17**(1980), 31–32.

[3] J.L. Brown, "On the expansion of the bivariate Gaussian probability density using results of nonlinear theory", *IEEE Trans. on Inf. Theory*, (1968), 158–159.

[4] R. Cranley and T.N.L. Patterson, "Randomization of number theoretic

methods for multiple integration", *SIAM J. on Numer. Anal.* **13**(1976), 904–914.

[5] P.J. Davis and P. Rabinowitz, *Methods of numerical integration*, Academic Press, New York, 1975.

[6] I. Deák and B. Bene, "Random number generation: a bibliography", MTA SZTAKI Working Paper MO/5 (1979), pp.54.

[7] I. Deák, "Three digit accurate multiple normal probabilities", *Numer. Math.* **35**(1980), 369–380.

[8] I. Deák, "An economical method for random number generation and a normal generator", *Computing* **27**(1981), 113–121.

[9] I. Deák, "Procedures for the computation of normal distribution function values and probabilities of rectangles, MTA SZTAKI Working Paper MO/48 (1983), 88–36.

[10] I. Deák, "Computing probabilities of rectangles in case of multinormal distribution", *Math. of Comp.* (forthcoming).

[11] U. Dieter and J.H. Ahrens, "Uniform random numbers. Nonuniform random numbers", Technische Hochschule, Graz. (1974), 380 pages.

[12] T.G. Donelly, "Bivariate normal distribution", *Comm. ACM* **16**(1973), 638.

[13] Z. Drezner, "Computation of the bivariate normal integral", *Math. Comp.* **32**(1978), 277–279.

[14] J.E. Dutt, "A representation of multivariate normal probability integral transforms", *Biometrika* **60**(1973), 637–645.

[15] S.M. Ermakov, "Monte Carlo methods and related problems", (in Russian) Izd. Nauka, Moscow, 1971.

[16] S.M. Ermakov and V.G. Zolotukhin, "Polynominal approximations and the Monte Carlo method", in: Theory of Probability and Applications, *Amer. Math. Soc. Transl.* **5**(1960), 428–431.

[17] J.B. Ewbank, B.L. Foote, and H.J. Kumin, "A method for the solution of the distribution problem of stochastic linear programming", *SIAM J. Appl. Math.* **26**(1974), 225–238.

[18] B.L. Fox, "Counterparts of variance reduction techniques for quasi Monte Carlo integration", manuscript, 1983.

[19] W. Freeden, "On integral formulas of the (unit) sphere and their application to numerical computation of integrals", *Computing* **25**(1980), 131–146.

[20] J.H. Friedman and M.H. Wright, "A nested partitioning procedure for numerical multiple integration", *ACM TOMS* **7**(1981), 76–92.

[21] A.C. Genz and A.A. Malik, "An imbedded family of fully symmetric numerical integration rules", *SIAM J. Numer. Anal.* **20**(1983), 580–588.

[22] I.J. Good and R.A. Gaskins, "The Centroid method of numerical integration", *Numer. Math.* **16**(1971), 343–359.

[23] S. Haber, "Stochastic quadrature formulas", *Math. of Comp.* **23**(1969), 751–764.

[24] S. Haber, "Numerical evaluation of multiple integrals", *SIAM Review* **12** (1970), 481–526.

[25] J.M. Halton, "A retrospective and prospective survey of the Monte Carlo method", *SIAM Review* **12**(1970), 1–63.

[26] J.M. Hammersley and D.C. Handscomb, *Monte Carlo Methods*, Methuen. London. 1964.

[27] B.J. Hansotia, "Stochastic linear program with simple recourse: the equivalent deterministic convex program for the normal, exponential and Erlang cases", *Naval Res. Logist. Quart.* **27**(1980), 257–272.

[28] E. Hlawka, "Zur angenaherten Berechnung mehrfacher Integrale", *Monatsh. Math.* **66**(1962), 140–151.

[29] D.K. Kahaner and M.B. Wells, "An experimental algorithm for n-dimensional adaptive quadrature", *ACM TOMS* **5**(1979), 86–96.

[30] P. Kall, "Computational methods for solving two-stage stochastic linear programming problems". *Zeitschrift für Ang. Math. und Physik* **30**(1979), 261–271.

[31] P. Kall, "Solving stochastic programming problem with recourse including error bounds", *Math. Operationsforsch. Statist. Ser. Optimization* **13**(1982), 431–447.

[32] P. Keast and J.N. Lyness, "On the structure of fully symmetric multidimensional quadrature rules", *SIAM J. Numer. Anal.* **16**(1979), 11–29.

[33] P. Keast and J.C. Diaz, "Fully symmetric integration formulas for the surface of the sphere in s dimension", *SIAM J. Num. Anal.* **20**(1983), 406–419.

[34] N.M. Korobov, "The approximate computation of multiple integrals", *Dokl. Akad. Nank SSSR* **124**(1959), 1207–1210.

[35] D.P. Laurie, "Sharper error estimates in adaptive quadrature", *BIT* **23** (1983), 258–261.

[36] V.I. Lebedev, "On cubature formulas of spheres", *J. Comp. Math. and Math. Physics* **16**(1976), 293–306.

[37] F. Mantel and P. Rabinowitz, "The application of integer programming to the computation of fully symmetric integration formulas in two and three dimensions", *SIAM J. Numer. Anal.* **14**(1977), 391–425.

[38] R.C. Milton, "Computer evaluation of the multivariate normal integral", *Technometrics* **14**(1972), 881–889.

[39] H. Niederreiter, "On a number theoretical integration method", *Aequationes Math.* **8**(1972), 304–311.

[40] H. Niederreiter, "Application of Diophantine approximations to numerical integration", in: Diophantine Approximations and its Applications. Academic Press, New York, London, 1973, (ed. C.F. Osgood) 129–199.

[41] H. Niederreiter, "Quasi-Monte Carlo methods and pseudo-random numbers", *Bull. Amer. Math. Soc.* **84**(1978), 957–1041.

[42] H. Niederreiter, "Existence of good lattice points in the sense of Hlawka", *Mh. Math.* **86**(1978), 203–219.

[43] H. Niederreiter, "Multidimensional integration using pseudo-random numbers", (this volume).

[44] A. Prékopa, "Contributions to the theory of stochastic programming", *Math. Programming* 4(1973), 202–221.

[45] A. Prékopa, S. Ganczer, I. Deák, and K. Patyi, "The STABIL stochastic programming model and its experimental application to the electrical energy sector of the Hungarian economy", in: *Stochastic Programming* (ed. M.A.H. Dempster) Oxford, 1980, 369–385.

[46] R. Scherer, "Über fehlerschranken bei produkt-kubatur", *ZAMM* 60)1980), T-315–T-317.

[47] I.A. Stegun and R. Zucker, "Automatic Computing Methods for special functions", *J. of Research NBS* 86(1981), 661–686.

[48] B. Strazicky, "On an algorithm for solution of the two-stage stochastic programming problem", *Methods of OR* **XIX**(1974), 142–156.

[49] A.H. Stroud, *Approximate calculation of multiple integrals*, Prentice-Hall, Englewood Cliffs, N.J., 1971.

[50] M. Sugihara and K. Murota, "A note on Haselgrove's method for numerical integration", *Math. of Comp.* 39(1982), 549–554.

[51] A.G. Sukharev, "On the problem of constructing optimal quadratures for multivariate functions", *Cybernetics* 7–11 (in Russian).

[52] T. Szántai, "A method for computing the value of the multinormal distribution function and its gradient", (in Hungarian) *Alkalmaz. Mat. Lapok.* 2(1976), 27–39.

[53] R. Terras, "Algorithms for some integrals of Bessel functions and multivariate Gaussian integrals", *J. Comp. Physics* 41(1981), 192–199.

[54] T. Tsuda, "Numerical integration of functions of very many variables", *Numer. Math.* 20(1973), 377–391.

[55] R. Wets, "The distribution problem and its relations to other problems in stochastic programming", in: *Stochastic programming* (ed. M.A.H. Dempster), 1980, 245–262.

[56] R. Wets, "Stochastic Programming: solution techniques and approximation schemes", in: *Mathematical programming - the state-of-the-art*, (Eds. A. Bachem, M. Groetschel, B. Korte) Springer, Berlin, 1983, 566–603.

[57] R.Wets, "Approximation schemes", 1983 (this volume).

[58] S. Yakowitz, J.E. Krimmel and F. Szidarovszky, "Weighted Monte Carlo integration", *SIAM J. Numer. Anal.* 15(1978), 1289–1300.

[59] S.K. Zaremba, "The mathematical basis of Monte Carlo and quasi-Monte Carlo methods", *SIAM Review* 10(1968), 303–314.

[60] S.K. Zaremba (editor), *Applications of Number Theory to Numerical Analysis*. Academic Press, New York, 1972.

[61] S.K. Zaremba, "On Cartesian products of good lattices", *Math. Comp.* 30(1976), 546–551.

CHAPTER 8

STOCHASTIC INTEGER PROGRAMMING

A. R. Kan and L. Stougie

8.1   Introduction.

This short chapter on *stochastic integer programming* will be quite different in nature from the preceding ones. To a large extent, this difference reflects the way in which current research traditions in integer programming differ from those in other areas of mathematical programming.

Initially, integer programming was concerned with a simple and yet fundamental extension of the generic *linear programming* model

$$\text{minimize} \quad cx \tag{8.1}$$

$$\text{subject to} \quad Ax = b \tag{8.2}$$

$$x \geq 0 \tag{8.3}$$

obtained by adding the constraint

$$x \in \mathbb{Z}^n.$$

Methods to solve this generic *integer programming* problem were sought in the hope that their efficiency would match the efficiency of the *simplex method* for linear programming. Since virtually every optimization problem encountered in practice turned out to allow formulation as an integer program, such a method would be a truly formidable solution tool.

Rapidly, however, it appeared that the great generality of integer programming comes at a price: neither the cutting plane approach pioneered by Gomory, the branch-and-bound approach first proposed by Land and Doig nor any other method proposed in the sixties turned out to be able to solve any but the smallest problems within reasonable time. Even today, when linear programming problems with thousands of variables are solved on a routine basis, integer programming problems with 80 or 100 variables may already present insurmountable problems.

For a while, optimists could keep hoping that some totally new approach, some brilliant fresh idea could provide a breakthrough to a truly efficient integer programming method. *Computational complexity theory*, however, put an end to that illusion in the early seventies, by showing that the computational differences encountered in solving integer programming problems are likely to be caused by the inherent complexity of the problem and not by the intellectual limitations of the researchers studying it. More precisely, if we associate the notion of an easy or well-solved problem with the existence of an algorithm whose running time increases at most *polynomially* with problem size, then the general integer programming problem is highly unlikely to be easy in this sense: it belongs to a class of notoriously difficult combinatorial optimization problems, the *NP-hard* problems, for which strong evidence suggests that any solution method has *superpolynomially* increasing running time in the worst case. For integer programming, an enumerative approach such as branch and bound, in which the (exponentially large) set of feasible solutions to (8.1), (8.2), (8.3) is implicitly or explicitly enumerated, provides a good example of such a method.

That, fortunately, is only part of the story. A more encouraging implication of the complexity results mentioned above is that the road to computational success for integer programming problems is through the exploitation of *special structure*. Methods that solve any integer program are very unlikely to be efficient, and in this respect that situation is very different from linear programming. But if specially designed solution methods are used that exploit the particular features of the model at hand, then the outlook is much brighter. We notice that for certain important subclasses of integer programming problems, such as network flow, shortest path and matching problems *polynomial time algorithms* have been designed implying that these problems belong to the above mentioned class of well-solved problems. Even if the problem in question is not easy in the formal sense it still pays to investigate if its special structure allows for sharper bounds, faster enumeration schemes or tighter cutting planes. In doing so, one may well end up with an enumerative solution method whose empirical behaviour is completely satisfactory.

Much of the above discussion carries over to stochastic integer programming. From the generic (two-stage) *stochastic linear programming problem.*

$$\text{minimize} \quad cx + \mathrm{E}(\min \mathbf{q}y | \mathbf{W}y = \mathbf{T}x - \mathbf{p}, y \geq 0)$$
$$\text{subject to} \quad Ax = b$$
$$x \geq 0$$

where *random variables* are boldfaced, it is easy to derive the generic (two-stage) *stochastic integer programming problem:*

$$\text{minimize} \quad cx + \mathrm{E}(\min \mathbf{q}y | \mathbf{W}y = \mathbf{T}x - \mathbf{p}, y \geq 0, y \in \mathbb{Z}^k) \quad (8.4)$$
$$\text{subject to} \quad Ax = b \quad (8.5)$$
$$x \geq 0 \quad (8.6)$$
$$x \in \mathbb{Z}^n. \quad (8.7)$$

However, since both the general stochastic linear programming problem and, as we have seen, the general integer programming problem enjoy a well-deserved reputation for computational intractability, so far hardly anybody has been tempted to consider methods to solve (8.4), (8.5), (8.6), (8.7) in full generality. There is no difficulty in principle: one could, for instance, write out the

equivalent deterministic program as in Chapter , Section , and solve the
resulting large integer programming problem, perhaps by exploiting the special
structure (though in the case of integer programming it is not so obvious how
to do that). But the resulting method is not likely to be of great computational
efficiency.

Many of the difficulties inherent to the general stochastic integer programming problem already show up when we consider what theoretical features of
linear programming contribute to the success of stochastic linear programming
codes. Take, for example, the pleasant properties of *parametric linear programming* that lead to convexity properties for stochastic linear programming. A
small example will already show how much less well behaved parametric integer
programs can be. Consider the (deterministic) function

$$z(x) = 1 - x + \max\{y | 0 \leq y \leq x, \quad y \in \mathbb{Z}\}. \tag{8.8}$$

Its graph is depicted in Figure 8.1, and it shows the peculiar discontinuities and
nonconvexities that integer programming gives rise to.

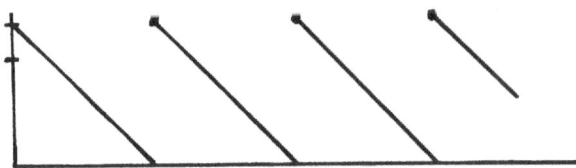

Figure 8.1

If the integrality constraints appear only at the first stage, then the expected
optimal second stage costs are still convex in the first stage decision variables
and the problem can be dealt with by fairly conventional means. The nonconvexities in the two-stage objective function induced by integrality constraints at
the second stage cause more fundamental problems. Of course, in stochastic integer programming one usually deals with a weighted sum of ill-behaved second
stage functions such as (8.8), the smoothing-out effect of which may eliminate
discontinuity. But convexity or concavity cannot be guaranteed under reasonably general conditions. For instance, let us define

$$Z(x) = 1 - x + \mathrm{E}[\max\{y | 0 \leq y \leq x + \beta, y \in \mathbb{Z}\}],$$

where the random variable β is uniformly distributed over the interval $[0, \delta]$
with $\delta < 1$. Simple calculations yield that for $k = 1, 2, \ldots$

$$Z(x) = \begin{cases} k - x, & k - 1 \leq x \leq k - \delta \\ (k - x)(1 + \frac{1}{\delta}) - 1, & k - \delta \leq x \leq k \end{cases}$$

The graph of Z is depicted in Figure 8.2. Due to the continuity of the distribution function of β. Z is a continuous, but still nonconvex function. General results on the shape of objective functions of two-stage decision problems are derived in Stougie [14].

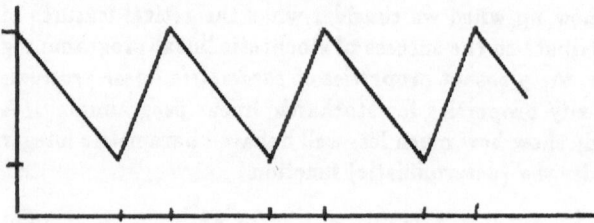

Figure 8.2

So, as in the case of deterministic integer programming, we turn to the exploitation of special structure as the last hope for some computational progress. Indeed, this is what most of the (limited) research efforts in the area have focused on. The above discussion suggests that an appropriate first step should be to obtain more insight in the behaviour of the *distribution problem solution* for these specially structured problems, and this has turned out to be an unexpectedly fruitful area of research. Natural probabilistic extensions of some traditional combinatorial optimization problems turn out to have the surprising property that the random variable corresponding to their optimal solution value converges in some stochastic sense to a simple analytical function of problem parameters when the problem size increases. These results are discussed in more detail in Section 8.2, which is devoted to the integer stochastic programming distribution problem.

In Section 8.3, we shall see how results on the distribution problem find application in the construction of solution methods for the two-stage decision problem. In fact, if the second stage problem is one of those for which an asymptotic closed form for the optimal solution value is known, then it is intuitively obvious that a *heuristic* of good asymptotic properties can be based on using the closed form expression in an approximation to the original objective (8.4). Results of this nature, together with a brief examination of the possibilities for an *optimization* method in contrast to *approximation (heuristic)* methods, can be found in Section 8.3.

By their very nature, the available results on specially structured stochastic integer programming problems are to a large extent ad hoc and hence of limited general value. We have not attempted to provide an exhaustive survey of the area; for that, we refer to Stougie [14] and to the annotated bibliography Karp et al., [7]. In fact, we propose to illustrate the nature of the results obtained on a very simple but typical stochastic integer programming problem, that might be called *the machine investment problem*. The first stage of this problem involves the acquisition of a certain number of *identical machines* at cost c each, subject to probabilistic information about the *processing times* $p_j (j = 1, \ldots, n)$ of the jobs that will have to be executed on these machines in the second stage. The objective of the second stage decision is to minimize the *makespan* (i.e., the maximum sum of the processing times assigned to any one machine) of the resulting schedule. If we denote the minimum makespan value as a function of the number of machines m by $C_n^*(m)$, then the stochastic program is to minimize

$$Z_n(m) = cm + E C_n^*(m) \tag{8.9}$$

where m is constrained to be integer. The computation of $C_n^*(m)$ is itself a (NP-hard) combinatorial optimization problem. Thus, this simple example incorporates all the features characterizing the collection of stochastic integer programming problems that we shall be addressing here.

8.2 The distribution problem

As announced in Section 8.1, probabilistic versions of traditional combinatorial optimization problems sometimes have the remarkable property that their optimal solution value is asymptotic to a simple function of certain problem parameters.

The machine investment problem provides a striking example. Recall that the second stage corresponds to the minimization of makespan on m machines. Specifically, any feasible schedule must satisfy the restrictions that each machine processes at most one job at a time and each job is processed during on uninterrupted interval of length equal to its processing time. For this NP-hard optimization problem enumerative methods provide the only available solution tool. We are interested, however, in a probabilistic version of it as it appears to the first stage decision maker. Let us assume that the processing times of the jobs are independent identically distributed (i.i.d.) random variables with expected value μ. Intuition suggests that the minimal makespan for n sufficiently large will be relatively close to the lower bound achieved by dividing the total workload $\sum_{j=1}^{n} p_j$ evenly among the m machines. We will show that this intuition is correct. For the proof we rely on the above lower bound and on an upper bound provided by a heuristic solution of the problem. We assume that $E p_1^2 < \infty$. For the formal analysis we define the following random model of the problem. Let the processing times of a problem with n jobs be the first n elements of a random vector drawn from an infinite dimensional sample space Ω.

The heuristic that we use is a simple list scheduling rule: the jobs are placed in an arbitrary fixed order and at each step the next job on the list is assigned to the first available machine (see Figure 8.3). Let $C_n^H(m)$ denote the makespan under this heuristic for given m and given a realization of the processing times. Let L be the latest time that all machines are occupied and let job k be completed last. By the nature of list scheduling, $L \leq \sum_{j=1}^{n} p_j / m$.

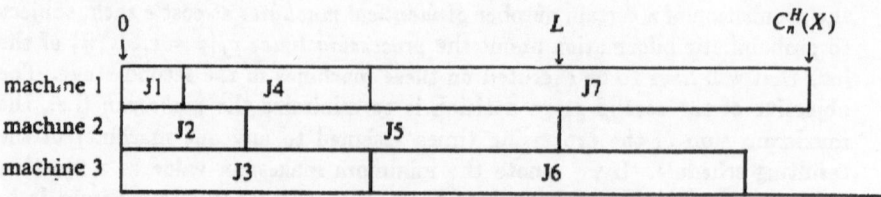

Figure 8.3   Illustration of the list scheduling heuristic.
Problem instance: $X = 3, n = 7, p = (1, 2, 4, 3, 5, 6, 7)$

Trivially, $p_k \leq \max_{j=1,\ldots,n} p_j = p_{\max}$. Therefore

$$C_n^H(m) \leq \sum_{j=1}^{n} p_j/m + p_{\max}.$$

This inequality combined with the lower bound $\sum_{j=1}^{n} p_j/m$ on the optimal makespan yields

$$\sum_{j=1}^{n} p_j/m \leq C_n^*(m) \leq C_n^H(m) \leq \sum_{j=1}^{n} p_j/m + p_{\max}.$$

Dividing this by $n\mu/m$ yields

$$\frac{\sum_{j=1}^{n} p_j - n\mu}{n\mu} + 1 \leq \frac{C_n^*(m)}{n\mu/m} \leq \frac{C_n^H(m)}{n\mu/m} \leq \frac{\sum_{j=1}^{n} p_j - n\mu}{n\mu} + 1 + \frac{m p_{\max}}{n\mu}. \quad (8.10)$$

Since μ is finite, the strong law of large numbers implies that

$$Pr\{ \lim_{n \to \infty} (\sum_{j=1}^{n} \mathbf{p}_j - n\mu)/n\mu = 0 \} = 1. \quad (8.11)$$

It remains to prove that

$$Pr\{ \lim_{n \to \infty} m\mathbf{p}_{\max}/n\mu = 0 \} = 1. \quad (8.12)$$

We note that the following lemma is proved in a.o. Feller, [3].

Lemma 8.1. *If $E\mathbf{p}_1^2 < \infty$. then*
(i) $\lim_{n \to \infty} \mathbf{p}_{\max}/\sqrt{n} = 0$ almost surely.
(ii) $\lim_{n \to \infty} E\mathbf{p}_{\max}/\sqrt{n} = 0$. $\square$

Therefore (8.12) holds for all values of m satisfying $m = 0(\sqrt{n})$. (8.10), (8.11) and (8.12) together imply that

$$Pr\{ \lim_{n \to \infty} \frac{C_n^*(m)}{n\mu/m} = 1 \} = 1. \quad (8.13)$$

Taking expectations in (8.10) and applying Lemma 8.1 (ii) implies that

$$\lim_{n\to\infty} \frac{\mathbf{EC}_n^*(m)}{n\mu/m} = 1.$$

If we assume that the common distribution function of the processing times has a positive derivative in 0, then it is even possible to prove that

$$\lim_{n\to\infty} \mathbf{C}_n^*(m) - \sum_{j=1}^{n} \mathbf{p}_j/m = 0$$

almost surely, and

$$\lim_{n\to\infty} E\mathbf{C}_n^*(m) - n\mu/m = 0. \tag{8.14}$$

The intuition behind these results is that under the above assumption there are enough jobs with very small processing times. These can be used for smoothing the differences in the execution times of the machines after having assigned the jobs with larger processing times. A rigorous proof of these results is far from easy (cf. Frenk and Rinnooy Kan,[5]). Result (8.13) is particularly illuminating, in that it shows how the optimal value of the second stage objective function (cf. (8.9)) can be written asymptotically as a simple function of the problem parameters n and μ, and of the first phase decision variable m. Even more pleasantly, we have seen that there exist simple scheduling heuristics whose solution values are also asymptotic to this same function. Similar asymptotic results are available for many other combinatorial problems. They can be broadly divided into three classes, in accordance with the different underlying probabilistic models.

(i)   Number problems

Here, randomness occurs in certain numerical parameters; typically, these are assumed to be i.i.d. random variables. The above scheduling problem provides a good example. The *linear assignment* problem

$$\text{minimize} \quad \sum_{i=1}^{n}\sum_{j=1}^{n} \mathbf{a}_{ij}x_{ij}$$

$$\text{subject to} \quad \sum_{i=1}^{n} x_{ij} = 1 \quad (j=1,\ldots,n)$$

$$\sum_{j=1}^{n} x_{ij} = 1 \quad (i=1,\ldots,n)$$

$$x_{ij} \in \{0,1\}$$

where the weights $\mathbf{a}_{ij}$ are i.i.d., is another one. Here, one can show under quite general conditions on the common distribution function F that the expected optimal solution value is asymptotic to $nF^{-1}(1/n)$ Frenk and Rinnooy Kan,

[5]. The proof amounts to showing that, with high probability of being correct, we may set all x_{ij} equal to zero, except those corresponding to the smallest weights for each i and each j. Thus with probability approaching 1 we obtain a feasible solution whose value is asymptotic to n times the value of the smallest order statistic; i.e., $nF^{-1}(1/n)$.

A third example is provided by the *knapsack* problem

$$\text{maximize} \quad \sum_{j=1}^{n} \mathbf{c}_j x_j$$

$$\text{subject to} \quad \sum_{i=1}^{n} \mathbf{a}_j x_j \leq bn$$

$$x_j \in \{0,1\}.$$

If the $\mathbf{c}_j$ and $\mathbf{a}_j$ are independent and uniformly distributed on $[0,1)$, then the expected optimal solution value is asymptotic to $n\sqrt{2b/3}$ if $0 < b < \frac{1}{6}$ and $n\left(-\frac{3}{2}b^2 + \frac{3}{2}b + \frac{1}{8}\right)$ if $\frac{1}{6} \leq b < \frac{1}{2}$ (if $b > \frac{1}{2}$, then asymptotically all x_j are equal to 1 with probability 1) Meanti et al.,[12]. To derive this result, one shows that the optimal solution value is asymptotic to the value of the linear relaxation, which is equal to $n\min_\lambda\{\mathbf{L}_n(\lambda)|\lambda \geq 0\}$ where

$$\mathbf{L}_n(\lambda) = \max\{\lambda + \frac{1}{n}\sum_{j=1}^{n}(\mathbf{c}_j - \lambda\mathbf{a}_j)x_j|0 \leq x_j \leq 1(j = 1,\ldots,n)\}$$

$$= \lambda b + \frac{1}{n}\sum_{j=1}^{n}(\mathbf{c}_j - \lambda\mathbf{a}_j)\mathbf{x}_j(\lambda)$$

with

$$\mathbf{x}_j(\lambda) = \begin{cases} 1 & \text{if } \mathbf{c}_j - \lambda\mathbf{a}_j \geq 0 \\ 0 & \text{otherwise.} \end{cases}$$

The strong law of large numbers implies that $\mathbf{L}_n(\lambda)$ converges almost surely to

$$L(\lambda) = \lambda b + E\mathbf{c}_1\mathbf{x}_1(\lambda) - \lambda E\mathbf{a}_1\mathbf{x}_1(\lambda)$$

and results from convex analysis can be used to show that $\min_\lambda\{\mathbf{L}_n(\lambda)|\lambda \geq 0\}$ converges (almost surely and in expectation) to the unique minimum of $L(\lambda)$. Elementary computations then yield a closed form expression for $L(\lambda)$ and through that the above result. As in most cases, this result is accompanied (and, indeed, derived through) a simple heuristic whose error disappears asymptotically.

(ii)   Euclidean problems

These problems can be formulated with respect to n points in the Euclidean plane; their probabilistic version then amounts to assuming these points to be distributed uniformly over (say) the unit square. The most famous example is the *traveling salesman* problem of finding the shortest tour through the points. The optimal solution value is asymptotic to $\beta\sqrt{n}$ with probability 1 (Steele, [13]), where in the case of the unit square $\beta \approx 0.765$. The proof of this result is extremely complicated, although it is not hard to appreciate the proportionality to $\sqrt{n}$ intuitively: for large n, the optimal tour through $4n$ points in a 4×4 square is 4 times as large as the optimal tour through n points in a 1×1 square, and scaling down the 4×4 square to a 1×1 one reduces the length by a factor of 2. Here, a simple *space partitioning* heuristic (Karp, [7]) achieves a solution value that is asymptotic to the optimal one. In one version, the unit square is partitioned into $s(n)$ equal size subsquares Q_i. The optimal local tours $T^*(Q_i)$ are computed and $s(n)$ points, one selected from each Q_i, are linked by a single global tour. This yields a Euclidean walk through all points which can be easily transformed into a tour of no greater length. For the analysis, ons shows that $\Sigma_i T^*(Q_i)$ exceeds the optimal tour length by no more than $\frac{3}{2}$ times the sum of the perimeters of the Q_i, which is an $0(\sqrt{s(n)})$ term. There are various ways to construct the global tour, so that its length adds no more then $0(\sqrt{s(n)})$ to the absolute error again. Then, by taking $s(n) = o(n)$ and invoking the above result, one sees that the relative error converges to 0 almost surely. Similar results, combined with similar heuristics, have been obtained for Euclidean location problems (cf. Zemel, [15]) and routing problems (cf. Marchetti Spaccamela et al.,[10] and Haimovich and Rinnooy Kan, [6]). For an overview, see Karp et al., [7].

(iii) Graph problems

Two natural models for random graphs, one in which each edge is present with probability p, and one in which m edges are scattered uniformly among the vertex pairs, provide the context for probabilistic versions of combinatorial optimization problems defined on graphs. The *maximum clique* problem of finding the size of the largest complete subgraph, is a particularly fine example: under the first probabilistic model, the maximum clique size is asymptotic to $2\ln n/\ln(1/p)$ (Matula, [11]). For results on other graph problems we refer once again to Karp et al., [7].

In all the above cases (with the exception of the linear assignment problem), it would be virtually impossible to solve every large instance of these optimization problems to optimality. Thus, if one wishes to have an exact solution to the distribution problem, this can only be achieved for small problem sizes. Given the parametric character of the necessary computations, dynamic programming is a natural tool to consider; as we shall see in the next section, it can occasionally be applied with reasonable success.

The asymptotic character of all the above results is one of their least attractive features. For some of them (in particular the number and graph problems), *speed of convergence* results provide additional information about the rate at which the objective function value converges to its limit. Especially for the Euclidean problems, however, such results are notoriously lacking.

8.3   Multi-stage decision problems

For the solution of integer multi-stage decision problems, the difficulty of which
has been discussed extensively in Section 8.1, heuristics can be designed in
which results of the type presented in Section 8.2 play a central role. When an
asymptotic characterization of the optimal value of the second stage problem
has been derived as in (8.13), it can be used as part of an *estimation* of the
overall two-stage objective function. As mentioned in Section 8.2, this estimate
is frequently a simple function of the first stage decision variables. Its minimization yields the heuristic first stage decision. We then require a heuristic
for solving the second stage problem, which is usually a NP-hard one. One
would like this heuristic to provide solutions of such quality that strong asymptotic optimality properties of the whole heuristic procedure are guaranteed.
Fortunately, simple heuristics frequently turn out to be good enough for these
purposes.

The combination of a good estimate of the total cost and a good approximation procedure for the solution of the second stage problem can be shown
rigidly to yield a guarantee for the asymptotic optimality of the resulting stochastic integer programming heuristic. More specifically, the relative error of
the heuristic, obtained by dividing the difference between the heuristic value
and the optimal value of the problem by the optimal value, can be shown to
converge stochastically to 0 with increasing problem size for a very general class
of models.

We illustrate the above ideas again with the example of the machine investment problem. The asymptotic characterization (8.14) of the optimal value
of the second stage scheduling problem allows us to estimate the overall cost of
the two-stage decision problem by the function

$$Z'_n(m) = cm + \frac{n\mu}{m}.$$

Minimization with respect to m of this unimodal convex function, subject to the
restriction that m is integral, produces a heuristic first stage decision m^{H_1} equal
to $\lfloor \sqrt{n\mu/c} \rfloor$ or to $\lceil \sqrt{n\mu/c} \rceil$, depending on which of these two values is more
favorable. For the solution of the second stage scheduling problem, we have
seen in Section 8.2 that the list scheduling rule yields a relative error that tends
to 0 almost surely if $m = 0(\sqrt{n})$. We note that $m^{H_1} \approx \sqrt{n\mu/c}$. Therefore, if
$C_n^{H_2}(m)$ denotes the makespan produced by the heuristic, the above makes it
easy to verify that

$$\lim_{n \to \infty} \frac{cm^{H_1} + EC_n^{H_2}(m^{H_1})}{cm^{H_1} + \frac{n\mu}{m^{H_1}}} = 1,$$

which establishes asymptotic optimality of the heuristic procedure as a whole.
A detailed description of the above result is given in Dempster et al., [2].

We can also compare the heuristic solution value with the optimal solution
value of the machine investment problem under the assumption that all information is available in advance, This problem can be formulated as finding a

function $\mathbf{m}_n^\circ : \Omega \to \mathbb{N}$ that gives for each realization of the random processing times a value m_n° for which

$$cm_n^\circ + C_n^*(m_n^\circ) = \min_{m \in \mathbb{N}} \{cm + C_n^*(m)\}.$$

The reasoning used for the justification of result (8.13) makes it easy to verify that almost surely

$$\lim_{n \to \infty} \frac{cm^{H_1} + \mathbf{C}^{H_2}(m^{H_1})}{cm_n^\circ + C_n^*(\mathbf{m}_n^\circ)} = 1.$$

This result implies that the relative error that can be attributed to imperfect information also tends to zero almost surely. This strong property of the heuristic was named *asymptotic clairvoyance* in Lenstra et al., [9].

In a similar way, heuristics of equal quality can be constructed for other two-stage decision problems of which the second stage problem allows asymptotic characterization of the optimal solution, such as vehicle routing problems (Marchetti Spaccamela et al., [10]) and location problems (Stougie, [14]) that are preceded by an investment decision. For instance, in the vehicle routing case, the objective is to minimize the sum of the cost of acquiring m vehicles at cost c each and the expected length of the longest route subsequently taken by any of the m vehicles to serve n customers from a common depot. By underestimating the latter by the expected cost of the shortest traveling salesman tour through all n customers divided by m (i.e. $\beta\sqrt{n}/m$), we again arrive at an asymptotically optimal heuristic. In the location problems the objective is to minimize the sum of establishing m depots and the expected sum of the distances from each of n customers to the nearest depots. A general framework for the design and analysis of such stochastic integer programming heuristics is presented in Lenstra et al., [9].

The remaining part of this section will be dedicated to optimization methods for stochastic integer programming. Only few results are available in this direction. Such methods have been designed for some two-stage decision problems, of which the stochastic parameters are assumed to have discrete distributions with only a small number of points with positive density. It is not surprising that the parametric relations between the various feasible solutions of these problems can efficiently be exploited by *dynamic programming* routines. This can again be illustrated through the machine investment problem. Let us assume that the processing times of the jobs can have only k possible values, $a_1, \ldots, a_k$. Let $C^*(m, n_1, \ldots, n_k)$ be the minimum makespan of a set of jobs consisting of n_i jobs with processing times a_i $(i = 1, \ldots, k)$. Any schedule and therefore also the optimal one can be split into a schedule of a subset of the jobs on a subset of the machines and the rest of the jobs on the rest of the machines. Based on this observation we derive the following recurrence relations:

$$C^*(1, n_1, \ldots, n_k) = \sum_{i=1}^k n_i a_i$$

and, for every m' satisfying $1 \leq m' \leq m$,

$$C^*(m, n_1, \ldots, n_k) =$$
$$\min_{\substack{0 \leq \ell_1 \leq n_1 \\ \vdots \\ 0 \leq \ell_k \leq n_k}} \max\{C^*(m', \ell_1, \ldots, \ell_k), C^*(m - m', n_1 - \ell_1, \ldots, n_k - \ell_k)\}$$

$$(8.15)$$

We can evaluate the objective function of the two-stage decision problem for each interesting value of m by computing the minimum makespan for each possible composition of the set of n jobs, weighing them with the corresponding probabilities, taking the weighted sum and adding cm. Solving the machine investment problem is now just a matter of selecting the minimum value.

This algorithm has a running time that, interestingly enough, is polynomial in the number n of jobs. It is, however, exponential in the number k of possible values of the processing times. This can be seen by setting m' in (8.15) equal to 1 and evaluating $C^*(m, n_1, \ldots, n_k)$ for all values of m ranging from 1 to n, and for all values of $n_1, \ldots, n_k$ satisfying $\sum_{i=1}^{k} n_i \leq n$. These are $O(n^{k+1})$ evaluations, each of which requires the solution of equality (8.15), which can be achieved by the comparison of $O(n^k)$ values, Hence the overall running time is $O(n^{2k+1})$. Obviously better versions and implementations are possible, and, in fact, the best running time bound that has been achieved is $O(n^{2k-1} \log n)$ (Lageweg et al., [8], elsewhere in this book). Therefore already for small values of k, only problems with a limited number of jobs can be solved.

In Lageweg et al., [8] the above dynamic programming routine is described in detail and tested. In the same paper similar routines are presented for a *capital budgeting problem* and for a *hierarchical bin packing problem*, for which at the first stage one has to decide upon the capacity of bins, in which items have to packed in the second stage, such that a minimum number of bins is required. In the location problems the objective is to minimize the sum of establishing m depots and the expected sum of the distances from each of n customers to the nearest depot. The computational results not only showed that the above dynamic programming routines do work satisfactorily but also yielded some insights into the shape of objective functions of integer two-stage decision problems involving discrete distributions. These confirmed earlier theoretical insights that were based on parametric analyses of deterministic integer programming problems (cf. Blair et al., [1]). To seek alternatives to these simple dynamic programming routines is but one of the many challenges that remain in the area of stochastic integer programming, an area which is only now starting to receive the attention that it deserves.

Acknowledgements

The first author was partially supported by NSF-grant ECS-831-6224.

References

[1] C. Blair and R.G. Jeroslow, "On the value function of an integer program", *Math. Progr.*, Vol. 23, No. 3(1979), 237–273.

[2] M.A.H. Dempster, M.L. Fisher, L. Jansen, B.J. Lageweg, J.K. Lenstra and A.H.G. Rinnooy Kan, "Analysis of heuristics for stochastic programming: results for hierarchical scheduling problems", *Math. Oper. Res.* 8(1983), 525–537.

[3] W. Feller, *An introduction to probability theory and its applications*, Vol. 1, 3rd edition, John Wiley and Sons, New York, 1968.

[4] J.B.G. Frenk and A.H.G. Rinnooy Kan, "The asymptotic optimality of the LPT rule", Technical report, Erasmus University, Rotterdam (1984).

[5] J.B.G. Frenk and A.H.G. Rinnooy Kan, "Order statistics and the linear assignment problem", Technical report, Erasmus University, Rotterdam (1985).

[6] M. Haimovich and A.H.G. Rinnooy Kan, "Bounds and heuristics for capacitated routing problems", *Math. Oper. Res.* (to appear).

[7] R.M. Karp, J.K. Lenstra, C. McDiarmid and A.H.G. Rinnooy Kan, "Probabilistic analysis", in: M. Ohegeartaigh, J.K. Lenstra and A.H.G. Rinnooy Kan, *Combinatorial optimization: annotated bibliographies*, Wiley, Chichester, England, 1984.

[8] B.J. Lageweg, J.K. Lenstra, A.H.G. Rinnooy Kan and L. Stougie, "Stochastic integer programming by dynamic programming", elsewhere in this book (1984).

[9] J.K. Lenstra, A.H.G. Rinnooy Kan, L. Stougie, "A framework for the design and analysis of hierarchical planning systems", *Ann. Oper. Res.* 1(1984), 23–42.

[10] A. Marchetti Spaccamela, A.H.G. Rinnooy Kan, L. Stougie, "Hierarchical vehicle routing", *Networks* 14(1984), 571–586.

[11] D.W. Matula, "Improved bounds on the chromatic number of a graph", Abstract, Department of Computer Science, Southern Methodist University, Dallas, TX (1983).

[12] M. Meanti, A.H.G. Rinnooy Kan, L. Stougie and C. Vercellis, "Asymptotic characterization of the optimal value of the Multi-Knapsack Problem", Technical report, Centre for Mathematics and Computer Science, Amsterdam, 1985.

[13] J.M. Steele, "Subadditive Euclidean functionals and nonlinear growth in geometric probability", *Ann. Probab.* 9(1981), 365–376.

[14] L. Stougie, "Design and analysis of solution methods for stochastic integer programming problems", forthcoming.

[15] E. Zemel, "Probabilistic analysis of geometric location problems", *Ann. of Oper. Res.* 1(1984).

PART III

Implementation

CHAPTER 9

A PROPOSED STANDARD INPUT FORMAT FOR
COMPUTER CODES WHICH SOLVE STOCHASTIC
PROGRAMS WITH RECOURSE

J. Edwards

Abstract

We explain our suggestions for standardizing input formats for computer codes
which solve stochastic programs with recourse. The main reason to set some
conventions is to allow programs implementing different methods of solution
to be used interchangeably. The general philosophy behind our design is a) to
remain fairly faithful to the *de facto* standard for the statement of LP problems established by IBM for use with MPSX and subsequently adopted by the
authors of MINOS, b) to provide sufficient flexibility so that a variety of problems may be expressed in the standard format, c) to allow problems originally
formulated as deterministic LP to be converted to stochastic problems with a
minimum of effort, d) to permit new options to be added as the need arises.

9.1 Introduction

In the latter half of 1984, the Adaptive Optimization project of the Systems
and Decision Sciences program at the International Institute for Applied Systems Analysis collected a number of computer programs written to solve various
problems in stochastic programming. Our goal was to organize these codes so
that they might be distributed on magnetic tape to researchers, who might
benefit from having several algorithms with which to experiment. However, we
came to realize that the process of tinkering with the various methods will be
greatly complicated because each program has its own format for input data.
We therefore developed a standard input format for stochastic programs with
recourse. To encourage and simplify its use, we based it on the input format
developed by IBM for the extended Mathematical Programming Subsystem
(MPSX) [1] and adopted by the authors of the Modular In-core Nonlinear Optimization System (MINOS) [2] and we wrote a number of low level subroutines
to read files written in the standard format.

9.2   The Problem

The general form of the stochastic program with recourse is taken to be

$$\text{minimize}\quad cx + Q(x)$$

$$\text{subject to}\quad A \left\{ \begin{array}{c} \leq \\ = \\ \geq \end{array} \right\} b$$

$$\ell \leq x \leq u$$

where

$$Q(x) = \mathrm{E}_\chi \left\{ \min_{y \in Y} q(y, \chi) \,|\, \mathrm{T}(\chi)x + \mathrm{W}(\chi)y = p(\chi) \right\},$$

x and y denote the decision and recourse variables, respectively, χ denotes an event, $\mathrm{T}(\chi)$ and $\mathrm{W}(\chi)$ denote the technology and recourse matrices, respectively, and E_χ denotes expectation. In subsequent references to the technology matrix, the recourse matrix, the stochastic right hand side, $p(\chi)$, and the penalty function, $q(y, \chi)$, we omit the arguments y and χ.

9.3   Organization of the Data: Control, Core, and Stochastics Files

The data required by a program written to solve the stochastic program in (1) can be divided logically into three files: a control file, a "core" file, and a "stochastics" file. Roughly speaking, the control file contains any data particular to the program and the core and stochastics files contain the data that define the problem.

As its name implies, the control file contains any information that is used to guide the execution of the program. For example, the control file might include a limit on the number of steps permitted and a tolerance for convergence if the algorithm implemented in the program were iterative in nature, file name and unit number assignments if the program required several files, or upper limits on the amount of storage needed if the program allocated array space "dynamically". The control file also contains any information that must be read before the program profitably can read the contents of the matrices and vectors that appear in the problem, e.g., the dimensions of those structures. Because the contents of the control file depend heavily on the algorithm employed and the manner in which it is implemented, we have not included a standard format for control files. Indeed, the rigid structure of the format we propose (particularly its strict use of specific columns as field delimiters) makes it unsuitable for application to files whose contents are liable to change frequently.

The core file contains the bounds on the decision vector, x, the contents of the matrices by which it is multiplied, and the contents and ranges of the rows of the deterministic right hand side vector, b. The core file for a stochastic LP thus corresponds in large measure to the data file that MPSX or MINOS would require to solve the equivalent nonstochastic LP (i.e., the same problem with $Q(x)$ removed).

The stochastics file defines the technology matrix, T, the distribution of the rows of the stochastic right hand side vector, p, the contents of the recourse matri $\langle$, W, and the function q. We have chosen to partition the input in this fashion so that a problem originally formulated as a linear program and expressed in standard MPSX format may be augmented later by a stochastics file, thereby permitting certain elements (e.g., the right hand side) to be stochastic.

9.4 Overview of the Standard Input Format

The proposed format is quite similar to the MPSX format, which is described on pages 199 through 209 of [1], although there are some differences. As in the MPSX format, each data file contains a number of sections, some of which are optional. A "header line" (or "header") marks the beginning of each section*. Most sections contain data lines. A data line is divided into six fields, some of which may be empty. Specific columns delineate field boundaries. There are three name fields, two numeric fields, and a code field. The columns that constitute these fields are

- columns 2 and 3: code field
- columns 5 through 12: first name field
- columns 15 through 22: second name field
- columns 25 through 36: first numeric field
- columns 40 through 47: third name field
- columns 50 through 61: second numeric field

(all column ranges are inclusive). Comment lines contain an asterisk (*) in the first column and may appear anywhere.

Unlike the MPSX format, names may contain imbedded blanks or leading blanks (although this last is not recommended). The contents of the name fields are interpreted as character strings, so names may begin with a digit. All lower case letters in the code and name fields are translated to their upper case equivalents. Values in the numeric fields must contain a decimal point. The MPSX convention concerning comments following a dollar sign ($) in the first column of the second or third name fields has not been adopted as part of the standard format.

Following are descriptions of each of the data files. Each description contains a list of the sections that constitute the corresponding data file. These sections must appear in the data file in the same order as they appear in the list, although sections marked "optional" need not appear at all.

---

* ([1], p. 199) uses the term "indicator card" rather than header.

9.5　The Core File

The core file specifies

- the linear portion of the objective, c,
- the contents of the constraint matrix, A, and possibly the contents of the technology matrix, T, and of the recourse matrix, W,
- the deterministic right hand side, b,
- the bounds on the decision vector, x, and
- the ranges on the right hand side.

The core file contains the following sections: NAME, ROWS, COLUMNS, RHS, RANGES, BOUNDS, and ENDATA. These sections assume more or less the same role in the standard format as they do in the MPSX format. Therefore, we give only an abbreviated description of these sections and note differences between the standard format and the MPSX format.

(1) **NAME** - This is an informative header line (the section contains no data lines). The user may enter any characters desired in columns 15 through 72 (the MPSX format restricts names to eight alphanumeric characters).

(2) **ROWS** - As in the MPSX format, this section specifies the names of the rows of A, the name of the row in the COLUMNS section that contains the elements of c, and the type of constraint (equality or inequality) represented by each row. In some cases, this section also specifies the names of the rows of T. Rows formed by a linear combination of two other rows (type "D" rows) and scaling of rows (use of the "'SCALE"' keyword) are supported in the MPSX format but are not permitted in the standard format.

(3) **COLUMNS** - As in the MPSX format, this section specifies the names of the columns of A and of c and contains the values of the nonzero elements of A and of c. In some cases, this section also specifies the names of the columns of W, contains the nonzero elements of W, and/or contains the nonzero elements of T. Scaling of columns (use of the "'SCALE"' keyword) is supported in the MPSX format but is not permitted in the standard format.

(4) **RHS** - This section specifies the names of the rows of b and contains the values of the nonzero elements of b. This section is identical to its counterpart in the MPSX format.

(5) **RANGES** (optional) - This section specifies the ranges on the rows of b. This section is identical to its counterpart in the MPSX format.

(6) **BOUNDS** (optional) - This section specifies the bounds on the rows of the decision vector, x. This section is identical to its counterpart in the MPSX format.

(7) **ENDATA** - This line marks the end of the core file (the section contains no data lines) and is identical to its counterpart in the MPSX format.

9.6 The Stochastics File

The stochastics file specifies

- the contents of the technology matrix, T,
- the distribution of the stochastic right hand side, p,
- the contents of the recourse matrix, W, and
- the form of the penalty function, q.

The stochastics file contains the following sections: NAME, TECHNOLOGY, DISTRIBUTIONS, RECOURSE, OBJECTIVES, and ENDATA. After the OB-JECTIVES section additional sections may appear containing data particular to a given algorithm. A program should read only those sections it needs from the file and should ignore the rest.

Most sections may take one of several forms, and the user must enter the name of one of them beginning in column 15 of the header line. A description of each of the sections, the forms they may assume, and their contents follows.

(1) **NAME** - This is an informative header line (the section contains no data lines). The user may enter any characters desired in columns 15 through 72.

(2) **TECHNOLOGY** - This section specifies the contents of T. The section may take one of the forms whose names follow:

DETERMINISTIC (the elements of T follow) - The technology matrix is given by the data following the section header. The format of the data is identical to that of the COLUMNS section of the core file, i.e., the contents of the matrix are specified in column order. The first name field on a line (columns 5 through 12) contains the name of the column. The remaining name/numeric field pairs (columns 15 through 22/25 through 36 and 40 through 47/50 through 61) specify a row name and the contents of the matrix at the position given by the row and column names. The row names form a subset of the row names in the ROWS section of the core file.

CORE (the elements of T appear in the core file) - The data consists of a list of names which form a subset of the names specified in the ROWS section of the core file. The contents of these rows (as specified in the COLUMNS section of the core file) constitute the technology matrix. One name appears per line, in the first name field (columns 5 through 12).

STOCHASTIC (the elements of T are supplied by a subroutine) - The data consists of a list of the names of the rows of the technology matrix. Each row name has associated with it one or more column names. The column names specify the active columns within the given row and form a subset of the column names specified in the COLUMNS section of the core file. The values for the technology matrix do not appear in either data file but are supplied by a subroutine written by the user. The row names appear in the first name field of a line (columns 5 through 12) and the other two name fields (columns 15 through 22 and 40 through 47) are available for the column names.

NONE (no data) - There is no data. The user must decide where and how to obtain the necessary values.

(3) **DISTRIBUTIONS** - This section specifies the distribution of the rows of p. The section may take one of the forms whose names follow:

DISCRETE (each row is independently distributed) - Each row of p may take one of a fixed number of values. The data for this form consists of a number of "definitions", which are analogous to the "vectors" in the RANGES and BOUNDS sections of the core file (see [1]). Each definition specifies the distribution of every row of p and consists of a number of sets of entries of the form "defname rowname value probability". Within a given definition, there is one such set for each of the rows named in the TECHNOLOGY section. "defname" is the name of the definition to which the entry belongs; it occupies the first name field on a line (columns 5 through 12). "rowname" is the name of the row associated with the entry; it occupies the second name field on a line (columns 15 through 22). "value" and "probability" are a value for the row and its likelihood, respectively. They occupy the first and second numeric fields (columns 25 through 36 and 50 through 61), respectively.

The sum of the probabilities for a given row must be unity. The values specified for a given row must be distinct. Entries for different rows or different definitions must not be mixed together in the input file.

As an example, let the T matrix have two rows, TROW1 and TROW2, and define two distributions for the rows of p as follows:

$$
\text{Row 1} = \begin{cases} 1 \\ 2 \\ 3 \\ 4 \end{cases} \text{ with probability } \begin{cases} 0.4 \\ 0.2 \\ 0.2 \\ 0.2 \end{cases} \tag{1}
$$

$$
\text{Row 2} = \begin{cases} 8 \\ 9 \\ 0 \end{cases} \text{ with probability } \begin{cases} 0.6 \\ 0.3 \\ 0.1 \end{cases}
$$

and

$$
\text{Row 1} = \begin{cases} 2 \\ 4 \end{cases} \text{ with probability } \begin{cases} 0.5 \\ 0.5 \end{cases} \tag{2}
$$

$$
\text{Row 2} = 2 \text{ with probability } 1.0.
$$

The contents of the name and numeric fields for these distributions are shown in Table 9.1. The user specifies which is the desired definition (our definition names "DIST1" and "DIST2" were chosen arbitrarily) when the appropriate input utility is called. Note that every value contains a decimal point.

Table 9.1   Contents of a sample DISCRETE DISTRIBUTIONS section.

First Name Field	Second Name Field	First Numeric Field	Second Numeric Field
DIST1	TROW1	1.0	0.4
DIST1	TROW1	2.0	0.2
DIST1	TROW1	3.0	0.2
DIST1	TROW1	4.0	0.2
DIST1	TROW2	8.0	0.6
DIST1	TROW2	9.0	0.3
DIST1	TROW2	0.0	0.1
DIST2	TROW1	2.0	0.5
DIST2	TROW1	4.0	0.5
DIST2	TROW2	2.0	1.0

SIMULATION (the rows are supplied by a subroutine) - There are no data lines in this case. The program obtains its values from a subroutine written by the user.

PIECEWISE (piecewise constant pdf) - Each row of p takes a value within one of a finite number of ranges. Within a range, all values are equally likely. However, within a set of ranges, all ranges are *not* equally likely. The data for this form consists of a number of "definitions", which are analogous to the "vectors" in the RANGES and BOUNDS sections of the core file (see [1]). Each definition specifies the distribution of every row of p and consists of a number of sets of entries of three lines each. Within a given definition, there is one such set for each of the rows named in the TECHNOLOGY section. Each three line entry within a set describes a range for the row associated with the set. The first line in an entry contains the letters "PC" in the code field (columns 2 and 3), the name of the definition to which the entry belongs in the first name field (columns 5 through 12), the name of the row with which this range is associated in the second name field (columns 15 through 22), and the probability that the row takes a value within the range in the first numeric field (columns 25 through 36). The second and third lines in an entry specify the upper and lower bounds of the range. For both bounds, the code field contains the letters "BD", the first name field contains the name of the definition to which the entry belongs, the second name field contains the name of the row with which the range is associated, and the first numeric field contains the bound value.

The sum of the probabilities for the ranges for a given row must be unity. Entries for different rows, different ranges, or different definitions must not be mixed together in the input file.

As an example, let the T matrix have two rows, TROW1 and TROW2, and define two distributions for the rows of p as follows:

$$\text{Row 1 in } \left\{ \begin{array}{c} [1,2] \\ [3,4] \end{array} \right. \text{ with probability } \left\{ \begin{array}{c} 0.6 \\ 0.4 \end{array} \right. \tag{1}$$

$$\text{Row 2 in } \left\{ \begin{array}{c} [5,7] \\ [1,3] \\ [0,1] \end{array} \right. \text{ with probability } \left\{ \begin{array}{c} 0.7 \\ 0.1 \\ 0.2 \end{array} \right.$$

and

$$\text{Row 1 in } \left\{ \begin{array}{c} [2,4] \\ [5,9] \end{array} \right. \text{ with probability } \left\{ \begin{array}{c} 0.5 \\ 0.5 \end{array} \right. \tag{2}$$

The contents of the code, name and numeric fields for these distributions are shown in Table 9.2. The user specifies which is the desired definition (our definition names "DIST1" and "DIST2" were chosen arbitrarily) when the appropriate input utility is called. Note that every value contains a decimal point.

SCENARIOS (the value of p is defined by a sample of vectors) - The p vector may take one of a finite number of values. The data for this form consists of a

Table 9.2   Contents of a sample PIECEWISE DISTRIBUTIONS section.

Code Field	First Name Field	Second Name Field	First Numeric Field
PC	DIST1	TROW1	0.6
BD	DIST1	TROW1	1.0
BD	DIST1	TROW1	2.0
PC	DIST1	TROW1	0.4
BD	DIST1	TROW1	3.0
BD	DIST1	TROW1	4.0
PC	DIST1	TROW2	0.7
BD	DIST1	TROW2	5.0
BD	DIST1	TROW2	7.0
PC	DIST1	TROW2	0.1
BD	DIST1	TROW2	1.0
BD	DIST1	TROW2	3.0
PC	DIST1	TROW2	0.2
BD	DIST1	TROW2	0.0
BD	DIST1	TROW2	1.0
PC	DIST2	TROW1	0.5
BD	DIST2	TROW1	2.0
BD	DIST2	TROW1	4.0
PC	DIST2	TROW1	0.5
BD	DIST2	TROW1	5.0
BD	DIST2	TROW1	9.0
PC	DIST2	TROW2	1.0
BD	DIST2	TROW2	2.0
BD	DIST2	TROW2	3.0

number of "definitions", which are analogous to the "vectors" in the RANGES and BOUNDS sections of the core file (see [1]). Each definition provides a sample of vectors and consists of sets of entries giving a value for p and the probability that p takes that value. The first line in each entry contains the letters "SC" in the code field (columns 2 and 3), the name of the definition to which the entry belongs in the first name field (columns 5 through 12), a name identifying the scenario in the second name field (columns 15 through 22), and the probability that p takes the value associated with this scenario in the first numeric field (columns 25 through 36). Subsequent lines specify the values that the rows of p assume under the scenario. There must be one of these lines for each row named in the TECHNOLOGY section. The code field of these lines contains the letters "RV", the first name field contains the name of the definition to which the entry belongs, the second name field contains the name of the row whose value the line specifies, and the first numeric field contains the value.

The sum of the probabilities for the scenarios in a given definition must be unity. Entries for different scenarios or different definitions must not be mixed together in the input file.

As an example, let the T matrix have two rows, TROW1 and TROW2, and define two distributions of the vector p as follows:

$$\text{Vector } = \begin{cases} \begin{bmatrix} 12 \\ 34 \\ 56 \end{bmatrix} \text{ with probability } \begin{cases} 0.5 \\ 0.3 \\ 0.2 \end{cases} \end{cases} \qquad (1)$$

and

$$\text{Vector } = \begin{cases} \begin{bmatrix} 12 \\ 15 \end{bmatrix} \text{ with probability } \begin{cases} 0.5 \\ 0.5 \end{cases} \end{cases} \qquad (2)$$

The contents of the code, name and numeric fields for these distributions are shown in Table 9.3. The user specifies which is the desired definition (our definition names SAMP1 and SAMP2 were chosen arbitrarily) when the appropriate input utility is called. The scenario names SCEN1, SCEN2, and SCEN3 where chosen arbitrarily. Note that every value contains a decimal point.

Table 9.3 Contents of a sample SCENARIOS DISTRIBUTIONS section.

Code Field	First Name Field	Second Name Field	First Numeric Field
SC	SAMP1	SCEN1	0.5
RV	SAMP1	TROW1	1.0
RV	SAMP1	TROW2	2.0
SC	SAMP1	SCEN2	0.3
RV	SAMP1	TROW1	3.0
RV	SAMP1	TROW2	4.0
SC	SAMP1	SCEN3	0.2
RV	SAMP1	TROW1	5.0
RV	SAMP1	TROW2	5.0
SC	SAMP2	SCEN1	0.5
RV	SAMP2	TROW1	1.0
RV	SAMP2	TROW2	2.0
SC	SAMP2	SCEN2	0.5
RV	SAMP2	TROW1	1.0
RV	SAMP2	TROW2	5.0

NONE (no data) - There is no data. The user must decide where and how to obtain the necessary values.

(4) **RECOURSE** - This section specifies the contents of W. The section may take one of the forms whose names follow:

SIMPLE (simple recourse) - There are no data lines in this case. The recourse matrix is assumed to be $[I, -I]$, where I has rank equal to the number of rows in the technology matrix.

DETERMINISTIC (the elements of W follow) - The recourse matrix is given by the data following the section header. The format of the data is identical to that of the COLUMNS section of the core file, i.e., the contents of the matrix are specified in column order. The first name field on a line (columns 5 through 12) contains the name of the column. The remaining name/numeric field pairs (columns 15 through 22/25 through 36 and 40 through 47/50 through 61) specify a row name and the contents of the matrix at the position given by the row and column names. The row names form a subset of the row names in the TECHNOLOGY section.

CORE (the elements of W appear in the core file) - The data consists of a list of names which form a subset of the column names specified in the COLUMNS section of the core file. The contents of those columns (as specified in the COLUMNS section of the core file) constitute the recourse matrix. One name appears per line, in the first name field (columns 5 through 12).

STOCHASTIC (the elements of W are supplied by a subroutine) - The data consists of a list of the names of the rows of the recourse matrix. Associated with each name is one or more column names. These column names specify the active columns within the given row and form a subset of the column names specified in the COLUMNS section of the core file. The values for the recourse matrix do not appear in either data file but are supplied by a subroutine written by the user. The row names appear in the first name field of a line (columns 5 through 12) and the other two name fields (columns 15 through 22 and 40 through 47) are available for the column names.

NONE (no data) - There is no data. The user must decide where and how to obtain the necessary values.

(5)   **OBJECTIVES** - This section specifies the form of q. The section may take one of the forms whose names follow:

LINEAR (q is a linear function) - The recourse objective is given by $q(y) = qy$, where q is given by the data following the section header. The data for this form consists of a number of "definitions", which are analogous to the "vectors" in the RANGES and BOUNDS sections of the core file (see [1]). Each definition specifies the elements of q and consists of entries of the form "defname name value", where "defname" is the name of the definition to which the entry belongs, "name" is the name of a column of W (or of a row of T; see below) and "value" is the value for the corresponding row of q. "defname" occupies the first name field on a line (columns 5 through 12), "name" occupies the second name field (columns 15 through 22) and "value" occupies the first numeric field (columns 25 through 36).

Entries for different definitions must not be mixed together in the input file.

As an example, let the W matrix have two columns, WCOL1 and WCOL2, and define two vectors q as follows:

$$q = [79] \tag{1}$$

and

$$q = [33] \tag{2}$$

The contents of the name and numeric fields for these vectors are shown in Table 9.4. The user specifies which is the desired definition (our definition names "VEC1" and "VEC2" were chosen arbitrarily) when the appropriate input utility is called. Note that every value contains a decimal point.

Table 9.4   Contents of a sample LINEAR OBJECTIVES section.

First Name Field	Second Name Field	First Numeric Field
VEC1	WCOL1	7.0
VEC1	WCOL2	9.0
VEC2	WCOL1	3.0
VEC2	WCOL2	3.0

PIECEWISE (q is two-piece linear) - The recourse objective is assumed to be two-piece continuous about zero, i.e.

$$q(y) = \sum_i q_i(y_i) \text{ with } q_i(y_i) = \begin{cases} -q_i^+ y_i, & y_i \leq 0 \\ q_i^- y_i & y_i \geq 0 \end{cases}$$

The data for this form consists of a number of "definitions", which are analogous to the "vectors" in the RANGES and BOUNDS sections of the core file (see [1]). Each definition specifies the values of q_i^+ and q_i^- for all i and consists of entries of the form "defname name value value", where "defname" is the name of the definition to which the entry belongs, "name" is the name of a column of W (or of a row of T; see below), the first value gives the corresponding value of q^+, and the second value gives the corresponding value of q^-. The names occupy the first and second name fields on a line (columns 5 through 12 and 15 through 22) and the values occupy the first and second numeric fields (columns 25 through 36 and 50 through 61).

Entries for different definitions must not be mixed together in the input file.

As an example, let the W matrix have two columns, WCOL1 and WCOL2, and define two vectors q as follows:

$$q = \left[\left\{ \begin{array}{l} -2, \ y_1 \leq 0 \\ 5, \ y_1 \leq 0 \end{array} \right. \quad \left\{ \begin{array}{l} -3, \ y_2 \leq 0 \\ 7, \ y_2 \leq 0 \end{array} \right. \right] \tag{1}$$

and

$$q = \left[\left\{ \begin{array}{l} -5, \ y_1 \leq 0 \\ 3, \ y_1 \leq 0 \end{array} \right. \quad \left\{ \begin{array}{l} -9, \ y_2 \leq 0 \\ 2, \ y_2 \leq 0 \end{array} \right. \right] \tag{2}$$

The contents of the name and numeric fields for these vectors are shown in Table 9.5. The user specifies which is the desired definition (our definition names VEC1 and VEC2 were chosen arbitrarily) when the appropriate input utility is called. Note that every value contains a decimal point and that the values of q_i^+ are positive.

Table 9.5 Contents of a sample OBJECTIVES (PIECEWISE) section.

First Name Field	Second Name Field	First Numeric Field	Second Numeric Field
VEC1	WCOL1	2.0	5.0
VEC1	WCOL2	3.0	7.0
VEC2	WCOL1	5.0	3.0
VEC2	WCOL2	9.0	2.0

NONE (no data) - There is no data. The user must decide where and how to obtain the necessary values.

Note - if the recourse matrix is simple (i.e., if there are no column names for W), row names of T are substituted for column names of W in the OBJECTIVES section.

(6) **ENDATA** - This line marks the end of the stochastics file (the section contains no data lines).

It is clear that we have covered only a few of the possibilities for most of the above sections. However, the format is such that new forms can be added as the need arises.

References

[1] IBM (International Business Machines, Inc.), (1972) *Mathematical Programming Subsystem - Extended (MPSX) and Generalized Upper Bounding (GUB) Program Description* document number SH20-0968-1

[2] B.A. Murtagh and M.A. Saunders (1977), *MINOS - A Large-Scale Nonlinear Programming System (For Problems with Linear Constraints) - User's Guide*, Technical Report SOL 77-9, Systems Optimization Laboratory, Department of Operations Research, Stanford University

CHAPTER 10

A COMPUTER CODE FOR SOLUTION OF PROBABILISTIC-CONSTRAINED STOCHASTIC PROGRAMMING PROBLEMS

T. Szantai

10.1 Introduction

The theory of logarithmic concave measures was developed by A. Prékopa [1, 2]. Due to this theory it became possible to handle joint probabilistic constraints in the stochastic programming problems. These constraints are of the form

$$P(d_i x \geq \beta_i, \qquad i = 1, \ldots, s) \geq p, \qquad (10.1)$$

where the random variables $\beta_1, \ldots, \beta_s$ have a logconcave joint distribution. For the calculation of the probability value (10.1) one can apply multi-dimensional integration techniques. Unfortunately these methods have an extremely slow convergence in higher dimensions. In these cases only Monte Carlo methods are applicable and this is the reason why the probabilistic-constrained stochastic programming problems of this type can not be solved efficiently by standard nonlinear programming codes. In the last ten years many test problems have been solved and many real applications have been worked out. All of these works required development an individual computer code suitable for the special problem to be solved.

In this paper we give a short description of a computer code which intends to solve a relatively wide class of probabilistic-constrained stochastic programming problems. In the last section we also give the results for some simple test problems.

The computer code is contained in the collection of experimental computer codes assembled by the Adaptation and Optimization (ADO) project of the Systems and Decision Sciences (SDS) program at the International Institute for Applied Systems Analysis (IIASA). This collection is available on computer tape to researchers. The tape contains a User's Manual for each program as well.

10.2   The Solution Method

We solve probabilistic-constrained stochastic programming problems of the form

$$\text{minimize} \quad c_1 x_1 + \cdots + c_n x_n a$$
$$\text{subject to} \quad Ax = b$$
$$x \geq 0 \tag{10.2}$$
$$\text{and} \quad P(DX \geq \beta) \geq p,$$

where A is a known $m \times n$ matrix, D is a known $s \times n$ matrix, b is known and of the appropriate dimension, p is the prescribed probability level, and $\beta_1, \ldots, \beta_s$ have joint normal probability distribution with expected values

$$E(\beta_1) = \mu_1, \ldots, E(\beta_s) = \mu_s,$$

with variances

$$D^2(\beta_1) = \sigma_1^2, \ldots, D^2(\beta_s) = \sigma_s^2,$$

and with the correlation matrix

$$R = \begin{bmatrix} 1 & r_{12} & \cdots & r_{1s} \\ r_{21} & 1 & \cdots & r_{2s} \\ \cdots & \cdots & \cdots & \cdots \\ r_{s1} & r_{s2} & \cdots & 1 \end{bmatrix}.$$

In problem (10.2) the linear constraints may include inequalities as well and explicit upper bounds on the variables can be specified.

For the solution of problem (10.2) we apply Veinott's supporting hyperplane algorithm. This algorithm solves general nonlinear programming problems and it is especially practical when the problem has just one nonlinear constraint above the possibly large number of linear constraints. A complete description of the algorithm is given in Veinott [4]. Here we give only details which are related to the stochastic feature of the problem.

To obtain a starting point in the interior of the feasible domain one can solve the linear programming problem

$$\text{minimize} \quad \sum_{i=1}^{n} (d_{i1} x_1 + \ldots + d_{in} x_n - \mu_i)/\sigma_i$$
$$\text{subject to} \quad Ax = b \tag{10.3}$$
$$Dx \geq \mu + t\sigma$$
$$x \geq 0$$

where d_{ij} is the element of D in the i-th row and j-th column and t is a constant. The value of parameter t should be chosen based on the desired probability level, p. For high probabilities the value 3 is recommended. If the optimal solution of the linear programming problem (10.3) turns out not to be an interior point of the feasible domain, i.e. it does not satisfy the probabilistic constraint, one should try to solve problem (10.3) with a larger value of the parameter t. Of course when choosing a relatively large parameter value t it may be that the linear programming problem (10.3) will not have any feasible solution. Experience shows that the selection of an appropriate parameter value is not difficult.

To obtain a starting point outside the feasible domain, the program solves the linear programming problem

$$\text{minimize} \quad Ax = b$$
$$\text{subject to} \quad x \geq 0 \tag{10.4}$$

In the case of an unbounded objective, one must provide additional constraints on the variables which do not disturb the probabilistic constraint.

To find the boundary point of the probabilistic constraint at each iteration we use an interval bisection algorithm with a sophisticated stopping rule. Let denote x_{in} the actual point in the interior of the feasible domain and x_{out} the point outside the feasible domain. We want to determine the value λ for which

$$x_\lambda = x_{\text{out}} + \lambda(x_{\text{in}} - x_{\text{out}}), \quad 0 < \lambda < 1$$

and

$$P(d_i x_\lambda \geq \beta_i, \quad i = 1, \ldots, s) = p.$$

In an earlier paper (see Szántai [3]) we published a method for constructing good lower and upper bounds on the probability values of type (10.1). This method is based on the so called Bonferroni inequalities. First of all one can reduce the size of the uncertainty interval by means of these bounds. Let us denote

$$P_{\text{lower}}(d_i x \geq \beta_i, \quad i = 1, \ldots, s)$$

and

$$P_{\text{upper}}(d_i x \geq \beta_i, \quad i = 1, \ldots, s)$$

the lower and upper bounds of the probability value (10.1). Then we can find first the values λ_{lower} and λ_{upper} for which

$$P_{\text{lower}}(d_i x_{\lambda_{\text{lower}}} \geq \beta_i, \quad i = 1, \ldots, s) = p$$

and

$$P_{\text{upper}}(d_i x_{\lambda_{\text{upper}}} \geq \beta_i, \quad i = 1, \ldots, s) = p.$$

It is clear that we may restrict the search on the interval $(\lambda_{\text{lower}}, \lambda_{\text{upper}})$ instead of the interval $(0,1)$.

We calculate the probability values by Monte Carlo simulation. Whereas we apply a variance reduction technique (see Szántai [3]) the calculation of the probability value (10.1) involves some errors. So we should take special care to stop outside of the feasible domain rather than inside. For this purpose we apply a modified stopping rule in the interval bisection algorithm which is as follows:

1. If $P(d_i x_{\lambda_{half}} \geq \beta_i, \quad i = 1, \ldots, s) \geq p + \varepsilon$ then let $\lambda_{upper} = \lambda_{half}$ and repeat the bisection.

2. If $P(d_i x_{\lambda_{half}} \geq \beta_i, \quad i = 1, \ldots, s) \leq p - 2\varepsilon$ then let $\lambda_{lower} = \lambda_{half}$ and repeat the bisection.

3. If $p - 2\varepsilon < P(d_i x_{\lambda_{half}} \geq \beta_i, \quad i = 1, \ldots, s) \leq p - \varepsilon$ then stop, $x_{\lambda_{half}}$ is a boundary point with the prescribed tolerance.

4. If $p - \varepsilon < P(d_i x_{\lambda_{half}} \geq \beta_i, \quad i = 1, \ldots, s) < p + \varepsilon$ then make a new, more accurate evaluation of the probability value (i.e. use more random numbers in the Monte Carlo simulation). Now

   (a) If $P_{new}(d_i x_{\lambda_{half}} \geq \beta_i, \quad i = 1, \ldots, s) > p$ then let $\lambda_{upper} = \lambda_{half}$ and repeat the bisection.

   (b) If $P_{new}(d_i x_{\lambda_{half}} \geq \beta_i, \quad i = 1, \ldots, s) \leq p$ then stop, $x_{\lambda_{half}}$ is a boundary point with the prescribed tolerance.

Here ε is the prescribed tolerance, $x_{\lambda_{half}}$ the point of the actual search interval and P_{new} the more accurate probability value. The four cases are illustrated on Figure 10.1.

Case 1.

$p + \varepsilon$

p Case 4.

$p - \varepsilon$

Case 3.

$p - 2\varepsilon$

Case 2.

Figure 10.1 The stopping rule illustrated.

For constructing the supporting hyperplane it is necessary to calculate the gradient vector of the probability (10.1) as a function of the variables x at the actual boundary point. The partial derivatives of the probability (10.1) can be expressed by means of the conditional probabilities. As in the case of normal distribution the conditional distributions are normal too, and we can apply the same Monte Carlo simulation for the gradient vector calculation as before.

The supporting hyperplane algorithm stops when for the actual point outside the feasible domain

$$P(d_i x_{out} \geq \beta_i, \quad i = 1, \ldots, s) \geq p - \varepsilon.$$

In this case we accept the last boundary point as the optimal solution of stochastic programming problem (10.2).

10.3 A Test Problem

Let us consider a coffee company marketing three different blends of coffee No. 1, 2 and 3. The coffee company has developed a rigid set of requirements for

		No. 1	No. 2	No. 3
	acidity	≤ 3.5	≤ 4.0	≤ 5.0
	caffeine	≤ 2.8	≤ 2.2	≤ 2.4
	liquoring value	≥ 7.0	≥ 6.0	≥ 5.0
	hardness	≤ 2.5	≤ 3.0	≤ 7.8
each of its 3 blends:	aroma	≥ 7.0	≥ 5.0	≥ 4.0

Forecasts indicate that the demands for the company's three blends during the coming month will be as follows:

blend No. 1.   3,000 pounds
blend No. 2.   40,000 pounds
blend No. 3.   20,000 pounds

On the first day of a particular month the company found that its available supply of green coffees was limited to eight different types as indicated in the following table. According to this table, these coffees vary according to (1) price, (2) quantity available, and (3) taste characteristics.

green coffee type	price per pound	available supply in pounds	acidity (pH)	percent caffeine content	liquoring value	hardness index	aroma index
type 1	0.35	24,000	4.0	1.8	6	2	8
type 2	0.20	74,000	4.5	1.0	5	7	4
type 3	0.44	5,000	3.0	3.0	8	2	7
type 4	0.41	20,000	4.0	2.0	6	2	7
type 5	0.36	5,000	3.5	1.5	6	3	9
type 6	0.34	4,000	3.6	1.1	6	4	7
type 7	0.36	5,000	3.2	1.4	6	3	8
type 8	0.19	100,000	5.1	1.7	5	9	1

The company is confronted with the problem of determining an optimum combination of available green coffees for next month's roasting operation. We may regard the demands for the company's 3 blends during the coming month as normally distributed random variables with expected values equal to the forecasts listed above. Then the company should determine an optimum combination of available green coffees so that the random demands will be met with a prescribed probability. Let x_{ij} be the amount of i-th type green coffee in the blend j. Then after some scaling we get the stochastic programming problem: minimize

$$
\begin{aligned}
(350x_{11} + 200x_{21} + 440x_{31} + 410x_{41} + 360x_{51} + 340x_{61} + 360x_{71} + 190x_{81} \\
350x_{12} + 200x_{22} + 440x_{32} + 410x_{42} + 360x_{52} + 340x_{62} + 360x_{72} + 190x_{82} \\
350x_{13} + 200x_{23} + 440x_{33} + 410x_{43} + 360x_{53} + 340x_{63} + 360x_{73} + 190x_{83})
\end{aligned}
$$

$$
\begin{array}{ll}
x_{11} + x_{12} + x_{13} \leq 25 & x_{51} + x_{52} + x_{53} \leq 5 \\
x_{21} + x_{22} + x_{23} \leq 75 & x_{61} + x_{62} + x_{63} \leq 4 \\
x_{31} + x_{32} + x_{33} \leq 5   & x_{71} + x_{72} + x_{73} \leq 5 \\
x_{41} + x_{42} + x_{43} \leq 20 & x_{81} + x_{82} + x_{83} \leq 100
\end{array}
$$

$0.5x_{11}$	$+x_{21}$	$-0.5x_{31}$	$+0.5x_{41}$		$+0.1x_{61}$	$-0.3x_{71}$	$+1.6x_{81}$	≤ 0
$-x_{11}$	$-1.8x_{21}$	$+0.2x_{31}$	$-0.8x_{41}$	$-1.5x_{51}$	$-1.7x_{61}$	$-1.4x_{71}$	$-1.1x_{81}$	≤ 0
$-x_{11}$	$-2x_{21}$	$+x_{31}$	$-x_{41}$	$-x_{51}$	$-x_{61}$	$-x_{71}$	$-2x_{81}$	≥ 0
$-0.5x_{11}$	$+4.5x_{21}$	$-0.5x_{31}$	$-0.5x_{41}$	$+0.5x_{51}$	$+1.5x_{61}$	$+0.5x_{71}$	$+6.5x_{81}$	≤ 0
x_{11}	$-3x_{21}$			$+2x_{51}$		x_{71}	$-6x_{81}$	≥ 0
	$0.5x_{22}$	$-x_{32}$		$-0.5x_{52}$	$-0.4x_{62}$	$-0.8x_{72}$	$+1.1x_{82}$	≤ 0
$-0.4x_{12}$	$-1.2x_{22}$	$+0.8x_{32}$	$-0.2x_{42}$	$-0.7x_{52}$	$-1.1x_{62}$	$-0.8x_{72}$	$-0.5x_{82}$	≤ 0
	$-x_{22}$	$+2x_{32}$					$-x_{82}$	≥ 0
$-x_{12}$	$+4x_{22}$	$-x_{32}$	$-x_{42}$		$+x_{62}$		$+6x_{82}$	≤ 0
$3x_{12}$	$-x_{22}$	$+2x_{32}$	$+2x_{42}$	$+4x_{52}$	$+2x_{62}$	$+3x_{72}$	$-4x_{82}$	≥ 0
$-x_{13}$	$-0.5x_{23}$	$-2x_{33}$	$-x_{43}$	$-1.5x_{53}$	$-1.4x_{63}$	$-1.8x_{73}$	$+0.1x_{83}$	≤ 0
$-0.6x_{13}$	$-1.4x_{23}$	$+0.6x_{33}$	$-0.4x_{43}$	$-0.9x_{53}$	$-1.3x_{63}$	$-x_{73}$	$-0.7x_{83}$	≤ 0
x_{13}		$+3x_{33}$	$+x_{43}$	$+x_{53}$	$+x_{63}$	$+x_{73}$		≥ 0
$-5.8x_{13}$	$-0.8x_{23}$	$-5.8x_{33}$	$-5.8x_{43}$	$-4.8x_{53}$	$-3.8x_{63}$	$-4.8x_{73}$	$+1.2x_{83}$	≤ 0
$4x_{13}$		$+3x_{33}$	$+3x_{43}$	$+5x_{53}$	$+3x_{63}$	$+4x_{73}$	$-3x_{83}$	≥ 0

$$
P\begin{pmatrix}
x_{11} & +x_{21} & +x_{31} & +x_{41} & +x_{51} & +x_{61} & +x_{71} & +x_{81} & \geq \beta_1 \\
x_{12} & +x_{22} & +x_{32} & +x_{42} & +x_{52} & +x_{62} & +x_{72} & +x_{82} & \geq \beta_2 \\
x_{13} & +x_{23} & +x_{33} & +x_{43} & +x_{53} & +x_{63} & +x_{73} & +x_{83} & \geq \beta_3
\end{pmatrix} \geq p,
$$

where the random variables $\beta_1, \beta_2, \beta_3$ are normally distributed with expected values

$$
E(\beta_1) = 3, \qquad E(\beta_2) = 40, \qquad E(\beta_3) = 20,
$$

with variances

$$
D^2(\beta_1) = 0.25, \qquad D^2(\beta_2) = 25, \qquad D^2(\beta_3) = 9
$$

and with three different correlation matrices (in three different groups of the test problems):

$$
R_1 = \begin{bmatrix} 1 & 0.1 & 0.1 \\ 0.1 & 1 & 0.9 \\ 0.1 & 0.9 & 1 \end{bmatrix}, \quad
R_2 = \begin{bmatrix} 1 & 0 & 0 \\ 0 & 1 & 0 \\ 0 & 0 & 1 \end{bmatrix}, \quad
R_3 = \begin{bmatrix} 1 & 0.1 & 0.1 \\ 0.1 & 1 & -0.9 \\ 0.1 & -0.9 & 1 \end{bmatrix}.
$$

Some results concerning the test problems are:

	probability level	optimal value
1. *Positive correlations* (R_1)		
deterministic problem	0.228	18500.0
stochastic problem No. 1	0.9	22564.0
stochastic problem No. 2	0.95	23603.6
stochastic problem No. 3	0.99	25500.6
2. *Independent case* (R_2)		
deterministic problem	0.125	18500.0
stochastic problem No. 1	0.9	22949.4
stochastic problem No. 2	0.95	23866.6
stochastic problem No. 3	0.99	25639.8
3. *Negative correlations* (R_3)		
deterministic problem	0.051	18500.0
stochastic problem No. 1	0.9	22961.6
stochastic problem No. 2	0.95	23885.2
stochastic problem No. 3	0.99	25680.6

In the above list the deterministic problem always means the linear programming problem with the forecasted demands. Its optimal solution has different probability levels according to the correlation matrices.

References

[1] A. Prékopa, "On Logarithmic Concave Measures with Application to Stochastic Programming", *Acta Scientiarum Mathematicarum* **32**(1971), 301–316.

[2] A. Prékopa, "Contributions to the Theory of Stochastic Programming", *Mathematical Programming* **4**(1973), 202–221.

[3] T. Szántai, "Evaluation of a Special Multivariate Gamma Distribution Function", *Mathematical Programming Study*, (1985) to appear.

[4] A.F. Veinott, "Supporting Hyperplane Method", *Operations Research* **15** (1967), 147–152.

CHAPTER 11

CONDITIONAL PROBABILITY AND CONDITIONAL EXPECTATION OF A RANDOM VECTOR

H. Gassmann

Abstract

Some problems in stochastic programming require the computation of conditional information on a multivariate random variable over an n-dimensional rectangle. For continuous distributions this involves a multidimensional integration and is thus a very hard problem. This paper describes various approximation methods in the case of the multivariate normal distribution along with numerical evidence of their performance. The extension to more general sets and other distributions such as the multi-gamma are discussed as well.

11.1 Introduction

Stochastic programming problems of the form

$$\min_x E_\xi \varphi(x, \xi),$$

where ξ is a random vector on some probability space (Ω, S, P) have been used extensively in the literature (see e.g. [2],[23],[24] and the references cited therein).

In principle, the above is a nonlinear programming problem and could be solved by ordinary NLP techniques. The reason why this is not done is that the evaluation of the objective function—and *a fortiori* of derivatives if they exist—is often extremely costly, since taking the expectation on ξ amounts to a multidimensional integration or sometimes a finite sum with a large number of terms.

Frequently $\varphi(x, \xi)$ is convex such that error bounds based on Jensen's inequality and on the Edmundson-Madansky inequality [10],[15] are available, and it is these bounds one works with rather than the function itself. Estimates are usually of the form

$$\sum_{i=1}^{I} p_i \varphi(x, \overline{\xi}^i) \leq E_x i \varphi(x, \xi) \leq \sum_{i=1}^{I} p_i u^i, \tag{11.1}$$

where p_i and $\overline{\xi}^i$ are the conditional probability and conditional mean of ξ given that $\xi \in A_i$. The set $\{A_i : i = 1, \ldots, I\}$ forms a partition of the sample space into polyhedral sets (bounded or unbounded), while u^i is some upper bound on the conditional expectation of $\varphi(x,\xi)$ given $\xi \in A_i$.

If ξ is a continuous random vector with distribution F, then $p_i = \int_{A_i} dF$, $\bar{\xi}^i = (\bar{\xi}_1^i, \ldots, \bar{\xi}_n^i) = \frac{1}{p_i}(q_1^i, \ldots, q_n^i)$, where $q_k^i = \int_{A_i} s_k dF(s)$. Evaluating these multidimensional integrals is a nontrivial problem in its own right, and it is this integration problem that the paper is concerned with. The main emphasis is on the multivariate normal distribution, and several probabilistic methods are developed in Sections 11.2-11.5 to compute p_i and q_k^i under the assumption that the partition $\{A_i\}$ consists of n-dimensional rectangles of the form

$$A_i = \prod_{j=1}^n [a_j^i, b_j^i].$$

Section 11.2 describes some trivial cases and estimation by a simple Monte-Carlo method. Deák's decomposition is presented in Section 11.3, Szántai's Bonferroni-type approach appears in Section 11.4. The two techniques are combined in Section 11.5 into a hybrid method which attempts to exploit the advantages of both. Numerical results are given in Section 11.6 to contrast the performance of these methods on a small number of sample problems. Section 11.7 discusses briefly some of the problems encountered when forming the quotient $\bar{\xi}_k^i = q_k^i/p_i$. In Section 11.8 we describe extensions of the various techniques to general polyhedral sets and a modification of Szántai's method to treat other multivariate distributions.

11.2   Multivariate Normal Distribution; Simple Monte-Carlo

Arguably the most commonly used continuous multivariate distribution is the multivariate normal distribution [13] whose density f is given by

$$f(z) = \frac{1}{(2pi)^{n/2}|\Sigma|^{1/2}} e^{-(z-\mu)'\Sigma^{-1}(z-\mu)}, \tag{11.2}$$

where n is the dimension of the random vector z, μ its mean and Σ its co-variance matrix, assumed symmetric and positive definite. The multivariate normal distribution possesses some attractive properties which will be used in the description of some of the methods in this paper. It is well known, for instance, that if $x \sim N(\mu, \Sigma)$, then $y = Cx \sim N(C\mu, C\Sigma C')$ for an arbitrary matrix C, the only proviso being that the product Cx be well defined. To simplify some of the presentation we shall assume given an n-dimensional rectangle $A = \prod_{i=1}^n [a_i, b_i]$, and a random vector $z \sim N(0, \Sigma)$ where Σ is a correlation matrix, i.e. diag $\Sigma = (1, 1, \ldots, 1)$. This does not constitute a loss of generality since z can always be standardized by the linear transformation $z \to y$ defined by $y_i = (z_i - \mu_i)/\sqrt{\sigma_{ii}}$.

We shall denote

$$p = \int_A f(z)dz, \quad q_k = \int_A z_k f(z)dz, \tag{11.3}$$

where f is the normal density as in (11.2). Before discussing the problem in full generality, we shall give some special cases for which the solution is easy.

1. If $n = 1$, there are no problems. The conditional probability $p = \frac{1}{\sqrt{2\pi}}$ $\int_{a_1}^{b_1} e^{-z^2/2} dz$ can be found for example by expanding the integrand into a power series. There are efficient and reliable routines in almost every mathematical software package which will perform the computation. The numerator q_1 is even easier to obtain since the corresponding integral can be solved analytically. Thus

$$q_1 = \int_{a_1}^{b_1} \frac{z_1}{\sqrt{2\pi}} e^{-z_1^2/2} dz_1 = \frac{1}{\sqrt{2\pi}} (e^{-a_1^2/2} - e^{-b_1^2/2}).$$

2. For $n = 2$, the answer is obtained almost as easily. The integral for p can be developed into a power series [6], similar to the one-dimensional case, and commercial software exists which performs the evaluation. In order to calculate q_1 and similarly q_2 it is possible to exploit the fact that $h(t) = te^{-t^2/2}$ permits analytical integration. Thus one may complete the square in the exponent, exchange the order of integration, and simplify. This gives (details are in [9])

$$
\begin{aligned}
q_1 = \frac{1}{\sqrt{2\pi}} \Bigg(& e^{-a_1^2/2}\left(\Phi\left(\frac{b_2 - \sigma_{12}a_1}{\sqrt{1-\sigma_{12}^2}}\right) - \Phi\left(\frac{a_2 - \sigma_{12}a_1}{\sqrt{1-\sigma_{12}^2}}\right)\right) \\
& - e^{-b_1^2/2}\left(\Phi\left(\frac{b_2 - \sigma_{12}b_1}{\sqrt{1-\sigma_{12}^2}}\right) - \Phi\left(\frac{a_2 - \sigma_{12}b_1}{\sqrt{1-\sigma_{12}^2}}\right)\right) \\
& + \sigma_{12}e^{-a_2^2/2}\left(\Phi\left(\frac{b_1 - \sigma_{12}a_2}{\sqrt{1-\sigma_{12}^2}}\right) - \Phi\left(\frac{a_1 - \sigma_{12}a_2}{\sqrt{1-\sigma_{12}^2}}\right)\right) \\
& - \sigma_{12}e^{-b_2^2/2}\left(\Phi\left(\frac{b_1 - \sigma_{12}b_2}{\sqrt{1-\sigma_{12}^2}}\right) - \Phi\left(\frac{a_1 - \sigma_{12}b_2}{\sqrt{1-\sigma_{12}^2}}\right)\right) \Bigg).
\end{aligned}
$$

where $\Phi(t)$ is the standard normal distribution function in one dimension.

3. The last trivial case occurs when z has independent components. In this situation, $\Sigma = I$, and the problem at hand can be reduced to separate applications of the one-dimensional computation as follows:

$$p = \int_A \frac{1}{(2\pi)^n/2} e^{-\sum_i z_i^2/2} dz = \prod_{i=1}^n \int_{a_i}^{b_i} \frac{1}{\sqrt{2\pi}} e^{-z_i^2/2} dz_i,$$

$$q_k = \int_A \frac{z_k}{(2\pi)^n/2} e^{-\sum_i z_i^2/2} dz = \frac{1}{\sqrt{2\pi}} (e^{-a_k^2/2} - e^{-b_k^2/2})$$

$$\prod_{i \neq k} \frac{1}{\sqrt{2\pi}} \int_{a_i}^{b_i} e^{-z_i^2/2} dz_i.$$

Difficulties arise for $n \geq 3$ if the components of z are correlated. While series expansions do exist [7,8], they converge slowly, and it is usually best to estimate the integrals by resorting to some sampling technique. The simple Monte-Carlo method [12] consists in generating an independent sample $\{z^1, \ldots, z^N\}$ of size N from the distribution of z, counting all instances for which the sample point lies in the rectangle A, ignoring the others, and forming the estimator

$$\hat{p} = \frac{1}{N} \sum_{j=1}^{N} 1_A(z^j),$$

where 1_A denotes the indicator function of the rectangle A. It can be shown that $\hat{p}$ is unbiased, that is the expected value of (the random variable) $\hat{p}$ is equal to p.

Similarly one has the estimators $\hat{q}_k = \frac{1}{N} \sum_{j=1}^{N} z_k^j 1_A(z^j)$, which can be shown to be unbiased for q_k. Unfortunately, since the estimators are random variables, any performance guarantee can only be formulated in probability, and an individual estimator may be far from the true value, even if its variance is small. Moreover, the variance of $\hat{p}$, and similarly of $\hat{q}_k$, is proportional to the inverse of the sample size, which necessitates a rather large sample if any meaningful accuracy requirement has to be satisfied.

For this reason much effort has been invested in finding variance reduction schemes. The most popular device is based on "antithetic variables", that is, whenever z^j is a point generated in the sample, one also includes the point $-z^j$. Other, more powerful methods will be described in the following sections.

A comment should be made here on how to construct the sample points. Since $y = Cx \sim N(C\mu, C\Sigma C')$ whenever $x \sim N(\mu, \Sigma)$, it suffices to generate n independent univariate standard normal deviates w_i^j, $i = 1, \ldots, n$ and to calculate $z^j = Lw^j$ for some matrix L such that $LL' = \Sigma$. An attractive choice for L in this setting is the Choleski decomposition [20] of Σ, because it can be computed efficiently and because its triangular structure may reduce the computational effort necessary in calculating z^j.

11.3 Deák's Method

Deák [3,4] describes a more efficient method for finding the conditional probability p, based on the decomposition

$$z = \lambda L v, \tag{11.4}$$

where λ is chi-distributed with n degrees of freedom, and v is uniformly distributed on S^n, the unit sphere in R^n. Here λ can be interpreted as the length of the vector z, v as its direction. L is taken as the Choleski decomposition, as in the simple Monte-Carlo method. Then

$$p = \int_A f(z)\,dz = \int_{S^n} \int_{r_1(v)}^{r_2(v)} d\chi_n(\lambda)\,dU(v),$$

where $r_1(v) = \min\{r : r \ge 0, a \le rLv \le b\}$,
$\qquad\quad r_2(v) = \max\{r : r \ge 0, a \le rLv \le b\}$.

For an illustration of this idea when $n = 2$, see Figure 11.1.

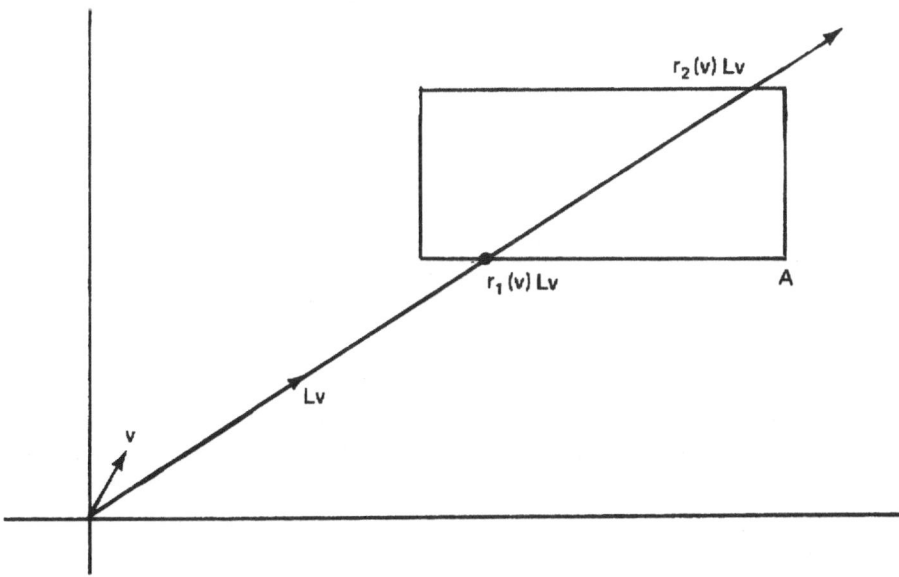

Figure 11.1 Deák's Method in Two Dimensions

Sampling is then performed on v only while the one-dimensional integral $\int_{r_1(v)}^{r_2(v)} d\chi_n(\lambda)$ is calculated explicitly.

This gives $\hat{p} = \frac{1}{N} \sum_{j=1}^{N} \mu^j$, where $\mu^j = \int_{r_1(w^j)}^{r_2(v^j)} d\chi_n(\lambda)$. Again one could use simple Monte-Carlo sampling from the uniform distribution on S^n, or the method of antithetic variables for some reduction in the variance. Deák presents a different idea which seams to work quite well in practice.

Instead of an independent sample of v^j, Deák advocates the use of "orthonormal variates", that is, starting from a random system $\{v_1^j, \ldots, v_n^j\}$ of orthonormal directions one forms all linear combinations $d^j = \frac{1}{\sqrt{k}} \sum_{\ell=1}^{k} (-1)^{n_\ell} v_{i_\ell}^j$ where $n_\ell = 0$ or 1 for all ℓ and $i_{\ell_1} \neq i_{\ell_2}$ if $\ell_1 \neq \ell_2$.

This method has the advantage of creating a large number of random points, namely $2^k \binom{n}{k}$, with comparably little effort. The larger sample size has a dramatic effect on the variance of the estimator $\hat{p}$, and Deák reports a "coefficient of efficiency" of up to 1000 [4]. Efficiency is measured as $\sigma_0^2 t_0 / \sigma_k^2 t_k$, where σ_0^2 is the approximate variance of the simple Monte-Carlo estimator for a fixed N, σ_k^2 is the approximate variance of Deák's estimator for the same N, forming linear combinations of k directions, t_0 and t_k are the respective computation times. This particular measure is used because for both methods the variance is roughly proportional to $1/N$, that is $\sigma_i^2 t_i$ is approximately constant. The parameter k can in principle be chosen arbitrarily from the set $\{1, 2, \ldots, n\}$, but the maximum value of $2^k \binom{n}{k}$ occurs when $k = \lfloor \frac{2n+1}{3} \rfloor$, and the computational complexity increases very fast. Deák reports best results for $k = 2, 3, 4$.

To adapt the method for computing $\hat{q}_i$, one observes that

$$
\begin{aligned}
\int_A z_i f(z) dz &= \int_{S^n} \int_{r_1(v)}^{r_2(v)} \lambda L_i v d\chi_n(\lambda) dU(v) \\
&= \int_{S^n} L_i v \int_{r_1(v)}^{r_2(v)} \lambda d\chi_n(\lambda) dU(v) \\
&= \beta \int_{S^n} L_i v \int_{r_1(v)}^{r_2(v)} d\chi_{n+1}(\lambda) dU(v),
\end{aligned}
\tag{11.5}
$$

where $\beta = \frac{\Gamma(\frac{n+1}{2})\sqrt{2}}{\Gamma(\frac{n}{2})}$.

This results in the estimator $\hat{q}_i = \frac{\beta}{N} \sum_{j=1}^{N} L_i v^j \tilde{\mu}^j$, with $\tilde{\mu}^j$ computed in analogy with μ^j on the previous page. It should be clear that $\hat{q}_i$ and $\hat{p}$ can all be computed simultaneously from the sample, and the $r_\ell(v)$ need to be determined just once.

11.4 Szántai's Procedure

Szántai [22] uses a completely different approach, based on a Bonferroni-type decomposition of the sample space R^n. It is not hard to see that

$$\int_A \psi = \int_{R^n} \psi - \sum_i \int_{\overline{A}_i} \psi + \sum_{i<j} \int_{\overline{A}_i \cap \overline{A}_j} \psi$$
$$- \sum_{i<j<k} \int_{\overline{A}_i \cap \overline{A}_j \cap \overline{A}_k} \psi \pm \ldots + (-1)^n \int_{\bigcap_{i=1}^n \overline{A}_i} \psi \tag{11.6}$$

where $\overline{A}_i = \{x : a_i > x_i \text{ or } b_i < x_i\}$ and ψ is an arbitrary integrand. Using the fact that $\int_{\overline{A}_i} dF$ and $\int_{\overline{A}_i \cap \overline{A}_j} dF$ can be calculated easily (we are in the case $n = 1$ or $n = 2$), and using simple Monte-Carlo simulation, one has available three different unbiased estimates for p, namely

(i) $\hat{p}^{(1)}$ - sampling directly from $\int_A dF$,

(ii) $\hat{p}^{(2)}$ - calculating explicitly $\int_{R^n} dF - \sum_i \int_{\overline{A}_i} dF = \sum_i \int_{a_i}^{b_i} dF_i + 1 - n$ and sampling from the rest,

(iii) $\hat{p}^{(3)}$ - calculating $\int_{R^n} dF - \sum_i \int_{\overline{A}_i} dF + \sum_{i<j} \int_{\overline{A}_i \cap \overline{A}_j} dF$

$$= \sum_{i<j} \int_{a_i}^{b_i} \int_{a_j}^{b_j} dF_{ij} + (2-n) \sum_i \int_{a_i}^{b_i} dF_i + \frac{(n-1)(n-2)}{2}$$

and sampling from the rest.

Szántai describes a way to condense the tail of the expansion in (11.6) into a single expression which involves only $i(j)$, the number of constraints $a_i \leq z_i^j \leq b_i$ which are violated by a given sample point z^j. In fact, one obtains

$$\hat{p}^{(2)} = \int_{R^n} dF - \sum_i \int_{\overline{A}_i} dF + \frac{1}{N} \sum_{j=1}^N \max\{0, i(j) - 1\}$$

$$\hat{p}^{(3)} = \int_{R^n} dF - \sum_i \int_{\overline{A}_i} dF + \sum_{i<j} \int_{\overline{A}_i \cap \overline{A}_j} dF \tag{11.7}$$

$$- \frac{1}{N} \sum_{j=1}^N \frac{\max\{0, i(j) - 1\}(i(j) - 2)}{2}$$

Finally, the covariance structure of the estimators is determined to form yet another unbiased estimator $\hat{p}^{(4)} = \lambda_1 \hat{p}^{(1)} + \lambda_2 \hat{p}^{(2)} + (1 - \lambda_1 - \lambda_2) \hat{p}^{(3)}$, where the weights λ_1 and λ_2 are chosen so as to minimize the variance of $\hat{p}^{(4)}$. (This minimization can be carried out analytically.) Szántai reports improvements in efficiency which are of the same order of magnitude as those in [4].

In order to adapt Szántai's method for calculation of the q_k it is necessary to evaluate expressions of the form $\int_{A_i} x_k dF$, $\int_{A_i \cap A_j} x_k dF$. This can be done by a procedure quite similar to the method presented earlier for the case $n = 2$, namely by completing the square in the exponent of the integrand. This yields the formula

$$
\int_{A_i \cap A_j} x_k dF = \frac{1}{\sqrt{2\pi}} \left[\sigma_{ik} e^{-a_i^2/2} \left(\Phi \left(\frac{b_j - \sigma_{ij} a^i}{\sqrt{1 - \sigma_{ij}^2}} \right) - \left(\Phi \frac{a_j - \sigma_{ij} a_i}{\sqrt{1 - \sigma_{ij}^2}} \right) \right) \right.
$$

$$
- \sigma_{ik} e^{-b_i^2/2} \left(\Phi \left(\frac{b_j - \sigma_{ij} b_i}{\sqrt{1 - \sigma_{ij}^2}} \right) - \Phi \left(\frac{a_j - \sigma_{ij} b_i}{\sqrt{1 - \sigma_{ij}^2}} \right) \right)
$$

$$
+ \sigma_{jk} e^{-a_j^2/2} \left(\Phi \left(\frac{b_i - \sigma_{ij} a_j}{\sqrt{1 - \sigma_{ij}^2}} \right) - \Phi \left(\frac{a_i - \sigma_{ij} a_j}{\sqrt{1 - \sigma_{ij}^2}} \right) \right)
$$

$$
\left. - \sigma_{jk} e^{-b_j^2/2} \left(\Phi \left(\frac{b_i - \sigma_{ij} b_j}{\sqrt{1 - \sigma_{ij}^2}} \right) - \Phi \left(\frac{a_i - \sigma_{ij} b_j}{\sqrt{1 - \sigma_{ij}^2}} \right) \right) \right]
$$

$$(11.8)$$

In the same manner as before one arrives at three unbiased estimators for q_k, namely

(i) Sampling directly from $\int_A z_k dF$,

(ii) calculating explicitly $\int_{\mathbb{R}^n} z_k dF - \sum_i \int_{\overline{A}_i} z_k dF = \sum_i \int_{A_i} z_k dF$ and sampling from the rest,

(iii) calculating explicitly $\int_{\mathbb{R}^n} z_k dF - \sum_i \int_{\overline{A}_i} z_k dF + \sum_{i<j} \int_{\overline{A}_i \cap \overline{A}_j} z_k dF = \sum_{i<j} \int_{A_i \cap A_j} z_k dF + (2-n) + \sum_i \int_{A_i} z_k dF$ and sampling from the rest.

Once more the affine combination of least variance is formed. Numerical results appear in Section 11.6.

11.5. A Hybrid Method

Deák's method uses a decomposition of the random variable, while Szántai's procedure can be thought of as a decomposition of the sample space R^n. Both authors report impressive acceleration on the computing time, and it seems natural to try to combine the two approaches, exploiting all the desirable features simultaneously. The derivation proceeds exactly as in the previous section, but now Deák's decomposition is to be used in sampling $\hat{p}^{(1)}, \hat{p}^{(2)}, \hat{p}^{(3)}$ instead of simple Monte-Carlo. Unfortunately this means that Formula (11.7) is no longer available. Instead one writes

$$
\sum_{i<j} \int_{\overline{A}_i \cap \overline{A}_j} dF(z) - \sum_{i<j<k} \int_{\overline{A}_i \cap \overline{A}_j \cap \overline{A}_k} dF(z) + \cdots + (-1)^n \int_{\bigcap_{i=1}^n \overline{A}_i} dF(z)
$$

$$
= \sum_{\ell=2}^n (\ell-1) \int_{B_\ell} dF(z) = \sum_{\ell=2}^n (\ell-1) \int_{S^n} \int_{C_\ell(v)} d\chi_n(\lambda) dU(v)
$$

$$
= \int_{S^n} \sum_{\ell=2}^n (\ell-1) \int_{C_\ell(v)} d\chi_n(\lambda) dU(v),
$$

(11.9)

where $B_\ell = \{z : a_i \leq z_i \leq b_i \text{ is violated for exactly } \ell \text{ indices } i\}$,

$C_\ell(v) = \{\lambda : a_i \leq \lambda L_i v \leq b_i \text{ is violated for exactly } \ell \text{ indices } i\}$.

In a completely analogous fashion one obtains

$$
\sum_{i<j<k} \int_{\overline{A}_i \cap \overline{A}_j \cap \overline{A}_k} dF - \cdots + (-1)^{n+1} \int_{\cap_i \overline{A}_i} dF
$$

$$
= \int_{S^n} \sum_{\ell=3}^n \binom{\ell-1}{2} \int_{C_\ell(v)} d\chi_n(\lambda) dU(v).
$$

(11.10)

In order to estimate the quantities in (11.9) and (11.10), a sample is generated from the uniform distribution on S^n. Since the event $L_i v = 0$ has probability zero, its occurrence can be ignored, which makes it possible to determine $C_\ell(v)$ in the following way.

Let critical values ℓ_i, u_i be defined by $\ell_i = \min\{a_i/L_i v, b_i/L_i v\}$, $u_i = \max\{a_i/L_i v, b_i/L_i v\}$ and assume without loss of generality that the vectors ℓ, u are arranged in ascending order. Further set $\ell_0 = u_0 = -\infty$, $\ell_{n+1} = u_{n+1} = +\infty$. If λ, i, j are such that $\ell_{i-1} \leq \lambda < \ell_i$ and $u_{j-1} < \lambda \leq u_j$, then it is not hard to see that exactly $n + j - i$ inequalities are violated, i.e. $\lambda \in C_{n+j-i}(v)$. Moreover, as λ decreases, the number of inequalities violated increases or decreases by 1, whenever the next critical value encountered is from the vector ℓ or u, respectively; none of the inequalities hold if $\lambda > u_n$. Finally $\lambda L v \in A$ if and only if all inequalities hold, i.e. if $\ell_n \leq \lambda \leq u_1$.

This suggests the following algorithmic procedure:

Step 0: (Initialize)

Determine ℓ, u.

Set $nviol < -n$, $nhigh < -n$, $nlow < -n$.

Set $upper \leftarrow +\infty, \nu_1 \leftarrow 0, \nu_2 \leftarrow 0, \nu_3 \leftarrow 0$.

Step 1: (Process the next interval)

Set $lower \leftarrow \max\{u_{nhigh}, \ell_{nlow}\}, \alpha \leftarrow \int_{lower}^{upper} d\chi_n(\lambda)$.

Set $i_2 \leftarrow nviol - 1, i_3 \leftarrow \binom{nviol-1}{2}$.

If $nviol = 0$, set $\nu_1 \leftarrow \nu_1 + \alpha$.

If $nviol \geq 2$, set $\nu_2 \leftarrow \nu_2 + i_2\alpha, \nu_3 \leftarrow \nu_3 - i_3\alpha$.

Step 2: (Update)

If $u_{nhigh} > \ell_{nlow}$, set $nviol \leftarrow nviol - 1, nhigh \leftarrow nhigh - 1$.

If $u_{nhigh} = \ell_{nlow}$, set $nhigh \leftarrow nhigh - 1, nlow \leftarrow nlow - 1$.

If $u_{nhigh} < \ell_{nlow}$, set $nviol \leftarrow nviol + 1, nlow \leftarrow nlow - 1$.

Set $upper \leftarrow lower$.

If $upper > 0$, go to 1.

Step 3: (Generate another sample point and repeat.)

For given sample size N this defines three estimators

$$\widehat{p}^{(1)} = \frac{1}{N}\sum_{j=1}^{N}\nu_1^j, \widehat{p}^{(2)} = \frac{1}{N}\sum_{j=1}^{N}\nu_2^j, \widehat{p}^{(3)} = \frac{1}{N}\sum_{j=1}^{N}\nu_3^j.$$

Similarly, there are the estimators

$$\widehat{q}_i^{(1)} = \frac{\beta}{N}\sum_{j=1}^{N}L_i v^j \overline{\nu}_1^j, \widehat{q}_i^{(2)} = \frac{\beta}{N}\sum_{j=1}^{N}L_i v^j \overline{\nu}_2^j, \widehat{q}_i^{(3)} = \frac{\beta}{N}\sum_{j=1}^{N}L_i v^j \overline{\nu}_3^j.$$

where $\overline{\nu}_\ell^j$ is obtained by the same algorithm using $\overline{\alpha} = \int_{lower}^{upper} d\chi_{n+1}(\lambda)$ in place of α. The constant β is as in (11.5). We explicitly remark that the critical values ℓ, u have to be calculated just once for each sample point.

From these triples of estimators one can then form Szántai's estimator $\widehat{q}_i^{(4)} = \lambda_1^i \widehat{q}_i^{(1)} + \lambda_2^i \widehat{q}_i^{(2)} + \lambda_3^i \widehat{q}_i^{(3)}$ which is the affine combination of least variance. (The weights λ_k^i may differ from component to component.)

11.6 Numerical Results

The following page gives an overview of the performance of the various methods of sections 2-5 when applied to a small number of low-dimensional problems. The abbreviations used in the table are

SMC · Simple Monte Carlo
AV · Antithetic Variables
DMC · Simple Monte Carlo applied to decomposition (11.4)
DAV · Antithetic Variables applied to decomposition (11.4)
De k · Deák's method, using k vectors at a time
Sz · Szántai's method
Hy k · Hybrid method, using k vectors at a time.

For each method we report a weighted efficiency rating for the estimator $\hat{p}$, namely the quantity $(\sigma_{100}^2 t_{100} + \sigma_{200}^2 t_{200} + \sigma_{500}^2 t_{500} + \sigma_{1000}^2 t_{1000}) \cdot 10^6$, where σ_x^2 is the variance and t_x is the CPU time for sample size x. Results for the $\hat{q}_i$ were quite similar and had to be omitted to save space. All computations were performed on the Amdahl 460 V/8 computer at the University of British Columbia Computing Centre.

Simple Monte-Carlo is expectedly worst for all four problems, while the hybrid method is characterized by extremely slow computing times. This may be due in part to the sorting algorithm which was used in determining the vectors ℓ and u of critical values. Up to 15% of total CPU time was spent in the sorting routine VSRTA of the IMSL library which uses a quicksort algorithm. A self-contained merge-insertion algorithm [14] might do better in some instances. Moreover, the one-dimensional integration could be accelerated greatly by computing only a table of reference values and interpolating whenever a new value is needed.

Szántai's procedure is comparable in performance to Deák's method, except on problem 4 where it is markedly inferior. It is interesting to note that problem 4 is also the problem with the smallest conditional probability. On problem 5, all 1000 sample points generated for Szántai's method lie outside the region of interest. Thus the sample variance is zero, although the estimate is inaccurate. The hybrid method clearly outperforms all competitors on problems 3 and 5 which have the highest conditional probability.

Further testing is clearly indicated, but at this stage it seems best to use Deák's method whenever the conditional probability is expected to be small and the hybrid method if the conditional probability is expected to be large. The best value for the parameter k in each case seems to be close to $n/2$.

Table 11.1 Comparison of Efficiency

N of	1	2	3	4	5	6
Dimension	3	3	4	5	10	10
Probability	0.028	0.20	0.91	0.0064	0.97	0.037
SMC	20.67	130.4	72.0	14.15	63.2	160.0
AV	11.80	86.0	44.4	5.76	54.2	81.2
DMC	9.27	39.4	53.6	3.80	26.0	15.6
DAV	5.31	10.69	51.0	2.92	18.8	7.23
De 1	3.52	7.29	4.09	2.60	20.4	6.49
De 2	1.93	2.37	3.58	0.76	18.0	2.08
De 3	3.02	5.39	3.83	0.61	—	—
Sz	11.60	13.71	1.09	10.09	0*	15.2
Hy 1	9.12	5.77	0.29	11.31	0.002	12.4
Hy 2	4.54	4.74	0.07	5.54	0.0001	11.2
Hy 3	8.06	8.40	0.18	5.37	—	—

*No random point was generated within the region of interest.

11.7 Conditional Expectation

Up to now the emphasis has been on determining "good" estimators $\hat{p}$ and $\hat{q}_k$ for the numerator and denominator of Formula (11.3). The real interest, however, lies in the ratio q_k/p, so it is natural to ask about statistical properties of the quotient $\hat{q}_k/\hat{p}$ for the various estimators, in particular about its mean and its variance.

The first unpleasant surprise is the fact that the quotients are not unbiased [16]. Using a Taylor expansion about the point q_k/p, it is not hard to verify that

$$E\left(\frac{\hat{q}_k}{\hat{p}}\right) = \frac{q_k}{p} + \frac{1}{p^2}\left(\sigma_{00}\frac{q_k}{p} - \sigma_{k0}\right) + 0(N^{-2}),   \tag{11.11}$$

where $\sigma_{00} = \text{var}(\hat{p})$, $\sigma_{k0} = \text{cov}(\hat{q}_k, \hat{p})$. The bias can be reduced (but not eliminated) by forming the estimators

$$\hat{r}_k = \frac{\hat{q}_k}{\hat{p}} - \frac{1}{\hat{p}^2}\left[\text{Var}(\hat{p})\frac{\hat{q}_k}{\hat{p}} - \text{Cov}(\hat{q}_k, \hat{p})\right]   \tag{11.12}$$

Expanding this expression into a Taylor series shows that $E(\hat{r}_k) = q_k/p + 0(N^{-2})$.

It should be noted that in formula (11.11) the true variance and covariance are used, while the quantities appearing in (11.12) are their estimates obtained from the sample. Further improvements may be effected by retaining higher order terms in the Taylor expansion as well as higher order sample moments. The possible gain is not easily assessed, however, and storage and computation requirements would be increased considerably.

Approximate formulas for the variance of the different estimators can also be obtained by Taylor expansion, namely

$$\text{Var}\left(\frac{\widehat{q_k}}{\widehat{p}}\right) = \frac{1}{p^2}\left[\sigma_{kk} - 2\sigma_{k0}\frac{q_k}{p} + \sigma_{00}\left(\frac{q_k}{p}\right)^2\right] + 0(N^{-2})$$

and

$$\text{Var}(\widehat{r}_k) = \frac{1}{p^2}\left[\sigma_{kk} - \sigma_{00}\left(\frac{q_k}{p}\right)^2\right] + 0(N^{-2}).$$

This variance inflation poses a serious problem, in particular if p is small, and cannot be eliminated even if p is known with certainty. On the other hand, rectangles of little mass contribute little to the overall bounds in formula (11.1), and perhaps the accuracy requirements may be relaxed a little in this case.

11.8 Extensions

It may sometimes be desirable to find conditional information on polyhedral sets other than n-dimensional rectangles. For instance, a decomposition of the sample space into (possibly unbounded) simplices improves the computation of the bounds in (11.1) and may be preferred in some situations.

Let therefore a (non-empty) polyhedral set $A := \{x : a \leq Tx \leq b\}$ be given, where some or all components of the vector a may be set to $-\infty$, similarly b may contain certain components equal to $+\infty$. We will assume that the rows of T are pairwise linearly independent and again seek to determine the quantities $p = \int_A dF, q_k = \int_A z_k dF$, where $z \sim N(0, \Sigma)$. This problem is very similar to the previous problem of n-dimensional rectangles, since the quantity $y = Tx$ is normally distributed with mean 0 and covariance matrix $\overline{\Sigma} = T\Sigma T'$, which may be singular if the number of constraints exceeds n. Thus $\overline{\Sigma}$ may not have a Choleski decomposition; instead the Choleski decomposition of Σ should be used to form the matrix TL. From then on it is smooth sailing, all the techniques and methods of sections 2-5 will go through provided L is replaced by TL in all formulas. Sampling should still be done from the n-dimensional normal distribution and the uniform distribution on S^n where appropriate.

Other multivariate distributions for which conditional information may be of interest include the multilognormal [13] and multigamma [18] distributions.

A random vector $z = (z_1, \ldots, z_n)$ has a multilognormal distribution if and only if the random vector $\ln z = (\ln z_1, \ldots, \ln z_n)$ has a multinormal distribution. Hence it is possible to work with the vector $\ln z$ instead, and all the previous results can be used.

Another interesting distribution is the multigamma distribution which has seen some application in chance-constrained stochastic programming problems [17]. By a suitable scaling transformation an arbitrary univariable gamma distribution can be reduced to standard form which is defined by the density

$$f_\theta(z) = \begin{cases} \frac{1}{\Gamma(\theta)}z^{\theta-1}e^{-z} & \text{if } z > 0 \\ 0 & \text{if } z \leq 0 \end{cases}$$

Hence we can assume without loss of generality that the multivariate gamma distribution is such that all the one-dimensional marginals have standard form. We shall write $z \sim \Gamma(\theta)$ if z has a standard gamma distribution with parameter θ. A well-known fact about the standard gamma distribution is its additivity, more precisely, if $z_1 \sim \Gamma(\theta_1), z_2 \sim \Gamma(\theta_2)$, and z_1 is independent of z_2 then $z_1 + z_2 \sim \Gamma(\theta_1 + \theta_2)$.

Deák's method no longer applies to the computation of the conditional probabilities, since the decomposition of formula (11.4) is not available any more, but Szántai's method can still be used, and it is easily adapted for computation of the q_k as well. To that end one needs to develop algorithms for the evaluation of univariate and bivariate gamma distributions in formula (11.6). Once this has been done, the quantities p, q_k can be estimated by a simple Monte-Carlo method in three different ways and the combination of least variance can be found.

The univariate case is easy: there are efficient library routines available to compute

$$\int_0^{a_i} \frac{1}{\Gamma(\theta)} z^{\theta-1} e^{-z} dz,$$

and for the conditional expectation one notes that

$$\int_0^{a_i} \frac{z}{\Gamma(\theta)} z^{\theta-1} e^{-z} dz = \int_0^{a_i} \frac{\theta}{\theta\Gamma(\theta)} z^\theta e^{-z} dz = \theta \int_0^{a_i} \frac{1}{\Gamma(\theta+1)} z^{(\theta+1)-1} e^{-z} dz.$$

For the bivariate distribution $F(a_1, a_2) = \int_0^{a_1} \int_0^{a_2} f(z_1, z_2) dz_2 dz_1$, Szántai uses the decomposition of z_1, z_2 into three independent standard gamma distributions x_1, x_2, x_3 with suitable parameters $\theta_1, \theta_2, \theta_3$, respectively, in the form

$$
\begin{aligned}
z_1 &= x_1 + x_2 \\
z_2 &= x_1 \qquad + x_3
\end{aligned}
$$

and obtains the series expansion

$$f(z_1, z_2) = f_{\theta_1+\theta_2}(z_1) f_{\theta_1+\theta_3}(z_2) \{1+$$
$$\sum_{r=1}^\infty r! \frac{\Gamma(\theta_1+r)\Gamma(\theta_1+\theta_2)\Gamma(\theta_1+\theta_3)}{\Gamma(\theta_1)\Gamma(\theta_1+\theta_2+r)\Gamma(\theta_1+\theta_3+r)} L_r^{\theta_1+\theta_2-1}(z_1) L_r^{\theta_1+\theta_3 \cdot 1}(z_2)\},$$
$$F(a_1, a_2) = F_{\theta_1+\theta_2}(a_1) F_{\theta_1+\theta_3}(a_2) +$$
$$\sum_{r=1}^\infty C(\theta_1, \theta_2, \theta_3, r) f_{\theta_1+\theta_2+1}(a_1) f_{\theta_1+\theta_3+1}(a_2) L_{r-1}^{\theta_1+\theta_2}(a_1) L_{r-1}^{\theta_1+\theta_3}(a_2),$$

where F_θ is the standard gamma distribution with parameter θ,

$$C(\theta_1, \theta_2, \theta_3, r) =$$
$$\frac{(r-1)!}{r} \frac{(\theta_1+r-1)(\theta_1+r-2)\ldots\theta_1}{(\theta_1+\theta_2+r-1)\ldots(\theta_1+\theta_2+1)(\theta_1+\theta_3+r-1)\ldots(\theta_1+\theta_3+1)}.$$

$L_r^\theta(\cdot)$ are the Laguerre polynomials [19] defined by $L_0^\theta(x) = 1$, $L_1^\theta(x) = \theta + 1 - x$ and the recursion

$$(r+1)L_{r+1}^\theta(x) - (2r + \theta + 1 - x)L_r^\theta(x) + (r + \theta)L_{r-1}^\theta(x) = 0, \quad r = 1, 2, 3, \ldots$$

A conditioning argument can be used to derive expressions for

$$\int_0^{a_1} \int_0^{a_2} z_1 f(z_1, z_2)\, dz_2\, dz_1, \quad \int_0^{a_1} \int_0^{a_2} z_2 f(z_1, z_2)\, dz_2\, dz_1,$$

based on the fact that the random variables z_1 and z_2 are conditionally independent given the value of x_1. Therefore one obtains

$$\int_0^{a_1} \int_0^{a_2} z_1 f(z_1, z_2)\, dz_2\, dz_1 = \int_0^{a_1} \int_0^{a_2} z_1 f(z_1, z_2) \int_0^\infty f_{\theta_1}(x_1)\, dx_1\, dz_2\, dz_1$$

$$= \int_0^\infty \int_0^{a_1} \int_0^{a_2} z_1 f(z_1, z_2) f_{\theta_1}(x_1)\, dz_2\, dz_1\, dx_1$$

$$= \int_0^{a_1 \wedge a_2} \int_{x_1}^{a_1} \int_{x_1}^{a_2} x_1 + (z_1 - x_1) f(x_1 + (z_1 - x_1)),$$

$$x_1 + (z_2 - x_1)) f_{\theta_1}(x_1)\, dz_2\, dz_1\, dx_1$$

$$= \int_0^{a_1 \wedge a_2} \int_0^{a_1 - x_1} \int_0^{a_2 - x_1} (x_1 + x_2) f_{\theta_2}(x_2) f_{\theta_3}(x_3) f_{\theta_1}(x_1)\, dx_3\, dx_2\, dx_1$$

$$= \int_0^{a_1 \wedge a_2} \int_0^{a_1 - x_1} \int_0^{a_2 - x_1} x_1 f_{\theta_1}(x_1) f_{\theta_2}(x_2) f_{\theta_3}(x_3)\, dx_3\, dx_2\, dx_1$$

$$+ \int_0^{a_1 \wedge a_2} \int_0^{a_1 - x_1} \int_0^{a_2 - x_1} x_2 f_{\theta_2}(x_2) f_{\theta_1}(x_1) f_{\theta_3}(x_3)\, dx_3\, dx_2\, dx_1$$

$$= \theta_1 \int_0^{a_1 \wedge a_2} \int_0^{a_1 - x_1} \int_0^{a_2 - x_1} f_{\theta_1 + 1}(x_1) f_{\theta_2}(x_2) f_{\theta_3}(x_3)\, dx_3\, dx_2\, dx_1$$

$$+ \theta_2 \int_0^{a_1 \wedge a_2} \int_0^{a_1 - x_1} \int_0^{a_2 - x_1} f_{\theta_2 + 1}(x_2) f_{\theta_1}(x_1) f_{\theta_3}(x_3)\, dx_3\, dx_2\, dx_1$$

$$= \theta_1 F'(a_1, a_2) + \theta_2 F''(a_1, a_2),$$

where F' is the distribution of the random vector

$$z_1' = x_1' + x_2$$
$$z_2' = x_1' + x_3, x_1' \sim \Gamma(\theta_1 + 1).$$

F'' is the distribution of

$$z_1 i'' = x_1 + x_2''$$
$$z_1'' = x_1 + x_3, x_2'' \sim \Gamma(\theta_2 + 1).$$

Similarly one can derive the formula

$$\int_0^{a_1} \int_0^{a_2} z_2 f(z_1, z_2)\, dz_2\, dz_1 = \theta_1 F''(a_1, a_2) + \theta_3 F'''(a_1, a_2),$$

where F'''' is the distribution of

$$z_1''' = x_1 + x_2$$
$$z_2''' = x_1 + x_3''', x_3''' \sim \Gamma(\theta_3 + 1).$$

The same idea can be used to derive values for expressions of the form

$$\int_0^{a_1} \int_0^{a_2} \int_0^{\infty} z_3 f(z_1, z_2, z_3) dz_3 dz_2 dz_1,$$

where $f(z_1, z_2, z_3)$ is the joint density of a trivariate gamma distribution which can be decomposed into independent standard gamma distributions $x_i \sim \Gamma(\theta_i)$ as follows:

$$\begin{array}{rcccccc} z_1 & = & x_1 & & +x_4 & +x_5 & & +x_7 \\ z_2 & = & & x_2 & +x_4 & \quad +x_6 & +x_7 . \\ z_3 & = & & x_3 & & +x_5 & +x_6 & +x_7 \end{array} \quad (11.13)$$

Conditioning on the values of x_4, x_5, x_6, x_7 and simplifying, one obtains

$$\int_0^{a_1} \int_0^{a_2} \int_0^{\infty} z_3 f(z_1, z_2, z_3) dz_3 dz_2 dz_1$$

$$= \theta_3 F_3(a_1, a_2, \infty) + \theta_5 F_5(a_1, a_2, \infty) + \theta_6 F_6(a_1, a_2, \infty) + \theta_7 F_7(a_1, a_2, \infty),$$

where F_i stands for the cumulative trivariate gamma distribution having the same decomposition structure as (11.13), when x_i is replaced by $x_i' \sim \Gamma(\theta_i + 1)$.

The integration in the z_3-direction is over the whole support of the random variable and can thus be suppressed by working with the two-dimensional marginal distributions of z_1, z_2. After suitably aggregating the components of z_i, this yields the expressions

$$F_3(a_1, a_2, \infty) = F_3'(a_1, a_2; \theta_4 + \theta_7, \theta_1 + \theta_5, \theta_2 + \theta_6),$$
$$F_5(a_1, a_2, \infty) = F_5'(a_1, a_2; \theta_4 + \theta_7, \theta_1 + \theta_5 + 1, \theta_2 + \theta_6),$$
$$F_6(a_1, a_2, \infty) = F_6'(a_1, a_2; \theta_4 + \theta_7, \theta_1 + \theta_5, \theta_2 + \theta_6 + 1),$$
$$F_7(a_1, a_2, \infty) = F_7'(a_1, a_2; \theta_4 + \theta_7 + 1, \theta_1 + \theta_5, \theta_2 + \theta_6),$$

where e.g. F_5' is the cumulative distribution of

$$z_1' = y_1' + y_2'$$
$$z_2' = y_1' + y_3', y_1' \sim \Gamma(\theta_4 + \theta_7), y_2' \sim \Gamma(\theta_1 + \theta_5 + 1), y_3' \sim \Gamma(\theta_2 + \theta_6).$$

Acknowledgement

This research was supported in part by NSERC grant # 67-7147.

References

[1] J.H. Ahrens and U. Dieter, "Computer Methods for Sampling from Gamma, Beta, Poisson and Binomial Distributions", *Computing* 12(1974), 223–246.

[2] J. Birge and R. Wets, "Designing Approximation Schemes for Stochastic Optimization Problems, in Particular for Stochastic Programs with Recourse", *Math. Programming Study* (to appear).

[3] I. Deák, "Computation of multiple normal probabilities", in P. Kall and A. Prékopa (eds.) *Recent Results in Stochastic Programming*, Springer Verlag, New York, 1980, 107–120.

[4] I. Deák, "Three digit accurate multiple normal probabilities", *Num. Math.* 35(1980), 369–380.

[5] I. Deák, "Procedures for the Computation of Normal Distribution Function Values and Probabilities of Rectangles", Working Paper No. MO/48, Computer and Automation Institute, The Hungarian Academy of Sciences, Budapest, 1983.

[6] T.G. Donnelly, "Bivariate normal distribution, algorithm 462", *Communications of the ACM* 16(1973), 638.

[7] J.E. Dutt, "A representation of multivariate normal probability integrals by integral transforms", *Biometrica* 60(1973), 637–645.

[8] C.K. Eagleson, "Polynomial expansions of bivariate distributions", *Ann. Math. Stat.* 35(1964), 1208–1215.

[9] H.I. Gassmann, "Means of a bivariate normal distribution over a rectangle: an efficient algorithm", Working Paper No. 971, Faculty of Commerce, The University of British Columbia, 1983.

[10] H.I. Gassmann, "Conditional Probability and Conditional Expectation of a Multivariate Normal Random Variable over a Rectangle", IIASA working paper WP-84-100, Laxenburg, Austria, 1984.

[11] H.I. Gassmann and W.T. Ziemba, "A tight upper bound for the expectation of a convex function of a multivariate random variable", *Math. Prog. Study* (to appear).

[12] J.M. Hammersley and D.C. Handscomb, *Monte Carlo Methods*, Methuen, London, 1964.

[13] N.L. Johnson and S. Kotz, *Distributions in Statistics*, Vol. 4, Wiley, New York, 1972.

[14] D.E. Knuth, *The Art of Computer Programming*, Vol. 3, Addison-Wesley, 1974.

[15] A. Madansky, "Bounds on the expectation of a convex function of a multivariate random variable", *Ann. Math. Stat.* 30(1959), 743–746.

[16] A.M. Mood, R. Greybill and C. Boes, *Introduction to the Theory of Statistics*, 3rd edition, McGraw-Hill, New York, 1974.

[17] A. Prékopa, S. Ganczer, I. De'ak and K. Patyi, "The STABIL stochastic programming model and its experimental application to the electrical energy sector of the Hungarian economy", in *Stochastic Programming*, M.A.H. Dempster (ed.), Academic Press, New York, 1980, 369–385.

[18] A. Prékopa and T. Szántai, "A new multivariate gamma distribution and its fitting to empirical data", *Water Resources Research* **14**(1978), 19–24.

[19] M. Spiegel, *Mathematical Handbook of Formulas and Tables*, McGraw-Hill, New York, 1968.

[20] J. Stoer, *Numerische Mathematik I*, Springer-Verlag, Berlin, 1972.

[21] R. Swaroop, J.D. Brownlow, G.R. Ashworth and W.R. Winter, "Bivariate Normal, Conditional and Rectangular Probabilities: A Computer Program with Applications", NASA Technical Memorandum # 81350, 1980.

[22] T. Szántai, "Evaluation of a special multigamma distribution function", *Math. Prog. Study* (to appear).

[23] R. Van Slyke and R. Wets, "*L*-shaped linear programs with applications to optimal control and stochastic programming", *SIAM J. Appl. Math.* **17**(1969), 638–663.

[24] R. Wets, "Stochastic programming: solution techniques and approximation schemes", in *Mathematical Programming: The State of the Art*, A. Bachem, M. Grötschel and B. Korte (eds.), Springer Verlag, 1982.

CHAPTER 12

AN *L*-SHAPED METHOD COMPUTER CODE FOR MULTI-STAGE STOCHASTIC LINEAR PROGRAMS

J.R. Birge

Abstract

A computer code implementing the *L*-shaped method of Van Slyke and Wets is described. The method is generalized to apply to problems with up to three periods and up to three hundred seventy-five different future scenarios. The main subroutines are described.

12.1 Introduction

The *L*-shaped method for two-stage stochastic linear programs was given by Van Slyke and Wets [20], see Chapter 3. It is an outer linearization procedure that approximates the convex objective term in the stochastic program by successively appending supporting hyperplanes. This paper describes a multi-stage implementation of this algorithm in which the supports are found by optimizing a nested sequence of problems. The mechanics of this algorithm and its convergence properties are described in Birge [4].

The method is a type of nested decomposition procedure that can be compared with inner linearization procedures such as those of Glassey [9, 10] and Ho and Manne [13]. It is also related to basis factorization approaches (Kall [14], Strazicky [19], see also the next chapter) and inner linearization of the dual (Dantzig and Madansky [6]).

The basic steps of the algorithm are described in Section 12.2. The main subroutines of the computer code are then given in Section 12.3. Significant variables and data structures are also described. Input and output formats are detailed in Section 12.4 along with examples of their form. Section 12.5 presents some observations and potential extensions.

12.2   Algorithm Description

The multi-stage stochastic linear program considered by the algorithm is

$$\text{minimize}\quad c_1 x_1 + E_{\xi_2}[\min c_2 x_2 + \cdots + E_{\xi_T}[\min C_T x_T]\cdots]$$

$$\text{subject to}\quad A_1 x_1 = b_1,$$
$$B_1 x_1 A_2 x_2 = \xi_2,$$
$$\vdots \tag{12.1}$$
$$B_{T-1} x_{T-1} + A_T x_T = \xi_T,$$
$$x_t \geq 0, t = 1,\ldots,T, \xi_t \in \Xi_t, t = 2,\ldots,T,$$

where c_t is a known vector in $\mathbb{R}^{n_t}$ for $t = 1,\ldots,T$, b_1 is a known vector in $\mathbb{R}^{m_1}$, ξ_t is a random m_2-vector defined on the probability space $(\Xi_t, \mathbf{F}_t, F_t)$ for $t = 1,\ldots,T$, and A_t and $Bsubt$ are correspondingly dimensioned known real-valued matrices. "E_{ξ_t}" denotes mathematical expectation with respect to ξ_t.

The L-shaped method of Van Slyke and Wets [20] applies to (1) when $T = 2$. Methods for the multi-stage problem have generally assumed a specific structure for the problem. Beale, et al. [2] and Ashford [1] for example, consider a multi-stage production problem and implement an appropriate approximation. The generalization of the L-shaped method implemented in the computer code described here and introduced in Birge [3],[4] does not, however, require any special structure except that the random vectors ξ_t are discretely distributed.

The algorithm is called the Nested Decomposition for Stochastic Programming Algorithm (NDSPA). It is based on the observation that given a realization ξ_t^j of the random vector in period t and given a solution $x_{t-1}^{a(j)}$ from period $t-1$, the decision problem at period t can be written (see Wets [21])

$$\text{minimize}\quad c_t x_t^j + Q_{t+1}(x_t^j) \tag{12.2}$$
$$\text{subject to}\quad A_t x_t^j = \xi_t^j + B_{t-1} x_{t-1}^{a(j)} \tag{12.2a}$$
$$D_t^{\ell,j} x_t^j \geq d_t^{\ell,j}, \ell = 1,\ldots,r_t^j, \tag{12.2b}$$
$$x_t \geq 0,$$

where $q_{t+1}(x_t)$ is a convex function, $D_t^{\ell,j} \in \mathbb{R}^{n_t}$ for all l, $r_t^j \leq m_{t+1}$, and (12.2b) is a feasibility cut, see Chapter 3.

Program (12.2) can then be solved using a *relaxed master problem*:

$$\text{minimize}\quad c_t x_t^j + \theta_t^j \tag{12.3}$$
$$\text{subject to}\quad A_t x_t^j = \xi_t^j + B_{t-1} x_{t-1}^{a(j)} \tag{12.3a}$$
$$D_t^{\ell,j} x_t^j \geq d_t^{\ell,j}, \ell = 1,\ldots,r_t^j \tag{12.3b}$$
$$E_t^{\ell,j} x_t^j + \theta_t^j \geq e_x^{\ell,j}, \ell = 1,\ldots,s_t^j \tag{12.3c}$$
$$x_t^j \geq 0. \tag{12.3d}$$

Program (12.3) is solved to obtain $(\overline{x}_t^j, \overline{\theta}_t^j)$. If $\overline{\theta}_t^j < Q_{t+1}(\overline{x}_t^j)$ then another *optimality cut* (12.3c) is added to (12.3) and (12.3) is resolved. If $xbar_t^j$ forces infeasibility in any future period then a *feasibility cut* (12.3b) is added to (12.3). This process is repeated until $\overline{\theta}_t^j \geq Q_{t+1}(\overline{x}_t^j)$. For the construction of feasibility and optimality cuts, see Chapter 3.

For implementation in multi-stage problems, it is assumed that there are a finite number K_t of *scenarios* in each period t. The scenarios consist of all possible realizations of the random vectors from periods 2 through t. For every period t scenario j, there corresponds a unique *ancestor* scenario $a(j)$ in period $t-1$ and, perhaps, several descendant scenarios $d(j)$ in period $t+1$. NDSPA solves (12.1) by first obtaining a feasible solution to (12.3) for all t and j and by then sequentially solving (12.2) using the relaxation in (12.3) from periods T to one.

NDSPA

Step 0.

Solve (12.3) for $t = 1$ (dropping the scenario index j) where $\theta_1 = 0$, $r_1 = s_1 = 0$ and (12.3a) is replaced by $A_1 x_1 = b_1$. Set $\theta_t^j = 0$ and $r_t^j = s_t^j = 0$ in (12.3) for all t and scenarios j at t. (The indices r_t^j and s_t^j are updated whenever a constraint (12.3b) or (12.3c) is added to (12.3)).

Step 1.

If (12.3) is infeasible for $t = 1$, STOP. The problem (12.1) is infeasible. Otherwise, let $\overline{x}_1$ be the current optimal solution of (12.3) for $t = 1$. Use $\overline{x}_1$ as in input in (12.3a) for $t = 2$. Solve (12.3) for $t = 2$ and all ξ_2^j, $j = 1, \ldots, K_2$. If any period two problem (12.3) is infeasible, then add a feasibility cut (12.3b) to (12.3) for $t = 1$, resolve (12.3) for $t = 1$, and return to 1. Otherwise let $t = 2$ and go to 2.

Step 2.

(a) Let the current period t optimal solutions be $\overline{x}_t^j$, $j = 1, \ldots, K_t$. Solve (12.3) for $t + 1$ and all $j = 1, \ldots, K_{t+1}$ using the ancestor solution $xbar_t^j$ in (12.3a).

(b) If any period $t+1$ problem is infeasible, add a feasibility cut (12.3b) to the corresponding ancestor period t problem and resolve that problem.

If the period t problem is infeasible, let $t = t - 1$.
If $t = 1$, go to 1.
Otherwise, return to 2.a.

Otherwise, all period $t + 1$ problems (12.3) are feasible.

If $t \leq T - 2$, let $t = t + 1$ and return to 2.a.
Otherwise ($t = T - 1$), remove the $\theta_r^j = 0$ restriction for all periods r and scenarios j at r. Let the current value of each θ_r^j be $\theta_r^j = -\infty$ if no constraints (12.3c) are present. Go to 3.

Step 3.

(a) Find $E_t^{\ell,j}$ and $e_t^{\ell,j}$ for a new constraint (12.3c) at each scenario t problem (12.3) using the current period $t+1$ solutions.

(b) If there exists j such that

$$\bar{\theta}_t^j < e_t^{\ell,j} - E_t^{\ell,j} \bar{x}_t^j, \qquad (12.4)$$

then add the new constraint (12.3c) to each period t problem (12.3) for which (12.4) holds. Solve each period t problem (12.3). Use the resulting solutions $(\bar{x}_t^j, \theta_t^j)$ to form (12.3a) for the corresponding descendant period $t+1$ problems (12.3) and resolve each period $t+1$ problem (12.3).

If $t < T-1$, let $t = t+1$ and go to 2.a.

Otherwise, return to 3.a.

Otherwise, $\bar{\theta}_t^j = e_t^{\ell,j} - E_t^{\ell,j} \bar{x}_t^j$ for all scenarios j at t.

If $t > 1$, let $t = t-1$ and return to 3.a. Otherwise, STOP. The current solutions $\bar{x}_\tau^j$, $\tau = 1, \ldots, T$ form an optimal solution of (12.1).

Steps 1 and 2 of NDSPA represent a *forward pass* to obtain feasibility in each scenario subproblem. Step 3 is a *backward pass* that solves (2) beginning with period T and passing backward to period 1. Unboundedness may be handled explicitly in the program following the procedure in Van Slyke and Wets [20] but in the computer code of NDSPA all variables are upper bounded and hence unboundedness is avoided. For period T, the computer code also has a special procedure for solving (3). It uses the *bunching* (see Wets [22] and Chapter 3, Section 3.4) method to look through all realizations of ξ_T and find those for which a given basis is optimal. This procedure is described in the next section and represents an alternative to the *sifting* procedure of Garstka and Rutenberg [8].

Experimental results using NDSPA have been encouraging. In Birge [3, 4], NDSPA is compared with a piecewise linear partitioning algorithm, a basis factorization procedure and the code MINOS (Murtagh and Saunders [17]) on a set of staircase test problems from Ho and Loute [12]. NDSPA consistently outperformed the other methods except on one problem in which its storage limitations were exceeded. In general, the results compared favorably with those of Kallberg and Kusy [15] and Kallberg, White, and Ziemba [16] for simple recourse problems. Each stochastic problem was solved in less than twice the time required to solve the deterministic problem with expectations substituted for the random variables.

12.3 The NDST3 Computer Code-Primary Subroutines

NDSPA has been coded in FORTRAN in a current version called NDST3. This code allows for three periods including three second period scenarios and one hundred twenty-five third period scenarios. Each scenario problem (12.3) is limited to three hundred fifty rows and six hundred columns. Within any scenario problem (12.3), there can be at most three thousand nonzero elements. Tolerances can be set in the BLOCK DATA section and in the example are set at 10^{-7} for zero tolerance, 10^{-5} for pivot tolerance, 10^{-4} for reduced cost tolerance and 10^{-10} for small tolerance. The linear programming sections of the code are from L PM-1 written by J.A. Tomlin (Pfefferkorn and Tomlin [18]).

Many variables in NDST3 have multiple subscripts. This questionable programming technique is used to make the scenario obvious. For example, $XLB(2,3,2,1)$ is the lower bound on the second variable in scenario 1 in period 3 with ancestor scenario 2 in period 2. In general, the last three subscripts of all variables with more than two subscripts are $(JCUR, JPER(2), JPER(3))$ where JCUR indicates the period of the scenario, JPER(2) indicates the period 2 ancestor scenario and JPER(3) denotes the period 3 scenario. This last period scenario is not used in the current version of NDST3 but has been used for a four period implementation. The current version is limited to three periods to avoid excessive storage requirements. The code can process four period problems if the period 3 index is incremented in all array definitions and sufficient memory is available. The subroutine SHIFTR, which manipulates data storage, must also be updated if the dimensions are changed.

The main variables in the code are stored in the blank common block. These variables and their descriptions follow.

Variable	Definition
$B(i,j,k,l)$	Current right-hand side element i in period j and scenario k, l
$X(i,j,k,l)$	Current value of variable basic in row i at period j and scenario k, l
$XLB(i,j,k,l)$ and $XUB(i,j,k,l)$	Lower and upper bounds of variable i at j, k, l
$XKSI(i,j,k,l)$	Current realization of random vector in row i at j, k, l
$YPI(i,j,k,l)$	Current dual variable value for row i at j, k, l
$NROW(j,k,l)$	Current number of rows at j, k, l
$NCOL(j,k,l)$	Current number of columns at j, k, l
$NELM(j,k,l)$	Current number of nonzero elements at j, k, l
$JH(i,j,k,l)$	Variable basic in row i at j, k, l
$KINBAS(i,j,k,l)$	Status (basic, nonbasic) of variable i at j, k, l
LA,IA,A	Linked lists of A_t matrix elements
LE,IE,E	Linked lists of elements in eta vector form of basis inverse
LBN,IBN,ABN	Linked lists of elements in B_t matrices

PROB(j,k,l)   Probability of scenario j, k, ℓ

The important variables in BLOCK 3 are

Variable	Definition
NND(i)	Number of scenarios in period i
NPASS	Number of passes from period t to $t-1$ or $t+1$
JPER(i)	Current scenario realization in period i
JCUR	Current period
JPASS	Indicator of forward or backward pass; JPASS = 1 for forward, JPASS = 2 for backward
NPER	Number of periods T

In BLOCK 4, the significant variables are

Variable	Definition
XTOPT	Value of $-e_T^{\ell,j} + E_T^{\ell,j}\bar{x}_T^j$ for checking for optimality
PRBY(i,j)	Probability of j-th realization of i-th random element in stochastic vector in period T
PRST(i,j,k)	Joint probability of i-th realization of first random element, j-th realization of second, and k-th realization of third for stochastic vector at t
CBST(i,j)	Value of j-th realization of i-th random element in stochastic vector at T
NCUR(i)	Current realization of i-th random element at T
IBST(i)	Row of i-th random element at T
NST	Number of random elements at T

The code NDST3 assumes that specific random vectors (with specific probabilities) are assigned for periods 2 through $T-1$ and that at period T the random vector includes NST independent random elements. The bunching approach can then be easily applied to these possibilities.

The main program in NDST3 organizes the algorithm and calls subroutines to implement the steps of NDSPA. The main routines called in this segment are:

INPUT	accepts all data input;
INCHK	echoes input;
NORMAL	solves the linear program in (12.3);
STRPRT	reports on current solution;
NDCOM	directs the algorithm for $t < T$;
PARSFT	controls the algorithm for $t = T$;
WRAPUP	writes output.

The main routine calls NORMAL to solve (12.3) if $t < T$ and then calls NDCOM to determine which problem to solve next. If $t = T$, PARSFT is called to solve (12.3) for period T and determine the next step of the algorithm. JCUR(t) is set equal to NPER+1 (i.e., T+1) whenever a terminating condition (infeasible or optimal) is met.

The following routines are all used by NORMAL in solving the linear program (12.3):

RHCHCK   checks now residuals;
BTRAN   performs backward transformation;
FORMC   forms objective function vector and checks feasibility;
PRICE   computes reduced costs and picks entering column;
CHUZR   performs minimum ratio test and determines leaving variable;
WRETA   forms new eta-vectors for product form of inverse;
SHIFTR   rearranges data storage;
INVERT   computes basis inverse using LU decomposition;
UNPACK(i) expands i-th column in A;
BUNPCK(i) expands i-th column in B;
SHFTE   shifts eta vectors around;
UPBETA   updates right-hand side and basis indicators.

NORMAL reinverts the basis every INVFRQ iterations or if the maximum row residual is greater than 10 times ZTOLZE. A maximum of ITRFRQ iterations is allowed.

The subroutine NDCOM handles all steps for NDSPA for $t < T$. The variable MSTAT is used to indicate infeasibility (QN) or feasibility (QF). If an infeasibility is found, then a feasibility cut is added in the subroutine FEASCT and t is set to $t-1$. If the current problem is feasible, then NDCOM determines the next subproblem to solve. If every scenario at period t has been solved, the NDCOM sets up problem (12.3) for period $t+1$. The subroutine BPRODX is called here to compute $B_t x_t^j$ and FRMRHS is called to find $\xi_{t+1}^{d(j)} + B_t x_t^j$.

If the algorithm has proceeded to the backward pass, the control shifts Again, if an infeasibility is found, then a feasibility cut is added to the corresponding ancestor scenario problem. First, any cuts (12.3b) or (12.3c) that are slack (satisfied as strict inequalities) are deleted in the subroutine DLETCT. This option saves on storage and does not affect convergence. NFLG $= 1$ signifies that the current problem (12.3) solution is optimal. For $t < T-1$, the code follows Step 2 of NDSPA and continues to $t+1$. If $t = T-1$ and condition (12.4) is not met, then NDCOM follows the iterations in Step 3 of NDSPA. If condition (4) is met in following this backwards iteration then an optimality cut (3.3) is placed on the corresponding ancestor scenario using the subroutine LKHDCT(K) where K is the preceding period. Optimality at period K is checked in the subroutine OPTCHK(K) which sets NFLG $= 1$ if (4) is not met.

Subroutine PARSFT performs Step 3 of NDSPA for $t = T$. It includes the variable JSTCH(i,j,k) that indicates the number of the basis found optimal for the alternative with realizations i, j, k for random elements 1, 2, and 3 respectively, in the last period. NCUR(i) is the current realization of the i-th random element and NXNF(i) is the realization of the i-th random element in the first infeasible basis found by the bunching procedure. NETND(i) keeps the number of eta vectors in the i-th basis and INFLG $= 0$ for no infeasibilities and

1 for an infeasibility found in passing through all alternative random vectors at T. YBX is a vector keeping $B_{T-1}\overline{x}_{T-1}^{a(j)}$.

The bunching procedure begins by calling NORMAL to obtain an optimal solution. On subsequent iterations, the procedure begins with the previous basis which is dual feasible and calls the subroutine DNORML which implements the dual simplex method. In either case, if an infeasibility is caught then a feasibility cut is made and control returns to the main program.

After having found a new optimal basis, the algorithm updates $E_{T-1}^{\ell,j}$ and $e_{T-1}^{\ell,j}$ and then loops through all right-hand side alternatives for which no feasible basis has yet been found. Since every scenario corresponds to the same objective function, an optimal basis for any scenario is dual feasible for all other scenarios. The appropriate right-hand side is set up and FTRAN is called to find the values of the basic variables. The subroutine DCHUZR is then called to determine a leaving (infeasible) variable. It returns IROWP = 0 for a feasible basis which is then optimal. If a leaving variable is found then DCHUZC is called to find an entering variable. If no entering variable is found then the current scenario is infeasible and control is returned to the main program. If an entering variable is found, then the current scenario is marked as the first scenario to check in the next bunching loop (if no scenario has been found infeasible for the current basis) and the next scenario is tested.

Whenever a scenario is found to be feasible for the current right-hand side then the values of $E_{T-1}^{\ell,j}$ and $e_{t-1}^{\ell,j}$ are updated, and the next scenario is chosen. When an optimal basis has been found for all period T scenarios then optimality is checked using the subroutine XOPTCK. NFLG = 1 is returned if (12.4) is not met and the algorithm proceeds back to period $T-1$. If (12.4) is met then a new optimality cut is added to the ancestor period $T-1$ problem.

The algorithm proceeds through these subroutines until optimality is found in NDCOM (for $T > 2$) or PARSFT($T = 2$) or until infeasibility is found in NDCOM. When one of these terminal conditions is reached, WRAPUP is called and the output described in the next section is produced.

12.4 Input and Output Formats

The input format for NDST3 basically follows the MPS standard for mathematical programs except in its splitting the data into periods. As a test problem we used, among others, SCAGR7.S2 which was adapted from the staircase test problems of Ho and Loute [12]. It contains two periods for the stochastic program, and, in the second period, there are three independent random variables with two values each. This leads to eight total scenarios.

The first row of the input contains five values used in program. Each is entered in I4 format, they are in order:

IFPROB   number of problem;
IOBJ   row of objective function (usually '1');
INVFRQ   iterations between matrix inversions;
ITRFRQ   total number of iterations allowed;
NPER   number of periods.

The next NPER rows contain the number of different right-hand side values (I4 format) for each period. The first and last periods have 1's because the first period is deterministic and the last period right-hand sides are input separately at the end of the program. The fourth row contains the probability of the first right-hand side value in F5.3 format. The next sections are ROWS, COLUMNS, and RHS sections for MPS format for all values in the first period set of constraints, $A_1 x_1 = b_1$. Following as ENDATA, lower bounds on all variables (excluding slacks) in 9F8.0 format and upper bounds in the same format are input. If an initial basis were entered then a section headed by BASIS and including columns and the corresponding row in the basis could be entered after the COLUMNS section. This format is discussed below as part of the output.

The next sections of the code include ROWS and COLUMNS sections to describe the matrix B_1 in (12.1). This is followed by an ENDATA and the probability of the next period's first right-hand side vector. The data for $A_2 x_2 = \xi_2^j$ would then be entered for each possible ξ_2^j and, if more periods were present, this would be followed in each case by the data for B_2 (possibly depending on j). This process of repeating the probability of ξ_t^j, giving the data for $A_t x_t = \xi_t^j$ and of then giving B_t repeats until all scenarios indicated in the command lines of the code have been input.

The last period scenario input is followed by a section marker STOCH which prompts the program to read in separate values and probabilities for random elements in the last period. For each random element, we must give the row name in columns 5–12, the value of the element is given in F12.4 format in columns 25–36 and the probability of that value is given in F12.4 format in columns 50–61. Each independent element is input with at most five values total.

Another version of NDST3, called NDST3.A, has also been developed at IIASA, Laxenburg, Austria. In this code, input follows the standard format set at IIASA except for the first line of input which contains the control parameters.

NDST3 writes two output files on devices 6 and 7. The first one contains most of the iteration and result information. The second contains only the variables that were basic in the optimal solution found by the program. That output may be inserted into an input file to provide the program with a starting basis. For detailed instructions about input/output format and the structure of this code, see NDSP User's Manual, Edwards [1985].

12.5   Extensions and Observations

As mentioned above, NDST3 can be easily expanded to handle larger problems and more scenarios. Some care, however, must be used in maintaining storage requirements within acceptable limits. Future versions of the code are planned to eliminate some redundancy and to enable more complex problems to be solved. Other planned options are to include the possibilities for some continuous distributions and to use approximating techniques from Birge and Wets [5] in achieving convergence within a predetermined tolerance. This has been implemented for a single random variable in a new code NDST4 and further refinements are planned.

The code has performed very well in general and in most situations significantly (by an order of magnitude) outperforms general purpose linear programming codes. The one problem in which it did not perform well, is one that required that a large number of feasibility cuts be added to the first period problem. These cuts were dense and, without deleting slack cuts, the problem required an excessive number of nonzero elements (i.e., more than three thousand). When slack cuts were deleted, the program obtained an unstable basis that did not generate a feasible first period solution. This may be a problem inherent in decomposition algorithms because of numerical error present in generating cuts. Two truely identical cuts may be generated that differ only in their error coefficients. This is the cutting plane analogy of the slow convergence characteristics observed in Dantzig-Wolfe decomposition (Ho [11]). When NDST3 was implemented so that the row residuals were checked on every iteration, this problem was solvable despite the instability. It then required 1379 simplex iterations compared to 1742 iterations for a simplex method implementation on the deterministic equivalent problem. No other problems tested required NDST3 to perform as many as half the number of iterations of the simplex method. The instability may therefore have caused slower convergence in this example. It appears that stability problems are rare but if further testing results in more of these difficulties, some testing of the integrity of cuts may have to be added.

References

[1] R.W. Ashford, "A Stochastic Programming Algorithm for Production Planning", Scicon report, Milton Keynes, 1982.

[2] E.M.L. Beale, J.J.H. Forrest, and C.J. Taylor, "Multi-time Period Stochastic Programming", in *Stochastic Programming*, M.A.H. Dempster (ed.), Academic Press, New York, 1980.

[3] J.R. Birge, "Solution Methods for Stochastic Dynamic Linear Programs", Technical Report SOL 80-29, Systems Optimization Laboratory, Stanford University, 1980.

[4] J.R. Birge, "Decomposition and Partitioning Methods for Multi-Stage Stochastic Linear Programs", Technical Report 82-6, Department of Industrial and Operations Engineering. The University of Michigan, 1982.

[5] J.R. Birge and R.J-B. Wets, "Designing Approximation Schemes for Stochastic Optimization Problems", IIASA Report, 1983.

[6] G.B. Dantzig and A. Madansky, "On the Solution of Two-Stage Linear Programs Under Uncertainty", in *Proceedings, 4th Berkeley Symposium on Math. Stat. and Prob.* U.C. Press, Berkeley.

[7] J. Edwards, "Documentation for the ADO/SDS collection of stochastic programming codes", WP85-02 IIASA, Laxenburg, Austria, 1985.

[8] S. Garstka and R. Rutenberg, "Computation in Discrete Stochastic Programming with Recourse", *Operations Research* **21**(1973), 112–122.

[9] C.R. Glassey, "Dynamic Linear Programs for Production Scheduling", *Operations Research* **19**(1971), 45–56.

[10] C.R. Glassey, "Nested Decomposition and Multi-Stage Linear Programs", *Management Science* **20**(1973), 282–292.

[11] J.K. Ho, "Convergence Behavior of Decomposition Algorithms for Linear Programs", Working Paper No. 179, College of Business Administration, The University of Tennessee (1984).

[12] J.K. Ho and E. Loute, "A Set of Staircase Linear Programming Test Problems", *Mathematical Programming* **20**(1981), 245–250.

[13] J.K. Ho and A.S. Manne, "Nested Decomposition for Dynamic Models", *Mathematical Programming* **6**(1974), 121–140.

[14] P. Kall, "Computational Methods for Solving Two-Stage Stochastic Linear Programming Problems", *Journal of Appl. Math. and Physics* **30**(1979), 261–271.

[15] J.G. Kallberg and M.I. Kusy, "A Stochastic Linear Program with Simple Recourse", Faculty of Commerce, The University of British Columbia, 1976.

[16] J.G. Kallberg, R.W. White, and W.T. Ziemba, "Short Term Financial Planning Under Uncertainty", *Management Science* **28**(1982), 670–682.

[17] B.A. Murtagh and M.A. Saunders, "Large Scale Linearly Constrained Optimization", *Mathematical Programming* **14**(1978), 41–72.

[18] C.E. Pfefferkorn and J.A. Tomlin, "Design of a Linear Programming System for IILIAC IV", Technical Report SOL 76-8, Systems Optimization Laboratory, Stanford University, 1976.

[19] B. Strazicky, "Some Results Concerning an Algorithm for the Discrete Recourse Problem", in *Stochastic Programming*, M.A.H. Dempster (ed.). Academic Press, New York, 1980.

[20] R. Van Slyke and R.J-B. Wets, "*L*-Shaped Linear Programs with Applications to Optimal Control and Stochastic Linear Programs", *SIAM Journal on Appl. Math.* **17**(1969), 638–663.

[21] R. J-B. Wets, "Programming Under Uncertainty: The Solution Set", *SIAM Journal on Appl. Math.* **14**(1966), 1143–1151.

[22] R. J-B. Wets, "Stochastic Programming: Solution Techniques and Approximation Schemes", in *Mathematical Programming: State-of-the-Art 1982*, A. Bachem, M. Groetschel, B. Korte (eds.). Springer-Verlag, Berlin, 1983.

CHAPTER 13

THE RELATIONSHIP BETWEEN THE L-SHAPED METHOD AND DUAL BASIS FACTORIZATION FOR STOCHASTIC LINEAR PROGRAMMING

J.R. Birge

Abstract

The basis factorization method of Strazicky for stochastic linear programs is shown to involve the same computational effort per iteration as the L-shaped method of Van Slyke and Wets. A variant of the factorization approach can then be found which is equivalent to the L-shaped method. The advantages of this decomposition approach over a standard factorization are discussed.

13.1 Introduction

We consider the problem

$$
\begin{aligned}
\text{minimize} \quad & cx + Q(x) \\
\text{subject to} \quad & Ax = b \\
& x \geq 0,
\end{aligned}
\tag{13.1}
$$

where $Q(x) = E_\xi[\min qy \text{ subject to } Wy = \xi - Tx, y \geq 0]$
and ξ is a random n_2-vector, where A is an $m_1 \times n_1$ real matrix, W is an $m_2 \times n_2$ real matrix, T is an $m_2 \times n_1$ real matrix, and c, q, and b are correspondingly dimensioned vectors. For $\xi \in \Xi = \{\xi^1, \xi^2, \ldots, \xi^N\}$, where $P(\xi = \xi^i) = p^i$, we have (13.1) is equivalent to

$$
\begin{aligned}
\text{minimize} \quad & cx + p^1 qy^1 + p^2 qy^2 + \cdots + p^N qy^N \\
\text{subject to} \quad & Ax = b \\
& Tx + Wy^1 = \xi^1 \\
& TxWy^2 = \xi^2 \\
& \quad \vdots \\
& Tx + Wv^N = \xi^N
\end{aligned}
\tag{13.2}
$$

The dual of (13.2) is

$$\text{maximize} \quad b^T \sigma + p^1 {\xi^1}^T \pi^1 + p^2 {\xi^2}^T \pi^2 + \ldots + p^N {\xi^N}^T \pi^N$$

$$\text{subject to} \quad A^T \sigma + p^1 T^T \pi^1 + p^2 T^T \pi^2 + \ldots + p^N T^T \pi^N \le c^T$$

$$W^T \pi^1 \le q^T$$

$$W^T \pi^2 \le q^T \tag{13.3}$$

$$\vdots$$

$$W^T \pi^N \le q^T$$

Kall [2] and Strazicky [3] observed that any feasible basis of (13.3) may be written as

$$\begin{bmatrix} B & Y \\ L & Z \end{bmatrix} \tag{13.4}$$

where B is a block diagonal matrix. For (13.3),

$$B = \begin{bmatrix} \tilde{W}_1 \tilde{I}_1 & & \\ & \tilde{W}_2 \tilde{I}_2 & \\ & & \tilde{W}_N^T \tilde{I}_N \end{bmatrix},$$

where $[\tilde{W}_i^T \tilde{I}_i]$ is an $n_2 \times n_2$ submatrix of $[W^T \ \ I]$ for all $i = 1, 2, \ldots, N$. Kall notes that we may reduce the size of $[\tilde{W}_i^T \tilde{I}_i]$ by taking an $m_2 \times m_2$ nonsingular submatrix $\widehat{W}_i^T$ from $\tilde{W}_i^T$. We have

$$\begin{bmatrix} \widehat{W}_i^T & 0 \\ \bar{W}_i^T & \bar{I}_i \end{bmatrix} \begin{bmatrix} \pi_1 \\ \bar{\rho}_i \end{bmatrix} = \begin{bmatrix} \hat{q} \\ \bar{q} \end{bmatrix} + \begin{bmatrix} \hat{I}_i \\ 0 \end{bmatrix} \hat{\rho}_i \tag{13.5}$$

or $(\widehat{W}_i^T)^{-1}\hat{q} + (\widehat{W}_i^T)^{-1}\hat{\rho}_i = \pi_i$, $\bar{\rho}_i = \bar{q} - (\bar{W}_i^T)(\widehat{W}_i^T)^{-1}\hat{q} - (\bar{W}_i^T)(\widehat{W}_i^T)^{-1}\hat{\rho}_i$. So, we can rewrite (13.5) as

$$[(\bar{W}_i^T)(\widehat{W}_i^T)^{-1}\bar{I}_i] \begin{bmatrix} \hat{\rho}_i \\ \bar{\rho}_i \end{bmatrix} = [\bar{q} - (\bar{W}_i^T)^1 \cdot (\widehat{W}_i^T)^{-1}\hat{q}]. \tag{13.6}$$

(13.6) substantially reduces the number of rows from (13.5) but it has a signif- icant drawback in terms of nonzero element storage. The sparse matrix W_i^{-T} may be transformed into a very dense matrix $(W_i^{-T})(\widehat{W}_i^T)^{-1}$. Kall uses this ma- trix in solving (13.3) and, therefore, must update the full $(n_2 - m_2) \times (n_2 - m_2)$ basis throughout the algorithm. Wets [5] has observed that $m_2 \times m_2$-matrices $\widehat{W}_i^T$ should be used as working bases so that updates only need to be performed within these sparse matrices instead of in the larger, dense $(n_2 - m_2) \times (n_2 - m_2)$ matrix in Kall's approach.

Wets also suggests that the algorithm may be made even more efficient by taking advantage of the repetition of similar bases among the $\widehat{W}_i^T$. LU decomposition and sparse updating may additionally be used to improve this approach. Wets then conjectures what is shown below: that this method involves the same computational effort per iteration as the L-shaped method of Van Slyke and Wets [4] and that a modification of the dual basis factorization method will follow the same path as the L-shaped method. This modification amounts to the dual decomposition procedure proposed by Dantzig and Madansky [1].

13.2 Discussion

We assume we have a feasible solution to (13.3), $(\sigma^0, \pi^{1,0}, \ldots, \pi^{N,0}, \lambda^0, \rho^{1,0}, \ldots, \rho^{N,0})$, where $(\lambda^0, \rho^{1,0}, \ldots, \rho^{N,0})$ are slack variables. We also assume that $\tilde{W}_i^T = W^T$ for all i so that only columns from A^T and the identity are basic in the first set of constraints. In the pricing operation, we solve

$$A^B x_B + A^N x_N = b,$$
$$I_N x_N = 0,$$

where I_N is the set of basic identity columns. We have $x_B = (A^B)^{-1} b$ and check for $x_B \geq 0$. If some $x_B(i) < 0$, then that column in A^B is replaced and the problem is solved again. If we restrict ourselves to only checking for primal feasibility in the x variables, then we are solving the dual problem

$$\text{maximize} \quad b^T \sigma$$

$$\text{subject to} \quad A^T \sigma \leq c^T - \sum_{i=1}^{N} p^i T^T \pi^{i,0},$$

or the primal problem

$$\text{minimize} \quad \left(c - \sum_{i=1}^{N} p^i \pi^{i,0} T\right) x$$
$$\text{subject to} \quad Ax = b \tag{13.7}$$
$$x \geq 0.$$

This is essentially the first step of the L-shaped method. The dual method involves the same steps of computing $A^B x_B = b$, $\pi A^B = c_B$ and $\rho = c_N - \pi A^N$ as in the primal method, so the computational effort is the same at each step. We note that this does not include pricing for y variables as would occur in the general dual method.

After all $x_B(i) \geq 0$ have been found, we let x^1 be the prices and we proceed to solve $\tilde{W}_i y^i = \xi^i - T x^1$ for all y^i. For every subproblem i, if $y^i(j) < 0$ then we choose a leaving column only from the identity columns in subproblem i. We relax dual feasibility in the first set of constraints. This process is equivalent to solving the subproblems

$$\text{minimize} \quad qy^i$$
$$\text{subject to} \quad Wy^i = \xi^i - Tx^1, \tag{13.8}$$
$$y^i \geq 0,$$

for all i as in the L-shaped method. We note again that the computations involved in a single iteration are the same in both methods except that we do not update the prices in the first set of constraints for the factorization method. We note also that solutions of (13.8) may be found quickly by finding all ξ^i for which a given basis is optimal.

After solving these problems, we obtain either an unbounded condition or all $y^i \geq 0$. In the former case, some subproblem (13.8) is infeasible. We then look at the column in (13.3) which gave the unboundedness condition. For $y_j^i < 0$, $y_j^i = (\widehat{W_i})^{-1}(j,\cdot) \cdot (\xi^i - Tx^1)$ and the column $-[(\widehat{W_i})(j,\cdot) \cdot \bar{W_i}]^T \leq 0$. We let $\pi = -(\widehat{W_i})(j,\cdot)$ and obtain

$$\pi(\xi^i - Tx^1) > 0, \tag{13.9}$$

and

$$\pi \bar{W_i} \leq 0. \tag{13.10}$$

(13.9) and (13.10) are the infeasibility conditions for (13.8) that we would find in the L-shaped method. In the dual method, we would choose a pivot from the first set of constraints so that we would force $\pi(\xi^i - Tx) \leq 0$. We introduce a new column in the main problem,

$$\begin{bmatrix} p^i(\xi^i \pi^i)^T \rho \\ (p^i T^T \pi^i)\rho \end{bmatrix}$$

where $\rho \geq 0$. The main problem is then

$$\text{maximize} \quad b^T \sigma + p^i(\xi^i \pi^i)^T \rho$$
$$\text{subject to} \quad A^T \sigma + (p^i T^T \pi^i)\rho \leq \bar{c}^T,$$
$$\rho \geq 0,$$

where $\bar{c}^T$ includes c^T and other fixed columns of π. This is equivalent to adding a constraint

$$(\pi^i T)x \leq \pi^i \xi,$$

as in the L-shaped algorithm. We next solve the main problem again and repeat.

If after solving the subproblems all $y^i \geq 0$, then either the problem is optimal or one of the first set of constraints in (13.3) has been violated. In this case, if we let $\theta = \sum_{i=1}^{N} p^i ((\xi^i)^T - (Tx_1)^T)\pi^{i,1}$, where $\pi^{i,1}$ is the initial solution of (13.3), then we have

$$\theta < \sum_{i=1}^{N} p_i ((\xi^i)^T - (Tx_1)^T)\pi^{i,2}$$

where $\pi^{i,2}$ is the optimal subproblem i solution. We observe that either $\pi^{i,1}$ or $\pi^{i,2}$ or linear combinations of these solutions may be used as solutions for the subproblems. We use this to obtain a substitute first period problem:

$$\text{maximize} \quad b^T\sigma + \lambda_1 (\sum_{i=1}^{N} p^i (\xi^i)^T \pi^{i,1}) + \lambda_2 (\sum_{i=1}^{N} p^i (\xi^i)^T \pi^{i,2})$$

$$\text{subject to} \quad A^T\sigma + \lambda_1 (\sum_{i=1}^{N} p^i (\xi^i)^T \pi^{i,1}) + \lambda_2 (\sum_{i=1}^{N} p^i (\xi^i)^T \pi^{i,2}) \le c^T \quad (13.11)$$

$$\lambda_1 + \lambda_2 = 1$$

$$\lambda_1, \lambda_2 \ge 0.$$

We solve problem (13.11) and repeat by adding a column for feasibility of the subproblems or by adding a column for choices of subproblem solutions as in (13.11). We note that these are the same steps as in the L-shaped decomposition method where $\theta < \sum_{i=1}^{N} p^i ((\xi^i)^T - (Tx_1)^T)\pi^i$ and a constraint on θ,

$$\left(\sum_{i=1}^{N} \pi^i T\right) x + \theta \ge \sum_{i=1}^{N} p^i (\xi^i)^T, \quad (13.12)$$

is added. The two methods with these specifications follow the same procedures for each iteration. We note also that these methods follow the same steps as Dantzig-Wolfe decomposition applied to the dual problem (13.3), (Dantzig-Mandansky [1], Van Slyke and Wets [4]).

13.3   Conclusion

We have shown that on each iteration of the L-shaped method, the number of steps is equivalent to that of the basis factorization method and that the L-shaped method may be viewed as a variant of the basis factorization approach. In general, however, the two methods will not follow the same path to optimality. By maintaining dual feasibility, the basis factorization restricts the path to optimality and requires more effort in checking for feasibility within the first set of constraints.

The decomposition variant of basis factorization also avoids two other problems inherent in the full factorization approach. For $X = B^{-1}Y$ in (13.4), the factorization approach uses the inverse of $(LX - Z)$ in performing simplex operations. X is composed of columns of B^{-1} since Y is composed of identity columns. The columns of B^{-1} need not be sparse and may be very dense, causing $(LX - Z)$ to be dense as well. The storage requirement for the nonzero elements of this $n_1 \times n_1$ matrix may be large.

Another difficulty in applying this factorization without decomposition is that, whenever an identity column in I_i in (13.5) is replaced, then $\widehat{W}_i^T$ must be changed and $(LX - Z)$ changes. This pivot alters the prices (x) for all other blocks $j \neq i$. Therefore, a pivot step is required for each new block into which this identity column enters. By fixing x in the decomposition, whenever a new matrix $\widehat{W}_i^T$ is introduced, all values ξ^j such that $y^j = (\widehat{W}_i^T)^{-1}(\xi^j - Tx) \geq 0$ can be found without performing separate pivot operations. For very large N, the standard factorization scheme may be forced through a long sequence of pivots, whereas the decomposition approach may change these bases quickly. For problems with large N, then, the decomposition variant above is probably the only tractable basis factorization method.

References

[1] G.B. Dantzig and A. Madansky, "On the solution of two-stage linear programs under uncertainty", in *Proceedings, 4th Berkeley Symposium on Math. Stat. and Prob.*, University of California Press, Berkeley, (1961), 165–176.

[2] P. Kall, "Computational methods for solving two-stage stochastic linear programming problems", *Journal of Appl. Math. and Physics* **30**(1979), 261–271.

[3] B. Strazicky, "Some results concerning an algorithm for the discrete recourse problem", in *Stochastic Programming*, M.A.H. Dempster (ed.), Academic Press, New York (1980), 263–271.

[4] R. Van Slyke and R. Wets, "L-shaped linear programs with applications to optimal control and stochastic linear programs", *SIAM Journal on App. Math.* **17**(1969), 638–663.

[5] R. J-B. Wets, "Stochastic programming: solution techniques and approximation schemes", in *Mathematical Programming: State-of-the-Art 1982*, Springer-Verlag, Berlin, 1982.

CHAPTER 14

DESIGN AND IMPLEMENTATION OF A STOCHASTIC PROGRAMMING OPTIMIZER WITH RECOURSE AND TENDERS

L. Nazareth

Abstract

This paper serves two purposes, to which we give equal emphasis. First, it describes an optimization system for solving large-scale stochastic linear programs with simple (i.e. decision-free in the second stage) recourse and stochastic right-hand-side elements. Second, it is a study of the means whereby large-scale Mathematical Programming Systems may be readily extended to handle certain forms of uncertainty, through post-optimal options akin to sensitivity or parametric analysis, which we term "recourse analysis". This latter theme (implicit throughout the paper) is explored in a proselytizing manner, in the concluding section.

14.1 Introduction

This paper is a sequel to Nazareth and Wets [21] and serves two purposes, to which we give equal emphasis. First, it describes an optimization system for solving a restricted but important class of large-scale stochastic linear programs with recourse. Second, it is a study and detailed illustration of the means whereby any large-scale Mathematical Programming System (MPS) designed for solving deterministic linear programs, could be readily extended to handle some forms of uncertainty, in particular, via post-optimal analysis options. This latter theme (implicit throughout the paper) is explored, in a proselytizing manner, in the concluding section.

The class of practical stochastic linear programs with which we are concerned (termed C1 problems in [21]) arise as a natural extension of the linear programming model as follows: given a linear program with matrix A, it is often the case that some of the components of the right-hand-side (exogenous) vector of resource availability or resource demand, are known only in probability and have been replaced (in the deterministic LP formulation) by some expected value. We seek to extend this linear program, using the recourse formulation. Rows of A corresponding to the stochastic right-hand-side are used to define the technology matrix T (we follow the notation and terminology of [21]) and the remaining rows of A are used to define the constraint matrix A, both A and T being typically large, sparse matrices. The recourse is assumed to be simple (i.e. *decision-free* in the second-stage problem) and specified in terms of costs (or penalties) on shortage and surplus. Furthermore, we restrict attention to the case where each component of the stochastic right-hand-side has a given discrete probability distribution. There are many applications for such a model, see Ziemba [27], and more complex stochastic linear programs with recourse can sometimes be solved by an iterative discretization or sampling procedure involving definition and solution of a sequence of C1 problems.

The above considerations are very much in the background of our implementation design, our choice of algorithms and of the more general issues which we wish to discuss regarding the extension of conventional Mathematical Programming Systems, so as to be able to handle at least some forms of uncertainty. Our optimization system is based primarily upon a version of Wolfe's generalized programming algorithm (see Dantzig [4]) given in Nazareth and Wets [21] Section 3.2.1 and, in more detail, in Nazareth [18]. It also includes a version of an algorithm based upon bounded variables (see Wets [25]) given in [21] Section 2.1 and, again in more detail, in [20]. Two simpler options, namely to solve an initial linear program and to permit some of its constraints to be "elastic" are also included to help get a recourse problem "off the ground". In our implementation (see Nazareth [19] for an overview of our overall approach) we have utilized current mathematical programming technology for specifying the problem (using standard MPS conventions [14] for the LP portion and a suitable extension to provide the added stochastic information), to represent the data internally (in packed data structures, space for which is dynamically allocated within a work storage array) and to implement our solution strategies (using an efficient and numerically stable implementation of the simplex method, namely the MINOS System of Murtagh and Saunders [15], [16]).

Finally, we want our design to mesh as naturally as possible with current Mathematical Programming Systems. In particular, we argue in the concluding section of our paper, that "recourse analysis" (simple recourse to start off with, but also more general forms of recourse) could be provided as a post-optimal analysis option in any large-scale MPS, to augment the options for parametric and sensitivity analysis that are now usually available.

14.2   Overview of the SPORT System

14.2.1   Problem

SPORT (pronounced SupPORT) is an acronym for *S*tochastic *P*rogramming *O*ptimizer with *R*ecourse and *T*enders. The current version solves large-scale stochastic programs with simple (decision-free in the second stage) recourse and discrete distribution of right-hand-side elements (termed C1 problems). The formal statement of such problems may be found in [21] (see (1.1) through (1.3) where $W = [I, -I]$ and where the right-hand-side $h(\omega)$ is the only stochastic quantity, with a known discrete distribution) and we shall not repeat here. Instead, we shall state the problem from the perspective emphasized in this paper, namely that of a given linear program in which inherent uncertainty in some of the right-hand-side (exogenous) elements is to be more fully taken into account. Consider therefore the linear program

$$\text{minimize} \quad cx$$
$$\text{subject to} \quad Ax = d \tag{14.1}$$
$$x \geq 0$$

where A is an $m \times n$ matrix (which is generally large and sparse), d is a given m-vector and c is a given n-vector. Some of the elements of d which correspond to demands (or available resources) may be, in reality, only known in probability and defined in (14.1) by taking some expected value. For simplicity, let us suppose that the corresponding "technology" constraints of (14.1) are the last m_2 constraints and let us denote them by $Tx = \overline{h}$, where T is an $m_2 \times n$ matrix. Let the remaining m_1 constraints be $Ax = b$ where A is an $m_1 \times n$ matrix and $d = \left(\begin{smallmatrix} b \\ h \end{smallmatrix}\right)$.

A useful extension to the LP model (14.1) is to permit the constraints $Tx = \overline{h}$ to be "elastic" (Tomlin [24]) by imposing a penalty q_i^+ on shortage in the i-th technology constraint when demand (corresponding to the right-hand-side element $\overline{h}_i$) exceeds the supply $(Tx)_i$, so that $y_i^+ = \overline{h}_i - (Tx)_i \geq 0$. Similarly let q_i^- be the penalty imposed on surplus (when the reverse of the earlier conditions holds) so that $y_i^- = (Tx)_i - \overline{h}_i \geq 0$. (The choice of notation q_i^+ for shortage and q_i^- for surplus is a little unfortunate, but is now standard.) Thus associated with the decision x for the "first-stage" or decision variables, we have a penalty of

$$Q_i(x, \overline{h}_i) = \begin{cases} q_i^+ \left(\overline{h}_i - (Tx)_i\right) & \text{when } \left(\overline{h}_i - (Tx)_i\right) \geq 0 \\ q_i^- \left((Tx)_i - \overline{h}_i\right) & \text{when } \left(\overline{h}_i - (Tx)_i\right) \leq 0. \end{cases}$$

To minimize first stage costs and all penalty costs we can formulate the extension of (14.1) as a problem with "elastic" constraints as follows:

$$\text{minimize} \quad cx + q^+ y^+ + q^- y^-$$
$$\text{subject to} \quad Ax = b$$
$$Tx + y^+ - y^- = \overline{h} \tag{14.2}$$
$$x \geq 0, y^+ \geq 0, y^- \geq 0$$

where q^+ and q^- are m_2-vectors with components q_i^+ and q_i^- respectively.

Unfortunately (14.2) does not address the uncertainty in the right-hand side vector, which so far has been replaced by $\bar{h}$. One way to address uncertainty is to compute the penalty cost associated with each *realization* of the random vector $h(\omega)$. Let us also define the "tender" or "bill of goods" associated with a decision x by $\chi = Tx$. Thus we have

$$Q_i(x, \bar{h}_i) \triangleq \psi_i(\chi_i, h_i(\omega)) = \begin{cases} q_i^+ (h_i(\omega) - \chi_i) & \text{when } (h_i(\omega) - \chi_i) \geq 0 \\ q_i^- (\chi_i - h_i(\omega)) & \text{when } (h_i(\omega) - \chi_i) \leq 0 \end{cases}$$

Let $\Psi(\chi) \triangleq E_\omega \left(\sum_{i=1}^{m_2} \psi_i(x_i, h_i(\omega)) \right) = \sum_{i=1}^{m_2} E_\omega \left(\psi_i(\chi_i, h_i(\omega)) \right) \triangleq \sum_{i=1}^{m_2} \Psi_i(\chi_i)$. We seek to minimize the cost of the decision cx and the expected value of the penalty costs. Thus we can formulate this extension of (14.1) as

$$\text{minimize} \quad cx + \sum_{i=1}^{m_2} \Psi_i(\chi_i)$$
$$\text{subject to} \quad Ax = b \tag{14.3}$$
$$Tx - \chi = 0$$
$$x \geq 0$$

For $C1$ problems it can be readily demonstrated (see, for example [25], [20]) that

$$\Psi_i(\chi_i) = \max_{\ell=0,\ldots,k_i} (s_{i\ell}\chi_i + e_{i\ell})$$

where $s_{i\ell}$ and $e_{i\ell}$ are defined from the probability distribution of $h_i(\cdot)$. Let this be given by values $h_{i1}, h_{i2}, \ldots, h_{ik_i}$ with $h_{i\ell} \leq h_{i,l+1}$, with associated probabilities $p_{i1}, p_{i2}, \ldots, p_{ik_i}$. Then, for $\ell = 0, \ldots, k_i$

$$s_{i\ell} = \left(\sum_{i=1}^{\ell} h_{it} \right) q_i - q_i^+$$
$$e_{i\ell} = q_i^+ \bar{h}_i - q_i \left(\sum_{t=1}^{\ell} h_{it} p_{it} \right) \tag{14.4}$$

where, by convention, $\sum_{i=1}^{0} = 0$, $q_i = (q_i^+ + q_i^-) > 0$ and $\bar{h}_i$ is the expected value of $h_i(\omega)$. Finally, using a theorem in [18], it is possible to state (14.3) in an *equivalent* form and in so doing also unify with (14.2) as follows:

$$\text{minimize} \quad cx + q^+ y^+ + q^- y^- + \sum_{i=1}^{m_2} \Psi_i(\chi_i)$$
$$\text{subject to} \quad Ax = b \tag{14.5}$$
$$Tx + y^+ - y^- - \chi = 0$$
$$x \geq 0, y^+ \geq 0, y^- \geq 0$$

For $\chi \triangleq \overline{h}$ we obtain (14.2) since $\Psi(\overline{h})$ is then a constant term. (14.5) is a piecewise-linear separable convex programming problem with which we shall be concerned henceforth. It makes possible both convenient implementation of the algorithms which we employ and the various options that we provide, as discussed in the next section.

14.2.2 Algorithms

The system is based primarily upon the Wolfe generalized programming approach as discussed in [21], Section 3.2.1. The particular algorithm implemented here termed ILSRDD (Inner Linearization—Simple Recourse— Discrete Distribution) is described, in detail, in [18]. The generalized programming approach was chosen because it proved effective in earlier experimental versions (see [18]) and because of its potential applicability to a wide class of stochastic programs (including problems with complete recourse and problems with probabilistic constraints, see [18]). We also include an alternative to ILSRDD. This is algorithm based upon problem redefinition and the introduction of bounded variables given by Wets [25] and implemented in the simpler form given in Nazareth and Wets [20]. The algorithm is termed BVSRDD (Bounded Variables–Simple Recourse–Discrete Distribution). This approach is much more limited in its range of possible application as we have discussed in [21], but we include it for the following reasons: (a) it is very convenient to have a second algorithm that works on basically the same input as ILSRDD, for purposes of comparisons of answers and validation of implementation. Two identical answers on a particular problem from two different algorithms are rather comforting in this world of uncertainty and although this is no guarantee of correctness, it provides some indication that an error (if any) is in the input data or its conversion into internal representations. (b) A fair amount of experience has been accumulated with an early implementation of this method for dense problems (see Kallberg & Kusy [11]) and a more advanced implementation (which handles sparsity) should be available. (When there are relatively few points in each distribution of $h_i(\cdot)$ then this may even be a quite efficient way to solve C1 problems. (c) The algorithm BVSRDD makes possible a simpler and more direct extension of a deterministic MPS when the aim is only to handle simple recourse.

Two further options are provided in order to be able to solve (14.5) with $\chi = \overline{h}$ (ELASTIC option) and in order to solve an initial linear program, equivalent to (14.5) with $\chi = \overline{h}$, $q_i^+ = q_i^- = \infty$ (MINOS option). Here, $\overline{h}$ denotes an arbitrary right-hand-side vector. Both of these options are useful as preliminaries to the recourse formulation.

14.2.3   Implementation

From a practical standpoint, the linear programs which we want to solve and extend are of the more general form:

$$\text{minimize} \quad cx$$

$$\text{subject to} \quad Ax \begin{pmatrix} \leq \\ = \\ \geq \end{pmatrix} b \tag{14.6}$$

$$\ell \leq x \leq u$$

where $\begin{pmatrix} \leq \\ = \\ \geq \end{pmatrix}$ indicates that constraints take one of three possible forms and u and ℓ are vectors of upper and lower bounds. Furthermore, we cannot usually expect the partition $A = \begin{pmatrix} A \\ T \end{pmatrix}$ with technology rows T coming last in the matrix A. In general, rows of A and rows of T will be interleaved in A. In addition, it is worthwhile to explicitly include a scale factor ρ to permit a weighting of the second-stage objective relative to the first (see [18]). Thus the practical problems which we seek to solve, are derived from (14.5) and (14.6) and take the form

$$\text{minimize} \quad cx + q^+ y^+ + q^- y^- + \rho \sum_{i=1}^{m_2} \Psi_i(\chi_i)$$

$$\text{subject to} \quad A^{(\alpha_i)} x (\leq = \geq) b_i, \qquad \alpha_i \in \Lambda \tag{14.7}$$

$$A^{\tau_i} x + y_i^+ - y_i^- - \chi_i = 0, \qquad \tau_i \in \Gamma$$

$$\ell \leq x \leq u, y^+, y^- \geq 0.$$

where A^{α_i}, $\alpha_i \in \Lambda$ defines the rows of A, A^{τ_i}, $\tau_i \in \Gamma$ defines the rows of T, and Λ and Γ are index sets with $|\Lambda| = m_1$, $|\Gamma| = m_2$ ($|\Lambda|$ denotes the number of indices in Λ).

Our system for solving recourse problems of the form (14.7) has three main phases:

Phase 1:   Problem Setup and Generation
Phase 2:   Specialized Setup and Solution
Phase 3:   Output

This is summarized in Figure 14.1. A design goal was that all algorithms work on essentially the same input and each ignore input data that is only required by the others, e.g. the limit on the number of cycles, which is only required by ILSRDD. The input is specified in the form of three files of information which are described in more detail in the next section. All that is *often* necessary to switch options is to change the algorithm card in the "control" file and check that enough work space has been provided for various items. The Problem Setup and Generation Phase results in the creation of two files required by MINOS—the SPECS file and the MPS file. The next main phase consists of reading in these files by MINOS, inserting additional columns into its packed data structures and finding the solution of the problem. Finally the Output Phase augments the solution output by MINOS with some additional information about the solution of the stochastic program with recourse.

The next three sections go into this in more detail.

14.3 Problem Setup and Generation

To be specific, we discuss this within the context of a very simple example. Consider the following product-mix example (due to J. Ho [10]). The problem has two products and three ingredients. We seek to minimize cost of production while maintaining the levels of fat and protein at acceptable levels, and not exceeding availability of ingredients. The demand for each product is a random variable with discrete distribution but in an LP formulation this must be replaced by some expected value. The problem is summarized as follows, where x_i, y_i, z_i denote the amount of each ingredient in product i $(i = 1, 2)$.

minimize subject to	x_1	$+2y_1$	$+3z_1$	$+x_2$	$+2y_2$	$+3z_2$		(OBJ)
Fat/Protein Content of Product 1:	$0.3x_1$	$+0.4y_1$	$+0.2z_1$				≥ 3.3	$(A3)$
Fat/Protein Content of Product 2:					$0.5y_2$	$+0.6z_2$	≤ 4.0	$(A4)$
Amount of Ingredient 1:	x_1			$+x_2$			≤ 15.0	$(A1)$
Amount of Ingredient 2:		y_1			$+y_2$		≤ 12.0	$(A2)$
Amount of Product 1:	x_1	$+y_1$	$+z_1$				$= \bar{h}_1$	$(T1)$
Amount of Product 2:				x_2	$+y_2$	$+z_2$	$= \bar{h}_2$	$(T2)$

$$x_i, y_i, z_i \geq 0, i = 1, 2$$

The penalties for under and over production are 2.0 and 1.0 units, respectively, for each product, and the probability distribution on demand $h(\cdot)$ is as follows:

	Product 1			Product 2		
Level	8.0	10.0	12.0	15.0	18.0	20.0
Probability	0.25	0.5	0.25	0.2	0.4	0.4

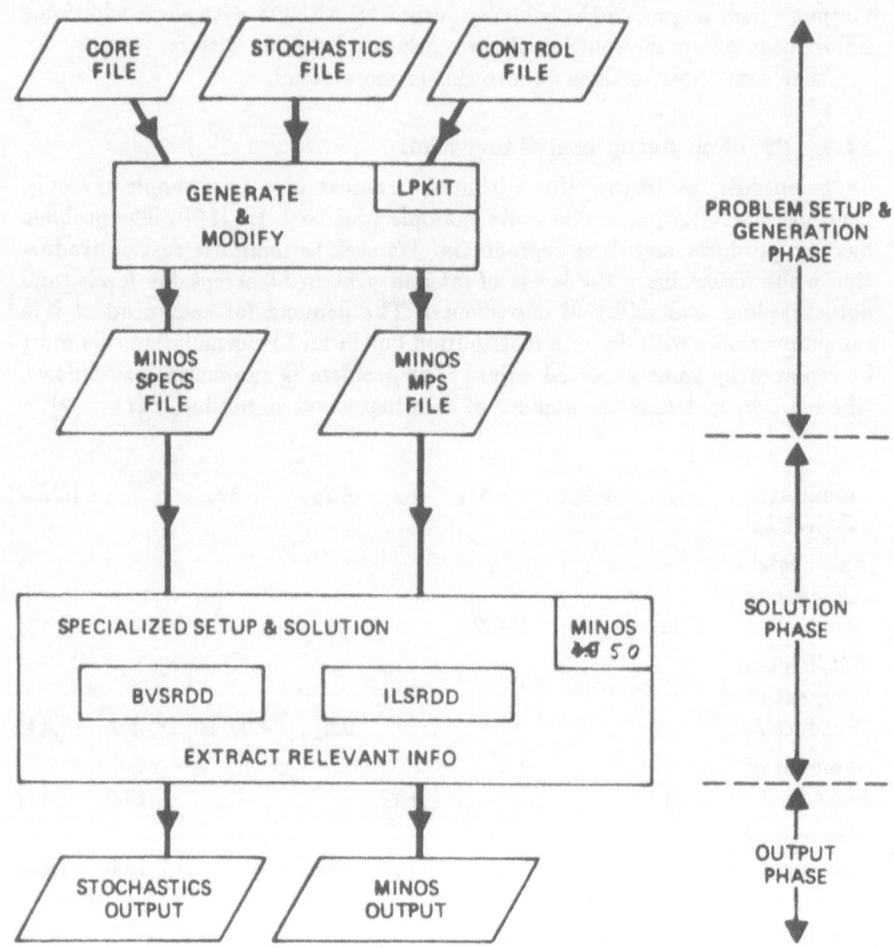

Figure 14.1   Overview

$\overline{h}_1 = 10.0$ and $\overline{h}_2 = 18.2$. The recourse function $\Psi(\chi)$ is defined in the usual way with $q^+ = (2.0, 2.0)$ and $q^- = (1.0, 1.0)$.

This simplified example will be quite adequate for purposes of illustration, and it can obviously be scaled up to a more realistic problem involving several products and ingredients.

14.3.1 Corefile

The input data corresponding to the decision variables x of the problem forms the "corefile". This specifies

- the names and types of each row of the problem
- the objective c
- the coefficients of A and T
- the deterministic right-hand-side elements
- the bounds on variables and ranges on rows

The "corefile" is specified in standard MPS format, see [14] and will often originate in a prior LP formulation. A and T can have interleaved rows and rows corresponding to T should normally be equality rows. However if these correspond to $\geq$ or $\leq$ rows i.e if there is no penalty on surplus or shortage, respectively, then provision is made in the system to change these to equality rows and a warning message is printed to that effect. This means that q_i^+ or q_i^- must be chosen appropriately at value zero. Note also that if there were non-zero elements in the right-hand-side vector corresponding to rows in the technology matrix they will be ignored by ILSRDD or BVSRDD and a message printed to this effect.

For our example, the corefile is given in Figure 14.2. (Slack variables were introduced explicitly in this case, but this is not necessary and could have been avoided by appropriate definition of row types.)

14.3.2 Stochastics File

The "stochastics" file specifies the information pertaining to the recourse problem. It gives:

- the row names identifying the technology matrix
- the probability distribution for each stochastic right-hand side
- the penalties q^+ and q^- on shortage and surplus
- the set of initial tenders for ILSRDD or the base tender for BVSRDD

An MPS-like format was designed for each of these items of information and is explained in the rest of this subsection. (An extension of this format is given in Edwards et al. [7].)

NAME   This is a header card. The user may enter any characters in columns 15 to 72.

TECHNOLOGY   The data consists of a list of names, one for each row in the technology matrix. These must be a subset of the list of rownames in the

```
NAME        LP
ROWS
        N OBJ
        E A1
        E A2
        E A3
        E A4
        E T1
        E T2
COLUMNS
            CLM1        OBJ         1.0         A1          1.0
            CLM1        A3          0.3         T1          1.0
            CLM2        OBJ         2.0         A2          1.0
            CLM2        A3          0.4         T1          1.0
            CLM3        OBJ         3.0         A3          0.2
            CLM3        T1          1.0
            CLM4        A3         -1.0
            CLM5        OBJ         1.0         A1          1.0
            CLM5        T2          1.0
            CLM6        OBJ         2.0         A2          1.0
            CLM6        A4          0.5         T2          1.0
            CLM7        OBJ         3.0         A4          0.6
            CLM7        T2          1.0
            CLM8        A4         -1.0
            CLM9        A1          1.0
            CLM10       A2          1.0
RHS
            RTH         A1         15.0         A2         12.0
            RTH         A3          3.3         A4          4.0
            RTH         T1         10.0         T2         18.2
BOUNDS
ENDATA
```

Figure 14.2   The corefile

"corefile". The submatrix corresponding to this set of rows in the COLUMNS section of the "corefile" defines the technology matrix. One name appears per line in columns 5 through 12.

DISTRIBUTION   The data consists of sets of entries of the form "rowname value probability". There is one such set for each of the rows named in the TECHNOLOGY section. "rowname" specifies the row associated with the entry (columns 5 through 12). "value" and "probability" specify the point and its associated probability. They occupy the first and second numeric fields (columns 25 through 36 and 50 through 61) respectively and must be specified as real numbers. The "rowname" repeats itself for each possible value associated with the row and the probabilities for this "rowname" must sum to unity.

OBJECTIVE   The data consists of entries of the form "name value value" where name is a rowname of T and the first value gives the value of q_i^+ and the second the value of q_i^- i.e the penalties on shortage and surplus respectively. The name occupies the first field (columns 5 through 12) and the values the first and second numeric fields (columns 25 through 36 and 50 through 61) respectively. They must be specified as real numbers.

TENDERS   The data consists of entries of the form "name rowname value" where name is the name associated with tender, "rowname" specifies the row associated with the entry and "value" is the level of the tender for this row. "name" repeats itself over all entries associated with the tender and there is one such "name" for each tender specified. "name" and "rowname" occupy the first two name fields (columns 5 through 12) and (15 through 22) respectively and "value" the first numeric field (columns 25 through 36). (If a set of these are provided for ILSRDD then the first one is used by BVRDD as its base tender, see Sec. 2.1 of [21].)

ENDATA   This card must be specified and flags the end of the "stochastics" file.

For our example the "stochastics" file is given in Figure 14.3.

```
NAME  TEST
TECHNOLOGY
          T1
          T2
DISTRIBUTION
          T1                        8.0                        0.25
          T1                       10.0                        0.5
          T1                       12.0                        0.25
          T2                       15.0                        0.2
          T2                       18.0                        0.4
          T2                       20.0                        0.4
OBJECTIVE
          T1                        2.0                        1.0
          T2                        2.0                        1.0
TENDERS
          TEND1      T1             8.0
          TEND1      T2            15.0
ENDATA
```

Figure 14.3   The stochastics file

14.3.3 Control File

The "control" file provides the information needed to guide the solution process. It gives:

- algorithm selected (generalized linear programming, bounded variable algorithm, elastic constraints or linear programming)
- input/output units for the files used by the system
- dimensioning information for various arrays within the system
- names of objective and right-hand-side vectors
- additional control parameters e.g. output level, cycle limit, etc.
- specification cards for MINOS

Our design here is similar to the MINOS SPECS file, but our format specification is more rigid and is based upon fields of four characters. Each main section is identified by a principal keyword which begins in column 1. Within each of these further options are identified by a second keyword which begins in column 5. Each of these options may have further suboptions and these are in turn identified by keywords beginning in column 9. The numerical strings or integers which provide the information that goes with a keyword are specified in a data field given by columns 23 through 30. Integers must of course be right justified. Only the first four characters (including blanks) of any keyword are significant.

The principal keywords, i.e. the keywords beginning in column 1, must be specified even when all defaults are selected.

The keywords are as follows:

BEGIN   This is a delimiter identifying the beginning of the control file

ALGORITHM   This identifies the selected algorithm. Options are ILSRDD, BVSRDD, ELASTIC or MINOS.

UNIT NUMBERS   The unit numbers are specified as follows:

CORE   unit number of "corefile". Default = 5

STOCHASTICS   unit number of "stochastics" file. Default = 7

SPECS   unit number of the MINOS SPECS file. Default = 8

MPS   unit numbers of the MINOS file specifying the matrix. Default = 9

DEBUG   unit number for debugging information. Default = 0 (no output)

LOG   unit number of the log file. Default = 0 (no output)

DIMENSIONS   This specifies information for setting up the work array

 ELEMENTS   an upper bound on the number of elements in the matrix (including space for input and generated tenders). Default = 1500

 ROWS   an upper bound on the number of rows (including technology). Default = 100

 TECHNOLOGY   an upper bound on the number of technology rows. Default = 20

 COLUMNS   an upper bound on the number of columns in the matrix (including tenders). Default = 300

PROBABILITIES   an upper bound on the number of discrete levels associated with each stochastic right-hand side. Default = 30

TENDERS   This provides information on tenders as follows:

INPUT   an upper bound on the number specified in the "stochastics" file. Default = 1

GENERATED   an upper bound on the number of tenders saved. Used in the round robin strategy. Default = 20

ELEMENTS   an upper bound on the total number of tender elements. Default = 2000

Note: One must be careful about specifying these quantities.

SELECTORS

OBJECTIVE   name of the objective row—up to 8 characters (must be provided)

RHS   name of the right-hand-side vector—up to 8 characters (must be provided)

BOUNDS   name of the bounds vector—up to 8 characters

RANGES   name of the ranges vector—up to 8 characters

CONTROL OPTIONS

OUTPUT   output level. Options are 1, 2 or 3, which provide increasingly verbose output. Default = 2

CYCLE   limit on number of tenders generated. Default = 1

SCALE   scale factor (see (14.1)), expressed as a percentage $(\rho = \text{SCALE}/100)$. Default = 100.

MINOS SPECIFICATIONS   Here one specifies any MINOS options which are then echoed into the MINOS SPECS file.

END   Delimiter indicating the end of the control section

In our example the "control" file is given in Figure 14.4.

14.3.4   Implementation of Problem Setup

This is done using some modules from LPKIT (see Nazareth [17]) suitably modified to suit our purposes. Additional routines have been written to set up information specified in the "stochastics" file into packed data structures and to generate the MINOS SPECS and MPS files.

14.4.   Specialized Setup and Solution

This part of the implementation is built around MINOS Version 5.0 whose outermost routines MINOS1 and MINOS2 were modified for our purposes. In particular, the PHANTOM COLUMNS option of MINOS (simply a device to provide some "elbow-room" in the data structures holding the problem) is extensively used in order to complete the setup of the recourse problem in the packed data structures used by the MINOS system.

```
BEGIN
ALGORITHM                    ILSRDD
UNIT NUMBERS
          CORE FILE              10
          STOCHASTICS FILE       11
          SPECS FILE             12
          MPS FILE               13
          DEBUG FILE             14
          LOG FILE               14
DIMENSIONS
          ELEMENTS              700
          ROWS                   10
          COLUMNS                40
          PROBABILITIES          20
          TENDERS
                    INPUT         1
                    GENERATED    10
                    ELEMENTS     99
SELECTORS
          OBJECTIVE             OBJ
          RHS                   RTH
CONTROL OPTIONS
          OUTPUT                 2
          CYCLE LIMIT            8
          SCALE FACTOR         100
END
```

Figure 14.4.   The control file

14.4.1   ILSRDD

The master program is defined by expression (3.7) in [21] with $W \triangleq [I, -I]$ and the obvious extension to match expression (14.7) in this paper. MINOS 5.0 sets up the A and T matrices in packed data structures from the MPS file which was generated in the previous phase. Then our modifications to subroutine MINOS2 pack in the additional columns corresponding to tenders. Other routines developed by us, which are called within the subroutine MINOS2, implement the generalized linear programming algorithm in coordination with the solution of each master program by MINOS 5.0. The detailed algorithm is given in [18].

14.4.2 BVSRDD

This is an implementation of the bounded variable method of Wets [25] in the form given in [21], Section 2.1. Further details of the algorithm may be found in [20]. There is a danger of performing a large number of pivot operations when the probability distribution of each right-hand-side element has many points (the so-called epsilon-to-death problem) but the associated computational effort is alleviated by the way in which MINOS updates its basis matrix representation. It is possible to improve the implementation (a) by using some of the acceleration techniques discussed in Wets [25] which, in effect, carry out several basis changes at the same time, (b) by specifying a good starting basis from the special structure in (14.7).

In contrast to ILSRDD, implementation is much more straightforward because only an initial linear program must be set up.

14.4.3 ELASTIC

This option implements the linear program (14.2) (see Section 14.2.1 of this paper), thereby permitting the "technology rows" to be elastic. The row names defining the technology rows and the penalties q^+ and q^- are defined by the stochastics file. Other data in this file is ignored.

14.4.4 MINOS

This simply provides the preliminary option of solving an initial linear program. The data in the stochastics file is not required here.

14.5 Output Phase

The output consists of two parts:

(a) MINOS output in standard MPS format. For a description of this see Murtagh & Saunders [16].

(b) SPORT output. This gives the first-stage and second-stage costs the optimal tender, the dual multipliers (prices) associated with the technology rows in the optimal solution and the probability levels of the equivalent chance-constrained program.

For the earlier example the output is given in Figure 14.5.

Figure 14.5 The output for the earlier example

14.6   Testing

The program has been exercised on several test problems as follows:

(a) The product-mix example of Section 14.3 due to J. Ho. This is a "toy" problem with 5 rows of which 2 are technology rows and 6 first-stage decision variables.

(b) The test problem given by Kallberg & Kusy [11]. This too is a "toy" problem with 3 rows of which 2 are technology rows and 6 first-stage decision variables. (Documented in King [12].)

(c) The test problem given by Cleef [3]. This has 9 rows of which 6 are technology rows and 16 first-stage decision variables. (Documented in King [12].)

(d) The problem of allocating aircraft to routes given in Dantzig [4]. This has 9 rows of which 5 are technology rows and 29 first-stage decision variables. (Documented in King [12].)

(e) A discretized version of the stochastic transportation problem given by Qi [23] formulated as a standard stochastic linear program with simple recourse. This has 78 rows of which 44 are technology rows and 1496 first-stage decision variables.

The bank asset and liability model given by Kusy & Ziemba [13] and a full-scale version of problem (d) above both provide good illustrations of the practical applications for which our program is designed.

14.7   Sportsmanship

The current system can be applied to a wider range of problems than would appear at first sight. For example when the stochastic linear program has stochastic technology matrices with a few discrete probability levels (which are independent of the right-hand-side distribution) say, $T_1, \ldots, T_t$ with probabilities $p'_1, \ldots, p'_t$, then we can express this as an equivalent problem

$$\text{minimize} \quad cx + p'_1 q^+ y_1^+ + p'_1 q^- y_1^- + \cdots + p'_t q^+ y_t^+ + p_t q^- y_t^-$$

$$
\begin{aligned}
\text{subject to} \quad & Ax && = b \\
& T_1 x + [I, -I]\binom{y_1^+}{y_1^-} && = h(\omega) \\
& \quad \vdots \qquad\qquad \ddots && \qquad \vdots \\
& T_t x \quad \cdots \quad \cdots + / \ [I, -I]\binom{y_t^+}{y_t^-} && = h(\omega) \\
& x, y_j^+, y_j^- \geq 0
\end{aligned}
\tag{14.8}
$$

Let us treat T defined by

$$
T = \begin{bmatrix} T_1 \\ \vdots \\ T_t \end{bmatrix}
$$

as a technology matrix in the usual way. Then we can set up the problem so that it can be solved by the system, as described earlier, with appropriate definition of penalties and distribution determined by (14.8).

In some situations the underlying probability distribution of $h(\cdot)$ is only known implicitly through a simulation model involving the random elements ω. Nazareth [18] discusses how the system can be extended to this case (see, in particular, Section 3.2 of [18] for some numerical experiments).

When the probability distribution of $h(\cdot)$ is not discrete, SPORT 2.0 can be used in conjunction with some iterative discretization procedure and computation of error bounds (see, for example, [26]).

When a more complex penalty structure is imposed on the second stage, program modifications would be required. This could, in many cases, be done fairly easily.

14.8 Availability

The Fortran implementation described here, SPORT 2.0 (pronounced SupPORT Version 2.0) was developed for use at IIASA on the VAX 11/780 (under the UNIX operating system). It uses MINOS 5.0 (the latest documented version), which is available in-house. Using the terminology in Nazareth [19], the current version of our system is a level-2 implementation, designed for algorithmic experimentation and for problem solving by an experienced user (one expected to be familiar both with his problem and with the implemented algorithm).

To use SPORT 2.0 at another site, it would be necessary to obtain MINOS 5.0 independently from Stanford University and to *substitute* our set of Fortran routines for the two MINOS 5.0 files MI00MAIN and MI10MACH. (Note that SPORT 2.0 will not run with versions of MINOS below 5.0.)

An earlier version of our system, designed for MINOS 4.9, SPORT 1.1, is available on the SDS/ADO tape, which is a collection of a number of routines for stochastic programming. This version provides readable Fortran and a manual (see Edwards [6]) to document our implementation. Note that it is *not* executable, since MINOS 4.9 is not included with it.

In order to obtain a copy of SPORT 2.0, please contact the author of this article at either of the following addresses: IIASA, System and Decision Sciences, A-2361, Laxenburg, Austria or CDSS, P.O. Box 4908, Berkeley, California 94704, USA.

14.9 Stochastic Programming with Recourse as a Form of Post-optimal Analysis in a Mathematical Programming System

Many large-scale Mathematical Programming Systems (e.g., MPSX/370 [1]) provide options for performing *parametric* and *sensitivity analysis* in the optimal solution of a linear program and for repeated (and efficient) reoptimization through a dual simplex procedure, when the right-hand-side is changed. (For MINOS, post-optimal analysis routines have been developed by Dobrowski, et al [5].)

A common approach for handling uncertainty in the right-hand-side is to use *scenario analysis*, which is indeed greatly facilitated by the above post-optimal options. Ermoliev and Wets [8] characterize this approach to dealing with uncertainty as being "seriously flawed" and explain why as follows: "Although it (scenario analysis) can identify 'optimal' solutions for each scenario (that specifies some values for the unknown parameters), it does not provide any clue as to how these 'optimal' solutions should be combined to produce a merely reasonable decision." Another approach that has been utilized by mathematical programmers as discussed in Section 14.2.1 is to introduce *elastic constraints* by defining penalties on shortage and surplus for a given right-hand-side. This, as we have noted, is in the spirit of the recourse model, but it does not yet address the stochastic aspect of the right-hand-side elements.

One aim of our paper has been to demonstrate (hopefully convincingly) that *recourse analysis* could be introduced in a very natural way as a post-optimal analysis option in an MPS and that its implementation is not substantially more difficult than that of other post-optimal analysis options currently provided within them. It could be argued, of course, since problem (14.7) can be directly expressed as a linear program, that it could be left up to the user to set up this linear program, create the appropriate MPS file and solve it in the conventional way. This is to impose upon him or her a laborious and error prone task. To do so would be as unreasonable as requiring that the user implement his own post-optimal parametric and sensitivity analysis. Another approach is to use an extended LP system based upon piecewise-linear (separable) programming (see Fourer [9]) to solve (14.5) or (14.7). Unfortunately such systems are not available as general purpose software. Thus it is necessary to fall back upon the more conventional mathematical programming systems.

The particular implementation described in earlier sections of this paper was developed for MINOS (specifically Version 5.0) in its linear programming mode, but an implementation for another large-scale linear programming system (MPS) could be patterned along rather similar lines (see, in particular, Figure 14.1). This would require the following:

(a) Firstly, augmentation of the standard MPS description of a linear program (which may be formulated and solved as a first step) by some standardized description of the stochastic information. A format along similar lines to Section 14.3.2 would be quite appropriate. Note that this does *not* conflict with the trend toward high-level modeling systems for defining mathematical programming problems (see, for example, the GAMS System of

Brooke, et al. [2]). MPS format (and its extension to stochastic problems) primarily serves the purpose of formalizing the *interface* to optimization codes and indeed MPS format continues to play this role in systems like GAMS. (With regard to the third "control" file of Figure 14.1, note that this is specific to the MINOS implementation and would obviously vary with different MPS systems.)

(b) Secondly, set up of one or more linear programming problems corresponding to (14.7) by augmenting internal data structures. The more straightforward implementation (because it involves only one augmentation) is to use some version of the bounded variable method of Wets [25] as in BVSRDD (see Section 14.4.2.). Assuming that a deterministic version of the problem has already been solved, the additional columns could be inserted directly into the packed data representation used by the MPS from the stochastic information supplied as described in (a) above, and the problem reoptimized. (It would be wasteful to generate a fresh MPS file for (14.7).) In MPSX/370, the augmentation and reoptimization could be done through the Extended Control Language (see [1]). The difficulty with the bounded variable approach arises when the distribution has many points, for example, when it is obtained by discretizing a continuous distribution. See the discussion in Section 14.4.2. Also it does not generalize to nonsimple recourse. The alternative is to implement the generalized linear programming approach, again directly inserting the added columns into internal data structures and solving a sequence of linear programs, each starting off where the previous one left off (as in ILSRDD, Section 14.4.1). As we have seen, implementation required modification only of the outermost level of MINOS and we believe this would be true for other MPS systems as well. The ILSRDD algorithm is very efficient in this context and as we may note, the approach applies to more general forms of recourse.

(c) Thirdly, the output of the solution in an appropriate way, again done most conveniently through access to the internal data structure.

To summarize, the mathematical programming field is ripe for incorporating some forms of stochastic programming with recourse into current large-scale MPS systems. We have provided a detailed illustration of how it can be done for one currently available MPS and how it could (possibly even should) be done for other systems.

Acknowledgements

It is a pleasure to thank Michael Saunders for making MINOS 5.0 (and an earlier version) available to us and for his very helpful advice on its usage in this context.

References

[1] M. Benichou, J.M. Gauthier, G. Hentges and G. Ribiere, "The efficient solution of large-scale linear programming problems - some algorithmic techniques and computational results", *Mathematical Programming* 13(1977), 280-322.

[2] A. Brooke, A. Drud and A. Meeraus, "High level modelling systems and nonlinear programming", Development Research Department, World Bank. (Presented at 12-th International Symposium on Mathematical Programming, Boston, Mass. 1985).

[3] H. Cleef, "A solution procedure for the two-stage stochastic program with simple recourse", *Z. Operations Research* 35(1981), 1-13.

[4] G.B. Dantzig, *Linear Programming and Extensions*, Princeton University Press, 1963.

[5] G. Dobrowolski, T. Rys, J. Hadjuk and A. Korytowski, "POSTAN 2- Extended postoptimal analysis package for MINOS", in A. Lewandowski and A. Wierzbicki eds., *Theory, Software and Test Examples for Decision Support Systems*, IIASA Collaborative Volume (1985), 142-176.

[6] J. Edwards, "Documentation for the ADO/SDS collection of stochastic programming codes", WP-85-02 (1985), IIASA, Laxenburg, Austria.

[7] J. Edwards, J. Birge, A. King and L. Nazareth, "A standard input format for computer codes which solve stochastic programs with recourse and a library of utilities to simplify its use", WP-85-03 (1985), IIASA, Laxenburg, Austria (appears, in part, in this volume).

[8] Y. Ermoliev and R. J-B. Wets, "Numerical techniques for stochastic optimization problems", PP-84-04 (1984), IIASA, Laxenburg, Austria (in this volume).

[9] R. Fourer, "Piecewise-linear programming", Technical Report, Dept. of Ind. Eng. and Management Sciences, Northwestern University, Evanston, IL, USA (1985).

[10] J. Ho, private communication (1983).

[11] J. Kallberg and M. Kusy, "Code Instruction for S.L.P.R., a stochastic linear program with simple recourse",, Technical Report, University of British Columbia (1976).

[12] A.J. King, "Stochastic programming problems: examples from the literature", Department of Applied Mathematics, University of Washington, Seattle, USA (1985) (appears in this volume).

[13] M. Kusy and W. Ziemba, "A bank asset and liability management model", CP-83-59, IIASA, Laxenburg, Austria (1983).

[14] IBM Document No. H20-0476-2, "Mathematical Programming System 360 Version 2, Linear and Separable Programming - User's Manual", IBM Corporation, New York.

[15] B.A. Murtagh and M.A. Saunders, "Large-scale linearly constrained optimization", *Mathematical Programming* 14(1978), 41-72.

[16] B.A. Murtagh and M.A. Saunders, "MINOS 5.0 User's Guide", Report No. SOL 83-20, Systems Optimization Laboratory, Stanford University (1983).

[17] J.L. Nazareth, "Implementation aids for optimization algorithms that solve sequences of linear programs by the revised simplex method, WP-82-107 IIASA, Laxenburg, Austria (1982).

[18] J.L. Nazareth, "Algorithms based upon generalized linear programming for stochastic programs with recourse", in: F. Archetti, ed., *Proceedings of IFIP International Workshop on Stochastic Programming: Algorithms and Applications*, Springer-Verlag (to appear).

[19] J.L. Nazareth, "Hierarchical implementation of optimizations methods", in: P. Boggs, R. Byrd and B. Schnabel, eds., *Numerical Optimization, 1984* SIAM, Philadelphia, 199–210 (1985).

[20] J.L. Nazareth and R.J-B. Wets, "Algorithms for stochastic programs: the case of non-stochastic tenders", A. Prékopa and R.J-B. Wets, eds., *Mathematical Programming Study*, (to appear).

[21] J.L. Nazareth and R.J-B. Wets, "Nonlinear programming techniques applied to stochastic programs with recourse", WP-85-62, IIASA, Laxenburg, Austria (1985) (appears in this volume).

[22] S.C. Parikh, Lecture notes on stochastic programming, unpublished, University of California, Berkeley (1968).

[23] L. Qi, Ph.D. Dissertation, Computer Science Department, University of Wisconsin, Madison (1983).

[24] J. Tomlin, private communication

[25] R. J-B. Wets, "Solving stochastic programs with simple recourse", *Stochastics* 10(1983), 219–242.

[26] R. J-B. Wets, "Stochastic programming: solution techniques and approximation schemes", in A. Bachem, M. Groetschel and B. Korte eds., *Mathematical Programming: The State of the Art*, Springer-Verlag, Berlin (1983) 566–603.

[27] W.T. Ziemba, "Stochastic programs with simple recourse", in: P. Hammer and G. Zoutendijk, eds., *Mathematical Programming in Theory and Practice*, North-Holland, Amsterdam (1974) 213–274.

CHAPTER 15

AN IMPLEMENTATION OF THE LAGRANGIAN FINITE-GENERATION METHOD

A.J. King

15.1 Introduction

An experimental code of the Lagrangian finite generation technique has been developed at IIASA for solving stochastic quadratic programs with simple recourse [1]:

$$\text{find} \quad x \in R^n \text{ to } \textit{maximize}:$$

$$\sum_{j=1}^{n} c_j x_j - \frac{1}{2}\sum_{j=1}^{n} d_j x_j^2 - E\{\sum_{i=1}^{\ell} e_i \theta(q_i^-, q_i^+; e_i^{-1}\mathbf{v}_i)\}$$

(SQP) subject to
$$-r_j^- \leq x_j \leq r_j^+ \qquad j = 1,\ldots,n$$

$$\sum_{j=1}^{n} a_{ij}x_j \leq b_i \qquad i = 1,\ldots,m$$

$$\mathbf{v}_i = \sum_{j=1}^{n} \mathbf{t}_{ij}x_j - \mathbf{h}_i \qquad i = 1,\ldots,\ell$$

where θ is a piecewise linear-quadratic function given by:

$$\theta(q^-, q^+; \tau) = \sup_{-1 \leq \sigma \leq q^+} [\sigma\tau - \frac{1}{2}\sigma^2] = \begin{cases} -q^-\tau - \frac{1}{2}(q^-)^2 & \text{if } \tau < -q^- \\ \frac{1}{2}\tau^2 & \text{if } -q^- \leq \tau \leq q^+ \\ -q^-\tau + \frac{1}{2}(q^+)^2 & \text{if } \tau > q^+ \end{cases}$$

the quantities $\mathbf{t}_{ij}, \mathbf{h}_i$ are square summable random variables and the other coefficients are fixed (nonstochastic) with $d_j \geq 0$ and $e_i \geq 0$.

The algorithm generates a sequence of points $\{\overline{x}^\mu, \mu = 1,\ldots\}$ that converge at a linear rate to the optimal solution, by solving at each step a modified version (SQP$_\mu$) of the original problem obtained by adding to the objective of (SQP) a proximal term [2]. More precisely we modify (SQP) by changing the linear and quadratic coefficients as follows

$$c_j^\mu = c_j + s_\mu^{-1}\overline{x}_j^\mu \qquad j = 1,\ldots,n$$
$$d_j^\mu = d_j + s_\mu^{-1} \qquad j = 1,\ldots,n$$

which has the effect of adding the proximal term $\frac{1}{2}s_\mu^{-1}\|x - \overline{x}^\mu\|^2$ to the objective. To solve (SQP_μ) we proceed by way of the *dual* [1], [3]:

$$\text{find} \quad \mathbf{z} \in L^2(\Omega, F, P : R^\ell), y \in R^m \text{ to minimize :}$$

$$\sum_{i=1}^{m} y_i b_i + \sum_{i=1}^{\ell} E\{\mathbf{z}_i h_i\} + \frac{1}{2}\sum_{i=1}^{\ell} E\{e_i \mathbf{z}_i^2\}$$

$$+ \sum_{j=1}^{n} d_j^\mu \theta\left(r_j^-, r_j^+ ; w_j/d_j^\mu\right)$$

(DQP_μ)

$$\text{subject to} \quad -q_j^- \leq z_j \leq q_j^+ \text{ a.s.} \quad i = 1,\ldots,\ell$$

$$0 \leq y_i \quad i = 1,\ldots,m$$

$$w_j = c_j^\mu - \sum_{i=1}^{m} y_i a_{ij} - \sum_{i=1}^{\ell} E\{\mathbf{z}_i t_{ij}\} \quad j = 1,\ldots,n$$

The properties of this problem: the appearance of the integrals in the objective constraints, and the simple nature of the boundary constraints on $\mathbf{z}$, permit us to solve this dual problem by a finite generation technique whereby we replace minimization over $\mathbf{Z} = \Pi_i[-q_i^-, q_i^+]$ with minimization over the convex set generated by a certain collection of elements $\mathbf{Z}^\nu = \{\boldsymbol{\xi}', \ldots, \boldsymbol{\xi}^{N^\nu}\}$, which turns out to be ordinary quadratic-programming. We then use the information gained by solving DQP_μ over co $\mathbf{Z}^\nu$ to generate a new collection $\mathbf{Z}^{\nu+1}$, and in this way obtain a sequence $\{\widehat{\chi}^\nu = 1,\ldots\}$ (the dual variables to DQP_μ) which converge at a linear rate to the optimal solution of SQP_μ [1].

The Lagrangian finite generation method requires that the random quantities $(\mathbf{h}, \mathbf{t})$ have finite discrete support. Of course it is not a restriction in the sense that some sort of discrete approximation scheme is needed to carry out the integrations. Discretization of measures for the solution of stochastic optimization problems is currently a very active research area. We will describe some of the work in this direction below in Section (15.6.2).

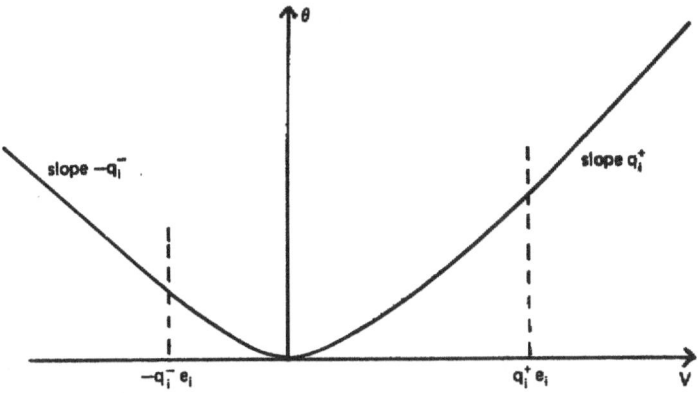

Figure 15.1   Graph of $e_i\theta\left(q_i^-,q_i^+;e_i^{-1}v_i\right)$

15.2   Discussion of SQP as a Model for Recourse Problems

The most important feature of SQP is the recourse penalty term for the first stage variables which takes the following form:

This can be viewed as a linear recourse penalty with a quadratic transition and is a generalization of the function $e_i q_i \theta\left(e_i^{-1} v_i\right)$ in [3]. The role of the piecewise linear quadratic penalty in the problem SQP is identical to that of the piecewise linear penalty in the stochastic linear program with recourse.

The usual statement of the stochastic linear program with simple recourse is as follows [4]:

$$\text{choose}\quad x \in R^n \text{ to } maximize:$$

$$\sum_{i=1}^{n} c_j x_j - \sum_{i=1}^{\ell} E\{q_i^+, v_i^+ + q_i v_i^-\}$$

(SLP)

$$\text{subject to}\quad -r_j^- \leq x_j \leq r_j^+ \qquad j=1,\ldots,n$$

$$\sum_{j=1}^{n} a_{ij}x_j \leq b_i \qquad i=1,\ldots,m$$

$$v_i^+ - v_i^- = \sum_{j=1}^{n} t_{ij}x_j - h_i \qquad i=1,\ldots,\ell$$

$$v_i^+ \geq 0, v_i^- \geq 0 \text{ a.s.} \qquad i=1,\ldots,\ell$$

With this formulation it is easy to see that if we take $d_j = 0$ and write down the limiting version of SQP as $e_i \to 0$, then we obtain SLP.

The reader should note that it is possible to solve SLP with the present algorithm—but in general the rate of convergence will not be linear. It is also possible to specify any value whatsoever to q_i^-, q_i^+ providing of course that $-q_i^- < q_i^+$. An important special case is the setting $q_i^- = 0, q_i^+ > 0$ giving a linear quadratic penalty of the form:

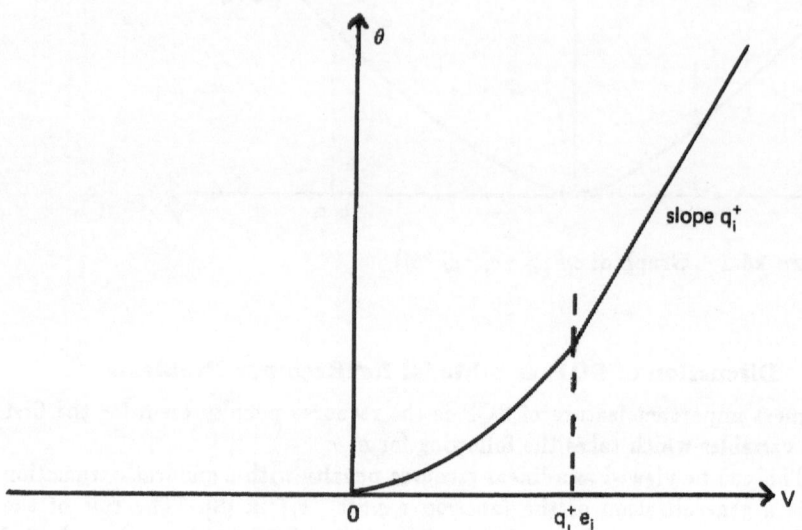

Figure 15.2   Graph of $e_i \theta \left(0, q_i^+ ; e_i^{-1} v_i \right)$

This type of penalty finds application in a variety of problems in resource management problems where the concern is to achieve $E\{v_i\} \leq 0$ and simultaneously to reduce the variance of the v_i above 0. The flexibility of the linear-quadratic penalty allows the decision maker to find, through the process of adjusting the various parameters, decision vectors $\bar{x}$ giving second stage outcomes v_i with certain desirable combinations of expectation and variance. An application of this model to the Lake Balaton eutrophication problem is discussed elsewhere in this volume, [9]. A second important case is where one or both of q_i^-, q_i^+ are infinite. This would give a purely quadratic recourse penalty in the appropriate direction. i.e., a one or two sided least-squares problem. Of course the same comments apply to r_j^-, r_j^+. We treat the case where $q_i^- = q_i^+ > 0$ separately in the next section as an important potential application of these to a numerical optimization problem in statistics.

15.3   Application of SQP to Robust Statistical Estimation

An important problem in applied statistics is to estimate the parameters x of a linear model $y = Tx$, there T is a given (possibly stochastic) matrix, from observations $\{\mathbf{y}_k, k = 1, \ldots, N\}$ which may be distorted by a noise term of unknown distribution. One formulation of this problem is the least squares model as originally proposed by Gauss:

choose x to *minimize*:

(LS)
$$\frac{1}{N} \sum_{k=1}^{N} |\mathbf{y}_k - Tx|^2.$$

Robust estimation in general is concerned with techniques of assessing and reducing the influence of the given set of observations upon the estimation of the parameter x (cf. [5]). One such technique is to modify the LS problem, reducing the influence of outliers in the sample by the use of the function:

$$\rho(\tau) = \begin{cases} |\tau - \frac{1}{2}| & \text{if } |\tau| \geq 1 \\ \frac{1}{2}\tau^2 & \text{if } |\tau| < 1 \end{cases}$$

giving the model (robust least squares):

choose (x, σ) to *minimize*:

(RLS)
$$\frac{1}{N} \sum_{k=1}^{N} \rho[\sigma^{-1}(\mathbf{y}_k - Tx)]$$

Except for the appearance of the σ as one of the variables involved in the minimization, the problem RLS can easily be seen as a particular interpretation of the model SQP, where σ corresponds to the e_i, we take $q_i^- = q_i^+ = 1$, and set $c_j = d_j = 0$. (In practice, σ—which is called the "nuisance parameter" for obvious reasons!— is usually held fixed in the solution of RLS.)

Thus we can derive in an analogous fashion to section 1 a sequence of problems RLS_μ whose dual:

choose $x \in R^N$ to *minimize*:

$$\frac{1}{N} \sum_{k=1}^{N} z_k y_k + \frac{1}{2} \frac{1}{N} \sum_{k=1}^{N} \sigma z_k^2 + \sum_{j=1}^{n} d_j^\mu \theta(r_j^-, r_j^+; w_j/d_j^\mu)$$

subject to $-1 \le z_k \le 1 \qquad k = 1, \ldots, N$

$$w_j = c_j^\mu - \frac{1}{N} \sum_{k=1}^{N} z_k T_j \qquad j = 1, \ldots, n$$

can be solved by the finite generation technique. In this formulation we have introduced primal constraints of the form $-r_j^- \le x_j \le r_j^+$, but this is not a serious problem since x_j can always be taken to be bounded in practice. As an additional feature, of course, we can include linear inequality constraints into the model RLS if it is required.

15.4 The Lagrangian Finite Generation Technique

In the paper [1], the authors develop a technique for solving a class of stochastic quadratic programs of which SQP_μ is a special example. The key idea is to approximate the dual problem DQP_μ by a sequence of quadratic subproblems which correspond to maximizing the dual objective over the convex hull of finitely many dual feasible solutions. The technique can be summarized in the following way:

1. find $(\hat{\chi}^\nu, \hat{z}^\nu)$ saddlepoint of $L^\mu(\chi, z)$ over $X \times \text{co } Z^\nu$

2. find $z^\nu \in \underset{z \in Z}{\text{argmax}} \, L^\mu(\hat{\chi}^\nu, z)$

3. determine $Z^{\nu+1} = \{\xi_1, \ldots, \xi_{N\nu+1}\} \supset \{\hat{z}^\nu, z^\nu\}$, return to step 1 with $\nu = \nu + 1$

where L^μ is the augmented Lagrangian associated with the primal-dual pair $\{\mathrm{SQP}_\mu, \mathrm{DQP}_\mu\}$.

The saddlepoint in step 1 is found by solving DQP_μ over the convex polytope co Z^ν:

choose $\hat\lambda \in R^{\ell \times N_\nu}, x \in R^n$ to *maximize* :

$$\sum_{j=1}^m y_i b_i + \sum_{i=1}^\ell \sum_{k=1}^{N^\nu} \lambda_i^k hexp_i^k + \tfrac{1}{2} \sum_{i=1}^\ell \sum_{k=1}^{N^\nu} \sum_{k'=1}^{N^\nu} \lambda_i^k \lambda_i^{k'} qexp_i^{kk'}$$

$$\sum_{j=1}^n [r_j^- w_j^0 + r_j^+ w_j^2 + \tfrac{1}{2} d_j^\mu (w_j^1)^2]$$

subject to $\lambda_i^k \geq 0 \quad i = 1,\ldots,\ell \quad k = 1,\ldots,N^\nu$

$$\sum_{k=1}^{N^\nu} \lambda_i^k \leq 1 \quad i = 1,\ldots,\ell$$

(EQ)

$y_i \geq 0 \quad i = 1,\ldots,m$

$$w_i^1 + w_j^2 - w_j^0 = \sum_{i=1}^\mu y_i a_{ij} - \sum_{i=1}^\ell \sum_{k=1}^{N^\nu} \lambda_i^k texp_{ij}^k, \quad j = 1,\ldots,n$$

$w_j^2 \geq 0, w_j^1$ free, $w_j^0 \geq 0 \quad j = 1,\ldots,n$

where $hexp_i^k = E\{\mathbf{h}_i \boldsymbol{\xi}_i^k\} \quad i = 1,\ldots,\ell \quad k = 1,\ldots,N^\nu$

$texp_{ij}^k = E\{\boldsymbol{\xi}_i^k \mathbf{t}_i\} \quad i = 1,\ldots,\ell \quad k = 1,\ldots,N^\nu$

$qexp_i^{kk'} = E\{e_i \boldsymbol{\xi}_i^k \boldsymbol{\xi}_i^{k'}\} \quad i = 1,\ldots,\ell \quad k = 1,\ldots,N^\nu$

$k' = 1,\ldots,N^\nu$

This is ordinary quadratic programming which can be solved by any one of a number of reliable codes, for example MINOS [6]. The dual multipliers for the n equality constraints give the primal decision vector $\hat\chi^\nu$, and the element $\hat{\mathbf{z}}^\nu = \sum_k \hat\gamma^k \boldsymbol{\xi}^k$ gives the dual half of the saddlepoint for step 1.

The maximization of the second step (over the whole space **Z**) can be obtained in closed form:

$$\mathbf{z}_i^{\nu} = \theta' \left(q_i^-, q_i^+; e_i^{-1} \left[\sum_{j-1}^{n} \mathbf{t}_{ij} \widehat{\chi}_j^{\nu} - \mathbf{h}_i \right] \right)$$

where θ' is the derivation of the piecewise linear-quadratic function θ,

$$\theta'(q^-, q^+; \tau) = \begin{cases} -1^- & \text{if } \tau < -\overline{q} \\ \tau & \text{if } -1^- \le \tau \le q^+ \\ q^+ & \text{if } \tau \ge q^+ \end{cases} .$$

There are a number of ways to generate the set $\mathbf{Z}^{\nu+1}$ in step 3 [**1**, section 3]. In the implementation at IIASA we set

$$\mathbf{Z}^{\nu+1} = \mathbf{Z}^0 \bigcup \{\widehat{\mathbf{z}}^{\nu}, \mathbf{z}^{\nu}, \ldots, \mathbf{z}^{\nu-\overline{\nu}-1}\},$$

where $\overline{\nu}$ is a predetermined maximum number of finite elements and $\mathbf{Z}^0$ is some initial fixed set. The finite generation method can be interpreted as a cutting plane technique, so although we are only required in theory to set $\mathbf{Z}^{\nu+1} = \{\widehat{\mathbf{z}}^{\nu}, \mathbf{z}^{\nu}\}$ one expects and in fact one obtains better results if more "cuts" are included.

Following this intuitive line of reasoning, we note also that it would be advantageous to include the element $\overline{\mathbf{z}}^{\mu}$ obtained by solving the preceding augmented Lagrangian $L^{\mu-1}$ whose saddlepoint is denoted by $(\overline{x}^{\mu}, \overline{z}^{\mu})$. The effect of including the element $\overline{z}^{\mu}$ in the initial set $\mathbf{Z}^0$ for at least the first few iterations of the finite generation method is quite dramatic as can be seen in the following example. This is a product mix problem in the form:

$$\text{maximize} \quad \sum_{j=1}^{4} c_j x_j - \sum_{i=1}^{2} E\{e_i \theta(0, q_q; e_i^{-1} \mathbf{v}_i)\}$$

$$\text{subject to} \quad 0 \le x_j \le r_j \qquad j = 1, \ldots, 4^i$$

$$\mathbf{v}_i = \sum_{j=1}^{4} \mathbf{t}_{ij} x_j - \mathbf{h}_i, \quad i = 1, 2.$$

The **t** matrix entries are all independent uniform and the **h** vector entries are independent normal. (For details see the Product Mix Problem [**10**]. In this example the algorithm halts when the relative duality gap, the normalized difference between the dual and primal, is less than 10^{-3}.

Example 1: Product mix problems

outer loop step		number of inner loop iterations (QP's)	duality gap
$\mu = 1$		4	0.8×10^{-1}
$\mu = 2$	(a) with $\bar{x}^{\mu}$	2	0.7×10^{-2}
	(b) without $\bar{x}^{\mu}$	2	0.5×10^{-2}
$\mu = 3$	(a) with $\bar{x}^{\mu}$	2	0.5×10^{-3}
	(b) without $\bar{x}^{\mu}$	2	0.5×10^{-2}
$\mu = 4$	(a) with $\bar{x}^{\mu}$	–	——
	(b) without $\bar{x}^{\mu}$	2	0.5×10^{-2}

15.5 Implementation

In this section we describe the details which transform the theory of the previous section into a practicable numerical method. The computer implementation of the finite generation method at IIASA is presently in an experimental and developmental stage. We shall keep the discussion focussed on the algorithm itself and the important numerical details which must be considered in any implementation.

Here is a rough outline of the complete algorithm. Details are described in the discussion which follows.

LAGRANGIAN FINITE GENERATION ALGORITHM

0. Initialization

    Read in data $A, b, c, d, e, q^-, q^+, r^-, r^+; n, m, \ell$

    Choose $\bar{x}^1, \bar{z}^1$

    Set $\mu = 1$

1. Outer Loop (Augmented Lagrangian cycle)

    Set $s_{\mu} = \sigma s_{\mu - 1}$

    Set $d^{\mu} = d + s_{\mu}^{-1}, c^{\mu} = c + s_{\mu}^{-1} \bar{x}^{\mu}$

    Determine $\mathbf{Z}^1 = \{\xi_1, \ldots, \xi_{N1}\}$ and $\bar{z}^{\mu} \in Z^1$

    Set $\nu = 1$

     $\hat{\chi}^1 = \bar{x}^{\mu}$

2. Inner Loop (Finite generation method)

    (a) Calculate $z_i^{\nu} = \theta' \left(q^-, q^+; e_i^{-1} \left[\sum_{j=1}^{n} t_{ij} \hat{\xi}_j^{\nu} - h_i \right] \right)$

         $i = 1, \ldots, \ell$

    Put $z^{\nu} \in Z^{\nu}$

    For each element $\xi^k \in Z^{\nu}, \quad k = 1, \ldots, N^{\nu}$ calculate:

$$hexp(i, k) = E\{h_i \xi_i^k\} \quad i = 1, \ldots, \ell \quad k = 1, \ldots, N^{\nu}$$

$$texp(i, j, k) = E\{t_{ij} \xi_i^k\} \quad i = 1, \ldots, \ell$$

$$qexp(i,k,kt) = E\{e_i \boldsymbol{\xi}_i^k \boldsymbol{\xi}_i^{k\prime}\} \quad i = 1,\ldots,\ell j = 1,\ldots,n$$

(b) Solve (QP) using MINOS; obtain solutions: $\widehat{\lambda}, \widehat{\chi}^\nu$

Set
$$\widehat{\boldsymbol{x}}^\nu = \sum_{k=1}^{N_\nu} \widehat{\lambda}^k \boldsymbol{\xi}^k$$

3. Stopping Criteria

(a) test $\widehat{\chi}^\nu$:

If inner loop not converged

Determine $\mathbf{Z}^{\nu+1} : \widehat{\boldsymbol{x}}^\mu \in \mathbf{Z}^{\nu+1}$

Set $x^{\nu+1} = \widehat{\boldsymbol{\xi}}^\nu$

Return to 2. with $\nu = \nu + 1$

If inner loop converged

Set $\overline{x}^{\mu+1} = \widehat{\boldsymbol{x}}^\mu$

                  $\overline{x}^{\mu+1} = \widehat{\chi}^\nu$

(b) test $\overline{x}^{\mu+1}$

If outer loop not converged, return to step 1 with $\mu = \mu + 1$

If outer loop converged, then stop.

Step 0. The initial guess $\overline{x}^1$ serves merely to provide a starting point for the augmented Lagrangian procedure. One could just as well set $\overline{x}^1 \equiv 0$. In the current implementation, $\overline{x}^1$ is the solution of the deterministic quadratic program obtained when the random variables $(\mathbf{h,t})$ are replaced by their expected values in the problem (SP). The initial "guess" $\overline{z}^1$ is included only for reasons of symmetry; the current implementation ignores it. However $\overline{z}^1$ will become important when the algorithm utilizes the full primal-dual augmented Lagrangian, as discussed below under "future developments".

Step 1. The primary purpose of this step is to update the augmented Lagrangian:

$$L^\mu(x,\boldsymbol{z}) = L^\mu(x,\boldsymbol{z}) - \frac{1}{2s_\mu}\|x - \overline{x}^\mu\|^2$$

The factor σ used to update s_μ is usually set between 1 and 2; for theoretical reasons we need $\sigma \geq 1$.

Step 2. This is the finite generation method. The method consists of two optimization problems

(1.) find saddlepoint $(\widehat{\chi}^\nu, \widehat{\boldsymbol{x}}^\nu)$ of $L^\mu(x,\boldsymbol{z})$ over $X \times \mathbf{Z}^\nu$

(2.) find $\boldsymbol{z}^\nu \in \operatorname*{argmin}_{\boldsymbol{z} \in \mathbf{Z}} L^\mu(\widehat{\chi}^\nu, \boldsymbol{z})$.

We have apparently inverted the order of the optimizations. In fact the present arrangement just serves to calculate the initial finite element $\boldsymbol{z}^1$ from $\widehat{\boldsymbol{x}}^\mu$ without unnecessary duplication of codes. The second minimization (2) is achieved in closed form, just as it appears in part (a) of this step.

The calculation of the integrals $hexp, texp, qexp$ in preparation for solving (QP) is dependent on the particular representation of the random variables $(\mathbf{h}, \mathbf{t})$ to be discussed in Section 15.6.2.

The saddlepoint problem of the finite generation method, in the form of the dual quadratic program (QP) (see above, Section 4), is solved by MINOS. The proper files and subroutines for utilizing the MINOS software are automatically generated by the computer codes; and the relevant solutions $\hat{\lambda}_i^k$ and the dual multipliers $\hat{\chi}_j^\nu$ (for the w_j equations) are passed from MINOS directly back to the algorithm. The pair $(\hat{\chi}^\nu, \hat{x}^\nu)$ is the saddlepoint for the first optimization problem (1) in the finite generation method.

Step 3. There are two sequences being generated $\{\hat{\chi}^\nu\}$ in the "inner loop" and $\{\overline{x}^\mu\}$ in the "outer loop", and we must specify stopping criteria for both. Of course in each case we specify a certain number of maximum iterations, typically 10 is the maximum for both, and once this maximum is reached we stop and make do with what we have. In the case of the inner loop we pass the last obtained (suboptimal) $\hat{\chi}^\nu$ on as the next proximal point $\overline{x}^{\mu+1}$ and hope for a better result on the next outer loop iteration; in the case of the outer loop we stop with a warning that the final $\overline{x}^\mu$ is not optimal.

In each case we test the relative norm of the difference of the successive iterates

$$\|\overline{x}^\mu - \overline{x}^{\mu-1}\|/\|\overline{x}^\mu\| < chieps$$
$$\|\hat{\chi}^\nu - \hat{\chi}^{\nu-1}\|/\|\hat{\chi}^\nu\| < chieps$$

The threshold "chieps" is a parameter chosen by the user. We have found that $chieps = 10^{-5}$ gives good results.

Finally there are two criteria based on the duality gaps for the respective problems. In the case of the inner loop it is possible to derive a criterion which ensures that the estimates $\overline{x}^{\mu+1} = \hat{\chi}^\nu$, while if not precisely the saddlepoint of L^μ, represents a good step in the sequence $\{\overline{x}^\mu\}$, i.e. it gives a linear rate of convergence. From [2] we know that to obtain the linear rate it is sufficient to choose $\overline{x}^{\mu+1}$ so that

$$\|\overline{x}^{\mu+1} - M_\mu(\overline{x}^\mu)\|^2 \le \frac{\delta_\mu^2}{2\theta_\mu}\|\overline{x}^{\mu+1} - \overline{x}^\mu\|^2$$

where $M_\mu(\overline{x}^\mu)$ is the primal half of the true saddlepoint for L^μ, and $\{\delta_\mu\}$ is a nonnegative sequence satisfying $\sum \delta_\mu < \infty$. From [1, theorem 3] it is easy to derive an inequality forcing this criterion which in our case turns out to be

$$\frac{1}{2}\frac{\delta_\mu^2}{\theta_\mu}\|\overline{x}^{\mu+1} - \overline{x}^\mu\|^2$$

$$\le \sum_{i=1}^{\ell} E\{\overline{x}^{\mu+1}(\overline{\mathbf{h}}_i - \sum_{j=1}^{n} \mathbf{t}_{ij}\overline{x}_j^{\mu+1}) + \tfrac{1}{2}(\overline{x}^{\mu+1})^2\}$$

$$+ \sum_{i=1}^{\ell} E\left\{e_i\theta\left(q_i^-, q_i^+; e_i^{-1}\left[\sum_{j=1}^{n} \mathbf{t}_{ij}\overline{x}_j^{\mu+1} - \mathbf{h}_i\right]\right)\right\}$$

These quantities are all available to the algorithm once a candidate $\left(\overline{x}^{\mu+1}, \overline{z}^{\mu+1}\right)$ $= \left(\widehat{\chi}^{\nu}, \widehat{z}^{\nu}\right)$ has been selected from step 2. This criterion turns out to be quite a convenient one, in practice often being satisfied before $\{\widehat{\chi}^{\nu}\}$ satisfies the relative norm test.

For both sequences $\{\widehat{\chi}^{\nu}\}$ and $\{\overline{x}^{\mu}\}$ we formulate in a straightforward manner a stopping criterion based on the duality gap between the primals and their respective duals. We calculate

$$\text{dual gap (1)} = \left(\text{value of SQP at } \overline{x}^{\mu+1}\right) - \text{value of DQP at } \overline{z}^{\mu+1}\right)$$
$$\text{dual gap (2)} = \left(\text{value of SQP}_\mu \text{ at } \widehat{\chi}^{\nu}\right) - \left(\text{value of DQP}_\mu \text{ at } \widehat{z}^{\nu}\right).$$

Then if dual gap (2) is small enough, typically we specify a tolerance of 10^{-6}, we say that $\{\widehat{\chi}^{\nu}\}$ has converged and set $\overline{x}^{\mu+1} = \widehat{\chi}^{\nu}$. If dual gap (1) is small enough (usually we set the tolerance at 10^{-3}) then we say that $\overline{x}^{\mu+1}$ "solves" SQP and stop.

15.6 Further Development

To this date the program has been tested on several problems, and performs quite satisfactorily on even a fairly complex problem such as the Lake Balaton eutrophication model where literally hundreds of variations have been successfully solved. The Balaton problem was modelled in the form DQP by Somlyódy and Wets [8], with $c_j = d_j = 0$, consisting of 35 deterministic constraints, 56 decision variables, 4 stochastic constraints developed from 15 independent random variables with a mixture of normal, log-normal and three-parameter-gamma distributions. This problem is now solved routinely by our codes. Similar experiences with other (smaller) problems verify that the method is quite reliable.

Typical formulations of the Balaton problem will require the solution of between 5 and 20 of the quadratic programs (QP). The amount of work performed depends on several factors; among these the principal ones appear to be the setting of the quadratic parameters s_μ, d_j, e_i.

The level and type of refinement of the discrete approximations to the measures is also an important feature. The current development program for the algorithm is centered on testing various approaches to improve the algorithm by modifying the quadratic parameters, as well as on various discretization schemes for improving the approximations to the probability measures.

15.6.1   Full implementation of proximal point algorithm

The theory of the finite generation technique does not require that the problem SQP be strongly quadratic in order to obtain convergence of the sequence $\{\hat{\chi}^\nu\}$, but if strong quadraticity is present in both primal and dual variables then we obtain linear convergence [1]. In problems where $d_j = 0$, the proximal term provides the quadratic behavior in the primal variables. In exactly the same way we can introduce a dual proximal term to give quadratic behavior in the dual variables when $e_i = 0$; this is achieved by setting

$$e_i^\mu = e_i + s_\mu^{-1} \quad \text{to replace } e_i,$$
$$\mathbf{h}_i^\mu = \mathbf{h}^{-\mu}/s_\mu \quad \text{to replace } h_i,$$

where we recall that $(\overline{x}^\mu, \overline{z}^\mu)$ is the approximate saddlepoint of $L^{\mu-1}$. This has the effect of adding the term $\frac{1}{2}s_\mu^{-1}E\{\|z - \overline{z}^\mu\|^2\}$ to the objective of the *dual* of SQP$_\mu$. Thus we can have a primal, a dual, or a primal-dual implementation of the proximal point method. The same theory holds in all cases [2]. However it is conceivable that one would like to omit the proximal point algorithm in one or both sets of variables. The next stage of the algorithm will include facilities for making these kinds of choices.

Thus it will be possible to solve even the completely linear problem, SLP, either directly (without proximal terms) or sequentially (with proximal terms). Ideally one should introduce the proximal terms only in cases where the finite generation method converges poorly, or is unstable. And of course one would like to predict the consequences of introducing these terms; for example, one would like to know the optimal setting of s_μ for a given problem SQP. The basic result is that there is a tradeoff: the higher s_μ the faster $\{\overline{x}^\mu\}$ converges, the lower s_μ the better $\{\hat{x}^\nu\}$ converges [1, theorem 5]. This effect is mediated through the quadratic form in SQP. The influence of these forms, and hence also the setting of s^μ, is quite dramatic as can be seen from the following runs of the (modified) Lake Balaton problem where the e_i varied between 0.5 and 50.

Example 2. Lake Balaton Problem

Outer Loop Step	$e = 50$ inner inner loop steps	duality gap	$e = 5.0$ inner inner loop steps	duality gap	$e = 0.5$ inner inner loop steps	duality gap
$\mu = 1$	2	0.8×10^{-1}	2	0.6×10^{-2}	3	0.3×10^{-2}
$\mu = 2$	1	0.3×10^{-1}	4	0.3×10^{-2}	3	0.3×10^{-3}
$\mu = 3$	1	0.7×10^{-2}	2	1.0×10^{-3}		(converged)
$\mu = 4$	1	1.3×10^{-2}	2	0.6×10^{-3}		
$\mu = 5$	1	0.8×10^{-2}	2	0.3×10^{-3}		
$\mu = 6$	1	0.6×10^{-2}		(converged)		
$\mu = 7$	1	0.5×10^{-2}				
$\mu = 8$	2	0.3×10^{-2}				
$\mu = 9$	1	0.3×10^{-2}				
$\mu = 10$	7	0.3×10^{-2}				
$\mu = 11$	2	0.3×10^{-2}				

(does not converge)

15.6.2 Discretization schemes for the probability measures

The augmented Lagrangian techniques coupled with the finite generation method constitute an effective algorithm for solving the problem SQP if the probability measure is discrete. As originally constituted the implementation of the algorithm at IIASA used Monte Carlo techniques to generate a sample of the random variables $(\mathbf{h}, \mathbf{t})$ and employed the sample as a discrete "empirical measure". In this way by increasing the size of the sample we generated a sequence of discrete measures p^{N_s} which converge in distribution to the true measure P. Then by epi-convergence arguments, see [7], the solutions $\bar{x}^{N_s}$ to the corresponding problems (SQP_{N_s}) converge to an optimal solution of SQP.

The implementation of this "simulation" scheme is quite straightforward. One simply stores, for each sample point $\omega \in \{1, \ldots, N_s\}$ values for the Monte Carlo simulations of the random variables $\mathbf{h}, \mathbf{t}$, i.e.,

$$\mathbf{h}_i \leftrightarrow h_i(\omega), \omega \in \{1, \ldots, N_s\}$$
$$\mathbf{t}_{ij} \leftrightarrow t_{ij}(\omega), \omega \in \{1, \ldots, N_s\}.$$

The finite elements $\boldsymbol{\xi}^k, k = 1, \ldots, N_\nu$ are also represented in this way, viz

$$\boldsymbol{\xi}_i^k \leftrightarrow \boldsymbol{\xi}_i^k(\omega), \omega \in \{1, \ldots, N_s\}$$

We calculate the integrals as follows:

$$E\{\mathbf{h}_i \boldsymbol{\xi}_i^k\} = \frac{1}{N_s} \sum_{\omega=1}^{N_s} h_i(\omega) \xi_i^k(\omega), \text{etc.}$$

Originally we believed that by beginning with a small sample, say N_1, and solving SQP_{N-1}, then this solution would be a good starting point for a larger sample $N_2 > N_1$. thus we envisioned a sequence of problems $\{\mathrm{SQP}_{N_s}\}$ each one using the last solution as a starting point. But in fact no advantage seems to be gained over using the initial point given by the deterministic problem with random quantities replaced by expectations.

A more promising approach to discretizing the probability measure is to take advantage of the convexity present in the problem and devise a discretization scheme based on conditional expectations, obtaining a discrete probability measure P_c. By solving the resulting problem SQP_c we obtain a solution $\bar{x}^c$ giving an optimal value which is a valid upper bound for the true value of (SQP) at $\bar{x}^c$ and hence also an upper bound for the true maximum of SQP [7, section 4].

The implementation of this "conditional expectations" representation is slightly more involved. Here we present the case where q is fixed (deterministic), and t_{ij}, h_i, are all independent. For each random variable we have constructed a partition of its support (a subset of the real line) and then we have calculated the conditional expectations $tcexp_{ij}(k)$ and $hcexp_i(k)k = 1, \ldots, npart$. We represent t_{ij}, h_i by the collection of discrete random variables which take values $tcexp_{ij}(k)$, $hcexp_{ij}(k)$ with probabilities $tprob_{ij}(k)$ and $hprob_i(k)$. Now let

$$\Gamma = \{\gamma = (\gamma_0, \gamma_1, \ldots, \gamma_n)|\gamma_k \in \{1, \ldots, npart\}\}$$

i.e. Γ is the set of all $(n+1)$-permutations of $npart$ letters. We represent the finite elements $\xi^k, k = 1, \ldots, N_\nu$ in the following way

$$\xi^k \leftrightarrow \xi^k(\gamma), \gamma \in \Gamma.$$

Thus, for example,

$$\xi_i^k = \theta'(q_i^-, q_i^+; e_i^{-1}(\sum_{j=1}^{n} t_{ij}x_j^k - h_i))$$

is calculated as

$$\xi_i^k(\gamma) = \theta'(q_i^-, q_i^+; e_i^{-1}(\sum_{j=1}^{n} tcexp_{ij}(\gamma_j)x_j^k - hcexp_i(\gamma_0))),$$

and the integrals are calculated as

$$E\{h_i\xi_i^k\} = \sum_{\gamma \in \Gamma} hcexp_i(\gamma_0)\xi_i^k(\gamma)hprob_i(\gamma_0) \prod_{j=1}^{n} tprob_{ij}(\gamma_j)$$

$$E\{t_i\xi_i^k\} = \sum_{\gamma \in \Gamma} tcexp_{ij}(\gamma_0)\xi_i^k(\gamma)hprob_i(\gamma_j) \prod_{j=1}^{n} tprob_{ij}(\gamma_j)$$

$$E\{e_i\xi_i^k\xi_i^{k'}\} = \sum_{\gamma \in \Gamma} e_i\xi_i^k(\gamma)\xi_i^{k'}(\gamma)hprob_i(\gamma_0) \prod_{j=1}^{n} tprob_{ij}(\gamma_j)$$

(We need only use $(n+1)$-permutations as opposed to $\ell(n+1)$-permutations because the problem is *separable* in the $i = 1, \ldots, \ell$ stochastic second stage constraints.)

There are some minor advantages to this implementation. Note that we do not have to keep in memory values of $tcexp_i$, $hcexp_i$ for each $\gamma \in \Gamma$ but only for each $k = 1, \ldots, npart$. Since $|\Gamma|$ can be quite large there is a considerable saving of memory allocation and access time over the Monte Carlo simulation implementation, where typically we would take $N_s \approx |\Gamma|$ (in a small problem). Furthermore, the resulting problem for the conditional expectations scheme can be stated in the standard input format [11].

The major advantage is, of course, that we have a valid upper bound. It is possible to combine this discretization with a lower bounding approach which utilizes the fact that the function

$$(\mathbf{h}, \mathbf{t}) \rightarrow -e_i \theta\left(q_i^-, q_i^+; e_i^{-1}\left(\sum_{j=1}^{n} t_{ij} x_j - h_i\right)\right)$$

is concave for fixed x, and then develops a measure on the extreme points of the partitions of the supports of the random variables assuming they are compact. One then develops a sequence of partitions, narrowing the gap between upper and lower bounds until a suitable tolerance is attained. For the case where $\mathbf{t}$ is fixed (deterministic), there is an optimal partitioning scheme [7]. However in the case where $\mathbf{t}$ is stochastic it is not yet clear what is the correct way to proceed.

References

[1] R.T. Rockafellar and R.J-B. Wets, "A Lagrangian finite generation technique for solving linear-quadratic problems in stochastic programming," CP-84-XX International Institute for Applied Systems Analysis, Laxenburg, Austria (1984).

[2] R.T. Rockafellar, "Augmented Lagrangians and applications of the proximal point algorithm in convex programming," *Math. of Oper. Res.* Vol. 1, No. 2 (1976).

[3] R.T. Rockafellar and R.J-B. Wets, "A dual solution procedure for quadratic stochastic programs with simple recourse," in: *Numerical Methods* V. Pereya and A. Reinoza (eds.), (Lecture Notes in Math. 1005, Springer-Verlag, Berlin) (1983).

[4] R.J-B. Wets, "Stochastic programming: solution techniques and approximation schemes," in: *Mathematical Programming: The State-of-the-Art* A. Bachau et al., (eds.), Springer-Verlag, Berlin (1983).

[5] P.J. Huber, *Robust Statistics*, Wiley, New York (1981).

[6] B.A. Murlagh and M.A. Saunders, "Large scale linearly constrained optimization," *Mathematical Programming* **14**(1978), pp. 41–72.

[7] J. Birge and R.J-B. Wets, "Designing approximation schemes for stochastic optimization problems, in particular for stochastic programs with recourse," WP-83-111, IIASA, Laxenburg, Austria (1983).

[8] L. Somlyódy and R.J-B. Wets, "Stochastic optimization models for lake eutrophication management," CP-85-16, IIASA, Laxenburg, Austria (1985).

[9] A.J. King, R.T. Rockafellar, L. Somlyódy and R.J-B. Wets, "Lake eutrophication management: the Lake Balaton project," in *Numerical Techniques for Stochastic Programming*, Y. Ermoliev and R.J-B. Wets, eds., Springer-Verlag (1986).

[10] A.J. King, "Stochastic programming problems: examples from the literature," *ibid*, (1986).

[11] J. Edwards, J. Birge, A. King, L. Nazareth, "A standard input format for computer codes which solve stochastic programs with recourse and a library of utilities to simplify its use," WP-85-03, IIASA, Laxenburg, Austria (1985).

CHAPTER 16

STOCHASTIC QUASIGRADIENT METHODS AND THEIR IMPLEMENTATION

A. Gaivoronski

16.1 Introduction

This paper discusses various stochastic quasigradient methods (see [1], [2]) and considers their computer implementation. It is based on experience gained both at the V. Glushkov Institute of Cybernetics in Kiev and at IIASA.

We are concerned here mainly with questions of implementation, such as the best way to choose step directions and step sizes, and therefore little attention will be paid to theoretical aspects such as convergence theorems and their proofs. Readers interested in the theoretical side are referred to [1],[2].

The paper is divided into five sections. After introducing the main problem in Section 16.1, we discuss the various ways of choosing the step size and step direction in Sections 16.2 and 16.3. A detailed description of an interactive stochastic optimization package (STO) currently available at IIASA is given in Section 16.4. This package represents one possible implementation of the methods described in the previous sections. Finally, Section 16.5 deals with the solution of some test problems using this package. These problems were brought to our attention by other IIASA projects and collaborating institutions and include a facility location problem, a water resources management problem, and the problem of choosing the parameters in a closed loop control law for a stochastic dynamical system with delay.

We are mainly concerned with the problem

$$\min\{F(x) : x \in X\}, F(x) = \mathrm{E}_\omega f(x,\omega), \qquad (16.1)$$

where x represents the variables to be chosen optimally, X is a set of constraints, and ω is a random variable belonging to some probabilistic space (Ω, B, P). Here B is a Borel field and P is a probabilistic measure.

There are currently two main approaches to this problem. In the first, we take the mathematical expectation in (16.1), which leads to multidimensional integration and involves the use of various approximation schemes [3-6]. This reduces problem (16.1) to a special kind of nonlinear programming problem which allows the application of deterministic optimization techniques. In this paper we concentrate on the second approach, in which we consider a very limited number of observations of random function $f(x,\omega)$ at each iteration in order to determine the direction of the next step. The resulting errors are smoothed out until the optimization process terminates (which happens when the step size becomes sufficiently small). This approach was pioneered in [7],[8].

We assume that set X is defined in such a way that the projection operation $x \rightarrow \pi_X(x)$ is comparatively inexpensive from a computational point of view, where $\pi_X(x) = \underset{z \in X}{\text{argmin}} \|x - z\|$. For instance, if X is defined by linear constraints, then projection is reduced to a quadratic programming problem which, although challenging if large scale, can nevertheless be solved in a finite number of iterations. In this case it is possible to implement a stochastic quasigradient algorithm of the following type:

$$x^{s+1} = \pi_X(x^s - \rho_s v^s). \tag{16.2}$$

Here x^s is the current approximation of the optimal solution, ρ_s is the step size, and v^s is a random step direction. This step direction may, for instance, be a statistical estimate of the gradient (or subgradient in the nondifferentiable case) of function $F(x)$: then $v^s \equiv \xi^s$ such that

$$E(\xi^s | x^1, x^2, \ldots, x^s) = F_x(x^s) + a^s, \tag{16.3}$$

where a^s decreases as the number of iterations increases, and the vector v^s is called a *stochastic quasigradient* of function $F(x)$. Usually $\rho_s \rightarrow 0$ as $s \rightarrow \infty$ and therefore $\|x^{s+1} - x^s\| \rightarrow 0$ from (16.2). This suggests that we should take x^s as the initial point for the solution of the projection problem at iteration number $s + 1$, thus reducing considerably the computational effort needed to solve the quadratic programming problem at each step $s = 1, 2, \ldots$. Algorithm (16.2)–(16.3) can also cope with problems with more general constraints formulated in terms of mathematical expectations

$$E_\omega f^i(x, \omega) \geq 0, i = \overline{1, m}$$

by making use of penalty functions or the Lagrangian (for details see [1],[2]).

The principal peculiarity of such methods is their nonmonotonicity, which may sometimes show itself in highly oscillatory behavior. In this case it is difficult to judge whether the algorithm has already approached a neighborhood of the optimal point or not, since exact values of the objective function are not available. The best way of dealing with such difficulties seems to be to use an interactive procedure to choose the step sizes and step directions, especially if it does not take much time to make one observation. More reasons for adopting an interactive approach and details of the implementation are given in the following sections.

Another characteristic of the algorithms described here is their pattern of convergence. Because of the probabilistic nature of the problem, their asymptotic rate of convergence is extremely slow and may be represented by

$$\|x^* - x^s\| \sim \frac{C}{\sqrt{k}}. \tag{16.4}$$

Here x^* is the optimal point to which sequence x^s converges and k is the number of observations of random parameters ω, which in many cases is proportional to the number of iterations. In deterministic optimization a superlinear asymptotic convergence rate is generally expected; a rate such as (16.4) would be considered as nonconvergence. But no algorithm can do asymptotically any better than this for stochastic problem (16.1) in the presence of nondegenerate random disturbances, and therefore the aim is to reach some *neighborhood* of the solution rather than to find the precise value of the solution itself. Algorithm (16.2)–(16.3) is quite good enough for this purpose.

16.2 Choice of Step Direction

In this section we shall discuss different ways of choosing the step direction in algorithm (16.2) and some closely related algorithms. We shall first discuss methods which are based on observations made at the current point x^s or in its immediate vicinity. More general ways are then presented which take into account observations made at previous points.

16.2.1 Gradients of random function $f(x,\omega)$

The simplest case arises when it is possible to obtain gradients (or subgradients in the nondifferentiable case) of function $f(x,\omega)$ at fixed values of x and ω. In this case we can simply take

$$\xi^s = f_x(x^s, \omega^s), \qquad (16.5)$$

where ω^s is an observation of random parameter ω made at step number s. If both the observation of random parameters and the evaluation of gradients are computationally inexpensive then it is possible to take the average of some specified number N of gradient observations:

$$\xi^s = \frac{1}{N} \sum_{i=1}^{N} f_x(x^s, \omega^i, s). \qquad (16.6)$$

These observations can be selected in two ways. The first is to choose the ω^i, s according to their probability distribution. If we do not know the form of the distribution function (as, for example, in Monte-Carlo simulation models) this is the only option. However, in this case the influence of low-probability high-cost events may not be properly taken into account. In addition, the asymptotic error of the gradient estimate ξ^s is approximately proportional to $1/\sqrt{N}$. The second approach may be used when we know the distribution of the random parameters ω. In this case many other estimates can be derived; the use of pseudo-random numbers* in particular may lead to an asymptotic error approximately proportional to $\log(N)/N$, which is considerably less than

---

* A concept which arose from the use of quasi-Monte-Carlo techniques in multidimensional integration [9].

in the purely random case. However, more theoretical research and more computational experience áre necessary before we can assess the true value of this approach. The main question here is whether the increase in the speed of convergence is sufficient to compensate for the additional computational effort required for more exact estimàtions of the $F_x(x^s)$.

Unfortunately, our theoretical knowledge concerning the asymptotic behavior of processes of type (16.2) tells us little about the optimal number of samples, even for relatively well-studied cases. For instance, what would be the optimal number N of observations for the case in which function $F(x)$ is differentiable and there are no constraints? In this case we can establish both asymptotic normality and the value of the asymptotic variance. If, additionally, ρ_s C/s then the total number of observations required to obtain a given asymptotic variance is the same for all $N \ll s$. If $s\rho_s \to \infty$ then the wait-and-see approach is asymptotically superior as long as $N \ll s$.

However, there is strong evidence that in cŏnstrained and/or nondifferentiable cases the value of N should be chosen adaptively. A very simple example provides some insight into the problem. Suppose that $x \in R^1$, $X = [a,\infty)$, $F(x) = x$, $f_x(x^s,\omega^s) = 1 + \omega^s$, where the ω^s, $s = 1,2,\ldots$, are independent random variables with zero mean. The obvious solution of this problem is $x = a$. Suppose for simplicity that $\rho_s \equiv \rho$. This will not alter our argument greatly because ρ_s usually changes very slowly for large s. In this case method (16.2),(16.5) will be of the form:

$$x^{s+1} = x^s - \rho(1 + \omega^s) + \tau_s,$$

$$\tau_s = \max\{0, a - x^s + \rho(1 + \omega^s)\}.$$

Method (16.2),(16.6) requires us to choose a step size N times greater than ρ; otherwise its performance would be inferior to that of method (16.2),(16.5) (unless the initial point is in the immediate vicinity of the minimum). Method (16.2),(16.6) then becomes

$$z^{s+1} = z^s - N\rho\left(1 + \frac{1}{N}\sum_{i=1}^{N}\omega^{i,s}\right) + \theta_s,$$

$$\theta_s = \max\{0, a - z^s + N\rho\left(1 + \frac{1}{N}\sum_{i=1}^{N}\omega^i, s\right)\}.$$

In order to compare the two methods we shall let s in the last equation denote the number of observations rather than the number of iterations and renumber the observations ω^i, s. The process

$$y^k = y^0 - \rho\sum_{i=0}^{k-1}(1 + \omega^i) + \sum_{i=0}^{k-1}\chi_i, \tag{16.7}$$

$$\chi_i = \begin{cases} 0 & \text{if } i \neq \mathbb{N} \text{ for } \ell = 1, 2, \ldots \text{ or } a < y^i - \rho(1 + \omega^i) \\ a - y^i + \rho(1 + \omega^i) & \text{otherwise} \end{cases}$$

has the property that $y^{\mathbb{N}} = z^s$ and therefore it is sufficient to compare y^k with x^k for $k = \mathbb{N}$, where

$$x^k = x^0 - \rho \sum_{i=0}^{k-1} (1 + \omega^i) + \sum_{i=0}^{k-1} \tau_i. \tag{16.8}$$

Suppose that $x^0 = y^0 \neq a$. Then if $t_x^1 = \min\{k : x^k = a\}$ represents the time at which process x^k first encounters the optimal point and $t_z^1 = \min\{\ell : y^{\mathbb{N}} = a\}$ represents the time of the corresponding encounter of process z^k with the optimal point, it is clear that $t_x^1 \leq t_z^1$ because from (16.7) and (16.8) we have that $y^k = x^k$ for $k \leq t_x^1$. This means that algorithm (16.2),(16.5) will get from some remote initial point to the vicinity of the optimal point faster then algorithm (16.2),(16.6) with $N > 1$. Now let us take $x^0 = y^0 = a$. Then (16.7) and (16.8) imply that $\chi_k = 0$ for $k < N$ while τ_k may differ from zero. Therefore in this case $x^N \geq y^N = z^1$ and the performance of algorithm (16.2),(16.6) with $N > 1$ becomes superior to that of algorithm (16.2),(16.5) after reaching the vicinity of the optimal point. This simple example demonstrates several important properties of constrained stochastic optimization problems, although more work is necessary before we can make any firm theoretical recommendations concerning the choice of the number of samples N. Above all, an appropriate definition of the rate of convergence is needed: recent results by Kushner [10] may be useful in this regard.

A rather general adaptive way of changing the number N would be to begin with a small value of N for the first few iterations ($N = 1$, for example), and increase N if additional tests show that the current point is in the vicinity of the optimum. The following averaging procedure has been shown to be useful in tests of this type:

$$v^{s+1} = (1 - \alpha_s)v^s + \alpha_s \xi^s, 0 \leq \alpha_s \leq 1, \tag{16.9}$$

where ξ^s is defined by (16.5) or (16.6). It can be shown (see [1], [2]) that $\|v^s - F_x(x^s)\| \to 0$ under rather general conditions, which include $\rho_s/\alpha_s \to 0$. The decision as to whether to change N may then be based on the value of $r_s = \|x^s - \pi_X(x^s - v^s)\|$. One possibility is to estimate ξ^s and its empirical variance at the same time:

$$\sigma_N^s = \frac{1}{N} \sum_{i=1}^{N} [f_x(x^s, \omega^{i,s}) - \xi^s]^2$$

and choose N such that $\sigma_N^s \leq \beta r_s$, where the value of β is set before beginning the iterations. In practice it is sufficient to consider a constant $\alpha_s \equiv \alpha \sim 0.01 - -0.05$, where the greater the randomness, the smaller the value of α. Our empirical recommendation for the initial value of N is $\sigma_N^0 \sim 0.1 \max_{x_1, x_2 \in X} \|x_1 - x_2\|$.

This method can be used to increase the number of samples per iteration automatically. Another possibility is to alter the value of N interactively; this is one of the options implemented in the interactive package STO, which has recently been developed at IIASA. Numerical experiments conducted with this package show that in problems where $f_x(x,\omega)$ has a high variance, choosing a value of N greater than one can bring about considerable improvements in performance.

The method described above uses increasingly precise estimates of the gradient, and therefore shares some of the features of the approximation techniques developed in [3–6] for solving stochastic programming problems. All of the remarks made here concerning sampling are also valid for the other methods of choosing ξ^s described below.

However, it is not always possible to use observations of the gradient $f_x(x,\omega)$ of the random function to compute a stochastic quasigradient. In many cases the analytic expression of $f_x(x,\omega)$ is not known, and even if it is, it may be difficult to create a subroutine to evaluate it, especially for large-scale problems. In this case it is necessary to use a method which relies only on observations of $f(x,\omega)$.

16.2.2 Finite-difference approximations

If function $F(x)$ is differentiable, one possibility is to use forward finite differences:

$$\xi^s = \sum_{i=1}^n \frac{f(x^s + \delta_s e_i, \omega^s_{i,1}) - f(x^s, \omega^s_{i,2})}{\delta_s} e_i, \tag{16.10}$$

or central finite differences:

$$\xi^s = \sum_{i=1}^n \frac{f(x^s + \delta_s e_i, \omega^s_{i,1}) - f(x^s - \delta_s e_i, \omega^s_{i,2})}{2\delta_s} e_i, \tag{16.11}$$

where the e_i are unit basis vectors from R^n. The most important question here is the value of δ_s. In order to ensure convergence with probability one it is sufficient to take any sequence δ_s such that $\sum_{i=1}^\infty \rho_i^2/\delta_i^2 < \infty$. If it is possible to take $\omega^s_{i,2} = \omega^s_{i,1}$ then any $\delta_s \to 0$ will do. However, the method may reach the vicinity of the optimal point much faster if δ_s is chosen adaptively. On the first few iterations δ_s should be large, decreasing as the current point approaches the optimal point. The main reason for this is that taking a large step δ_s when the current point is far from the solution may smooth out the randomness to some extent, and may also overcome some of the problems (such as curved valleys) caused by the erratic behavior of the deterministic function $F(x)$. One possible way of implementing such a strategy in an unconstrained case is given below.

(i) Take a large initial value of δ_s, such as $\delta_s \sim 0.1 \max_{x_1, x_2 \in X} \|x_1 - x_2\|$.

(ii) Proceed with iterations (16.2), where ξ^s is determined using (16.10) or (16.11). While doing this, compute an estimate of the gradient v^s from (16.9).

(iii) Take

$$\delta_{s+1} = \begin{cases} \delta_s & \text{if } v^s \geq \beta_1 \delta_s \\ \beta_2 \delta_s & \text{otherwise} \end{cases},$$

where the values of β_1 and β_2 should be chosen before beginning the iterative process.

It can be shown that this process converges when $\omega_{i,1}^s \equiv \omega_{i,2}^s$, although it will also produce a good approximation to the solution even if this requirement is not met. Estimate (16.9) is not the only possibility—in fact, any of the estimates of algorithm performance given in Section 16.3 would do.

Another strategy is to relate changes in the finite-difference approximation step to changes in the step size. This is especially advisable if the step size is also chosen adaptively (see Section 3). In the simplest case one may fix $\beta_1 > 0$ before starting and choose $\delta_s = \beta_1 \rho_s$, which, although contrary to theoretical recommendations, will nevertheless bring the current point reasonably close to the optimal point. To obtain a more precise solution it is necessary to reduce β_1 during the course of the iterations. This may be done either automatically or interactively; both of these options are currently available in the stochastic optimization package STO.

Finite-difference algorithms (16.10) and (16.11) have one major disadvantage, and this is that the stochastic quasigradient variance increases as δ_s decreases. This means that finite-difference algorithms converge more slowly than algorithms which use gradients (16.5). There are two ways of overcoming this problem. Firstly, if it is possible to make observations of function $f(x, \omega)$ for various values of x and fixed ω, it is a good idea to take the same values of ω for the differences (i.e., $\omega_{i,1}^s = \omega_{i,2}^s$) when δ_s is small because this reduces the variance of the estimates quite considerably. Another way of avoiding this increase in the variance is to increase the number of samples used to obtain ξ^s when approaching the optimal point, i.e., to use finite-difference analogues of (16.6). If there exists a $\gamma > 0$ such that $N_s \delta_s^2 > \gamma$, where N_s is the number of samples taken at step number s, then the variance of ξ^s remains bounded.

It is sometimes useful to normalize the ξ^s, especially when the variance is large.

Another disadvantage of the finite-difference approach is that it requires $n + 1$ evaluations of the objective function for forward differences and $2n$ for central differences, where n is the dimension of vector x. This may not be acceptable in large-scale problems and in cases where function evaluation is computationally expensive. In this situation a stochastic quasigradient can be computed using some analogue of random search techniques.

16.2.3   Analogues of random search methods

When it is not feasible to compute $n+1$ values of the objective function at each iteration, the following approach (which has some things in common with the random search techniques developed for deterministic optimization problems) may be used:

$$\xi^s = \sum_{i=1}^{M_s} \frac{f(x^s + \delta_s h_i, \omega_{i,1}^s) - f(x^s, \omega_{i,2}^s)}{\delta_s} h_i. \qquad (16.12)$$

Here the h_i are vectors distributed uniformly on the unit sphere, M_s is the number of random points and δ_s is the step taken in the random search. The choice of M_s is determined by the computational facilities available, although it is advisable to increase M_s as δ_s decreases. This method of choosing ξ^s has much in common with finite-difference schemes, and the statements made above about the choice of δ_s in the finite-difference case also hold for (16.12).

16.2.4   Smoothing the objective function

Methods of choosing ξ^s which rely on finite-difference or random search techniques are only appropriate when the objective function $F(x)$ is differentiable. The use of similar procedures in the nondifferentiable case would require some smoothing of the objective function. Suppose that the function $F(x)$ is not differentiable but satisfies the Lipschitz condition, and consider the function

$$F(x,r) = \int F(x+y)\mathrm{d}H(y,r), \qquad (16.13)$$

where $H(y,r)$ is a probability measure with support in a ball of radius r centered at zero. We shall assume for simplicity that $H(y,r)$ has nonzero density inside this ball. The function $F(x,r)$ is differentiable and $F(x,r) \rightarrow F(x)$ uniformly over every compact set as $r \rightarrow 0$. It is now possible to minimize the nonsmooth function $F(x)$ by computing stochastic quasigradients for smooth functions $F(x,r)$ and find the optimal solution of the initial problem by letting $r \rightarrow 0$. This idea was proposed in [11] and studied further in [12]. It is not actually necessary to calculate the integral in (16.13)—it is sufficient to compute ξ^s using equations (16.10)–(16.12), but at point $x^s + y^s$ rather than point x^s, where y^s is a random variable distributed according to $H(y,r_s)$. In this case (16.10) becomes:

$$\xi^s = \sum_{i=1}^{n} \frac{f(x^s + y^s + \delta_s e_i, \omega_{i,1}^s) - f(x^s + y^s, \omega_{i,2}^s)}{\delta_s} e_i. \qquad (16.14)$$

The most commonly used distribution $H(y,r)$ is uniform distribution on an n-dimensional cube of side r. If we want to have convergence with probability one we should choose r_s such that $\delta_s/r_s \rightarrow 0$ and $(r_s - r_{s+1})/\rho_s \rightarrow 0$. In practical computations it is also advisable to choose the smoothing parameter r_s in a similar way to δ_s, using one of the adaptive procedures discussed above.

Smoothing also has beneficial side effects in that it improves the behavior of the deterministic function $F(x)$. In the case where $F(x)$ may be written as the sum of two functions, one with a distinct global minimum and the other with highly oscillatory behavior, smoothing may help to overcome the influence of the oscillations, which may otherwise lead the process to local minima far from the global one. Thus it can sometimes be useful to smooth the objective function even if we can obtain a gradient $f_x(x,\omega)$. In this case we should take a large value for the smoothing parameter r_s on the first few iterations, decreasing it as we approach the optimal point. The points at which r_s should be decreased may be determined using the values of additional estimates, such as those described below in Section 16.3 or given by (16.9). Everything said about the choice of the finite-difference parameter δ_s is also valid for the choice of the smoothing parameter, including the connection between the step size and the smoothing parameter and the possibility of interactive control of r_s. The only difference is that a decrease in r_s does not lead to an increase in the variance of ξ^s and that it is preferable to have $\delta_s < r_s$. This is also reflected in the stochastic optimization software developed at IIASA.

All of the methods discussed so far use only the information available at the current point or in its immediate vicinity. We shall now discuss some more general ways of choosing the step direction which take into account the information obtained at previous points.

16.2.5 Averaging over preceding iterations

The definition of a stochastic quasigradient given in (16.3) allows us to use information obtained at previous points as the iterations proceed; this information may sometimes lead to faster convergence to the vicinity of the optimal point. One possible way of using such information is to average the stochastic quasigradients obtained in preceding iterations via a procedure such as (16.9). The v^s obtained in this way may then be used in method (16.2). This is another way of smoothing out randomness and neutralizing such characteristics of deterministic behavior as curved valleys and oscillations. Methods of this type may be viewed as stochastic analogues of conjugate gradient methods and were first proposed in [13]. We can choose ξ^s according to any of (16.5), (16.6), (16.10), (16.11), (16.12), or (16.14). Since $v^s \rightarrow F_x(x^s)$ under rather general conditions (see [1], [2]), method (16.9) can be considered as an alternative to method (16.6) for deriving precise estimates of gradient $F_x(x)$. This method has an advantage over (16.6) in that it provides a natural way of using rough estimates of $F_x(x^s)$ on the first few iterations and then gradually increasing the accuracy as the current point approaches the optimal point. In this case (16.9) can be incorporated in the adaptive procedures used to choose the smoothing parameter and the step in the finite-difference approximation.

However, it is not necessary to always take $\alpha_s \rightarrow 0$, because we have convergence for any $0 \leq \alpha_s \leq 1$. Sometimes it is even advantageous to take $\alpha_s \equiv \alpha =$ constant, because in this case more emphasis is placed on information obtained in recent iterations. In general, the greater the randomness, the smaller the value of α that should be taken. Another averaging technique is given by

$$v^{s+1} = \frac{1}{M_s} \sum_{i=s-M_s+1}^{s} \xi^s, \qquad (16.15)$$

where M_s is the size of the memory, which may be fixed.

16.2.6   Using second-order information

There is strong evidence that in some cases setting

$$v^s = A_s \xi^s \qquad (16.16)$$

may bring about considerable improvements in performance. Here ξ^s can be chosen in any of the ways discussed above. Matrix A_s should be positive definite and take into account both the second-order behavior of function $F(x)$ and the structure of the random part of the problem. One possible way of obtaining second-order information is to use analogues of quasi-Newton methods to update matrix A_s. To implement this approach, which was proposed by Wets in [3], it is necessary to have $\|\xi^s - F_x(x^s)\| \to 0$.

16.3.   Choice of Step Size

The simplest way of choosing the step-size sequence in (16.2) is to do it before starting the iterative process. Convergence theory suggests that any series with the properties:

$$\rho_s > 0, \sum_{s=1}^{\infty} \rho_s = \infty, \sum_{s=1}^{\infty} \rho_s^2 < \infty. \qquad (16.17)$$

can be used as a sequence of step sizes. In addition, it may be necessary to take into account relations between the step size and such things as the smoothing parameter or the step in a finite-difference approximation. Relations of this type have been briefly described in the preceding sections. In most cases the choice $\rho_s \sim C/s$, which obviously satisfies (16.17), provides the best possible asymptotic rate of convergence. However, since we are mainly concerned with reaching the vicinity of the solution, rule (16.17) is of limited use because a wide variety of sequences can be modified to satisfy it. The other disadvantage of choosing the step-size sequence in advance is that this approach does not make any use of the valuable information which accumulates during solution. These "programmed" methods thus perform relatively badly in the majority of cases.

The best strategy therefore seems to be to choose the step size using an interactive method. It is assumed that the user can monitor the progress of the optimization process and can intervene to change the value of the step size or other parameters. This decision should be based on the behavior of the estimates $\widehat{F}(x^s)$ of the current value of the objective function. The estimates may be very rough and are generally calculated using only one observation per iteration, as in the following example:

$$\widehat{F}^s = \frac{1}{s} \sum_{i=1}^{s} f\left(x^i, \omega^i\right). \tag{16.18}$$

It appears that although the observations $f(x^s, \omega^s)$ may vary greatly, the $\widehat{F}^s$ display much more regular behavior. Monitoring the behavior of some components of the vector x^s in addition to the $\widehat{F}^s$ also seems to be useful. One possible implementation of the interactive approach may proceed along the following lines:

(i) The user first chooses the value of the step size and keeps it constant for a number of iterations (usually 10–20). During this period the values of the estimate $\widehat{F}^s$ and some of the components of the vector x^s are displayed, possibly with some additional information.

(ii) The user decides on a new value for the step size using the available information. Three different cases may occur:

  – The current step size is too large. In this case both the values of the estimate $\widehat{F}^s$ and the values of the monitored components of x^s exhibit random jumps. It is necessary to decrease the step size.

  – The current step size is just right. In this case the estimates decrease steadily and some of the monitored components of the current vector x^s also exhibit regular behavior (steadily decrease or increase). This means that the user may keep the step size constant until oscillations occur in the estimate $\widehat{F}^s$ and/or in the components of the current vector x^s.

  – The current step size is too small. In this case the estimate $\widehat{F}^s$ will begin to change slowly, or simply fluctuate, after the first few iterations, while the change in x^s is negligible. It is necessary to increase the step size.

(iii) Continue with the iterations, periodically performing step (ii), until changes in the step size no longer result in any distinct trend in either the function estimate or the current vector x^s, which will oscillate around some point. This will indicate that the current point is close to the solution.

This method of choosing the step size requires an experienced user, but we have found that the necessary skills are quickly developed by trial and error. The main reasons for adopting an interactive approach may be summarized as follows:

  – Interactive methods make the best use of the information which accumulates during the optimization process.

  – Because the precise value of the objective function is not available, it is impossible to use the rules for changing the step size developed in deterministic optimization (e.g., line searches).

- Stochastic effects make it extremely difficult to define formally when the
  step size is "too big" or "too small"; theoretical research has not thrown
  any light on this problem.

The main disadvantage of the interactive approach is that much of the
user's time is wasted if it takes the computer a long time to make one observation
$f(x^s, \omega^s)$. For this reason a great effort has been made to develop *automatic
adaptive* ways of choosing the step size, in which the value of the step size
is chosen on the basis of information obtained at all or some of the previous
points x^i, $i = \overline{1, s}$. Methods of this type are considered in [14–20]. The approach
described in the following sections involves the estimate of some measures of
algorithm performance which we denote by $\Phi^i(xbar^s, u^s)$, where $\overline{x}^s$ represents
the whole sequence $\{x^1, x^2, \ldots, x^s\}$ and u^s the set of parameters used in the
estimate. In general, algorithm performance measures are attempts to formalize
the notions of "oscillatory behavior" and "regular behavior" used in interactive
step-size regulation, and possess one or more of the following properties:

- the algorithm performance measure is quite large when the algorithm exhibits distinct regular behavior, i.e., when the estimates of the function
  value decrease or the components of the current vector x^s show a distinct
  trend;
- the algorithm performance measure becomes small and even changes its
  sign if the estimates of the current function value stop improving or if the
  current point starts to oscillate chaotically;
- the algorithm performance measure is large far from the solution and small
  in the immediate vicinity of the optimal point.

Automatic adaptive methods for choosing the step size begin with some reasonably large value of the step size, which is kept constant as long as the value
of the algorithm performance measure remains high, and then decreases when
the performance measure becomes less than some prescribed value. The behavior of the algorithm usually becomes regular again after a decrease in the
step size, and the value of the performance measure increases; after a number of iterations oscillations set in and the value of the performance measure
once again decreases. This is a sign that it is time to decrease the step size.
A rather general convergence result concerning such adaptive piecewise-linear
methods of changing the step size is given in [18]. However, in many cases it
is difficult to determine how close the current point is to the optimal point using only one such measure—a more reliable decision can be made using several
of the measures described below. Unfortunately, it is not possible to come to
any general conclusions as to which performance measure is the "best" for all
stochastic optimization problems. Moreover, both the values of the parameters
used to estimate the performance measure and the value of the performance
measure at which the step size should be decreased are different for different
problems. Therefore if we fix these parameters once and for all we may achieve
the same poor performance as if we had chosen the whole sequence of step sizes
prior to the optimization process. Thus, it is necessary to tune the parame-

ters of automatic adaptive methods to different classes of problems, and the interactive approach can be very useful here. An experienced user would have little difficulty in using the values of the performance measures to determine the correct points at which to change the step size, and in learning what type of performance measure behavior requires an increase or a decrease in the step size. The interactive approach is of particular use if one iteration is not very time-consuming and there are a number of similar problems to be solved. In this case the user can identify the most valuable measures of performance in the first few runs, fix their parameters and incorporate this knowledge in automatic adaptive step-size selection methods for the remaining problems.

Although interactive methods usually provide the quickest means of reaching the solution, they cannot always be implemented, and in this case automatic adaptive methods prove to be very useful. The stochastic optimization package STO developed at IIASA and the Kiev stochastic and nondifferentiable optimization package NDO both give the user the choice between automatic adaptive methods and interactive methods of determining the step size. Below we describe some particular measures of algorithm performance and methods of choosing the step size.

The main indicators used to evaluate the performance of an algorithm are estimates of such things as the value of the objective function and its gradient. The averaging procedure (16.9) may be used to estimate the value of the gradient, as described earlier in this paper. The main advantage of this procedure is that it allows us to obtain estimates of the mean values of the random variables without extensive sampling at each iteration, since a very limited number of observations (usually only one) is made at each iteration. This estimate, although poor at the beginning, becomes more and more accurate as the iterations proceed. One example of such an estimate is (16.18), which is a special case of the more general formula

$$\widehat{F}^{s+1} = (1 - \gamma_s)\widehat{F}^s + \gamma_s f(x^s, \omega^s). \tag{16.19}$$

Any observation μ^s with the property

$$\mathrm{E}(\mu^s | x^1, x^2, \ldots, x^s) = F(x^s) + d_s \tag{16.20}$$

can be used instead of $f(x^s, \omega^s)$ in (16.19), where $d_s \to 0$. For example, (16.6) would do. In order to get $\lim_{s \to \infty} |\widehat{F}^s - F(x^s)| = 0$ it is necessary to have $\rho_s / \gamma_s \to 0$. However, estimate (16.18) assigns all observations of function values the same weight. This sometimes leads to considerable bias in the estimate for all the iterations the user can afford to run. Therefore for practical purposes it is sometimes more useful to adopt procedures of the type described in Section 2 for the estimation of gradients. These include estimate (16.19) with fixed $\gamma_s \equiv \gamma$, where $\gamma \sim 0.01 - -0.05$, and the method in which the average is taken over the preceding M_s iterations:

$$\widehat{F}^s = \frac{1}{M_s} \sum_{i=s-M_s+1}^{s} f(x^i, \omega^i). \tag{16.21}$$

Although these estimates do not converge asymptotically to $F(x^s)$, they place more emphasis on observations made at recent points. All of the estimates $\widehat{F}^s$ may also be used in an interactive mode to determine the step size, as described above. In addition, the values of the parameters used to determine the step size may also be chosen interactively. For example, the values of parameters b_1 and b_2 in

$$\rho_s = \frac{b_1}{b_2 + s}$$

can be made to depend on the behavior of $\widehat{F}^s$.

We shall now describe some automatic adaptive rules for choosing the step size. The important point as regards implementation is how to choose the initial value of the step size ρ_0. We suggest that the value of a stochastic quasigradient ξ^0 should first be computed at the initial point, and that the initial value of the step size should then be chosen such that

$$\rho_0 \ell \|\xi^0\| \sim D,$$

where $\ell \sim 10 - - 20$ and D is a rough estimate of the size of the domain in which we believe the optimal solution to be located. This means that it is possible to reach the vicinity of each point in this domain within the first 20 iterations or so.

16.3.1   Ratio of function estimate to the path length

Before beginning the iterations we choose the initial step size ρ_0, two positive constants α_1 and α_2, a sequence M_s and an integer $\widehat{M}$. After every $\widehat{M}$ iterations we revise the value of the step size in the following way:

(i) Compute the quantity

$$\Phi^1(\overline{x}^s, u^s) = \frac{\widehat{F}^{s-M_s} - \widehat{F}^s}{q(s, M_s)}. \tag{16.22}$$

Here the u^s are the averaging parameters used in the estimation of both $\widehat{F}^s$ and M_s, while $\overline{x}^s$ is again the whole sequence of points preceding x^s. The quantity

$$q(s, M_s) = \sum_{i=s-M_s}^{s-1} \|x^{i+1} - x^i\| \tag{16.23}$$

is the length of the path taken by the algorithm during the preceding M_s iterations. The function $\Phi^1(\overline{x}^s, u^s)$ is another example of a measure which can be used to assess algorithm performance.

(ii) Take a new value of the step size:

$$\rho_{s+1} = \begin{cases} \alpha_1 \rho_s & \text{if } \Phi^1(\overline{x}^s, u^s) \leq \alpha_2 \\ \rho_s & \text{otherwise} \end{cases}. \tag{16.24}$$

In this method the step size is changed at most once every $\widehat{M}$ iterations. This is essential because function Φ^1 changes slowly, and if its value is less than α_2 at iteration number s it is likely that the same will be true at iteration number $s + 1$. Therefore $\widehat{M}$ should lie in the range 5–20. This procedure can be modified in various ways, such as continuing for $\widehat{M}$ iterations with a fixed step size, then starting to compare values until inequality (16.24) is satisfied whereupon the step size is reduced. We then wait another $\widehat{M}$ iterations and repeat the procedure. Recommended values of α_1 and α_2 lie within the ranges 0.5–0.9 and 0.005–0.1, respectively. The number M_s may be chosen to be constant and equal to $\widehat{M}$. If we have a number of similar problems it is very useful to make the first run in a semi-automatic mode, i.e., to intervene in the optimization process to improve the values of parameters $\alpha_1, \alpha_2, \widehat{M}$ – the new values can then be used in a fully automatic mode to solve the remaining problems.

This algorithm is by no means convergent in the traditional sense, but it outperformed traditional choices like C/s in numerical experiments because it normally reaches the vicinity of the optimal point more quickly. However, it is possible to safeguard convergence by considering a second sequence C/s, where C is small, and switching to this sequence if the step size recommended by (16.24) falls below a certain value. This step size regulation was introduced in [15].

16.3.2 Use of gradient estimates

Take $\Phi^2 = \widehat{G}^s$ instead of $\Phi^1(\overline{x}^s, u^s)$ in (16.24), where $\widehat{G}^s$ is one of the gradient estimates discussed above, and the u^s represent all the parameters used, including averaging parameters and the frequency of changes in the step size.

16.3.3 Ratio of progress and path

The quantity $\|x^{s-M_s} - x^s\|$ represents the progress made by the algorithm between iteration number $s - M_s$ and iteration number s. If we keep the step size constant, the algorithm begans to oscillate chaotically after reaching some neighborhood of the optimal point. The smaller the value of the step size, the smaller the neighborhood at which this occurs, and thus the total path between iterations s and $s - M_s$ begins to grow compared with the distance between points x^{s-M_s} and x^s. This means that the function

$$\Phi^3(\overline{x}^s, u^s) = \frac{\|x^{s-M_s} - x^s\|}{\sum_{i=s-M_s}^{s-1} \|x^{i+1} - x^i\|} \tag{16.25}$$

can be used as a performance measure in equation (16.24).

16.3.4   Analogues of line search techniques

The decision as to whether (and how) to change the step size may be based on the values of the scalar product of adjacent step directions. If we have $(\xi^{s-1}, \xi^s) > 0$, then this may be a sign that regular behavior prevails over stochastic behavior, the function is decreasing in the step direction and the step size should be increased. Due to stochastic effects the function will very often increase rather than decrease, but in the long run the number of bad choices will be less than the number of correct decisions. Analogously, if this inequality does not hold then the step size should be decreased. The rule for changing the step size is thus basically as follows:

$$\rho_{s+1} = \begin{cases} \rho_s & \text{if } -\alpha_1 \leq (\xi^{s-1}, \xi^s) \leq \alpha_1 \\ \alpha_2 \rho_s & \text{if } (\xi^{s-1}, \xi^s) > \alpha_1 \\ \alpha_3 \rho_s & \text{if } (\xi^{s-1}, \xi^s) \leq -\alpha_1 \end{cases}, \qquad (16.26)$$

where the values of α_1, α_2, α_3 (recommended values $\alpha_1 \sim 0.4 - -0.8$, $1 < \alpha_2 \leq 1.3$ and $0.7 \leq \alpha_3 < 1$) should be chosen before starting the iterations. It is also advisable to have upper and lower bounds on the step size to avoid divergence. Sometimes it is convenient to normalize the vectors of step directions, i.e., $\|\xi^s\| = 1$. The lower bound may decrease as the iterations proceed. This method may also be applied to the choice of a vector step size, treating some (or all) variables or groups of variables separately. A number of different methods based on the use of scalar products of adjacent step directions to control the step size have been developed by Uriasiev [19], Pflug [16], and Ruszczynski and Syski [20].

16.4   IIASA Implementation

The interactive stochastic optimization package implemented at IIASA (STO) is based on the same ideas as the package for stochastic and nondifferentiable optimization developed in Kiev (NDO). It allows the user to choose between interactive and automatic modes and makes available the stochastic quasigradient methods described in Sections 2 and 3. In the interactive mode the program offers the user the opportunity to change the step parameters and the methods by which the step size and step direction are chosen *during the course of the iterations*. The user can also stop the iterative process and obtain a more precise estimate of the value of the objective function before continuing. The package is written in FORTRAN-77.

Before initiating the optimization process the user has to:

(i) Provide a subroutine UF which calculates the value of function $f(x, \omega)$ for fixed x and ω and, optionally, a subroutine UG which computes the gradient $f_x(x, \omega)$ of this function; the function evaluation subroutine should

```
FUNCTION UF(N,X)
DIMENSION X(N)
Calculation of f(x, ω)
RETURN
END
```

be of the form: Here N is the dimension of the vector of variables X. (Note that the implementation on the IIASA VAX actually requires the subroutine to be entered in lower-case letters rather than capitals.) A description of a subroutine which calculates a quasigradient is given later in this paper.

(ii) Compile these subroutines with the source code to obtain an executable module.

(iii) Provide at least one of the following additional data files:
- algorithm control file (used only in the noninteractive option)
- parameter file (used only in the interactive option)
- initial data file (should always be present)

All of these files are described in some detail later in the paper.

The optimization process can then begin. The program first asks the user a series of questions regarding the required mode (interactive or automatic), method of step size regulation, choice of step direction, etc. These questions appear on the monitor and should be answered from the keyboard or by reference to a data file. We shall represent the dialogue as follows:

Question? *Answer*

The first question is

interactive mode? reply yes or no   *yes/no*

To choose the interactive option the user should type in *yes* (or *y*); to select the automatic option he should answer *no* (or *n*). In the latter case the program would ask no further questions, but would read all the necessary information from the algorithm control file (which is usually numbered 2—under UNIX conventions its name is fort.2). The iterative process would then begin, terminating after 10,000 iterations if no other stopping criterion is fulfilled. The algorithm control file must contain answers to all of the following questions except those concerned either with dialogue during the iterations or with the parameter file (such questions are marked with an asterisk * below). This file is given a name only for ease of reference — the important thing for the user is its number.

Assume now that the user has chosen the interactive option by answering *yes* to the first question. The program then asks

parameter file?   *(number)*   *

The user should respond either with the number of the file of default parameters or with the number of the file in which the current values of the algorithm parameters are stored. The file of default parameters is provided with the program and has the name fort.12 (under UNIX conventions); thus, to refer the program to the default file the user should answer *12*. The purpose of this file is to help the user to set the values of algorithm parameters in the ensuing dialogue and also to store such improved values as may be discovered by the user through trial and error. If the user assigns the algorithm parameters any values other than those in the default file, the new values become the default values in subsequent runs of the program. This file is optional.

The program then asks

read parameter file? reply yes or no   *yes/no*   ∗

The answer *yes* implies that the file specified in the previous question exists, and that default parameter values are stored in this file. In this case, when asking the user about parameter values, the program will read the default option in the parameter file and reproduce it on the screen together with the question. If the user accepts this default value he should respond with *0* (zero); otherwise he should enter his own value, which will become the new default value.

The answer *no* means that no default values are available at the moment. In this case the program will form a new default file (labeled with the number given as an answer to the previous question); its contents will be based on the user's answers to future questions. This new default file, once formed, can be used in subsequent runs.

The next question is

number of variables?   *(number)*

to which the user should respond with the dimension of the vector of variables x. He is then asked

initial data file?   *(number)*

and should reply with the number of the initial data file. This file should contain the following elements (in exactly this order):

- The initial point, which should be a sequence of numbers separated by commas or other delimiters.
- Any additional data required by subroutines UF or UG if such data exists and the user chooses to put it in the initial data file (optional).
- Information about the constraints (described in more detail below)

The program then asks

step size regulation?   *is*

Here *is* is a positive integer from the set $\{1, 2, 3, 4, 6, 7\}$, where the different values of *is* correspond to different ways of choosing the step size. (The integer 5 is reserved for an option currently under development.)

is Definition

1 Adaptive automatic step size regulation (16.24) based on algorithm performance function (16.22) and function estimate (16.18).

2 Manual step size regulation based on algorithm performance function (16.22) and function estimate (16.18).

3 Adaptive automatic step size regulation (16.24) using algorithm performance measure (16.22) and a function estimate based on a finite number of previous observations (16.21).

4 Manual step size regulation based on the same estimates of algorithm performance as for *is* = 3.

6 Automatic step size regulation using algorithm performance measure (16.24) and function estimate (16.19) with fixed γ_s.

7 Manual step size regulation based on the same estimates of algorithm performance as for $is = 6$.

The difference between adaptive automatic and manual step size regulation (see $is = 1, 2$) is that in the first case the step size is chosen automatically, although the user may terminate the iterations at specified points and continue with another step size regulation, while in the second case the user changes the value of the step size himself. Both step size regulations are based on the same estimates of function value and algorithm performance.

The next question is

step direction? (5 figures)   *id1 id2 id3 id4 id5*

The user has to respond with five figures which specify various ways of choosing the step direction, e.g., 11111. We shall refer to these figures as *id1, id2, id3, id4* and *id5*. The subroutine which estimates the step direction makes some number of initial observations $\vec{\xi}^i$, s at each step; these are then averaged in some way to obtain the vector ξ^s, and the final step direction v^s is calculated using both ξ^s and values of v^i for $i < s$.

The value of *id1* specifies the nature of the initial observations $\vec{\xi}^i$, s.

id1 Definition

1 A direct observation of a stochastic quasigradient is available for $\vec{\xi}^i$, s and the user has to specify a subroutine UG to calculate it:
SUBROUTINE UG(N,X,G)
DIMENSION X(N),G(N)
Calculation of a stochastic quasigradient
RETURN
END
where G(N) is an observation of a stochastic quasigradient.
2 Central finite-difference approximation of the gradient as in (16.11).
3 The $\vec{\xi}^i$, s are calculated using random search techniques (16.12).
4 Forward finite-difference approximation of the initial observations $\vec{\xi}^i$, s as in (16.10).
5 Central finite-difference approximation of the gradient as in (16.11). All observations of the function used in one observation of $\vec{\xi}^i$, s are made with the same values of random parameters ω.
6 The $\vec{\xi}^i$, s are calculated using random search techniques (16.12). All observations of the function used in one observation of $\vec{\xi}^i$, s are made with the same values of random parameters ω.
7 Forward finite-difference approximation of the initial observations $\vec{\xi}^i$, s as in (16.10). All observations of the function used in one observation of $\vec{\xi}^i$, s are made with the same values of random parameters ω.

Note that for *id1*= 5, 6, 7 all observations of the function used in one observation of $\vec{\xi}^i$, s are made with the same values of random parameters ω. In this case the user should write a function UF which supports this feature as follows:

```
FUNCTION UF(N,X)
DIMENSION X(N)
COMMON/OMEG/LO,MO
If LO=1 and MO=1 then obtain new values
of random factors ω and set MO=0. Make
an observation of the function at point x.
RETURN
END
```

The second figure $id2$ determines the point at which observations are made:

$id2$ Definition

1 The initial direction is calculated at the current point x^s

2 The initial direction is calculated at a point chosen randomly from among those in the neighborhood of the current point x^s

The value of $id3$ defines the way in which the step in a finite-difference or random search approximation of $\overline{\xi}^i$, s is chosen:

$id3$ Definition

1 The approximation step is fixed. The observations of the objective function at point x^s originally used to obtain gradient observations $\overline{\xi}^i$, s are not used to update the estimate of the function employed for step size regulation.

2 The ratio δ_s/ρ_s of the step in the finite-difference approximation to the step size of the algorithm is fixed (see (16.10)–(16.12)). The observations of the objective function at point x^s originally used to obtain gradient observations $xibar^i$, s are not used to update the estimate of the function employed for step size regulation.

3 The approximation step is fixed. The observations described for $id3=1,2$ above *are* used to update the current estimate of the objective function.

4 The ratio δ_s/ρ_s of the step in the finite difference approximation to the step size of the algorithm is fixed (see (16.10)–(16.12)). The observations described for $id3=1,2$ above *are* used to update the current estimate of the objective function.

The fourth figure $id4$ defines the type of averaging used to obtain ξ^s from observations $\overline{\xi}^i$, s.

$id4$ Definition

1 No averaging, $xi^s = \overline{\xi}^i$, s, $i=1$.

2 Number of samples > 1.

The value of $id5$ specifies the way in which the final step direction v^s is obtained from previous values of v^s and from xi^s.

$id5$ Definition

1 No previous information is used. The final vector v^s is simply set equal to xi^s.

2 (16.9) is used.

3 A positive number n_3 is provided by the user. Set $k(s) = \max\{k : kn_3 + 1 \leq s\}$. Then the final direction v^s is computed from (16.15), where $M_s = s - k(s)n_3 + 1$.

4 No previous information is used. The final vector v^s is set equal to ξ^s and is normalized.

5 (16.9) is used. The final vector v^s is normalized.

6 A positive number n_3 is provided by the user. Let $k(s) = \max\{k : kn_3 + 1 \leq s\}$. Then the final direction v^s is computed from (16.15), where $M_s = s - k(s)n_3 + 1$. The final vector v^s is normalized.

The program then asks about the type of constraints present in the problem:

constraints? *(number)*

The answer (in the present implementation) must be 1,2,3 or 4. These values define the type of constraints present and correspond to the following options:

1 There are no constraints at all.

2 There are upper and lower bounds on the variables. The values of these bounds should be given at the end of the initial data file in the form of strings of numbers separated by commas or other delimiters. The string containing the upper bounds should come first.

3 There is one constraint $\sum_{i=1}^{n} a_i x_i \leq b$. The coefficients a_i should be given at the end of the initial data file. The string containing the coefficients of linear form comes first and then, on a separate line, the right-hand side.

4 There are general linear constraints $b_l \leq Ax \leq b_u$. In this case the program computes a projection on these constraints at each iteration, using the quadratic programming package SOL/QPSOL [21]. The previous point x^{s-1} is used as the initial approximation to the solution at iteration number s. The precision of projection also varies, being rough during the first few iterations and improving as the process proceeds. All of these facilities are intended to reduce the amount of computation required at each iteration. The following information should appear at the end of the initial data file (in exactly this order):

- upper bounds on variables x
- lower bounds on variables x
- upper bounds b_u on general linear constraints
- lower bounds b_l on general linear constraints
- number of nonzero elements in matrix A
- numbers of nonzero elements in the columns of matrix A
- nonzero elements of matrix A in increasing order of column number
- row numbers of nonzero elements, in the same order as the elements themselves

The next question is

termination condition? *(number)*

There is currently only one possible answer, which is 1. This means that the iterations terminate when the step size becomes smaller than some value specified by the user. Additional options are under development.

The program then asks the user whether the interactive mode is required *during* the iterations:

interactive mode during iterations? reply yes or no   *yes/no*

Note that the answer to this question should not be included in the algorithm control file for the completely noninteractive option (as indicated by the asterisk). If the user replies *yes* (or *y*), the program will allow the user to change the parameters of the algorithm and even the algorithm itself during the course of the iterations. If the answer is *no* (or *n*) the program will not communicate with the user during the iterations but will instead ask the following two questions:

number of iterations?   *(number)*

This is the number of iterations that should be performed before the process terminates (if it has not already been terminated by some other condition). It is necessary to put an answer to this question in the algorithm control file for the completely non-interactive option.

extra output? reply yes or no   *yes/no*

This is the program's way of asking the user whether information about the iterations should be saved. Note that these two questions do not appear if the user has chosen to run the program in the interactive mode during the iterations.

Now comes a group of questions about step direction parameters. These questions depend on the values of *id1*, *id2*, *id3*, *id4* and *id5* given previously (see the discussion of answers to the question step direction?).

If *id1* = 4, 5 then the question

number of random directions?   *(number)*

appears. The required answer is M_s from (16.12).

If *id2* = 2 the user is asked

relation between step size and neighborhood?   *(number)*

The answer is the ratio of the step size to the size of the neighborhood (of the current point) from which the observation point is chosen (i.e., r_s/ρ_s in the discussion of (16.13)).

If *id3* = 1, 3 and *id1!* = 1 the program asks

step in finite difference approximation?   *(number)*

The required answer is the value of step δ_s in the finite-difference or random search approximation (16.10)–(16.12) of the gradient observation. In this case δ_s is fixed. However, if *id3* = 2, 4 the question

relation between step in finite difference
approximation and step size?   *(number)*

appears. The answer is the ratio δ_s/ρ_s of the finite-difference approximation step to the algorithm step size.

If $id_4 = 2$ the program asks

number of samples? *(number)*

This is the number of samples taken at one point to obtain the averaged estimate (see, for instance, N in (16.6)).

The question

discount rate? *(number)*

appears if $id5 = 2, 5$. The required answer is the (fixed) value of α_s from (16.9). However, if $id5 = 3, 6$ the program asks

number of averaging steps? *(number)*

The user should respond with the value of n_3 (see earlier discussion of $id5$ options).

We now have a group of questions concerning the values of step size parameters. Which questions appear depends on the way in which the step size is being chosen (see earlier discussion of the question step size regulation?).

If the user has chosen automatic step size regulation ($is = 1, 3, 6$) he will be asked the following four questions:

initial step size? *(number)*

This is ρ_0.

multiplier? *(number)*

The required answer is α_1 from (16.24).

frequency of step size changes? *(number)*

The user should give the value of $\widehat{M}$ (see discussion of (16.24)).

lower bound on function decrease? *(number)*

This is α_2 from (16.24).

However, if the user has chosen to regulate the step size interactively ($is = 2, 4, 7$) he will only be asked

value of step size? *(number)*

The following questions appear only if there are general linear constraints, i.e., if the answer to the question constraints? is 4:

number of general linear constraints? *(number)*

correspondence between step size and
accuracy of projection? *(number)*

The answer to the first question is obvious but the second requires some explanation. In order to keep the amount of computation to a minimum, the accuracy τ_s of projection is linked to the value of the step size: $\tau_s = \varepsilon \rho_s$. This leads to only rough projection during the first few iterations (when the step size is large) and more precise projection as the current point approaches the optimal point. The required answer to the last question is the value of ε; recommended values lie in the range 0–1.

Another group of questions is concerned with the estimates of the objective function and also affects the choice of step size:

size of memory? *(number)*

The answer is M_s from (16.22), which in this implementation is fixed. If the step size regulation is defined by $is = 6, 7$ the program asks

multiplier for function averaging? *(number)*

The user should give the value of γ_s in (16.19), which is fixed.

With the answers to these questions the algorithm control file for the non-interactive option is complete. The rest of this section describes the ways in which the algorithm parameters and the algorithm itself may be modified during the course of the iterations. This may be done only if the answer to the question "interactive mode during iterations? reply yes or no" was *yes*. In this case the program will now perform the first iteration and produce a string of information something like this:

    1 0. 7505.826 7505.826 0. 1.000 100.458 109.575

Here the first number is the number of the current iteration, the second is the value of some algorithm performance measure (see (16.22), (16.25) for examples of such functions), the third is the estimate of the value of the objective function at the current point (see (16.18), (16.19), (16.21) for examples of such estimates), the fourth is an observation of $f(x^s, \omega^s)$, the fifth currently has no meaning and always contains 0, the sixth is the step size, and the rest are values of variables x_i^s (the default is that only the values of the first two such variables are displayed). After this string the following question will appear:

continue? reply "space",step,dir,var,estim,go,yes or no   •

This gives the user the opportunity to continue without any change, to alter the frequency of communication, to change the step size or step direction parameters, to display variables other than the first two, to stop at the current point and obtain a precise estimate of the value of the objective function, to switch from interactive to automatic mode, or to terminate the iterations and continue the solution with another algorithm. We shall now describe all of these options in some detail.

"space" If the user hits the space bar nothing will change and the program will perform another 10 iterations. The information about the process is displayed after each iteration; after the 10-th iteration the user is once again given the opportunity to make changes (the question "continue? reply "space",step..." appears).

step This means that the user wants to change the step size parameters (but not the step size regulation itself) and all the related questions will be repeated. Default or previous values of the step parameters will appear on the screen together with the questions.

dir This means that the user wants to change the step direction parameters (but not the way in which the step direction is chosen) and the questions concerned with this will be repeated. Default or previous values of the direction parameters will appear on the screen together with the questions.

var In this case the quantity and/or the selection of variables displayed on the screen may be changed. The following questions will appear:

number of printed variables? *(number)* ∗

i.e., if the user wants to print out the values of four variables rather than the default two, he answers *4*.

printed variables? *(number, number,....)* ∗

Here the user specifies which particular variables he wants displayed by giving the numbers of the chosen variables separated by commas. Questions concerning the frequency of communication will also appear here (see description of response *yes* below).

estim In this case the program will stop at the current point and estimate the value of the objective function. The following questions will appear:

number of observations? *(number)* ∗

i.e., the number of observations to be made, and

message frequency? *(number)* ∗

i.e., the number of observations after which the current estimate is displayed. The user is also asked for the point at which the estimate should be made:

what point? reply current, new or exit *current/new/exit* ∗

If the answer is *new* the program asks the question:

where to find new point? reply screen or file *screen/file* ∗

If the user wants to enter the new point from the keyboard he should reply *screen* (or *s*). He should then type the desired point on a new line, separating the components by commas. If, however, the new point is stored in some file the response should be *file* (or *f*) and the user is then asked

file number? *(number)* ∗

The answer is obviously the number of the file containing the new point. This new point is taken as the starting point for future iterations if the user answers *yes* to the following question:

replace current point by new? reply yes or no *yes/no* ∗

which appears when the estimation of the objective function at the new point has been completed. This facility makes it possible to exchange the current point for an arbitrary point chosen by the user and also to make precise estimations at arbitrary points. Finally, if the answer to the question "what point? reply current, new or exit" is *exit* the estimation procedure will end and the iterations will continue. **go** This means that the user does not want to continue in the interactive mode; he wants the process to proceed automatically. This is useful once the algorithm parameters have been established and also in the case when one iteration is very time-consuming. The user is then asked

number of iterations? *(number)* ∗

i.e., the total number of iterations before termination. After this the program has no more communication with the user and terminates after the specified number of iterations. *yes* In this case the frequency of communication can be changed. The following questions appear:

output frequency?   *(number)*   *

This is the number of iterations after which information about the process is displayed on the screen (the default value is 1, i.e., a string of information is printed after every iteration).

dialogue frequency?   *(number)*   *

This is the number of process information strings (see above) printed before the user is asked the question "continue? reply space,step,dir,var,estim,yes or no". The default is 10, i.e., the user is given ten strings of information about the process before he is asked whether he wishes to make any changes. *no* This means that the user wishes either to terminate the iterations or change the method. The program asks:

continue? reply "space",yes or no   *"space"/yes/no*   *

Here hitting the space bar means that the user wishes to proceed with the iterations using the same method, maybe returning to the initial point (see below); *yes* means he wishes to change the way in which the step size and/or step direction are chosen (the program will ask further questions about this—see below); *no* means that he wishes to terminate the iterations completely (some self-explanatory questions will then appear). If the user answers *"space"* or *yes* the program will ask

return to initial values? reply yes or no   *yes/no*   *

and the user should give the appropriate response.

The very first appearance of the question "continue? reply space,step,dir, var,estim,yes or no" is followed by the question

least value of step size? *(number)*   *

The answer is the least permissible value of the step size. If the current step size is less than this value then the iterations will terminate. In other cases the process terminates after 10,000 iterations with a question about whether to continue or not.

Everything that appears on the screen during the interactive dialogue automatically also goes to file number 15 (fort.15 in UNIX). This makes it possible to study the process after it has terminated.

This section provides some idea of the capabilities of the package of stochastic optimization subroutines STO available at IIASA. The implementation described here is the first version, and development of the second continues. This revised version will include methods for solving certain special problems, in particular problems with recourse, and new methods for step size regulation will be introduced.

16.5 Some Numerical Experiments

16.5.1 Facility location problem

We first consider a simple model of facility location in a stochastic environment. Suppose that we have to determine the amounts x_i of materials, facilities, etc., required at points $i = \overline{1,n}$ in order to meet a demand ω_i. The demand is random, and all we know is its distribution function $P\{\omega_1 \leq \overline{\omega}_1, \ldots, \omega_n \leq \overline{\omega}_n\} = H(\overline{\omega})$. The actual value $\omega = (\omega_1, \ldots, \omega_n)$ of the demand is not known when the decision concerning the $x = (x_1, \ldots, x_n)$ has to be made. Assume that we have made a decision x about the distribution of facilities and then found that the actual demand is ω. We have to pay for both oversupply and shortfalls, i.e., the penalty charged at the i-th location is $\psi_1^i(\omega_i - x_i)$ if $\omega_i \geq x_i$ and $\psi_2^i(x_i - \omega_i)$ if $\omega_i < x_i$, where the functions $\psi_1^i(y)$ and $\psi_2^i(y)$ are nondecreasing. In the simplest case these functions are linear and the total penalty for fixed x and ω is $\sum_{i=1}^n \max\{a_i(\omega_i - x_i), b_i(x_i - \omega_i)\}$, where $a_i \geq 0$, $b_i \geq 0$, $i = \overline{1,n}$. In most cases it is reasonable to select x in such a way that the average penalty is at a minimum, i.e., to minimize the following function:

$$F(x) = \mathrm{E}_\omega f(x,\omega) = \mathrm{E}_\omega \sum_{i=1}^n \max\{a_i(\omega_i - x_i), b_i(x_i - \omega_i)\} =$$

$$\int \sum_{i=1}^n \max\{a_i(\omega_i - x_i), b_i(x_i - \omega_i)\} \mathrm{d}H(\omega). \tag{16.27}$$

This approach can easily be generalized to deal with more complex facility location models (see [1],[15],[22]). The numerical experiment presented here is basically an application of the facility location model described above to the problem of high school location in Turin, Italy (see [15],[22]). In this example n is the number of districts in the city (16.23 in this case), ω_i is the number of students who want to attend schools in district i, and x_i is the capacity of schools in district i. It is assumed that a student living in district i will choose a school in district j with probability p_{ij}, where

$$p_{ij} = \frac{e^{-\lambda c_{ij}}}{\sum_{j=1}^n e^{-\lambda c_{ij}}}$$

and c_{ij} is proportional to the distance between districts i and j. The values of c_{ij} are taken from [15], as are the values of the parameters ($\lambda = 0.15$ and $a_i = b_i = 1.0$ for all i). The demand ω_i is assessed by assigning individual students to a school in a particular district on the basis of probabilities p_{ij}, thus simulating the student's choice of school. In order to reduce the amount of computation the number of students was scaled. Table 16.1 gives the resulting solution (the number of places that should be provided), together with the total number of students actually attending schools in each district.

Table 16.1 The solution of the problem of high school location in Turin, Italy [15],[22]

District	1	2	3	4	5	6	7	8
Number of students	14.0	13.0	15.0	11.0	14.0	14.0	11.0	12.0
Solution	17.9	13.0	18.9	19.0	16.0	13.9	10.8	10.2
District	9	10	11	12	13	14	15	16
Number of students	12.0	23.0	26.0	23.0	22.0	18.0	14.0	15.0
Solution	13.0	19.8	26.0	20.0	16.6	15.7	14.0	13.0
District	17	18	19	20	21	22	23	
Number of students	14.0	14.0	10.0	10.0	5.0	8.0	21.0	
Solution	13.0	15.7	10.0	10.1	5.0	10.3	17.0	

All real data was divided by a scaling factor of 100. We also have the constraint $\sum_{i=1}^{n} x_i = M$, where M is the total number of students in the city divided by 100 (339 in this case). Once ω has been obtained it is quite easy to calculate a stochastic quasigradient. We can use vector $\xi^s = (\xi_1^s, \xi_2^s, \ldots, \xi_n^s)$ in method (16.2), where

$$\xi_i^s = \begin{cases} -a_i & \text{if } \omega_i^s \geq x_i^s \\ b_i & \text{if } \omega_i^s < x_i^s \end{cases}.$$

Here ω_i^s is the demand in district i (calculated by simulating the students' behavior) at iteration number s, and x_i^s is the i-th component of the solution at this iteration. The initial point was obtained by assuming that each student goes to school in his native district. After extensive averaging, the value of the objective function at this point was found to be 74.2—the optimal value is 55.9. We shall first present results obtained using the interactive option for changing the step size, i.e., results obtained by giving the answer *2* to the question "step size regulation?" The step direction was specified as 11111, i.e, a direct observation of a stochastic quasigradient is available, this observation is made at the current point, the approximation step is fixed, there is no averaging, and no previous information is used. The size of the memory available for calculating the performance measure (16.22) was set at 10. Table 16.2 reproduces the information displayed on the monitor during the first 30 iterations.

Table 16.2 Information displayed during the first 30 iterations (facility location problem, interactive step size regulation)

Iter. no.	Performance measure	Estimate $\widehat{F}^s$ of $F(x^s)$	Observation of $f(x^s, \omega^s)$	Step size	x_4	x_{23}
2	-0.335	73.696	75.304	1.000	13.435	19.435
3	-0.172	73.739	73.826	1.000	14.565	18.565
4	-0.029	72.500	68.783	1.000	15.783	17.783
5	0.200	68.243	51.217	1.000	16.826	16.826
6	0.201	67.275	62.435	1.000	17.522	17.522
7	0.196	66.435	61.391	1.000	18.391	16.391
8	0.172	66.326	65.565	1.000	19.435	15.435
9	0.108	67.952	80.957	1.000	18.391	16.391
10	0.082	68.539	73.826	1.000	17.609	17.609
12	0.119	68.609	84.609	1.000	19.522	19.522
14	0.017	67.491	55.304	1.000	19.696	17.696
16	0.010	66.011	59.565	1.000	19.435	19.435
18	0.064	65.174	52.348	1.000	19.348	19.348
20	0.066	64.287	64.435	1.000	19.522	17.522
22	0.097	64.221	56.174	1.000	19.609	15.609
24	0.076	63.181	51.043	1.000	17.609	15.609
26	0.062	63.271	60.870	1.000	19.870	15.870
28	0.025	63.221	64.696	1.000	19.696	17.696
30	0.036	63.032	42.522	1.000	17.696	17.696

The observations of $f(x^s, \omega^s)$ given in Table 16.1 do not provide any clues as to whether the algorithm is improving the values of the objective function $F(x^s)$ or not. At first sight these observations appear to oscillate randomly between 40 and 80. By contrast, the estimates $\widehat{F}^s$ of the function $F(x^s)$ display much more stable behavior, generally decreasing during the first 22 iterations from 73 to 64 and then stabilizing around the values 63–64 with some small oscillations. Looking at the behavior of the two selected variables, we see that their values show a steady increase or decrease until iteration number 8 for x_4 and iteration number 5 for x_{23}. In later iterations both variables exhibit oscillatory behavior. The value of the performance measure during the first 4 iterations is negative, due to the instability of the initial estimates. It then begins to increase and reaches approximately 0.2, reflecting the regular behavior of the estimate $\widehat{F}_s$. After this it decreases in an oscillatory fashion to the range 0.03–0.06. All of this indicates that it is time to decrease the step size.

Table 16.3 Information displayed during iterations 31–59 (facility location problem, interactive step size regulation)

Iter. no.	Performance measure	Estimate $\hat{F}^s$ of $F(x^s)$	Observation of $f(x^s, \omega^s)$	Step size	x_4	x_{23}
31	0.045	62.379	42.783	0.500	18.087	17.087
33	0.025	62.295	62.783	0.500	18.261	16.261
35	0.052	61.652	52.609	0.500	19.391	16.391
37	0.063	61.565	46.957	0.500	19.348	16.348
39	0.079	61.318	52.261	0.500	19.261	17.261
41	0.050	61.211	68.174	0.500	19.174	16.174
43	0.051	60.815	51.304	0.500	18.261	16.261
45	0.070	60.452	57.913	0.500	17.304	16.304
47	0.059	60.279	45.652	0.500	17.348	15.348
49	0.035	60.277	64.957	0.500	18.391	15.391
51	0.043	60.104	61.739	0.500	18.652	14.652
53	0.017	60.133	64.696	0.500	18.565	14.565
55	0.017	60.240	67.043	0.500	18.652	14.652
57	−0.030	60.819	65.565	0.500	18.565	15.565
59	−0.052	61.189	85.391	0.500	18.609	16.609

After changing the step size, the estimates of $F(x^s)$ decreased steadily during iterations 31–51, and then started to increase during iterations 52–59 (see Table 16.3). The performance measure first increased, reaching a level of 0.05–0.07 between iterations 35 and 47 before dropping back to negative values. It is necessary to decrease the step size once again.

Table 16.4 Information displayed during iterations 62–80 (facility location problem, interactive step size regulation)

Iter. no.	Performance measure	Estimate $\hat{F}^s$ of $F(x^s)$	Observation of $f(x^s, \omega^s)$	Step size	x_4	x_{23}
62	−0.098	61.971	92.557	0.200	17.652	17.052
66	−0.067	61.684	46.713	0.200	18.104	17.504
70	0.013	61.353	61.026	0.200	18.226	16.826
74	0.087	61.167	55.739	0.200	17.861	16.461
78	0.061	60.832	58.104	0.200	18.296	16.896
80	0.020	61.001	78.557	0.200	18.348	17.348

We decided to stop after iteration number 80 (see Table 16.4) and estimate the value of the objective function at the current point. The average after the first 500 observations was 56.53, which shows that we are fairly close to the optimal solution. Note that this estimate is considerably lower than the value of $\hat{F}^s$ (61.0) given in the table. This is due to the fact that the estimate $\hat{F}^s$ is calculated from (16.18) including only one additional observation $f(x^s, \omega^s)$ per iteration, and it therefore includes observations made at early points which are clearly far from the optimum. Nevertheless, this estimate is still useful in determining the value of the step size because it reflects the general behavior of the algorithm. Subsequent iterations improved the value of the objective function only marginally (see Table 16.5).

Table 16.5 Information displayed during iterations 90–3070 (facility location problem, interactive step size regulation)

Iter. no.	Performance measure	Estimate $\widehat{F}^s$ of $F(x^s)$	Observation of $f(x^s, \omega^s)$	Step size	x_4	x_{23}
90	0.063	60.601	54.087	0.200	17.930	17.730
100	0.143	59.876	45.739	0.100	18.287	17.687
120	0.022	59.579	57.670	0.100	18.330	17.530
140	0.061	58.890	45.374	0.100	18.626	17.826
160	−0.011	59.161	56.278	0.100	19.226	17.626
180	0.319	58.761	44.744	0.020	19.379	17.299
200	0.008	58.608	49.144	0.020	19.237	17.277
300	0.317	57.847	43.322	0.020	18.946	17.146
400	−0.368	57.627	81.986	0.005	18.909	17.129
500	0.270	57.584	63.554	0.005	18.869	17.099
800	−0.830	57.012	58.455	0.001	18.967	17.017
1100	3.773	57.071	66.512	0.0003	18.980	17.000
1570	1.521	56.858	79.613	0.0001	18.983	16.998
2070	0.916	56.629	46.567	0.0001	18.975	16.998
2570	−0.874	56.603	71.741	0.0001	18.978	17.001
3070	0.118	56.425	55.729	0.0001	18.982	17.000

Our final estimate of the objective function was 56.0, which is close to the optimal solution.

The same results can be obtained by automatic regulation of the step size. In this case we give the answer *1* to the question "step size regulation?", i.e., adaptive automatic step size regulation (16.24) using function estimate (16.18). We also set

initial step size	1.0
multiplier	0.7
frequency of step size change	15
lower bound on function decrease	0.02
size of memory	15

(see the description of the step size parameters in Section 16.4). The results are presented in Table 16.6.

Table 16.6. Information displayed during iterations 2–1200 (facility location problem, adaptive automatic step size regulation)

Iter. no.	Performance measure	Estimate $\widehat{F}^s$ of $F(x^s)$	Observation of $f(x^s, \omega^s)$	Step size	x_4	x_{23}
2	3.663	77.826	60.261	1.000	10.739	20.739
4	1.590	72.522	57.739	1.000	12.739	18.739
6	1.091	69.232	54.522	1.000	14.826	20.826
8	0.892	65.457	48.174	1.000	14.826	18.826
10	0.736	63.609	56.087	1.000	16.913	18.913
15	0.453	64.980	65.652	1.000	18.130	18.130
20	0.071	64.435	58.522	1.000	17.522	19.522
30	0.023	64.304	49.652	1.000	19.783	15.783
50	0.007	61.951	49.391	1.000	17.609	15.609
70	0.017	61.563	68.696	0.790	15.104	15.104
100	0.017	60.593	90.195	0.490	18.665	18.245
150	0.017	60.246	65.349	0.240	20.166	16.855
200	0.054	59.526	48.282	0.082	19.657	17.223
300	0.036	59.277	50.012	0.028	19.131	17.248
400	−0.035	58.495	58.695	0.020	19.074	16.999
500	−0.100	58.440	63.486	0.010	18.903	16.986
600	0.143	57.936	36.450	0.007	18.913	16.984
700	0.446	57.683	47.760	0.003	18.955	16.998
800	−0.024	57.387	43.263	0.003	18.945	16.995
900	0.412	57.116	50.086	0.002	18.975	16.958
1000	0.430	57.006	43.503	0.001	18.947	16.969
1100	−0.063	56.726	76.801	0.001	18.969	16.997
1200	0.165	56.623	65.457	0.001	18.989	16.994

The value of the objective function at the final point (average of 4000 observations) is 56.2, which is close to the optimal value. The behavior of the algorithm was virtually the same as in the interactive case: quite a reasonable approximation of the optimal solution was obtained after 100–150 iterations, with little improvement being observed thereafter.

16.5.2 Control of water resources

This example is taken from work by A. Prékopa and T. Szantai. An extended description of the problem together with a solution obtained by reduction to a special type of nonlinear programming problem is given in [23]. Here we shall show how the problem can be solved using stochastic quasigradient methods. The basic aim is to control the level of water in Lake Balaton (a large, shallow lake in western Hungary). A certain volume of water ω_i flows into the lake from rivers, rainfall, etc., in time period i. This inflow varies randomly from one period to another, but it is possible to derive its probabilistic distribution from previous observations. The control parameter is the amount x_i of water released from the lake into the River Danube in each time period; the objective is to maximize the probability of the water level lying within specified bounds. It turns out that a reasonable control policy can be determined by considering only two consecutive periods of time, which in this example are measured in months. After appropriate transformations we arrive at the following problem (for details see [23]):

$$\max_{x_1, x_2} P\{Z(x_1, x_2)\}$$

$$0 \leq x_1 \leq R$$

$$0 \leq x_2 \leq R,$$

where the set $Z(x_1, x_2)$ is defined as follows:

$$Z(x_1, x_2) = \{(\omega_1, \omega_2) : a_1 \leq \omega_1 - x_1 \leq b_1, a_2 \leq \omega_2 - x_1 - x_2 \leq b_2\}.$$

Here a_i, b_i are respectively the lower and upper bounds on the "generalized" water level: in this particular example we took $a_1 = a_2 = -205$, $b_1 = b_2 = 95$, $R = 200$. The random water inputs ω_1 and ω_2 have a joint normal distribution $H(\omega_1, \omega_2)$ with expectations $E(\omega_1) = -28.07$, $E(\omega_2) = -59.43$ and covariance matrix

$$C = \begin{pmatrix} 3636.12 & 4660.51 \\ 4660.51 & 10121.36 \end{pmatrix}.$$

Let $\chi(x_1, x_2, \omega_1, \omega_2)$ denote the indicator function of the set $Z(x_1, x_2)$, i.e.,

$$\chi(x_1, x_2, \omega_1, \omega_2) = \begin{cases} 1 & \text{if } (\omega_1, \omega_2) \in Z(x_1, x_2) \\ 0 & \text{otherwise} \end{cases}.$$

The problem then becomes

$$\max_{x \in X} \int \chi(x_1, x_2, \omega_1, \omega_2) dH(\omega_1, \omega_2)$$

and can be solved using stochastic quasigradient methods. We took $(95,95)$ as the initial point; the value of the objective function at this point was 0.32. According to [23], the optimal solution is $(2,0)$, with an objective function value of 0.857. We decided to solve the problem using a finite-difference approximation of a stochastic quasigradient. Below we demonstrate how our interactive software package STO may be used to solve this problem, specifying interactive step size regulation (option 2) and step direction 21124, (i.e., taking a central finite-difference approximation of the gradient, calculating the step direction at the current point, with a fixed approximation step, a number of samples greater than 1, no previous information, and such that the step direction vector has unit norm).

The parameters were set at the following values:

step in finite difference approximation	10.0
number of samples	5
value of step size	10.0
size of memory	20

The results are given in Table 16.7.

Table 16.7 Information displayed during iterations 1–110 (water management problem, interactive step size regulation)

Iter. no.	Performance measure	Estimate $\widehat{F}^s$ of $F(x^s)$	Observation of $f(x^s, \omega^s)$	Step size	x_4	x_{23}
1	0.	0.	0.	10.000	102.071	102.071
2	1.000	0.	0.	10.000	102.071	102.071
4	0.025	0.250	1.000	10.000	106.543	93.127
6	0.011	0.333	0.	10.000	113.614	110.198
8	0.007	0.375	0.	10.000	106.543	113.127
10	0.006	0.400	0.	10.000	106.543	93.127
15	0.003	0.333	0.	10.000	83.944	101.254
20	0.002	0.350	0.	10.000	68.397	90.630
30	0.001	0.467	0.	10.000	18.240	93.229
40	0.000	0.475	1.000	10.000	48.678	63.727
50	0.000	0.500	1.000	10.000	41.277	29.097
60	0.000	0.567	1.000	10.000	0.	43.004
70	0.000	0.571	1.000	10.000	1.056	30.405
80	0.000	0.588	1.000	10.000	1.386	14.142
90	0.000	0.600	1.000	10.000	0.	24.142
100	0.000	0.610	1.000	10.000	7.071	20.000
110	0.000	0.609	1.000	10.000	10.000	0.

After iteration 110 we stopped and estimated the value of the function at the current point on the basis of 4000 observations—we obtained a value of 0.843, which is close to the optimal value. Subsequent iterations improved the value of the objective function only marginally (see Table 16.8).

Table 16.8 Information displayed during iterations 120–8090 (water management problem, interactive step size regulation)

Iter. no.	Performance measure	Estimate $\hat{F}^s$ of $F(x^s)$	Observation of $f(x^s, \omega^s)$	Step size	x_4	x_{23}
120	0.000	0.625	1.000	10.000	102.071	102.071
150	0.000	0.673	1.000	1.000	0.106	1.707
200	0.001	0.720	0.	1.000	2.707	6.309
390	0.005	0.792	1.000	0.100	3.071	7.835
590	−0.001	0.797	0.	0.100	1.787	8.110
1090	0.000	0.829	1.000	0.100	3.463	6.392
2090	0.000	0.845	1.000	0.100	0.383	5.538
3090	−0.005	0.852	0.	0.010	0.161	4.895
4090	−0.004	0.854	1.000	0.005	0.071	5.049
5090	0.004	0.856	1.000	0.005	0.064	4.955
6090	−0.002	0.855	1.000	0.005	0.106	4.980
7090	0.007	0.856	1.000	0.001	0.016	4.970
8090	0.005	0.855	1.000	0.001	0.020	4.985

After iteration 200 we changed the step in the finite-difference approximation to 1.0. The value of the objective function at the final point was 0.85, i.e., we had reached the optimal value. However, the values of the controls were far from the solution due to the flatness of the function around the optimum.

16.5.3 Determining the parameters in a closed loop control law for stochastic dynamical systems with delay

We have so far considered only static optimization problems. However, all of the techniques described above can also be applied to many classes of dynamical stochastic optimization problems. The example that we shall consider was suggested by A. Wierzbicki and is the problem of finding the optimal control parameters in a closed loop control law for a linear dynamical system disturbed by random noise. The state equations include response delay and may be written as follows:

$$z_{t+1} = a_t z_t + u_{t-k} + \omega_t, t = \overline{0, T}$$
$$z_0 = 1, u_{-i} = 0, i = \overline{0, k}, \tag{16.28}$$

where t is a discrete time, z_t is the state of the dynamical system at time t, u_t is the value of the control at time t, and ω_t is the random noise at time t. In this particular example the ω_t were taken to be distributed uniformly over the interval $[-b, b]$ and such that ω_i and ω_j are uncorrelated for $i \neq j$. However, neither this particular type of distribution nor these correlation properties are prerequisites for the use of the methods described in the preceding sections. The controls u_t were chosen according to the following closed loop control law:

$$u_t = x_1 \left(-z_t - x_2 \sum_{\tau=0}^{t} z_\tau \right), \tag{16.29}$$

where the decision parameters are $x_1 \geq 0$ and $x_2 \geq 0$.

The objective is to minimize the deviation of the state of the system from zero. We may therefore state the problem as follows: minimize the objective function

$$F(x_1, x_2) = \mathrm{E}_\omega \sum_{t=1}^{T} z_t^2 \qquad (16.30)$$

with respect to the control law parameters x_1 and x_2, subject to constraints (16.28) and (16.29) and nonnegativity constraints on x_1, x_2. We solved the problem with the following parameter values: time horizon $T = 100$, delay $k = 5$, state equation coefficient $a = 0.9$, bounds for random noise $b = 0.1$. With these values the optimal control parameters are $x_1 = 0.1$, $x_2 = 0$; the value of the objective function obtained after 10,000 observations was 4.52. It was discovered during preliminary runs that for $x_1 \geq 0.3$, $x_2 \geq 0.1$ the system becomes unstable and therefore these values were taken as upper bounds for the variables.

We set the initial point equal to the upper bounds $x_1^0 = 0.3$, $x_2^0 = 0.1$; the value of the objective function at this point (based on 3000 observations) was 422.56. We chose automatic step size regulation (option 1), i.e., the step size changes are based upon performance function (16.22). The step direction was specified as 71114, i.e., taking a forward finite-difference approximation of the gradient of the random objective function $f(x, \omega)$ with all observations of the function needed for one gradient evaluation made at the same value of the noise; with a fixed finite difference step and the finite-difference evaluation performed at the current point; without averaging; using no previous information and normalizing the resulting step direction. The parameters of the algorithm were as follows:

step in finite difference approximation	0.0001
initial step size	0.1
multiplier (for diminishing the step size)	0.85
frequency of step size change (actually the frequency with which the step size is reviewed)	15
lower bound on function decrease (the lowest value of performance function (16.22) which does not lead to a decrease in the step size)	0.09
size of memory (for evaluating (16.22))	15
least value of step size (stopping criterion)	0.000001

The results of the calculations are given in Table 16.9.

Table 16.9 Information displayed during iterations 1–120 (control law problem, automatic step size regulation)

Iter. no.	Performance measure	Estimate $\widehat{F}^s$ of $F(x^s)$	Observation of $f(x^s, \omega^s)$	Step size	x_4	x_{23}
1	0.	8.141	8.141	0.100	0.232	0.027
2	18.570	6.284	4.427	0.100	0.149	0.
3	12.231	5.695	4.517	0.100	0.054	0.
4	7.093	6.013	6.968	0.100	0.097	0.
5	6.727	5.450	3.199	0.100	0.075	0.
10	3.428	5.056	4.084	0.100	0.073	0.
15	2.416	4.759	4.254	0.100	0.103	0.099
20	0.421	4.733	4.214	0.100	0.029	0.
30	0.119	4.651	5.326	0.100	0.052	0.
40	0.058	4.615	4.896	0.100	0.050	0.
50	−0.012	4.631	5.143	0.085	0.071	0.
70	0.001	4.668	5.131	0.072	0.112	0.
90	0.005	4.665	4.943	0.061	0.076	0.059
100	0.042	4.621	3.481	0.052	0.076	0.
120	0.033	4.601	4.872	0.044	0.094	0.

Table 16.10. Information displayed during iterations 150–1500 (control law problem, automatic step size regulation)

Iter. no.	Performance measure	Estimate $\widehat{F}^s$ of $F(x^s)$	Observation of $f(x^s, \omega^s)$	Step size	x_4	x_{23}
150	0.044	4.517	3.776	0.032	0.102	0.000
170	0.084	4.485	4.234	0.023	0.101	0.
200	−0.015	4.473	5.224	0.017	0.101	0.
240	0.087	4.473	4.413	0.012	0.087	0.009
300	−0.155	4.503	4.478	0.006	0.095	0.
340	0.036	4.491	4.958	0.005	0.090	0.
400	0.089	4.501	4.973	0.002	0.093	0.000
440	−0.299	4.512	4.544	0.001	0.092	0.003
500	−0.131	4.512	3.571	0.001	0.098	0.
540	−0.416	4.502	4.437	0.001	0.101	0.
600	0.225	4.515	4.789	0.001	0.102	0.
640	0.710	4.508	3.704	0.001	0.101	0.
700	0.046	4.501	4.120	0.001	0.101	0.
800	0.079	4.517	4.633	0.000	0.100	0.000
900	−1.183	4.533	5.070	0.000	0.099	0.000
1000	2.700	4.534	4.860	0.000	0.099	0.000
1500	29.344	4.504	4.621	0.000	0.099	0.000

We stopped after iteration 120 to estimate the value of the objective function, which was calculated to be 4.54 after 3000 observations and is fairly close to the optimal value. Subsequent iterations improved the solution only marginally (see Table 16.10).

This example once again demonstrates the characteristic behavior of stochastic optimization algorithms: the neighborhood of the optimal solution is reached reasonably rapidly; oscillations then occur in this neighborhood and the current approximation to the optimal solution improves slowly.

The nature of stochastic quasigradient algorithms allows easy extension of model (16.28)–(16.30) to multivariable and nonlinear systems.

References

1. Yu. Ermoliev, *Methods of Stochastic Programming* (in Russian), Nauka, Moscow, 1976.

2. Yu. Ermoliev, "Stochastic quasigradient methods and their applications to systems optimization", *Stochastics*, **9**(1983), 1–36.

3. R. J-B. Wets, "Stochastic programming: solution techniques and approximation schemes", in *Mathematical Programming. The State of the Art*, Springer-Verlag, 1983.

4. L. Nazareth and R. J-B. Wets, "Algorithms for stochastic programs: the case of nonstochastic tenders", Working Paper WP-83-5, International Institute for Applied Systems Analysis, Laxenburg, Austria, 1983.

5. P. Kall, *Stochastic Linear Programming*, Springer-Verlag, Berlin, 1976.

6. A. Prékopa, "On probabilistic constrained programming", in H. Kuhn (Ed.), *Proceedings of the Princeton Symposium on Mathematical Programming*, pp. 113–138. Princeton University Press, Princeton, 1970.

7. H. Robbins and S. Monroe, "A stochastic approximation method", *Ann. Math. Stat.* **22**(1951), 400–407.

8. J. Kiefer and J. Wolfowitz, "Stochastic approximation of the maximum of a regression function", *Ann. Math. Stat.*, **23**(1952), 462–466.

9. B. Fox and H. Niedereiter, Lectures on quasi-Monte-Carlo methods given at IIASA, 1983.

10. H. Kushner, Asymptotic behavior of stochastic approximation and large deviations. Lecture given at IIASA, 1983.

11. Yu. Ermoliev and E. Nurminski, "Limit extremum problems", *Kibernetika* **4**(1973), 130–132.

12. A. Gupal, *Stochastic Methods for Solution of Nonsmooth Optimization Problems*, Naukova Dumka, Kiev, 1979.

13. A. Gupal and A. Basenov, "Stochastic analogue of conjugate gradient methods", *Kibernetika* **1**(1972), 79–83.

14. H. Kesten, "Accelerated stochastic approximation", *Ann. Math. Statist.* **29**(1958), 41–59.

15. Yu. Ermoliev, G. Leonardi and J. Vira, "The stochastic quasigradient method applied to a facility location problem", Working Paper WP-81-14, International Institute for Applied Systems Analysis, Laxenburg, Austria, 1981.

16. G. Pflug, "On the determination of the step size in stochastic quasigradient methods", Collaborative Paper CP-83-25, International Institute for Applied Systems Analysis, Laxenburg, Austria, 1983.

17. V. Fabian, "Stochastic approximation method", *Czechoslovakian Mathematical Journal* **10**(1960), 191–200.

18. N. Chepurnoi, Dissertation, V. Glushkov Institute of Cybernetics, Kiev, 1982.

19. S. Uriasiev, Dissertation, V. Glushkov Institute of Cybernetics, Kiev, 1982.

20. A. Ruszczynski and W. Syski, "A method of aggregate stochastic subgradients with on-line stepsize rules for convex stochastic optimization problems", *Mathematical Programming Study*, forthcoming.

21. P. Gill, W. Murray, M. Saunders, and M. Wright, "User's guide for SOL/-QPSOL: a FORTRAN package for quadratic programming", Technical Report SOL 83-7, Systems Optimization Laboratory, Stanford University, 1983.

22. Yu. Ermoliev and G. Leonardi, "Some proposals for stochastic facility location models", Working Paper WP-80-176, International Institute for Applied Systems Analysis, Laxenburg, Austria, 1980.

23. A. Prékopa and T. Szantai, "On optimal regulation of a storage level with application to the water level regulation of a lake", *European Journal of Operations Research*, **3**(1979), 175–189.

CHAPTER 17

STEPSIZE RULES, STOPPING TIMES AND THEIR IMPLEMENTATION IN STOCHASTIC QUASIGRADIENT ALGORITHMS

G.Ch. Pflug

1. Introduction

We consider the constrained optimization problem

$$f(x) = E_P(q(x, \cdot)) = min! \qquad x \in S \qquad (17.1)$$

where S is a closed, convex set of constraints $S \subseteq \mathbf{R}^k$. The symbol E_P or briefly E denotes the expectation with respect to the probability measure P which is defined on some measurable space $(\Omega, \mathcal{A})$

$$E_P(q(x, \xi)) = \int q(x, \xi) dP(\xi). \qquad (17.2)$$

There are, in principle, two different ways of attacking the problem (17.1):

(a) *Reduction to deterministic optimization*

The easiest situation arises if the integral (17.2) may be calculated analytically. In that case the problem (17.1) reduces to a deterministic constrained optimization problem. But even if there is no closed-form analytical representation of (17.2) the integral may be approximated with arbitrary accuracy. This may be done by approximating the probability measure P by a sequence $\tilde{P}_n$ such that $\tilde{P}_n \to P$ (in an appropriate sense) to guarantee that

$$\int q(x, \xi) d\tilde{P}_n(\xi) \to \int q(x, \xi) dP$$

and the first integrals are easy to calculate. Very often discrete measures are used for $\tilde{P}_n$. Another possibility is to calculate $E_P(q(x, \xi))$, directly by Monte Carlo or quasi Monte Carlo methods.

(b) *Stochastic quasigradient method*

For this group of methods it is not necessary to get good approximations of $E_P(q(x, \cdot))$–stochastic estimates suffice. If ξ is a random variable (random number or random vector) with distribution P then

$$\mathbf{Q}_x = q(x, \xi)$$

is a random variable with expectation $E(\mathbf{Q}_x) = f(x)$. A statistical approximation of the gradient $\nabla f(x)$ of the objective function f may be obtained by considering a difference approximation

$$(Y_x)_i = \frac{1}{2h}[q(x + he_i, \xi) - q(x - he_i, \xi)]$$

where $(Y)_i$ denotes the i-th component of the vector Y and e_i are the i-th unit vectors. Then, if $f(x)$ is twice differentiable

$$E_P(Y_x) = \nabla f(x) + 0(h^2).$$

Such a random vector Y_x is called a *stochastic quasigradient*, giving the method its name. Only the stochastic quasigradient (SQG) approach will be considered in this paper.

Sometimes there are even unbiased estimates of $\nabla f(x)$ available. This is e.g. the case if

$$x \mapsto q(x, \xi)$$

is differentiable in the $L^1(P)$-sense. This means that there is a vector of L^1-functions $\nabla g(x, \xi)$ such that

$$\int |q(x_1, \xi) - q(x_2, \xi) - (x_1 - x_2)'\nabla q(x, \xi)| dP(\xi) = o(\|x_1 - x_2\|). \quad (17.3)$$

In that case evidently

$$E(\nabla q(x, \xi)) = \nabla f(x).$$

It is important to notice that the following chain of implications holds

$$[q(x, \xi) \text{ differentiable for every } \xi \text{ and } L_1\text{-dominated}] \Longrightarrow$$
$$[q(x, \cdot) \text{ differentiable in the } L_1\text{-sense}] \Longrightarrow$$
$$[f(x) \text{ differentiable}]$$

The converse implications are not true as can be seen from the following examples.

Example (a).

Let

$$q(x, \xi) = \begin{cases} a(x - \xi) & \text{if } x \geq \xi \\ b(\xi - x) & \text{if } x < \xi \end{cases}$$

Such a specification is often encountered in economic applications where a denotes the surplus costs and b the shortage costs of a random demand ξ, x being the offer. $x \mapsto q(x, \xi)$ is not a differentiable function. However, if ξ is integrable then

$$\nabla q(x, \xi) = \begin{cases} a & \text{if } x \geq \xi \\ -b & \text{if } x < \xi \end{cases}$$

is the L^1-derivative of $q(\cdot, \cdot)$ as one can see immediately.

Example (b).

Let

$$q(x, \xi) = \begin{cases} 0 & \text{if } x + \xi \in A \\ 1 & \text{if } x + \xi \notin A \end{cases}$$

where A is some predetermined region in $\mathbb{R}^k$. Such problems arise in optimal control, if the probability that the control x plus noise ξ lies in the set A should be maximized. This function $q(x, \xi)$ is not L^1-differentiable although the function

$$E_P(q(x, \cdot))$$

is differentiable if P has a density w.r.t. Lebesgue measure.

A similar notion holds for subdifferentiable functions: A $\mathbb{R}^k$-valued random variable Y_x is called *stochastic subgradient* if

$$E_P(Y_x) \in \partial f(x)$$

This is again a weaker statement than the pointwise subdifferentiability of $x \mapsto q(x, \xi)$.

What concerns the smoothness properties of our problem (17.1) we may distinguish two cases:
(a) the functions $q(x, \cdot)$ are L^1-(sub)differentiable
(b) the functions $q(x, \cdot)$ are not L^1-(sub)differentiable but the expectations $f(x)$ are.

The stochastic quasigradient method uses a recursively defined stochastic sequence X_n to approximate the solution of (17.1):

$$X_{n+1} = \Pi_S(X_n - \rho_n Y_{X_n}) \tag{17.4}$$

where in case (a)

$$Y_x = \nabla q(x, \xi_n) \tag{17.4a}$$

and in case (b)

$$Y_x = \frac{q(x + h_n, \xi_n) - q(x - h_n, \xi_n)}{2h_n}. \tag{17.4b}$$

Here $\{\xi_n\}$ is a sequence of i.i.d. random variables with distribution P and Π_S denotes the projection onto the closed convex set S. The nonnegative constants ρ_n represent the stepsizes.

The use of algorithms of the form (17.4a) goes back to a pioneering paper by Robbins and Monro [17]. Kiefer and Wolfowitz studied for the first time stochastic minimization problems with difference approximations for the gradients. It is however important to notice that the iterative process (17.4a) is not a Kiefer-Wolfowitz process. This is so because we have used the same random element ξ_n two times in the definition of Y_x. Consequently, under some mild assumptions the variance of Y_x satisfies $\mathrm{Var}(Y_x) = 0(h_n^{-1})$. The original approach of Kiefer and Wolfowitz uses the quantities

$$Y_x = \frac{q(x + h_n, \boldsymbol{\xi}_n^{(1)}) - q(x - h_n, \boldsymbol{\xi}_n^{(2)})}{2h_n}. \tag{17.4c}$$

with $\boldsymbol{\xi}_n^{(1)}$ independent of $\boldsymbol{\xi}_n^{(2)}$. In that case $\mathrm{Var}(Y_x) = 0(h_n^{-2})$ only. If the randomness comes from generated random numbers then the use of (17.4b) guarantees a certain variance reduction which is impossible in the case when the randomness stems from measurement errors which was the situation considered by Kiefer and Wolfowitz.

In the following we shall restrict ourselves to consider the case (17.4a). The reason for doing so is that the behavior of (17.4b) depends much on the smoothness properties of $q(\cdot, \cdot)$. If this function is L^2-differentiable, then the variance of Y_x is bounded as in the case of (17.4a). Otherwise the variance of Y_x given by (17.4b) may increase with decreasing h_n. But although the convergence theorems for the case (b) procedure require that $h_n \to 0$ with increasing n it is reasonable to keep h_n away from zero in practical implementations. Otherwise numerical difficulties are encountered by forming the quotient in (17.4b). Thus for practical applications we may assume that the variance of Y_x is bounded anyway.

The convergence properties of the iterative sequence X_n given by (17.4) were studied by many authors. These properties include almost sure convergence of X_n to x^* the solution of (17.1). (Dvoretzky [3], Kushner and Clark [13], Ermoliev [4], Hiriart-Urruty [9] among others) rates of convergence (Schmetterer [20]), asymptotic laws (Blum [1], Fabian [7]) laws of iterated logarithms, etc.

We shall give here a simple but illustrative a.s. convergence result for random step sizes. Randomness does not mean here that a random line search is made but the stepsize ρ_n may depend on the information obtained up to the n-th step (i.e. an adaptive stepsize rule). Denote by x^* the solution of the problem (17.1) which is assumed to be unique. The σ-algebra generated by $\boldsymbol{\xi}_1, \boldsymbol{\xi}_2, \ldots, \boldsymbol{\xi}_{n-1}$ is denoted by $\mathcal{F}_n$. Moreover we shall assume that

(i) $\langle \nabla f(x), x - x^* \rangle \geq a \|x - x^*\|^2$
(ii) $\|\nabla f(x)\| \leq A + B\|x - x^*\|^2$
(iii) $\mathrm{Var}(Y_x) \leq C$
(iv) $\rho_n \geq 0, \rho_n$ is $\mathcal{F}_n -$ measurable

Theorem.
(i) Under the above conditions

$$\sum \rho_n = \infty \quad \text{a.s.} \qquad \sum \rho_n^2 < \infty \text{ a.s.}$$

implies that $X_n \to x^*$ a.s.

(ii) If $f(x)$ is convex and S is bounded then $\rho_n \to 0$ a.s. and $\sum \rho_n = \infty$ implies that $\overline{X}_n \to x^*$ a.s. where $\overline{X}_n$ is the weighted mean of the process

$$\overline{X}_n = \frac{\sum_{i=1}^n \rho_i X_i}{\sum_{i=1}^n \rho_i}$$

Miroszahmedov and Uryasev [14].

Outline of the proof. According to the assumptions

$$E(\|X_{n+1} - x^*\|^2|\mathcal{F}_n) \leq \|X_n - x^*\|^2 - 2\rho_n \langle \nabla f(X_n), X_n - x^* \rangle$$
$$+ \rho_n^2(A + B\|X_n - x^*\|^2) + \rho_n^2 C$$
$$= \|X_n - x^*\|^2(1 + \beta_n) - \eta_n + \mu_n \text{ (say)}$$

with $\beta_n, \eta_n, \mu_n \geq 0$ and the series $\sum \beta_n$ and $\sum \mu_n$ converge a.s. By a theorem due to Robbins and Sigmund [18] this implies

$$\|X_n - x^*\|^2 \text{ converges a.s. and}$$
$$\sum \rho_n \langle \nabla f(X_n), X_n - x^* \rangle < \infty \quad \text{a.s.} \tag{17.5}$$

Part (i) of the theorem follows, since (17.5) implies that

$$a \sum \rho_n \|X_n - x^*\|^2 < \infty \quad \text{a.s.}$$

which together with $\sum \rho_n = \infty$ and the a.s. convergence of $\|X_n - x^*\|^2$ gives

$$\|X_n - x^*\|^2 \to 0 \quad \text{a.s.}$$

In order to prove part (ii) of the theorem we introduce the notation $Y_n := Y_{X_n}$ and $Z_n := Y_n - E(Y_n)$.

By iterating the recursion we find that

$$0 \leq \|X_n - x^*\|^2 \leq \|X_0 - x^*\|^2 - 2\sum_{i=0}^n \rho_i \langle \nabla f(X_i), X_i - x^* \rangle$$
$$- 2\sum_{i=0}^n \rho_i \langle Z_i, x_i - x^* \rangle + \sum_{i=0}^n \rho_i^2 \|Y_i\|^2$$

Because of the convexity

$$\sum_{i=0}^n \rho_i \langle \nabla f(X_i), X_i - x^* \rangle \geq \sum_{i=0}^n \rho_i (f(X_i) - f(x^*))$$
$$\geq (f(\overline{X}_n) - f(x^*)) \sum_{i=1}^n \rho_i.$$

Hence

$$0 \leq f(\overline{X}_n) - f(x^*) \leq \frac{1}{2} \left(\sum_{i=1}^n \rho_i \right)^{-1} \left[\|X_0 - x^*\|^2 - 2 \sum_{i=0}^n \rho_i \langle Z_i, X_i - x^* \rangle \right.$$

$$\left. + \sum_{i=0}^n \rho_i^2 \|Y_i\|^2 \right]$$

(17.6)

It can easily be deduced from the assumptions that the right hand side of (17.6) converges to zero a.s. implying thus

$$\overline{X}_n \to x^* \text{ a.s.} \quad \square$$

The just proven theorem gives conditions for the stepsizes ρ_n which are so general that they cover a variety of cases. On the other hand, it does not tell us which stepsize rule is good, or even the best. All choices fulfilling $\sum \rho_n^2 < \infty$, $\sum \rho_n = \infty$ lead to a.s. convergence. A detailed study of such rules follows in the next section.

17.2   Stepsize rules and stopping times

Almost sure convergence results are only of limited importance for practical purposes. It is much more important to design a procedure which stops after a finite number of steps within a *neighborhood of the solution* x^* which has predetermined size. To put it more formally let $\| \cdot \|_D$ be a certain norm in $\mathbb{R}^k$ and α resp. ε two constants representing the desired confidence level and the size of a confidence region. An approximation procedure X_n is of practical use only in connection with a stopping time $\tau = \tau(\alpha, \varepsilon)$ such that X_τ, the process stopped at τ, satisfies

$$P_{x_0}\{\|X_\tau - x^*\|_D \leq \varepsilon\} \geq 1 - \alpha \qquad \forall x_0 \tag{17.7}$$

where P_{x_0} denotes the law of the process $\{X_n\}$ started at $X_0 = x_0$. Formula (17.7) is nothing else than the definition of a fixed width confidence region of level α.

Unfortunately exact level α confidence regions are difficult to obtain, even in the much simpler case of the sequential estimation of a mean value (see e.g. Chow and Robbins [2]). It is much easier to get *asymptotic level α confidence regions*. This is a family of stopping times $\{\tau_\varepsilon\}$ such that

$$\lim_{\varepsilon \to 0} P_{x_0}\{\|X_{\tau_\varepsilon} - x^*\|_D \leq \varepsilon\} \geq 1 - \alpha \qquad \forall x_0 \tag{17.8}$$

It may happen that the speed of convergence depends heavily on the starting value x_0. In that case the actual significance level of a confidence region may be arbitrarily low if the starting value was poorly chosen.

It is therefore useful to consider a uniform version of (17.8). In particular we call $\{\tau_\varepsilon\}$ a family of *uniform asymptotic level α confidence regions* if

$$\liminf_{\varepsilon \to 0} \inf_{x_0} P_{x_0}\{\|X_{\tau_\varepsilon} - x^*\|_D \leq \varepsilon\} \geq 1 - \alpha \qquad (17.9)$$

A stopping time fulfilling (17.9) is considered to be robust against bad influences of the starting value. In order to get error estimates some knowledge about the speed of convergence in (17.9) is of great help but usually difficult to obtain.

It is important to stress that stopping times must be seen in connection with stepsize rules. Typically a certain rule for determining the stepsizes leads to a certain asymptotic behavior of the process X_n which in turn is the basis for the definition of a stopping rule τ. On the other hand one may also define stepsizes on the basis of a sequence of increasing stopping times by changing the stepsizes (say by multiplication with 1/2) exactly at these times. Thus the interrelations between stepsizes and stopping times are rather close.

We shall now define some common stepsize rules and the pertaining stopping times. Recall that

(i) $\rho_n \mathcal{F}_n$-measurable $\qquad\qquad\qquad\qquad\qquad\qquad\qquad (17.10)$
(ii) $\sum \rho_n = \infty$ a.s.
(iii) $\sum \rho_n^2 < \infty$ a.s.

are the minimal conditions to guarantee convergence.

(a) *Deterministic stepsize rules (DSR)*

The simplest rule consists in taking $\{\rho_n\}$ as a sequence of nonrandom constants fulfilling (17.10), e.g.

$$\rho_n = \frac{\rho}{n^\beta} \frac{1}{2} < \beta \leq 1 \text{ or } \rho_n = \frac{\rho \log n}{n}.$$

The quickest rate of convergence is achieved by taking

$$\rho_n = \frac{\rho}{n} \qquad (17.11)$$

which is by far the most popular choice.

Many asymptotic results are known if the stepsizes are chosen according to (17.11): If the solution x^* is an interior point of S then

$$\sqrt{n}(X_n - x^*) \to N(0, \Sigma) \qquad (17.12)$$

where Σ is the solution of the matrix equation

$$\left(\rho A - \frac{I}{2}\right)\Sigma + \Sigma\left(\rho A' - \frac{I}{2}\right) = \rho^2 C \qquad (17.13)$$

with C being the covariance matrix of $q(x^*, \xi)$

$$C = \text{Cov}(q(x^*, \xi)) \qquad (17.14)$$

and A is the hessian of f at x^*

$$\nabla f(x) = A(x - x^*) + 0(\|x - x^*\|^2).\tag{17.15}$$

Here $\to$ denotes the convergence in distribution. Equation (17.13) may be made explicit for Σ either by writing

$$\Sigma = \rho^2 \int_0^\infty e^{u(\frac{I}{2} - \rho A)} C e^{u(\frac{I}{2} - \rho A')} du$$

or by introducing the vec operation (which transforms a matrix into a vector by putting the columns one above the other)

$$\text{vec}\,\Sigma = \left(I \otimes \left(\rho A - \frac{I}{2}\right) + \left(\rho A' - \frac{I}{2}\right) \otimes I\right)^{-1} \text{vec}\,C.$$

It is important to notice that the asymptotic distribution (17.12) is independent of the starting value x_0. There is even a much stronger result known. Consider the random function

$$Z_n(t) = \frac{[nt]}{n} X_{[nt]} \quad 0 \le t \le 1\tag{17.16}$$

($[x]$ denotes the integer part of x). The random process $Z_n(t)$ contain the whole information of the approximating sequence $X_1, X_2, \ldots, X_n$ up to time n. It may be shown that

$$Z_n(t) \to \int_{(0,1]} e^{(\ln u)(\rho A - I)} dW(tu)\tag{17.17}$$

Where W is a Gaussian process with statiionary independent increments in $\mathbb{R}^k$ processing the covariance matrix $\text{Cov}(W(1)) = C$ (see Walk [22]). Functional limit theorems of type (17.17)—sometimes called "invariance principles"—help very much to get a deeper insight into the pathwise behavior of the approximating process $\{X_n\}$. In particular large deviation results or laws of the iterated logarithm may be based on result (17.17).

Moreover a stopping time leading to asymptotic level α confidence regions may be derived from the asymptotic distribution (17.17). If A and C are known and the norm $\|\cdot\|_D$ is defined by $\|x\|_D = \sqrt{x'Dx}$ for a positive definite matrix D then

$$\tau_\varepsilon = \frac{\kappa_\alpha}{\varepsilon}\tag{17.18}$$

is a family of stopping times satisfying (17.8), i.e.

$$\lim_{\varepsilon \to 0} P\{\|X_{\tau_\varepsilon} - x^*\|_D \le \varepsilon\} \ge 1 - \alpha.$$

Here κ_α denotes the upper α-quantile of the distribution of

$$R = Z'DZ \text{ where } Z \sim N(0, \Sigma)\tag{17.19}$$

Unfortunately this distribution is tedious to calculate. There is however a good approximation by a Γ-distribution

$$R \sim \text{ approximately } \Gamma\left(\frac{\text{tr}^2(D\Sigma)}{2\,\text{tr}((D\Sigma)^2)}, \frac{2\,\text{tr}((D\Sigma)^2)}{\text{tr}(D\Sigma)}\right) \tag{17.20}$$

where $\Gamma(\alpha, \beta)$ has the density

$$\frac{x^{\alpha-1}\exp(-x/\beta)}{\beta^\alpha\Gamma(\alpha)} \qquad x > 0.$$

Approximation (17.20) is based on the comparison of the first two cumulants of the distributions (see Kendall and Stuart [10]). Notice that the Γ-distribution degenerates to a $\chi^2(\text{tr}(D\Sigma))$ distribution if $D\Sigma$ is idempotent.

If A and C are not known, they have to be estimated during the procedure. (A possible method of estimation is indicated in the last section.) Suppose that $\widehat{A}_n$ resp. $\widehat{C}_n$ are consistent estimates of A resp. C. Then $\widehat{\Sigma}_n$ given by

$$(\rho\widehat{A}_n - \frac{I}{2})\widehat{\Sigma}_n + \widehat{\Sigma}_n(\rho\widehat{A}'_n - \frac{I}{2}) = \rho^2\widehat{C}_n$$

consistently estimates Σ.

Let $\kappa_{\alpha,n}$ be the upper α-quantile of the Γ-distribution (17.20) where Σ is replaced by $\widehat{\Sigma}_n$. Then we may define the

(a') *deterministic stepsize stopping time (DST):*

$$\tau_\varepsilon = \inf\{n \,|\, \frac{\kappa_{\alpha,n}}{n} \le \varepsilon\} \tag{17.21}$$

By using the functional limit law (17.17) one may prove that τ_ε leads to an asymptotically unbiased confidence region, hence satisfies (17.8).

A quite similar result holds if the point of solution x^* lies on the boundary. Denote by K^* the tangent cone to S at the point x^* and suppose that H is the largest linear subspace contained in K^*. It may then be proved that the limit law of $\sqrt{n}(X_n - x^*)$ is again a normal distribution but concentrated on H (see Pflug [16]). Thus the constrained situation may be reduced to the unconstrained by considering only the projection of the Hessian matrix A and the covariance matrix C onto the subspace H. The situation is however different if $\dim H = 0$. In that case S is pointed at x^* and the asymptotics are different. This is however a rather unlikely case.

The big disadvantage of the rules (a) and (a') lies in the fact that the pertaining confidence regions are *not at all uniform* in the sense of definition (17.9). This is clear since everything was based on asymptotic formulas which do not reflect the influence of the starting value x_0. This is in fact the most important reason for making these rules such less competitive in practice.

The idea behind more elaborated stepsize rules is clear: Some information concerning the progress of the procedure should be gathered during the approximation process and should influence the actual stepsizes. Some possible ways of doing so are listed below.

(b) *The adaptive stepsize rule (ASR)*

This rule was formulated for the first time in Mirozahmedov and Uriasev [14]. Let Y_n denote the n-th stochastic gradient, i.e. $Y_n = Y_{X_n}$. The rule is to adapt ρ_n according to the inner product $\langle Y_n, Y_{n-1} \rangle$, i.e.

$$\rho_{n+1} = \rho_n \exp[a\rho_n \langle Y_{n+1}, Y_n \rangle - \delta\rho_n] \tag{17.22}$$

where a and δ are some fixed constants. The motivation for this choice comes from deterministic optimization since there a rule of the form

increase ρ_n if $\langle Y_{n+1}, Y_n \rangle > 0$
decrease ρ_n if $\langle Y_{n+1}, Y_n \rangle < 0$

leads to an optimal speed of convergence.

The term $-\delta\rho_n$ in the exponent of (17.22) is added to guarantee the convergence of ρ_n to zero. Mirozahmedov and Uriasev show that the assumptions of the convergence theorem part (ii) are fulfilled and hence

$$\overline{X}_n \to x^* \text{ a.s.}$$

The same rule but with $\delta = 0$ was studied by Rusczynski and Syski [19]. Some comments on this rule can be found in Section 17.3.

The stopping criterion pertaining to this rule is

(b′) *The adaptive stopping time (AST)*

$$\tau = \inf\{n | \rho_n \le \rho^*\} \tag{17.23}$$

This time does not lead to a confidence region with fixed size.

(c) *The decrease of objective function rule (DOSR)*

This rule is based on a recursive estimate $\widehat{f}_n$ of the objective function $E_p(q(x, \xi))$ namely

$$\widehat{f}_0 = q(X_0, \xi_0)$$
$$\widehat{f}_{n+1} = (1 - \beta_n)\widehat{f}_n + \beta_n q(X_{n+1}, \xi_{n+1}).$$

The constants β_n determine the degree of smoothing, e.g. $\beta_n = \beta$ (exponential smoothing) or $\beta_n = (n+1)^{-1}$ (arithmetic mean). The stopping rule itself employs

$$\rho_{n+1} = \begin{cases} \gamma_1 \rho_n & \text{if } \dfrac{\widehat{f}_{n-M_n} - \widehat{f}_n}{\sum_{i=n-M_n+1}^{n} \|X_i - X_{i-1}\|} \le \gamma_2 \\ \rho_n & \text{otherwise} \end{cases}$$

Here $\gamma_1 < 1$ and $\gamma_2 > 0$ are fixed constants and $\{M_n\}$ is an (increasing) sequence of nonnegative integers. Thus ρ_n stays either constant or decreases by a factor. The pertaining stopping time is

(c′) *The decrease of objective function stopping time (DOST).*

$$\tau_n = \inf\{n | \rho_n \leq \rho^*\}$$

Unfortunately there are no general properties known of this rule.

(d) *The ratio of progress stepsize rule (RPSR)*

This rule is similar to the above but measures the progress in the argument.

$$\rho_{n+1} = \begin{cases} \alpha_1 \rho_n & \text{if } \dfrac{\|X_n - X_{n-M_n}\|}{\sum_{i=n-M_n+1}^{n} \|X_i - X_{i-1}\|} \leq \alpha_2 \\ \rho_n & \text{otherwise .} \end{cases}$$

Again the stopping time is

(d′) *The ratio of progress stopping time (RPST)*

$$\tau_n = \inf\{n | \rho_n \leq \rho^*\}$$

Both preceding rules suffer from the defect that the last M_n steps (with M_n increasing) have to be kept in memory. They are described in Ermoliev and Gaivoronski [5].

(e) *The oscillation test stepsize rule (OTSR)*

This rule keeps the stepsize constant as long as some statistical test indicates that the behavior of the path is pure oscillation and no progress in the objective function is made. Then the step size is decreased by some factor.

Consider the procedure (17.4) with fixed stepsize

$$X_{n+1} = \Pi_S(X_n - \rho Y_{X_n}) \tag{17.24}$$

This is by construction a time-homogeneous Markov process. Under some weak regularity conditions this process is ergodic, i.e. it converges in law to the unique stationary measure of (17.24). Let $\tilde{X}_n^\rho$ be a stationary sequence of this Markovian process. It may be shown that if x^* lies in the interior of S then

$$\rho^{-\frac{1}{2}}(\tilde{X}_n^\rho - x^*) \to N(0, \Sigma) \tag{17.25}$$

as $\rho \to 0$ where Σ is a solution of

$$A\Sigma + \Sigma A' = C \tag{17.26}$$

with A resp. C given by (17.14) resp. (17.15). The similarity of (17.26) to (17.13) is interesting to notice. Again Σ may be calculated from A and C as

$$\Sigma = \int_0^\infty \exp(uA) \cdot C \cdot \exp(uA') du$$

or

$$\text{vec } \Sigma = (I \otimes A + A' \otimes I)^{-1} \text{vec } C.$$

Thus for small but constant ρ the process X_n converges in law to a normal distribution with mean x^* and covariance matrix $\rho\Sigma$. This corresponds to the well known fact that for fixed stepsize ρ the process approaches first some neighborhood of the solution and begins to oscillate around it afterwards. The OTSR makes a decision for decreasing the stepsize by testing whether this oscillatory behavior is already reached. As a test statistic we may use the inner product of subsequent gradients $V_n = (Y_n, Y_{n-1})$. If the sequence X_n is stationary and has the limiting distribution (17.25) then

$$E(V_n) = \rho \operatorname{tr}(A'A(I - \rho A)\Sigma) - \rho \operatorname{tr}(AC).$$

If A is symmetric and $\rho \ll 1$ this expression may be approximated by

$$E(V_n) = -\frac{1}{2}\rho \operatorname{tr}(AC).$$

If X_n is not yet oscillating $E(Y_n, Y_{n-1})$ is typically much larger. The unknown matrices A and C may be estimated consistently by $\hat{A}_n$ resp. $\hat{C}_n$. By equation (17.26) this leads also to an estimate $\hat{\Sigma}_n$ of Σ.

The OTSR is defined by a sequence of stopping times $\{\nu_n\}$ which are defined recursively by

$$\nu_0 = 0$$

$$\nu_{n+1} = \inf\{n | \frac{1}{n - \nu_n} \sum_{i=\nu_n+1}^{n} (Y_i, Y_{i-1}) \le \tag{17.27}$$

$$\le \rho_n \operatorname{tr}(\hat{A}_n'\hat{A}_n(I - \rho_n\hat{A}_n)\Sigma\hat{A}_n) - \rho_n \operatorname{tr}(\hat{A}_n\hat{C}_n) + \gamma_1\}$$

The stepsizes ρ_n are defined as

$$\rho_n = \rho_0 \cdot \gamma_2^i \text{ for } \nu_i \le n < \nu_i + 1 \tag{17.28}$$

Thus ρ_n is decreased by the factor $\gamma_2 < 1$ exactly at the times ν_i.

(e') *The oscillation test stopping time (OTST)*

A pertaining stopping time is also based on (17.25) employing the same ideas as were used in (17.20). Thus let $\kappa_{\alpha,n}$ be the upper α-quantile of the Γ-distribution (17.20) with Σ replaced by $\hat{\Sigma}_n$ then the stopping times τ_ε are

$$\tau_\varepsilon = \inf\{n | \frac{\kappa_{\alpha,n}}{n} \le \varepsilon\} \tag{17.29}$$

This family of stopping times leads to exact level α confidence regions. Moreover under mild assumptions these regions are uniform in the starting value and thus satisfy (17.9)

If the solution point x^* lies on the boundary then the result should be modified as in the DSR-case (a). Again the largest hyperplane H contained in K^*, the tangent cone to S at x^* carries the whole mass of the asymptotic distribution. By projecting everything onto the space H this case may be reduced to the unconstrained case. Details of the algorithm are presented in Section 17.4.

(f) *The inner product stepsize rule (IPSR)*

It has been pointed out in section (e) that the expectation of the inner product $V_n = \langle Y_n, Y_{n-1} \rangle$ of two subsequent gradients is negative, if the process is oscillating. This fact can be used for the definition of a very simple stepsize rule. Instead of comparing $E(V_n)$ with the asymptotically correct, but complicated expression given in formula (17.27) only the sign of $E(V_n)$ is considered.

More precisely the IPSR is defined by a sequence of stopping times $\{\nu_n\}$

$$\nu_0 = 0$$

$$\nu_{n+1} = \inf \left\{ n \Big| \frac{1}{n - \nu_n} \sum_{i=\nu_n+1}^{n} \langle Y_i, Y_{i-1} \rangle \leq 0 \right\}$$

The stepsizes are–as in the OTSR–defined as

$$\rho_n = \rho_0 \cdot \gamma^i \text{ for } \nu_i \leq n < \nu_{i+1}$$

It is evident that the IPSR decreases the stepsizes at an earlier stage than the OTSR. This is sometimes desirable since the more complicated estimations which are needed for the oscillation test rule are only valid for small ρ. Thus a good compromise is to begin with the simple inner-product rule which provides a fast convergence to a neighborhood of the solution. If the stepsizes ρ_n are small enough then the rule should be switched to OTSR. By such a procedure one avoids the very quick decrease of ρ_n in a later stage of the approximation process.

(g) *A review of other stepsize rules and stopping times*

There are many other rules known, some of which are restricted to the univariate case. Kesten [11] proposed for instance to choose

$$\rho_1 = 1$$

$$\rho_{n+1} = \begin{cases} \rho_n & \text{if } \operatorname{sgn} Y_{n-1} = \operatorname{sgn} Y_n \\ \frac{1}{m+1} & \text{if } \operatorname{sgn} Y_{n-1} \neq \operatorname{sgn} Y_n \text{ and } \rho_n = \frac{1}{m} \end{cases}$$

and showed a.s. convergence of this procedure.

Farell [8] considers also the univariate case and defines a stopping time of the following kind. Suppose it is known that the solution x^* lies in some interval

$x_0^{(1)} \leq x^* \leq x_0^{(2)}$. Then one may start two independent procedures $X_n^{(1)}$ and $X_n^{(2)}$ with initial points

$$X_0^{(1)} = x_0^{(1)}$$
$$X_0^{(2)} = x_0^{(2)}$$

and stop if for the first time

$$|X_n^{(1)} - x_n^{(2)}| \leq \varepsilon$$

This procedure leads to asymptotic confidence regions of variable size.

Fabian [6] accelerates the approximation procedure by doing a kind of line search. He takes additional observations of the objective function at $X_n + \frac{1}{n}Y_n$, $X_n + \frac{2}{n}Y_n, \ldots$ etc. and chooses

$$\rho_n = \frac{j}{n} \text{ where } j = \max\{i | q(X_n + \frac{i}{n}Y_n, \xi_n) < q(X_n, \xi_n)\}$$

He shows convergence but was unable to prove that this procedure is better than the DSR.

17.3 A comparison of different rules

It is rather impossible to give general statements about the superiority of one rule over another because detailed analysis of the performances of many rules has not been done. Therefore we restrict ourselves to compare them only for a very simple but basic stochastic approximation problem.

We assume that $S \equiv \mathbf{R}^k$, i.e. an unconstrained problem and $f(x) = \frac{1}{2}x'Ax$, a quadratic form with positive definite matrix A. The stochastic gradients are

$$\nabla g(x, \xi) = Ax + Z$$

where the errors $Z \sim N(0, C)$. Thus the procedure is

$$X_{n+1} = X_n - \rho_n A X_n - \rho_n Z_n \tag{17.30}$$

with $\{Z_n\}$ being a sequence of i.i.d. $N(0, C)$ random variables. Clearly the solution x^* equals zero in this example.

If $X_0 = x_0$ is the starting value then X_n may alternatively be represented as

$$X_n = \prod_{i=0}^{n-1}(I - \rho_i A)x_0 - \sum_{i=0}^{n-1} \rho_i \prod_{j>i}^{n-1}(I - \rho_j A)Z_i$$

$$= \prod_{i=0}^{n-1}(I - \rho_i A)x_0 - U_n \text{ (say)}$$

The first summand represents the influence of the starting value and U_n is the "error" term. Consider first the DSR-situation, i.e. $\rho_n = \frac{\ell}{n}$. Then

$$\sum_{i=0}^{n-1} \frac{\ell}{i} \prod_{j>i}^{n-1}(I - \frac{\ell}{j}A)Z_i$$

has a normal distribution with expectation zero and covariance matrix

$$\Sigma_n = \sum_{i=0}^{n-1} (\tfrac{\rho}{i})^2 \prod_{j>1}^{n-1} (I - \tfrac{\rho}{j}A)C \prod_{j>i}^{n-1} (I - \tfrac{\rho}{j}A')$$

which converges in accordance with (17.13) to a solution of

$$(\rho A - \tfrac{I}{2})\Sigma + \Sigma(\rho A' - \tfrac{I}{2}) = \rho^2 C.$$

On the other hand if ρ_n is kept constant $\rho_n \equiv \rho$ (like in the DOSR, RPSR and OTSR situation) then the error term

$$U_n = \sum_{i=0}^{n-1} \rho(I - \rho A)^{n-i-1} Z_i$$

converges in law to the autoregressive AR(1) process

$$\tilde{U}_n = \rho \sum_{i=0}^{\infty} (I - \rho A)^i Z_{n-i}$$

Thus the approximation process with constant stepsize may be represented as a sum of a component converging to zero and a stationary process. This is in accordance with (17.25).

The covariance matrix of $\tilde{U}_n$ satisfies

$$\mathrm{Cov}(\tilde{U}_n) = \rho^2 \sum_{i=0}^{\infty} (I - \rho A)^i C(I - \rho A') = \rho\Sigma + 0(\rho)$$

as $\rho \to 0$ where Σ is given by (17.26).

In a similar manner the gradient process $Y_n = AX_n + Z_n$ may be rewritten as

$$Y_n = A(I - \rho A)^n - \rho A \sum_{i=0}^{n-1} (I - \rho A)^{n-i-1} Z_i + Z_n$$

$$= A(I - \rho A)^n - W_n \text{ (say)}$$

where W_n converges in law to an autoregressive moving average process $\tilde{W}_n$

$$\tilde{W}_n = \rho A \sum_{i=0}^{\infty} (I - \rho A)^{i-1} Z_{n-1} - Z_n$$

Thus the expectation of $Y_n'Y_{n-1}$ is approximately

$$E(Y_n'Y_{n-1}) \approx x_0 A(I - \rho A)^n (I - \rho A')^n A' x_0$$

$$+ \mathrm{tr}\left(C\left(\rho^2 A^2 \sum_{i=0}^{\infty} (I - \rho A)^{2i-1} - \rho A\right)\right)$$

$$\approx x_0 A(I - \rho A)^n (I - \rho A')^n A' x_0 + \rho\,\mathrm{tr}(A'A\Sigma) - \rho\,\mathrm{tr}(AC)$$

(see (17.27)). Thus testing $Y_n'Y_{n-1}$ with the oscillation test (OTSR) is equivalent with testing whether the term $x_0 A(I - \rho A)^n(I - \rho A')^n A' x_0$ is already negligible. Since this criterion takes the influence of the starting value into account the resulting procedure leads to *uniform asymptotic confidence regions*. This is the main advantage of this method.

The DOSR compares progress in the objective function with progress in the argument. The objective function is estimated in our example by

$$\widehat{f}_n = \frac{1}{2n} \sum_{i=1}^{n} (X_i'AX_i + Z_i'AZ_i)$$

Thus the expectation of this estimate is

$$E(\widehat{f}_n) \approx \frac{1}{2}(\text{tr}(\rho A\Sigma) + \text{tr}(AC)).$$

On the other hand the expression $\|X_i - X_{i-1}\|^2$ has expectation

$$E(\|X_i - X_{i-1}\|^2) = \text{tr}(\rho A\Sigma) + \text{tr}(C)$$

so that for very small ρ

$$E(\|X_i - X_{i-1}\|) = 0(\text{tr}(C)^{1/2}).$$

Hence for very small ρ

$$\frac{\widehat{f}_n}{\sum_{i=1}^{n} \|X_i - X_{i-1}\|} \approx \frac{\text{tr}(AC)}{n\,\text{tr}(C)^{1/2}}$$

irrespectively of ρ. We see that the DOSR will take approximately the same number of steps between two consecutive stepsize reductions.

What concerns the ASR, it was shown by Ruszynski and Syski that for the sequence given by (17.22)

$$n\rho_n \to 1/\delta \quad \text{a.s.}$$

Thus this rule leads back to the DSR case, at least in an asymptotic sense.

Let us turn now to the case $\delta = 0$. If $C = 0$, i.e. the error term Z is zero, then the rule (17.22) reduces to

$$\log \rho_{n+1} = \log \rho_n + (X_n A^2 X_n - \rho_n X_n A^3 X_n).$$

Since $X_n A^2 X_n - \rho_n X_n A^2 X_n > 0$ if $\rho_n < \frac{1}{\lambda_{\max}}$ where $\lambda_{\max}$ denotes the maximal eigenvalue of A one can see that in this case ρ_n *does not converge to zero*. This results in an exponential speed of convergence of X_n. The weighted means $\overline{X}_n$ converge then with the rate $1/n$.

If the error terms are present, i.e. $C \neq 0$ it may be shown that

$$\frac{\rho_n}{\sqrt{n}} \quad \text{converges a.s.}$$

Thus the assumption $\Sigma\rho_n^2 < \infty$ is not fulfilled. Hence the procedure X_n is not convergent itself and the weighted means $\overline{X}_n$ converge with a speed $E(\overline{X}_n^2) = 0(\frac{\log n}{n})$.

17.4 The implementation of the oscillation test

The oscillation test routines were implemented by the author at IIASA. Some details of the implemented algorithms are given here.

Recall that the method consists in keeping the stepsize constant as long as the test rejects the hypothesis that the behavior of the path is already oscillation. If this hypothesis is strongly rejected then even an increase of the stepsize is advisable.

The method needs estimates for A, the hessian of the objective function at x^* and C, the covariance matrix of the errors.

An estimate for C is easily found. At each step we take *two* independent observations of the stochastic gradient

$$Y_n^{(1)} = \nabla q(X_n, \xi_n^{(1)})$$
$$Y_n^{(2)} = \nabla q(X_n, \xi_n^{(2)})$$

and calculate

$$Y_n = \tfrac{1}{2}(Y_n^{(1)} + Y_n^{(2)})$$
$$Z_n = \tfrac{1}{2}(Y_n^{(1)} - Y_n^{(2)})$$

The variance of Y_n is half of the variance of $Y_n^{(i)}$. This random variable is taken for determining X_{n+1}. The error variable Z_n is used for the estimation of C. As we know from the general considerations the asymptotic distribution is concentrated on the largest linear subspace contained in the tangent cone of S at x^*. Let K_n be the tangent cone of S at X_n. (If X_n is in the interior of S then $K \equiv \mathbf{R}^k$). Let H_n be the largest subspace contained in K_n and let $\overline{Z}_n$ be the projection of Z_n onto K_n. Then

$$\hat{C}_n = \frac{1}{n} \sum_{i=1}^{n} \overline{Z}_i \overline{Z}_i' \qquad (17.31)$$

is used as an estimate of C.

Next define $\Delta X_n = X_n - X_{n-1}$, $\Delta Y_n = Y_n - Y_{n-1}$. As $E(\Delta Y_n) = A\Delta X_n + 0(\|X_n - x^*\|^2 + \|\Delta X_n\|^2)$ we may construct an estimate of the relevant part of A as follows: Project ΔY_n and ΔX_n onto H_n to give $\overline{\Delta Y_n}$ and $\overline{\Delta X_n}$. The matrix $\hat{A}_n$ should satisfy

$$\overline{\Delta Y_n} \sim A\overline{\Delta X_n} \qquad (17.32)$$

Thus we may adjust $\hat{A}_n$ recursively to satisfy (17.32) by setting

$$\hat{A}_{n+1} = \hat{A}_n - \frac{1}{n}(\hat{A}_n\overline{\Delta X_n} - \overline{\Delta Y_n}) \cdot \frac{\overline{\Delta X_n}}{\|\overline{\Delta X_n}\|^2} \qquad (17.33)$$

It remains to solve th equation

$$\hat{A}_n \hat{\Sigma}_n + \hat{\Sigma}_n \hat{A}_n = \hat{C}_n$$

for the determination of the covariance matrix of X_n (see (17.26)). To use the indicated explicit formulas is however too time consuming. We use instead again a recursive way of solving (17.26), namely

$$\widehat{\Sigma}_{n+1} = \widehat{\Sigma}_n - \frac{1}{n}[\widehat{A}_n\widehat{\Sigma}_n + \widehat{\Sigma}_n\widehat{A}_n - \widehat{C}_n]. \tag{17.34}$$

It may be shown that if $\widehat{A} \to A$ (positive definite) and $\widehat{C}_n \to C$ (positive definite) then $\widehat{\Sigma}_n$ converges to a solution of (17.26). The oscillation test compares

$$\frac{1}{n - \nu_n} \sum_{i=\nu_n+1}^{n} (\overline{Y}_i, \overline{y}_{i+1}) \text{ with } \rho_n(\text{tr}(\widehat{A}_n'\widehat{A}_n\widehat{\Sigma}_n) - \text{tr}(\widehat{A}_n\widehat{C}_n)).$$

If the difference is smaller than γ_1 the hypothesis of oscillation is accepted and the stepsize is decreased (by a factor γ_2 – usually $\gamma_2 = \frac{1}{2}$).

If however the inner products are much larger than their asymptotic expectations, the stepsize is increased (by a factor $\gamma_3 > 1$).

The asymptotic confidence region can be found by looking at the distribution of the quadratic form

$$(X_n - x^*)'D(X_n - x^*)$$

which is approximately

$$\Gamma\left(\frac{\text{tr}^2(D\widehat{\Sigma}_n)}{2\,\text{tr}((D\widehat{\Sigma}_n)^2)}, \rho_n \frac{2\,\text{tr}((D\widehat{\Sigma}_n)^2)}{\text{tr}(D\widehat{\Sigma}_n)}\right)$$

(see (17.20)). If the upper α percentile of this distribution is smaller than ε then the whole procedure can be stopped and we know that

$$P\{\|X_r - x^*\|_D \leq \varepsilon\} \approx 1 - \alpha$$

if ε is small enough.

Sometimes it is not required to know $1 - \alpha$ confidence regions but the knowledge of the expectation of $(X_n - x^*)'D(X_n - x^*)$ suffices. Since

$$E((X_n - x^*)'D(X_n - x^*)) \approx \rho\,\text{tr}(D\Sigma_n)$$

this value can easily be calculated. If this value is smaller than some predetermined constant the whole procedure stops. Due to the careful testing and estimation the final value obtained is a very reliable one.

References

[1] J.R. Blum, "Multidimensional Stochastic Approximation Method", *Ann. Math. Statist.* **25**(1954), 737–744.

[2] Y.S. Chow and H. Robbins, "On the asymptotic theory of fixed-width sequential confidence intervals", *Ann. Math. Stat.* **36**(1965), 457–462.

[3] A. Dvoretzky, "On stochastic approximation", in *Proceedings of the Third Berkeley Symposium on Math. Statist. and Prob.*, University of California Press (1956).

[4] Yu. Ermoliev, "Methods of Stochastic Programming", *Nauka*. Moscow (1976). (in Russian)

[5] Yu. Ermoliev and A. Gaivoronski, *Stochastic Quasigradient Methods and their Implementation*, IIASA working paper WP-84-55 (1984).

[6] V. Fabian, "Stochastic approximation of minima with improved asymptotic speed", *Am. Math. Statist.* **38**(1967), 191–200.

[7] V. Fabian, "On asymptotic normality in stochastic approximation", *Ann. Math. Statist.* **39**(1968), 1327–1332.

[8] R.H. Farell, "Bounded length confidence intervals for the zero of a regression function", *Ann. Math. Statist.* **33**(1962), 237–247.

[9] Hiriart-Urruty, *Thèse*. Annales de l'Université de Clermont No. 58(1976), 12-ieme fasc.

[10] Kendall, M.G. and A. Stuart, *The Advanced Theory of Statistics*, Vol. I., Griffin, London (1963).

[11] H. Kesten, "Accelerated Stochastic Approximation", *Ann. Math. Statist.* **29**(1958), 41–59.

[12] H. Kiefer and J. Wolfowitz, "Stochastic estimation of the maximum of a regression function", *Ann. Math. Statist.* **23**(1952), 462–466.

[13] H. Kushner and S. Clark, *Stochastic Approximation Methods for Constrained and Unconstrained Systems*, App. Math. Sciences Vol. 26., Springer-Verlag, New York (1978).

[14] F. Mirozahmedov and S.P. Uryasev, "Adaptive stepsize regulation for stochastic optimization algorithm", *Zurnal vicisl. mat. i. mat. fiz.* **23**, **6**(1983), 1314–1325. (in Russian)

[15] G. Pflug, *On the determination of the step size in stochastic quasigradient methods*, IIASA-Collaborative paper CP-83-25(1983).

[16] G. Pflug, "Stochastic Minimization with constant step-size Asymptotic laws", to appear in *SIAM Journal of Control*.

[17] H. Robbins and S. Monro, "A stochastic approximation method", *Ann. Math. Statist.* **22**(1951), 400–407.

[18] H. Robbins and D.A. Siegmund, "Convergence theorem for nonnegative almost supermartingales and some applications", *Optimizing Methods in Statistics* (J.S. Rustagi, ed.), Academic Press (1971), 233–257.

[19] A. Ruszcynski and W. Syski, "A method of aggregate stochastic subgradients with on-line stepsize rules for convex stochastic optimization problems", to appear in *Mathematical Programming Study*.

[20] L. Schmetterer, "Stochastic approximation", *Proc. Fourth Berkeley Symp. Univ. California Press*, **Vol. I**, (1961), 587–609.

[21] R.L. Sielken, "Some Stopping Times for Stochastic Approximation Procedures", *Z. Wahrscheinlichkeitstheorie verw. Gebiete* **27**(1973), 79–86.

[22] H. Walk, "An Invariance Principle for the Robbins-Monro Process in a Hilbert Space", *Z. Warscheinlichkeitstheorie verw. Gebiete* **39**(1977), 135–150.

CHAPTER 18

ADAPTIVE STOCHASTIC QUASIGRADIENT PROCEDURES*

S. Urasiev

18.1 Introduction

In this chapter we deal with iterative algorithms for solving stochastic optimization problem

$$\min E f(x, \omega) \tag{18.1}$$

subject to constraints

$$x \in X \subset R^n$$

where x are variables to be chosen which take values in Euclidean space and ω are random parameters which belong to some probability space. Our main concern is the improvement of performance features of the stochastic quasigradient (SQG) method

$$x^{s+1} = \pi_X(x^s - \rho_s \xi^s) \tag{18.2}$$

where π_X is the projection operator on the set X, x^s-current approximation to solution, ρ_s is the stepsize and ξ^s is step direction, which roughly speaking, in average points to the direction of gradient of the function $Ef(x, \omega)$. Reader can find survey of such methods and further references in Chapter 6 (see also [1]). One of the main challenges which arise before implementor of SQG methods is appropriate selection of the stepsize ρ_s. Theory gives only very general guidelines:

$$\rho_s \longrightarrow 0, \sum_{s=0}^{\infty} \rho_s = \infty, \sum_{s=0}^{\infty} \rho_s^2 < \infty.$$

In papers written earlier on stochastic approximation [2], stepsize was chosen in advance to satisfy these conditions. For instance, $\rho_s = c/s$. In what follows, such choices which depend only on iteration number will be called programmed or off-lined rules. Unfortunately they lead to very slow convergence, although they assure in some sense optimal asymptotic rate. However, in practical computations SQG methods can be used to reach reasonable neighborhood of solution, not exact value of solution. For such purposes, asymptotic results are not relevant as well as programmed rules of choosing stepsize. In this chapter, adaptive or on-line rules for computing ρ_s are studied which exhibit much

---

* This chapter is based on the report presented at the International Conference on Stochastic Optimization, Kiev, 1984.

more satisfactory behavior. Such methods utilize information gathered during optimization process to make decision about current value of stepsize ρ_s. More specifically, ρ_s may depend on observations of random function $f(x^k, u^k)$ or stochastic quasigradient ξ^k in some or all preceding iterations $k \leq s$. Some on-line rules were proposed in [3]–[7]. This chapter is based on [5]–[7] and describes one particular adaptive SQG method in which stepsize increases or decreases depending on whether subsequent quasigradients point to the same or to the opposite directions.

This chapter consists of 5 sections. In Section 18.2 the adaptive SQG method is described, its convergence is analyzed in Section 18.3. Implementation details are discussed in Section 18.4, and the chapter ends in Section 18.5 with a description of some particular problems solved by algorithm together with results of numerical experiments.

18.2 Algorithm Description

In what follows we shall consider algorithm of type (18.2) for problem (18.1). It will be assumed that the process takes place in probability space (Ω, A, P) where A is σ-field and P- probability measure. Vector ξ^s from (18.2) is stochastic quasigradient, i.e.

$$E(\xi^s / B^s) = F_x(x^s) + b^s$$

where $F_x(x^s)$ is gradient of the function $F(x) = E_\omega f(x, \omega)$, conditions on b^s will be imposed later and B^s is σ-field defined by the process history, i.e., random variables $\{x^o, \ldots, x^s\}$. We shall keep in mind that x^s depends on random parameters from Ω, but will not specify this dependence explicitly.

We shall explain at first the idea of adaptive stepsize control informally. Here, for simplicity we shall assume that function $F(x)$ is smooth and $X = R^n$. It is quite naturally to choose step ρ_s to minimize $F(x)$ along direction ξ^s, i.e., such that function $\varphi_s(\rho)$ has minimum over ρ, where

$$\varphi_s(\rho) = E[F(x^s - \rho\xi^s)/x^s].$$

This is analogue of stepsize rules used extensively in deterministic optimization. It is easy to see that

$$\frac{\partial}{\partial \rho}\varphi_s(\rho)|_{\rho_s} = E\big[\frac{\partial}{\partial \rho}F(x^s - \rho\xi^s)|_{\rho=\rho_s}/x^s\big]$$
$$= -E\big[\langle\nabla F(x^s - \rho_s\xi^s), \xi^s\rangle/x^s\big]$$
$$= -E\big[\langle\nabla F(x^{s+1}), \xi^s\rangle/x^s\big]$$
$$= -E\big[\langle\xi^{s+1}, \xi^s > /x^s\big], \quad s = 0, 1 \ldots$$

Thus, $-\langle\xi^{s+1}, \xi^s\rangle$ is stochastic quasigradient of function $\varphi_s(\rho)$ in point ρ_s on iteration $s + 1$. To modify step ρ_s, let us use the following gradient procedure:

$$\rho_{s+1} = \rho_s + \lambda_s\langle\xi^{s+1}, \xi^s\rangle\lambda_s > 0, s = 0, 1 \ldots \tag{18.3}$$

The value of $\left(\xi^{s+1}, \xi^s\right)$ gives some information whether current value ρ^s exceeds minimum of function $\varphi_s(\rho)$ over ρ on the iteration s. If $\left(\xi^{s+1}, \xi^s\right) > 0$ then it is probable that minimum of $\varphi_s(\rho)$ is greater than ρ_s and it is necessary to increase stepsize and decrease it when sign is negative. This information is used to modify step ρ_s.

Naturally, decision based on this arguments will be subject to error due to stochastic phenomena. However, this errors will be smoothed out in the course of iterations. It is convenient to rewrite relation (18.3) in the following from

$$\rho_{s+1} = \rho_s a_s\left(\xi^{s+1}, \xi^s\right), a_s > 0, \quad s = 0, 1, \ldots, \tag{18.4}$$

(18.3) is the special case of (18.4) since for each λ_s, such that $\rho_{s+1} > 0, a_s$ can be selected respectively so that $\rho_s + 1$ computed by formulas (18.3), (18.4) coincide. In order to guarantee fulfillment of the convergence condition for SQG algorithms $\sum_{s=0}^{\infty} \rho_s = \infty$ (see Chapter 6), the value a_s is calculated by formula

$$a_s = a^{\rho_s}, a > 1, \quad s = 0, 1, \ldots,$$

Convergence of the algorithm (18.2) with the stepsize rule (18.4) can be established [7] in deterministic case, when $\xi^s = F_x(x^s)$ and $F(x)$ is a strongly convex function. For stochastic case, let us modify formula (18.4) as follows

$$\begin{aligned}
\rho_{s+1} &= \rho_s a^{\rho_s\langle\xi^{s+1}, \xi^s\rangle - \delta\rho_s} \\
&= \rho_s a^{\langle\xi^{s+1}, x^s - x^{s+1}\rangle - \delta\rho_s}, \delta > 0, \quad s = 0, 1, \ldots
\end{aligned} \tag{18.5}$$

Introduction of the term $\delta\rho_s$ guarantees fulfillment of one more convergence condition

$$\rho_s \longrightarrow 0 \text{ a.s. } s \longrightarrow \infty.$$

18.3 Convergence Analysis

Besides convergence of sequence x^s to the solution of problem (18.1), we are also interested in convergence of some convex combinations of this sequence. With sequence x^s generated by algorithm (18.2), (18.5), it will be associated the sequence

$$\bar{x}^s = \sum_{\ell=0}^{s} \rho_\ell x^\ell / \sum_{\ell=0}^{s} \rho_\ell \tag{18.6}$$

and the convergence of $\bar{x}^s$ to the solution will be studied. If such convergence does occur the initial sequence x^s will be called Cesaro convergent. The advantages of dealing with such convergence are the following:

- the sequence $\bar{x}^s$ displays much more regular behavior than original sequence x^s
- $\bar{x}^s$ can be computed with almost no additional effort in iterative way using the sequence $\bar{x}^s$.

– some convergence conditions can be relaxed for Cesaro convergence and in some cases x^s does not converge to the solution in ordinary sense, but is Cesaro convergent.

This type of convergence was used in [8],[9]. The following theorem from [7] gives conditions for Cesaro convergence of the method (18.2). We shall use the abbreviation a.s. for the words "almost sure".

Theorem 1. *Let $F(x)$ be a convex function defined on convex closed bounded set $X \subset R^n$,*

$$\max_{x,y \in X} \|x - y\| = C_1, \tag{18.7}$$

$$E\|\xi^s - F_s(x^s) - b^s\|^2 \leq C_2^2 \tag{18.8}$$

$$\overline{\lim_{s \to \infty}} \|b^s\| \leq \overline{b}, \tag{18.9}$$

$$\rho_s > 0 \text{ a.s.}, \quad s = 0, 1, \ldots, \tag{18.10}$$

$$E\rho_s^s < \infty, \quad s = 0, 1, \ldots, \tag{18.11}$$

$$\rho_s \longrightarrow 0 \text{ a.s. with } s \longrightarrow \infty, \tag{18.12}$$

$$\sum_0^\infty \rho_s = \infty \text{ a.s.}, \tag{18.13}$$

and at least one of the two following conditions is satisfied:

(1) *Step ρ_s depends only on $(x^0, \ldots, x^s, \xi^0, \ldots, \xi^{s-1})$ (it is measurable with respect to σ-algebra B_s induced by $(x^0, \ldots, x^s, \xi^0, \ldots, \xi^{s-1})$);*
(2) *$\rho_s \rho_{s-1}^{-1} \longrightarrow 1$ a.s., ρ_s depends only on $(x^0, \ldots, x^s, \xi^0, \ldots, \xi^s)$ (it is measurable with respect to σ-algebra induced by $(x^0, \ldots, x^s, \xi^0, \ldots, \xi^s)$).*

Then

$$\overline{\lim_{s \to \infty}} F(\overline{x}^s) - F(x^*) \leq \overline{b} C_1 \text{ a.s.}$$

where

$$x^* \in X^* = \{x^* : F(x^*) = \min_{x \in X} F(x)\}.$$

and $\overline{x}^s$ is defined by (18.6).

Corollary. *If $b^s \longrightarrow 0$ a.s. then all accumulating points of the sequence $\overline{x}^s$ are the solutions of problem (18.1).*

The main difference between conditions for Cesaro convergence and convergence in usual sense for SQG methods it that condition $\sum_{s=0}^\infty \rho_s^2 < \infty$, which is needed for normal convergence a.s., is not needed for Cesaro convergence. This makes verifying convergence conditions for adaptive SQG methods much easier.

Now we are prepared to give convergence results for adaptive SQG method (18.2), (18.5).

Theorem 2. *Let $f(x)$ be a convex (possibly nonsmooth) function defined on some vicinity of convex compact subset $X \subset R^n$. If the following conditions are satisfied*

$$\max \|x - y\| = C_1 \tag{18.14}$$

$$\sup \|\xi^s\| < C_2 \text{ a.s.} \tag{18.15}$$

$$\overline{\lim}_{s \to \infty} \|b^s\| \le \bar{b}, \tag{18.16}$$

$$\delta > C_1 \overline{\lim}_{s \to \infty} \inf_{h \in \partial F(x^s)} \|\xi^s - h\| \text{ a.s.} \tag{18.17}$$

where $\partial F(x)$ is the set of subgradients of $F(x)$ at point x. Then

$$\overline{\lim}_{s \to \infty} (F(\bar{x}^s) - \min_{x \in X} F(x)) \le \bar{b} C_1 \text{ a.s.}$$

i.e. if $\lim_{s \to \infty} b^s = 0$ then

$$F(x^s) - \min x \in X F(x) \longrightarrow 0 \text{ a.s.}$$

and all accumulating points of the sequence $\bar{x}^s$ are solutions of the problem (18.1) a.s.

Proof. Condition (18.10) of Theorem 1 follows directly from (18.5) since $\rho_0 > 0$ and $a > 0$. Here we shall give only an outline of the proof, which consists of checking conditions of theorem 1. We shall check here conditions (18.12)-(18.13) of the theorem 1 and assume $b^s = 0$ (for more details see [5]-[7]).

1. Let us show that condition (18.13) of Theorem 1 is satisfied, i.e. $\sum_{s=0}^{\infty} \rho_s = \infty$ a.s.. Assume the opposite, i.e. exists such constant K that probability of the event

$$A = \{\omega : \sum_{s=0}^{\infty} \rho_s \le K\}$$

is positive. From projection properties and (18.15), we get the estimate

$$\|x^{s+1} - x^s\| \le \|\rho_s \xi^s\| \le \rho_s C_2 \text{ a.s.} \tag{18.18}$$

Stepsize rule (18.5) together with (18.18) yields:

$$\rho_{s+1} = \rho_s a^{\langle \xi^{s+1}, x^s - x^{s+1} \rangle - \delta \rho_s} \ge \rho_s a^{-\|\xi^{s+1}\| \|x^s - x^{s+1}\| - \delta \rho_s}$$

$$\ge \rho_s a^{-(C_2^s + \delta)\rho_s} = \rho_s a^{-C_3 \rho_s}$$

where $C_3 = (C_2^2 + \delta)$. Therefore for $\omega \in A$ the following relation holds

$$\rho_{s+1} \ge \rho_s a^{-C_3 \rho_s} \ge \rho_0 a^{-C_3 \Sigma_0^s \rho_l} \ge \rho_0 a^{-C_3 K}$$

which implies $\sum_{s=0}^{\infty} \rho_s = \infty$ for $\omega \in A$ contradicting initial assumption. Therefore condition (18.13) is satisfied.

2. Consider now condition (18.12) and let us prove that $\rho_s \longrightarrow 0$ as $s \longrightarrow 0$ a.s. Denoting

$$C_s = \inf_{h \in \partial F(x^s)} \|\xi^s - h\|$$

we obtain the following estimate:

$$
\begin{aligned}
\langle \xi^{s+1}, x^s - x^{s+1} \rangle - \delta \rho_s &\leq \langle \widehat{F}_x(x^{s+1}), x^s - x^{s+1} \rangle \\
&\quad + \langle \xi^{s+1} - \widehat{F}_x(x^{s+1}), x^s - x^{s+1} \rangle - \delta \rho_s \\
&\leq F(x^s) - F(x^{s+1}) \\
&\quad + \|\xi^{s+1} - \widehat{F}_x(x^{s+1})\| \|x^s - x^{s+1}\| - \delta \rho_s \\
&\leq F(x^s) - F(x^{s+1}) \\
&\quad + C_2 \rho_s \|\xi^{s+1} - \widehat{F}_x(x^{s+1})\| - \delta \rho_s \text{ a.s.}
\end{aligned}
$$

Since $\widehat{F}_x(x^{s+1})$ in the last relation is an arbitrary vector belonging to set $\partial F(x^{s+1})$, we obtain

$$
\begin{aligned}
\langle \xi^{s+1}, x^s - x^{s+1} \rangle - \delta \rho_s &\leq F(x^s) - F(x^{s+1}) + C_1 \rho_s \inf_{h \in \partial F(x^s)} \|\xi^s - h\| - \delta \rho_s \\
&= F(x^s) - F(x^{s+1}) + (C_2 C_s - \delta) \rho_s \text{ a.s.}
\end{aligned}
$$

By substituting this estimate into (18.5), we obtain

$$
\begin{aligned}
\rho_{s+1} &\leq \rho_s a^{F(x^s) - F(x^{s+1}) + (C_2 C_s - \delta) \rho_s} \\
&\leq \rho_0 a^{\sum_{\ell=0}^{s}(F(x^\ell) - F(x^{\ell+1})) + \sum_{\ell=0}^{s}(C_2 C_s - \delta)\rho_\ell} \\
&= \rho_0 a^{F(x^0) - F(x^{s+1}) + \sum_{\ell=0}^{s}(C_2 C_s - \delta)\rho_\ell}
\end{aligned}
$$

Taking into consideration $\sum_{s=0}^{\infty} \rho_s = $ a.s. and relations (18.14), (18.17), we see that the expression in the exponent in the last relation tends to $-\infty$:

$$\lim_{s \to \infty} [f(x^0) - f(x^s)) + \sum_{\ell=0}^{s}(C_2 C_s - \delta)\rho_\ell] \longrightarrow -\infty \text{ a.s.}$$

Since $a > 1$, this implies $\rho_s \to 0$ a.s.

Now, we have to show that condition (18.2) of Theorem 1 is satisfied. The following relation is satisfied:

$$\frac{\rho_{s+1}}{\rho_s} = a^{\langle \xi^{s+1}, x^s - x^{s+1} \rangle - \delta \rho_s}.$$

Since $\rho_s \to 0$ a.s., then

$$\langle \xi^{s+1}, x^s - x^{s+1} \rangle - \delta \rho_s \longrightarrow 0 \text{ a.s. and}$$

$$\frac{\rho_{s+1}}{\rho_s} \longrightarrow 0 \text{ a.s.}$$

after all conditions of Theorem 1 are tested, the statement of this theorem follows from it.

18.4   Implementation Strategies

In this paragraph problems which arise during the implementation of stochastic quasigradient algorithm (18.2), (18.5) are discussed. Its implementation includes some heuristical elements. The implemented method can be used for the fast finding of good initial approximation in the vicinity of solution. The implemented algorithm described below performed essentially better than the method with programmed rule for step size selection. First, we shall present the algorithm and then discuss some of its features.

Algorithm. Set $s = 0$ at the beginning of the computation.

Step 1.   Computation of stochastic quasigradient ξ^s.

Step 2.   Averaging of the stochastic quasigradient norm $\|\xi^s\|$

$$G_s = G_{s-1} + (\|\xi^s\| - G_{s-1}) \cdot D.$$

At the beginning of the computation $G_1 = 0$.

Step 3.   The computation of the average current point drift

$$Q_s = G_s \rho_s$$

Step 4.   Check the stopping criterion: if $Q_s < Q_*$ or $s > s_*$, finish the computation, otherwise go to the next step.

Step 5.   The computation of scalar production T_s:

$$T_s = (\xi^s, x^{s-1} - x^s).$$

Step 6. Averaging of the T_s absolute value:

$$Z_s = Z_{s-1} + (|T_s| - Z_{s-1}) \cdot D.$$

At the beginning of the computation $Z_{-1} = 0$.

Step 7.   Rule for the step size ρ_s selection:

$$\rho_s = \rho_{s-1} R^{\frac{T_s}{Z_s}} \times \begin{cases} 1 & \text{if } t_s > 0 \\ U & \text{if } T_s \leq 0. \end{cases}$$

Step 8.   Reducing the step size change.

$$\rho_s = \begin{cases} 3\rho_{s-1} & \text{if } \rho_s \rho_{s-1}^{-1} > 3, \\ \frac{\rho_{s-1}}{4} & \text{if } \rho_s \rho_{s-1}^{-1} < 4^{-1}, \\ \rho_s & \text{otherwise.} \end{cases}$$

Step 9.   Finding the next approximation

$$x^{s+1} = x^s - \rho_s \xi^s.$$

Step 10. Projection on the feasible region X

$$x^{s+1} = \pi_X(x^{s+1})$$

Step 11. Take $s = s + 1$ and go to Step 1.

Two stopping criteria are implemented in the method. The first one is by the number of iterations. The second stopping criterion is by the value of the mean point trend which is equal to the product of the mean norm of the quasigradient ξ^s by the step size ρ_s. When the value of the shift becomes less than the threshold value Q_*, the method stops (steps 3,4). The step size control differs from theoretical one (18.5) in several aspects (step 7). For one thing, value T_s is divided by the averaged absolute value of T_s. Additional reduction of the step size by means of factor U, $0 < U \leq 1$ is introduced. The additional reduction takes place only if

$$T_s = \langle \xi^s, x^{s-1} - x^s \rangle \leq 0.$$

Since T_s/Z_s is some random value, step size ρ_s can increase or decrease, sometimes by too large a factor (step 7). In order that the next step does not differ too strongly from the preceding one ρ_{s-1}, some bounding coefficients are provided for increase or decrease of the step size (step 8).

Recommendations on the choice of the algorithm parameters. The following recommendations are obtained as a result of numerical experiments.

- The value of the mean change of step size $R(1 < R < 3)$ is usually set to $R = 2$.
- The value of the initial step size has no essential effect on the method convergence rate. However, if additional information is available, the initial value of the step size factor ρ_0 can be set approximately

$$\|x^0 - x^*\|(E(\|\xi^0\|))^{-1},$$

  where x^0- initial approximation, x^*- estimated location of extremum point;
- Parameter k defines averaging factor $D = \frac{1}{k}$ in the averaging formulas (Steps 2 and 6). Usually k is selected within the range $4 \leq k \leq 6$
- Parameter U (additional coefficient of step size reduction) is selected within the range $0.8 \leq U \leq 1$. With $k > 1$ coefficient U can be equal to 1 since step size decreases fast without additional decrease.
- The value of mean shift Q_* in stopping criterion is to be set approximately to the required solution accuracy for components of x.

18.5 Results of Numerical Experiments

Let us note firstly that it is advisable to average the values of variables and of the objective function during fixed number of the last iterations and take these quantities as the final approximation to the solution. The averaged value of coordinates x^s will now be designated as x and the averaged value $f(x^s, \omega^s)$ as $f(x)$.

Problem 1. The following problem is an example of multi-commodity facility location problem [7]. It is necessary to minimize

$$F(x) = E \sum_{i=1}^{5} \max\{a_i(x_i - \theta_i); b_i(\theta_i - x_i)\},$$

under constraints

$$
\begin{array}{rcrcrcrcrcl}
x_1 &+& x_2 &+& 2x_3 &+& 3x_4 &+& x_5 &=& 200 \\
x_1 &&&&&&&&& \le & 50 \\
&& x_2 &&&&&&& \le & 7 \\
&&&& x_3 &&&&& \le & 7 \\
&&&&&& x_4 &&& \le & 80 \\
&&&&&&&& x_5 & \le & 25
\end{array}
$$

$$x_i \ge 0, i = \overline{1,5}.$$

Here θ_i are random values uniformly distributed on intervals $[A_i, B_i]$, $i = 1, \ldots, 5$. Vectors $a = (a_1, \ldots, a_5)$, $b = (b_1, \ldots, b_5)$, $A = (A_1, \ldots, A_5)$, $B = (B_1, \ldots, B_5)$ are defined as follows:

$$
\begin{aligned}
A &= (0,0,0,0,0); & B &= (60,15,17,90,40); \\
a &= (1,0,3,1,2); & b &= (3,4,1,2,3).
\end{aligned}
$$

This problem allows analytical solution, which makes it possible to compare solution obtained by algorithm with exact one. The analytical form of the objective function is the following:

$$f(x) = \tfrac{1}{3}x_1^2 + \tfrac{2}{15}x_2^2 + \tfrac{2}{17}x_3^2 + \tfrac{1}{60}x_4^2 + \tfrac{1}{16}x_5^2$$
$$- 3x_1 - 4x_2 - x_3 - 2x_4 - 3x_5 + 278.5.$$

Stochastic quasigradient is computed by formula

$$\xi^s = (\xi_1^s, \ldots, \xi_5^s),$$

$$\xi_i^s = \begin{cases} a_1, & \text{if } x_i^s \ge \theta_i^s, \\ -b_1, & \text{if } x_i^s < \theta_i^s, \end{cases} \quad i = 1, \ldots, 5$$

The following exact solution was obtained using quadratic programming methods:

$$x^* = (41.88057; 7.00000; 2.48092; 41.27456; 22.33456),$$
$$f(x^*) = 98.100089.$$

Algorithm parameters are

$$R = 1.5; k = 4; U = 0.9; \rho_0 = 1.0.$$

Initial point is

$$x^0 = (0,0,0,0,0); f(x^0) = 278.5.$$

Step size on the 91st iteration $\rho_{91} = 0.1532$.

The results for averaged values of the coordinates and of the function for 91st to 100th iteration are as follows

$$\bar{x}_i = \frac{1}{10} \sum_{s=91}^{100} x_i^s, i = \overline{1,5}; \bar{x}_1 = 40.5485; \bar{x}_2 = 6.9981;$$

$$\bar{x}_j = 2.4381; \bar{x}_4 = 42.2561; \bar{x}_5 = 20.3561;$$

$$\bar{f}(x) = \frac{1}{10} \sum_{s=91}^{100} f(x^s, \theta^s) = 97.4185.$$

For comparison, below are given results of the solution of the same problem using the method with programmed control of step size. Initial approximation was the same. In asymptotically optimal [11] off-line step size rule $\rho_s = 1/\ell(s + a)$, parameter ℓ must be equal to the least eigenvalue of the objective function Hessian, i.e., $\ell = 1/30$. In this case we selected $a = 10$ and got approximately the same performance. However, our choice was based on exact information on objective function. If such information is not available, the off-line decision rule works in a much worse way.

Problem 2. A random locational equilibrium problem (Weber problem) [12]. The classical statement of Weber problem is as follows: given n points $\omega_i, i = \overline{1,n}$ in two-dimensional Euclidean space R^2, find a point $x \in R^2$ which minimizes the sum of distances $\|\omega_i - x\|$. In generalized statement of the problem [12] each point $\omega_i, i = \overline{1,n}$ is considered to be a random variable represented by some probability measure $\theta_i(\omega)$ over R^2. The problem now is to find the location of a point $x \in R^2$ which minimizes the weighted sum of expectations of distances between point x and points $\omega_i, i = \overline{1,n}$, i.e.

$$F(x) = \sum_{i=1}^{n} \beta_i \int \int_{R^2} \|x - \omega\| \theta_i(d\omega) \longrightarrow \min_{x \in R^2},$$

where $\beta_i > 0, i = \overline{1,n}$. The stochastic quasigradient at point x^s can be chosen as follows:

$$\xi^s = \sum_{i=1}^{n} \beta_i \gamma_i$$

where

$$\gamma_i = \begin{cases} \frac{x - \omega_i^s}{\|x - \omega_i^s\|} \\ 0 \qquad \text{otherwise} \end{cases}$$

and ω_i^s is distributed according to $\theta_i(\omega)$.

In this particular example, the number of destination points was chosen as $n = 30$, and $\theta_i, i = \overline{1,n}$ were taken as bivariate normal density functions whose means and standard deviations were generated randomly in the range 0–20. The weights β_j were also generated randomly in the range 0–10.

Exact value of the extremum is $x^* = (8,36;9.36)$ initial approximation $x^0 = (41,87)$. The results for averaged values of the variables x^s for 50-th to 60-th iteration are as follows

$$\bar{x} = \frac{1}{10} \sum_{s=51}^{60} x^s = (9.1, 10.2),$$

and for 190-th to 200-th are

$$\bar{x} = \frac{1}{10} \sum_{s=191}^{200} x^s = (8.9, 9.0).$$

If the initial approximation is $x^0 = (54, 30)$. The results for averaged values of the variables x^s for 20-th to 30-th iteration are as follows

$$\bar{x} = \frac{1}{10} \sum_{s=20}^{30} x^s = (8.0, 10.1),$$

and for 190-th to 200-th are

$$\bar{x} = \frac{1}{10} \sum_{s=190}^{200} x^s = (7.9, 9.7)$$

The following table contains detailed description of the problem.

x_1 means	3.02	6.07	9.77	16.26	6.12	14.80	7.24	7.52	15.91	13.57
	2.08	12.70	0.16	15.78	3.95	11.89	4.68	6.11	9.19	11.56
	12.43	19.98	15.33	18.20	7.84	1.16	4.54	17.48	10.78	1.45
x_2 means	7.63	6.62	15.40	10.83	4.85	17.14	2.20	9.30	17.30	14.60
	5.68	4.77	19.10	17.17	0.80	10.82	11.48	18.99	0.36	2.52
	10.00	1.93	11.39	16.41	16.21	2.09	16.69	8.70	12.04	2.93
x_1 devs.	18.65	18.95	0.45	13.50	17.55	1.12	18.42	1.59	15.65	9.49
	19.13	18.19	19.56	19.14	11.93	7.26	1.72	11.37	7.09	16.05
	15.62	4.31	15.44	1.40	5.82	8.56	16.72	5.29	10.36	12.49
x_2 devs.	3.77	15.79	8.68	6.29	7.97	9.23	5.81	3.17	17.91	7.02
	16.27	15.08	5.12	6.11	1.55	19.25	8.24	17.78	13.48	9.80
	5.49	15.13	7.07	16.83	15.86	9.90	19.44	16.35	0.37	15.31
weights	8.50	9.48	6.03	8.16	9.05	1.80	8.17	7.57	3.43	9.62
	2.87	3.77	4.34	4.88	0.11	2.13	7.75	1.64	5.75	6.12
	4.57	4.45	2.95	0.17	7.53	9.39	7.38	1.15	2.09	7.20

References

[1] Yu.M. Ermoliev, "Methods of stochastic Programming", Nauka, 1976, p.150 (in Russian).

[2] H. Robbins and S. Monro, "A stochastic approximation method", *Ann. Math. Statist.* **22**(1951), 400–407.

[3] H. Kesten, "Accelerated stochastic approximation", *Ann. Math. Statist.* **29**(1958), 41–59.

[4] G. Pflug, "On the determination of the step size in stochastic quasigradient methods". Collaborative Paper, CP-83-25, International Institute for Applied Systems Analysis, Laxenburg, Austria, 1983.

[5] S.P. Uriasiev, "Step regulation for direct methods of stochastic programming", *Kibernetika* **6**(1980), 85–87 (in Russian).

[6] S.P. Uriasiev, "Adaptive stepsize rules for stochastic optimization methods", Ph.D. thesis. Institute of Cybernetics, Kiev, 1983.

[7] F. Mirzoakhmedov and Urjasév, "Adaptive Step Size Control for Stochastic Optimization Algorithm", *Ih urn. vych.mat.i mat. fisiki*, **6**(1983), 1314–1325 (in Russian).

[8] R.E. Bruck, "On weak convergence of an Ergodic iteration the solution of variational inequalities for Monotone Operators in Hilbert Space".

CHAPTER 19

A NOTE ABOUT PROJECTIONS IN THE IMPLEMENTATION OF STOCHASTIC QUASIGRADIENT METHODS

R.T. Rockafellar and R.J-B Wets[*]

Given a stochastic optimization problem find $x \in X \subset R^n$ that minimizes $F(x) = E\{f(x,\xi)\}$ where $f : R^n x \Xi \to R$ is a real-valued-function, the quasi-gradient algorithm generates a sequence $\{x^1, x^2, \ldots\}$ of points of X (converging to the optimal solution with probability 1) through the recursion:

$$x^{\nu+1} := \underset{X}{\text{prj}} \left(x^\nu - \rho_\nu z^\nu \right)$$

where prj_X denotes the projection on X, $\{\rho_\nu, \nu = 1, \ldots\}$ is a sequence of positive scalars that tend to 0, and z^ν is a stochastic quasi-gradient of F at x^ν; see Chapter 5.

Unless X is a simple convex set, e.g. a rectangle or a ball, the projection operation may be too onerous to allow for a straightforward implementation of the iterative step; one would have to find at each step

$$x^{\nu+1} = \text{argmin}[\text{dist}^2 (x^\nu - \rho_\nu z^n, x) | x \in X],$$

which means solving a mathematical program with quadratic objective function. Therefore the implementations of the stochastic quasi-gradient method rely usually on various schemes to bypass this projection operation, through penalization or primal-dual methods, for example. There are however a few cases when it is possible to design a very effective subroutine to perform the projection operation.

We describe a simple method for projecting a point $\hat{y} \in R_+^n$ on a convex set X, assumed to be nonempty, that is the intersection of a rectangle $C \subset R^n$ and a set determined by a single linear or more generally by a separable nonlinear constraint of the type:

$$\sum_{j=1}^{n} a_j(x_j) \leq b, \tag{19.1}$$

where the a_j are convex differentiable functions such that for every $j = 1, \ldots, n$, the derivative a'_j of a $a_j(\cdot)$ is positive and bounded away from zero on C where

$$C := \{x \in R^n | \ell_j \leq x_j \leq u_j, \quad j = 1, \ldots, n\} \tag{19.2}$$

* Supported in part by grants of the National Science Foundation.

with $\ell_j = -\infty$ and $u_j = +\infty$ if x_j is not bounded below or above. We had to deal with such a case in connection with the model described in Chapter 22. (For related work, cf. [2]–[6].) Since the derivative of a convex function is a monotone nondecreasing function, the preceding condition on the derivative is satisfied if (and only if)

$$a'_j(\ell_j) > 0 \text{ if } \ell_j \text{ is finite} \tag{19.3}$$

or if $\ell_j = -\infty$

$$\lim_{\tau \to -\infty} a'_j(\tau) = a'_j(\ell_j) > 0.$$

Set $a'_j(u_j) = \lim_{\tau \to \infty} a'_j(\tau)$ if $u_j = +\infty$. In the special case when $a_j(\cdot)$ is linear, in which case we write

$$a_j(x_j) = a_j x_j, \tag{19.4}$$

this condition boils down to having $a_j > 0$.

The projection $\mathrm{prj}_X \hat{y}$ of $\hat{y}$ on X is the optimal solution of the (convex) nonlinear program

$$\text{find} \quad x \in C \subset R^n$$

$$\text{such that} \quad \sum_{j=1}^{n} a_j(x_j) \leq b \tag{19.5}$$

$$\text{and} \quad z = \frac{1}{2}\mathrm{dist}^2(\hat{y}, x) \text{ is minimized.}$$

Here "dist" is the Euclidean distance, i.e. the objective is the quadratic form

$$\mathrm{dist}^2(\hat{y}, x) = \sum_{j=1}^{n} x_j^2 - 2\sum_{j=1}^{n} \hat{y}_j x_j + \sum_{j=1}^{n} \hat{y}_j^2. \tag{19.6}$$

Since the feasible region

$$X = C \cap \{x | \sum_{j=1}^{n} a_j(x_j) \leq b\} \tag{19.7}$$

is a closed convex set, and the objective is an inf-compact (closed and bounded level sets) strictly convex function, the projection problem (19.5) is always solvable and it has a *unique* solution which is $\mathrm{prj}_X \hat{y}$.

Of course, it would be very easy to find the optimal solution of such a problem if there were no additional constraints besides $x \in C$. Our purpose is to show that with a single additional constraint it is possible to devise an algorithmic procedure for solving (19.5) that requires only marginally more work. This is achieved by constructing a (partial) dual to (19.5) whose solution gives us the (optimal) Lagrange multiplier λ^* to associate to the constraint $\Sigma_j a_j(x) \leq b$.

When this multiplier λ^* is known, then the theory of convex optimization allows us to replace (19.5) by the following separable convex optimization problem:

$$\text{find} \quad x \in C \subset R^n$$

$$\text{such that} \quad \sum_{j=1}^{n} [\frac{1}{2}(x_j - \hat{y}_j)^2 + \lambda' a_j(x_j)] \text{ is minimized.} \qquad (19.8)$$

The solution to such a problem yields $x^* = \text{prj}_X \hat{y}$, with

$$x_j^* = \begin{cases} \ell_j & \text{if } (\ell_j - \hat{y}_j) + \lambda^* a_j'(\ell_j) \geq 0, \\ u_j & \text{if } (u_j - \hat{y}_j) + \lambda^* a_j'(u_j) \leq 0, \\ x_j & \text{where } x_j + \lambda^* a_j'(x_j) = \hat{y}_j, \text{ otherwise.} \end{cases} \qquad (19.9)$$

In particular if $a_j(\cdot)$ is linear (19.4), then (19.9) becomes

$$x_j^* = \begin{cases} \ell_j & \text{if } (\ell_j - \hat{y}_j) + \lambda^* a_j \geq 0, \\ u_j & \text{if } (u_j - \hat{y}_j) + \lambda^* a_j \leq 0, \\ \hat{y}_j - \lambda^* a_j & \text{otherwise.} \end{cases} \qquad (19.10)$$

Thus all that is needed is an efficient procedure for finding λ^*. To do so let us consider the following convex optimization problem:

$$\text{find} \quad \lambda \in R_+$$

$$\text{such that} \quad g(\lambda) \text{ is maximized,} \qquad (19.11)$$

where

$$g(\lambda) = \min_{x \in C} \sum_{j=1}^{n} \frac{1}{2}(x_j - \hat{y}_j)^2 + \lambda a_j(x_j)] - \lambda b. \qquad (19.12)$$

In fact this problem is *dual* to our original problem (19.8). This claim can be substantiated by appealing to the general duality theory for convex optimization problems, cf. [7]; the Lagrangian generating (19.5) and (19.11) as a dual pair of problems is the function:

$$L(x, \lambda) = \begin{cases} \sum_{j=1}^{n} [\frac{1}{2}(x_j - \hat{y}_j)^2 + \lambda a_j(x_j)] - \lambda b & \text{if } x \in C, y \geq 0, \\ +\infty & \text{if } x \notin C, \lambda \geq 0, \\ -\infty & \text{if } \lambda < 0. \end{cases}$$

We can also argue directly as follows: define

$$\varphi(\eta) = \sup[\eta \lambda + g(\lambda) | \lambda \in R_+].$$

Note that $\varphi(0)$ is then the optimal value of (19.11). From (19.12) it follows that

$$\varphi(\eta) = \sup_{\lambda \geq 0} \lambda \left[\sum_{j=1}^{n} a_j(x_j) - b + \right] + \min_{x \in C} \frac{1}{2} \text{dist}^2(x, \hat{y})$$

and in particular for $\eta = 0$, since $X = C \cap \{x | \sum_{j=1}^{n} a_j(x_j) \leq b\}$ is nonempty, we obtain

$$\varphi(0) = \min_{x \in C} \frac{1}{2}[\text{dist}^2(x, \hat{y})] \text{ if } \sum_{j=1}^{n} a_j(x_j) \leq b$$

which is the optimal value of the projection problem (19.5). The equality of the optimal values implies in turn that if x^0 solves (19.5) and λ^0 solves (19.11) then from definition (19.12), we have

$$\lambda^0 \left(\sum_{j=1}^{n} a_j(x_j^0) - b \right) = 0 \tag{19.13}$$

Thus the multiplier λ^* that we seek, to substitute in (19.9), is the optimal solution of (19.11), the 1-dimensional optimization problem (on R). For any $\lambda \in R_+$, we can find an explicit expression, that yields the argmin of (19.11), similar to (19.9), namely

$$x_j(\lambda) = \begin{cases} \ell_j & \text{if } \lambda \geq \eta_j^+ = (\hat{y}_j - \ell_j)/a_j'(\ell_j), \\ u_j & \text{if } \lambda \leq \eta_j^- = (\hat{y}_j - u_j)/a_j'(u_j), \\ x_j & \text{if } \eta_j^- \leq \lambda \leq \eta_j^+ \\ & \text{where } x_j + \lambda a_j'(x_j) = \hat{y}_j. \end{cases} \tag{19.14}$$

Note that we have used the facts that a_j' is nonnegative and nondecreasing, so that $a_j'(\ell_j) \leq a_j'(u_j)$ and hence $\eta_j^- \leq \eta_j^+$ for all j. With

$$\begin{aligned} J^-(\lambda) &= \{j | \lambda < \eta_j^-\}, \\ J^+(\lambda) &= \{j | \lambda \geq \eta_j^+\}, \end{aligned} \tag{19.15}$$

and

$$J(\lambda) = \{j | \eta_j^- \leq \lambda < \eta_j^+\},$$

we have that

$$\begin{aligned} g(\lambda) = &\sum_{j \in J^-(\lambda)} [\frac{1}{2}(u_j - \hat{y}_j)^2 + \lambda a_j(u_j)] \\ &+ \sum_{j \in J^+(\lambda)} left[\frac{1}{2}(\ell_j - \hat{y}_j)^2 + \lambda a_j(\ell_j)] \\ &+ \sum_{j \in J(\lambda)} [\frac{1}{2}(x_j(\lambda) - \hat{y}_j)^2 + \lambda a_j(x_j(\lambda))] - \lambda b. \end{aligned} \tag{19.16}$$

The function g is concave: expression (19.12) gives us g as the sum of a linear function $(-b)\lambda$ and a min-function (of a collection of linear functions in λ). Thus the derivative, if it exists, is a monotone nonincreasing function of λ. Finding the maximum of g on R_+ corresponds to finding λ^* such that $g'(\lambda^*) = 0$, unless $g(0) \leq 0$ in which case $\lambda^* = 0$. Here, unless a'_j is pathological, we have that

$$g'(\lambda) = \sum_{j \in J^-(\lambda)} a_j(u^j) + \sum_{j \in J^+(\lambda)} a_j(\ell_j) - b$$
$$+ \sum_{j \in J(\lambda)} [(x_j(\lambda) - \hat{y}_j)x'_j(\lambda) + a_j(x_j(\lambda)) + \lambda a'_j(x_j(\lambda))x'_j(\lambda)]$$

and using the definition of $x_j(\lambda)$ when $j \in J(\lambda)$ this simplifies to

$$g'(\lambda) = \sum_{j \in J^-(\lambda)} a_j(u_j) + \sum_{j \in J^+(\lambda)} a_j(\ell_j) + \sum_{j \in J(\lambda)} a_j(x_j(\lambda)) - b. \qquad (19.17)$$

In the linear case, this becomes

$$g'(\lambda) = \sum_{j \in J^-(\lambda)} a_j u_j + \sum_{j \in J^+(\lambda)} a_j \ell_j + \sum_{j \in J(\lambda)} [a_j \hat{y}_j - a_j^2 \lambda] - b. \qquad (19.18)$$

To find $\lambda^* \in \text{argmax}[g(\lambda)|\lambda \in R_+]$, we propose the following procedure:

Step 0. Order $\{\eta_j^-, \eta_j^+, j = 1, \ldots, \eta\}$, say as $(\theta_1, \ldots, \theta_{2n})$, recording for each θ_i the corresponding label $(j, -)$ or $(j, +)$. (Ties correspond to an entry in the θ-vector repeated the appropriate number of times.)
Set $\theta^- = 0, \theta^+ = \theta_p$ with $p = \min(j|\theta_j > 0.)$
Construct $J^-(\theta^- = 0), J^+(0), J(0)$.
Compute

$$g'(0) = \sum_{j \in J^-(0)} a_j(u_j) + \sum_{j \in J^+(0)} a_j(\ell_j) + \sum_{j \in J(0)} a_j(\hat{y}_j) - b.$$

If $g'(0) \leq 0$, stop. Set $\lambda^* = 0$ and exit.
If $g'(0) > 0$, continue.

Step 1. Compute $g'(\theta^+)$ using (19.17) or (19.18).
If $g'(\theta^+) \leq 0$, then find $\lambda^* \in [\theta^-, \theta^+]$ such that $g'(\lambda^*) = 0$, exit.
If $g'(\theta^+) > 0$, continue.

Step 2. Set $p := p + 1, \theta^- := \theta^+, \theta^+ := \theta_p$
Adjust $J^-(\theta^-), J^+(\theta^-), J(\theta^-)$
Return to Step 1.

The algorithm clearly converges since it is a systematic search of a monotone nonincreasing function that eventually must reach the interval $[\alpha_p, \alpha_{p+1}]$ in which g' takes on the value 0; the problem is known to have a solution, see the preceding comments about duality.

In the linear case, all operations prescribed by the algorithm are simple and straightforward. The derivative $g'(\lambda)$ is given by (19.18). In Step 1, when $g'(\alpha^+) \leq 0, \lambda^*$ is given by the expression

$$\lambda^* = \beta/\gamma$$

where

$$\beta = \sum_{j \in J^-} a_j u_j + \sum_{j \in J^+} a_j \ell_j + \sum_{j \in J} a_j \hat{y}_j - b,$$

and

$$\gamma = \sum_{j \in J} a_j^2.$$

When the $a_j(\cdot)$ are nonlinear, the evaluation of $g'(\lambda)$ requires first the evaluation of $x_j(\lambda)$ for all $j \in J(\lambda)$. Also in Step 1 there may be difficulties in finding λ^* when $g'(\theta^+) \leq 0$. To begin with, let us consider the equations

$$x_j + \lambda a_j'(x_j) = \hat{y}_j. \tag{19.19}$$

Usually, there are many situations when it is easy to find a closed form expression for x_j as a function of λ. For example, if $a_j(z) = \alpha z^2 + \underline{z} + \gamma$ with $\alpha > 0$ (recall that $a_j(\cdot)$ is convex), then

$$x_j(\lambda) = (\hat{y}_j - \lambda \beta)/(1 + 2\alpha\lambda).$$

In general, however, even when an explicit expression for the derivative is available, we may have to resort to a numerical procedure for finding $x_j(\lambda)$. But here we are greatly aided by the following observation. For $\lambda \in [\bar{\jmath}^-, \eta_j^+]$ the function

$$z \mapsto \left(z + \lambda a_j'(z) - \hat{y}_j\right)$$

is monotone nondecreasing between ℓ_j and u_j with

$$(\ell_j - \hat{y}_j) + \lambda a_j'(\ell_j) < 0$$

and

$$(u_j - \hat{y}_j) + \lambda a_j'(u_j) \geq 0,$$

as follows from the definition of η_j^- and η_j^+, see (19.14). Thus a secant method [1], that we used in our implementation, is a very efficient procedure to find $x_j(\lambda)$.

We now turn to finding λ^* with $g'(\lambda^*) = 0$, knowing that

$$g'(\theta^-) > 0 \quad \text{and} \quad g'(\theta^+) \leq 0,$$

where g' is given by (19.17). The sets $J^-(\lambda)$, $J^+(\lambda)$ and $J(\lambda)$ remain fixed on this interval. Let

$$\beta = b - \sum_{j \in J^-} a_j(u_j) - \sum_{j \in J^+} a_j(\ell_j),$$

and

$$\gamma(\lambda) = \sum_{j \in J(\lambda)} a_j(x_j(\lambda)).$$

Note that from the definition of θ^- and θ^+ it follows that

$$\eta_j^- \leq \theta^- \leq \theta^+ \leq \eta_j^+, \text{ for all } j \in J.$$

Moreover, $\lambda \mapsto \gamma(\lambda)$ is a decreasing function with

$$\gamma(\theta^-) > \beta \text{ and } \gamma(\theta^+) \leq \beta.$$

We need to find λ^* such that $\gamma(\lambda^*) = \beta$. Unless we have some expression for $a_j(x_j(\lambda))$ that can be handled easily, we again need to rely on a numerical procedure, and in this case too the secant method suggests itself [1]. That is what we have used in our own implementation of the procedure.

This projection method is extremely efficient in the linear case but also produces very good results in the nonlinear case, in which case its efficiency is that of the secant method used in finding λ^* and $x_j(\lambda)$.

If there is more than one constraint, in addition to the upper and lower bounds, it may still be possible to use the procedure outlined here. For example it is possible to keep track of the active constraints, and when only one (or no) extra constraint is violated then we could use this procedure to obtain the projection, provided the projected point does not violate some other constraint. We should thus be able to cope with two or three extra constraints, resorting only once in a while to a general optimization procedure for solving (19.5).

Acknowledgment: We very much appreciate the comments of Dr. Richard Cottle (Stanford University) as well as his help on bibliographical questions.

References

[1] R.P. Brent, *Algorithms for Minimization with Derivatives*, Series in Automatic Computation, Prentice-Hall, New Jersey, 1978.

[2] G.R. Bitran and A.C. Hax, "Disaggregation and resource allocation using convex knapsack problems with bounded variables", *Management Science* **27**(1981) 431–441.

[3] R.W. Cottle and S.G. Duvall, *A Lagrangean relaxation algorithm for the constrained matrix problem*, Technical Report SOL 82-10, Systems Optimization Laboratory, Department of Operations Research, Stanford University, 1982.

[4] J.A. Ferland, B. Lemaire, and P. Robert, *Analytic solutions for nonlinear programs with one or two equality constraints*, Technical Report # 285, Departement d'Informatique et de Recherche Operationnelle, Université de Montréal, 1978.

[5] R. Helgason, J. Kennington and H. Lall, "A polynomial algorithm for a singly constrained quadratic program", *Mathematical Programming* **18** (1980) 338–343.

[6] R.K. McCord, "Minimization with one equality constraint and bounds on the variables", Ph.D. Thesis, Department of Operations Research, Stanford University, 1979.

[7] R.T. Rockafellar, *Conjugate Duality and Optimization*, Conference Series in Applied Mathematics 16, SIAM Publications, Philadelphia, 1974.

CHAPTER 20

DESCENT STOCHASTIC QUASIGRADIENT METHODS

K. Marti

20.1 Introduction

The FORTRAN-code "SEMI STOCHASTIC APPROXIMATION" can be applied in solving stochastic optimization problems of the following type

$$\text{minimize} \quad f(x)$$
$$\text{subject to} \quad x \in D, \tag{20.1}$$

where D is a closed convex subset of $\mathbb{R}^n$ and $F = F(x)$ is the convex mean value function defined by

$$F(x) = Eu(A(\omega)x - b(\omega)), \quad x \in \mathbb{R}^n. \tag{20.1.1}$$

Here $(A(\omega), b(\omega))$ is an $m \times (n+1)$ random matrix and u is a convex loss function on $\mathbb{R}^m$ such that the mean value $F(x)$ in (20.1.1) is real for every $x \in \mathbb{R}^n$. We suppose that the set D^* of optimal solutions x^* of (20.1) is nonempty.

Problems of the form (20.1) arise in many different connections, as e.g.

- Stochastic linear programming with recourse [7], [22]
- Portfolio optimization [9], [23]
- Error minimization and optimal design [2], [20]
- Statistical prediction [1]
- Optimal decision functions [5], [10].

Since the gradient (or subgradient) ∂F of F exists under weak assumptions and is given then by the formula

$$\partial F(x) = EA(\omega)'\partial u(A(\omega)x - b(\omega)), \tag{20.2}$$

where A' is the transpose of a matrix A and ∂u denotes the subgradient of u, our basic problem (20.1) could be attacked in principle by a gradient (or quasigradient) procedure of the type

$$x_{k+1} = P_D(x_k - \rho_k g_k), k = 1, 2, \ldots, \tag{20.3}$$

where $\rho_k > 0$ is a step size, $g_k \in \partial F(x_k)$ and P_D denotes the projection of $\mathbb{R}^n$ onto D.

However, in practice the computation of the gradient (subgradient) $\partial F(x_k)$ causes in general the following difficulties:

- Either formula (20.2) can not be evaluated at all because only a stochastic estimate Y_k of an element $g_k \in \partial F(x_k)$ is available [3], [21]. Hence, in this case we only have

$$Y_k = g_k + \text{ noise with some } g_k \in \partial F(x_k); \qquad (20.4.1)$$

- Or, though the integrand $A'\partial u(Ax - b)$ and the probability distribution $P_{(A(\cdot),b(\cdot))}$ of $A(\omega), b(\omega)$ in (20.2) is known, the numerical evaluation of this formula (20.2)—involving a multiple integral—is impossible in practice, since it takes too much computing time. In this case $\partial F(x_k)$ may be approximated by

$$Y_k \in A(\omega_k)'\partial u(A(\omega_k)x_k - b(\omega_k)), \qquad (20.4.2)$$

where $(A(\omega_k), b(\omega_k))$ is a realization of the random matrix $(A(\omega), b(\omega))$ generated independently of x_k by means of a pseudo random generator [11].

Consequently, in both cases (20.4.1) and (20.4.2) the gradient procedure can not be applied in practice.

It is therefore often replaced by the stochastic quasigradient method [3], [6]

$$X_{k+1} = P_D(X_k - \rho_k Y_k), k = 1, 2, \ldots, \qquad (20.5)$$

where the random direction is defined now as described by (20.4.1) or (20.4.2). Selecting *a priori* a sequence of step sizes $\rho_1, \rho_2, \ldots$ such that

$$\rho_k > 0, \sum_{k=1}^{\infty} \rho_k = +\infty, \sum_{k=1}^{\infty} \rho_k^2 < +\infty,$$

e.g. $\rho_k = \frac{c}{q+k}$ for some constants $c > 0$ and $q \in \mathbb{N} \cup \{0\}$, it is well known [19], [21] that the sequence of random iterates $X_1, X_2, \ldots$ generated by (20.5) converges with probability one to the set D^* of optimal solutions x^* of (20.1), provided that the approximates Y_k of $\partial F(x_k)$ fulfill a certain uniform second order integrability condition and D^* is a bounded set.

Unfortunately, due to their probabilistic nature, stochastic approximation procedures only have a very slow asymptotic rate of convergence of the type

$$E\|X_k - x^*\|^2 = 0(k^{-\lambda}),$$

where λ is a constant such that $0 < \lambda \leq 1$.

Moreover, the main disadvantage of stochastic quasigradient procedures (20.5) is their nonmonotonicity which sometimes may be displayed in a highly oscillatory behavior [4]. Hence, in many cases one does not know whether the algorithm has reached already a certain neighborhood of an optimal solution x^* or not.

For improving the convergence behavior of (20.5), several methods were suggested, e.g based on the adaptive selection of the step sizes ρ_k, see [8], or based on the use of second order information about F, see [18].

A further method—having a partial monotonicity property— is presented in the following.

20.2 Semi-Stochastic approximation

As was shown in several papers [10], [12], [14], [15], [17], for several classes U of convex loss functions u and several classes Π of distributions $P_{(A(\cdot),b(\cdot))}$ of the random matrix $(A(\omega), b(\omega))$, our minimization problem (20.1) has the following important

Property: (20.6)

At certain "nonefficient" or "nonstationary" points $x \in D$ there exists a deterministic (feasible) descent direction $h = h(x)$ of F which can be computed with less computing expenses than an element g_k of $\partial F(x_k)$. Moreover, $h(x)$ is stable with respect to variations of the loss of function $u \in U$.

Consequently, if at a certain iteration point X_k this property (20.6) holds, then clearly one will replace the stochastic direction $-Y_k$, which is a descent direction only in the mean, by the descent direction $h_k = h(X_k)$ of F available then at X_k with low computing expenses.

Hence, we obtain—as already described in [11], [13]—the following

Descent Stochastic Quasigradient Method

$$X_{k+1} = \begin{cases} P_D(X_k + \rho_k h_k), & \text{if (20.6) holds at } x_k \\ P_D(X_k - \rho_k Y_k), & \text{else.} \end{cases} \tag{20.7.1}$$

In many important applications this hybrid procedure has the important feature that property (20.6) is fulfilled, e.g. at every second iteration point X_k. Hence, in this case (20.9.1) has the more convenient form

$$X_{k+1} = \begin{cases} P_D(X_k + \rho_k h_k), & \text{if } k \in \mathbb{N}_1 \\ P_D(X_k - \rho_k Y_k), & \text{if } k \in \mathbb{N}_2, \end{cases} \tag{20.7.2}$$

where $\mathbb{N}_1, \mathbb{N}_2$ is a known decomposition of the set of integers $\mathbb{N}$, e.g. $\mathbb{N}_1 = \{1, 3, 5 \ldots\}$, $\mathbb{N}_2 = \{2, 4, 6, \ldots\}$. As was shown in [13], if the step sizes $\rho_1, \rho_2, \ldots$ are chosen such that

$$\rho_k > 0, \sum_{k \in \mathbb{N}_2} \rho_k < +\infty, \sum_{k=1}^{\infty} \rho_k^2 < +\infty,$$

then the semi-stochastic approximation procedure (20.7) converges with probability one to the set D^* of optimal solutions x^* of (20.1).

As expected, several numerical examples [11] show that the descent stochastic quasigradient method (20.7) has a much better convergence behavior than the pure stochastic quasigradient method. Especially, the highly oscillatory behavior of the random iterates $X_k, k = 1, 2, \ldots$, observed in (20.5) is damped very much by using also deterministic descent directions h_k in (20.7); moreover, the set D^* of optimal solutions is reached more exactly. In a recent paper [16] the rate of convergence of (20.7) could be estimated as follows.

Theorem 2.1. *Denote by $b_k = E\|X_k - x^*\|^2$ and $b_k^s = E\|X_k^s - x^*\|^2$ the mean square error of the descent stochastic quasigradient, pure stochastic quasigradient method, respectively.*

(a) If a fixed rate of stochastic and deterministic steps are taken in (20.7), then there are constants Q_1, Q_2 with $0 < Q_1 < 1, Q_1 < Q_2$ such that

$$Q_1 \cdot b_k^s \leq b_k \leq Q_2 \cdot b_k^s \text{ as } k - > \infty. \qquad (20.8)$$

Furthermore, Q_1, Q_2 are given by known formulas and $Q_2 < 1$ holds if $\frac{N}{M} < \gamma$, where N, M is the number of stochastic, deterministic steps, respectively, in one complete turn of iterations and *gamma* is a certain ratio depending on the parameters of the problem (20.1).

(b) If the stochastic steps in (20.7) are taken at a decreasing rate, then the speed of convergence is increased from $b_k^2 = 0\left(\frac{1}{k}\right)$ in the pure stochastic case to $b_k = 0\left(k^{-\lambda}\right)$ with a constant $1 < \lambda < 2$ in the semi-stochastic case.

20.3 Construction of deterministic descent direction

Up to now deterministic feasible descent directions may be constructed if the distributions $P_{(A(\cdot), b(\cdot))}$ are

- stable [12]
- invariant [15]
- discrete [14].

The following implementation is based on the assumption that $(A(\omega), b(\omega))$ has a $m(n + 1)$-dimensional

normal distribution with   (20.9)

mean $(\overline{A}, \overline{b})$ and   (20.9.1)

$$\text{covariance matrix } Q = \begin{pmatrix} Q_{11} & Q_{12} & \cdots & Q_{1m} \\ Q_{21} & Q_{22} & \cdots & Q_{2m} \\ \vdots & \vdots & & \vdots \\ Q_{m1} & Q_{m2} & \cdots & Q_{mm} \end{pmatrix}, \qquad (20.9.2)$$

where the $(n + 1) \times (n + 1)$ matrix Q_{ij} denotes the covariance matrix of the i-th and j-th row $(A_i(\omega), b_i(\omega)), (A_j(\omega), b_j(\omega))$, resp., of the random matrix $(A(\omega), b(\omega))$.

Besides (20.9) we still suppose:

The objective function F of (20.1) is not constant $\qquad(20.10)$
on arbitrary line segments $\overline{xy}$ of $\mathbf{R}^n$.

From assumption (20.9) follows that the random m-vector $A(\omega)x - b(\omega)$ has a normal distribution with mean $\overline{A}x - \overline{b}$ and covariance matrix

$$
Q_x = \begin{pmatrix}
\widehat{x}'Q_{11}\widehat{x} & \widehat{x}'Q_{12}\widehat{x} & \cdots & \widehat{x}'Q_{1m}\widehat{x} \\
\widehat{x}'Q_{21}\widehat{x} & \widehat{x}'Q_{22}\widehat{x} & \cdots & \widehat{x}'Q_{2m}\widehat{x} \\
\vdots & \vdots & & \vdots \\
\widehat{x}'Q_{m1}\widehat{x} & \widehat{x}'Q_{m2}\widehat{x} & \cdots & \widehat{x}'Q_{mm}\widehat{x}
\end{pmatrix}
$$

where $\widehat{x} = \left(\begin{smallmatrix} x \\ -1 \end{smallmatrix}\right)$.

The key for the construction of descent directions is now this

Theorem 3.1. *Suppose that assumptions (20.9) and (20.10) are fulfilled. If n-vectors $x, y \neq x$ are related according to the relations*

$$\overline{A}x = \overline{A}y \qquad(20.11.1)$$

$$Q_x - Q_y \text{ is positive semidefinite,} \qquad(20.11.2)$$

then $F(y) \le F(x)$ and $h = y - x$ is a descent direction of F at x. Moreover, if $x \in D$ and in addition to $(20.11.1)$ and $(20.11.2)$ we still have

$$y \in D, \qquad(20.11.3)$$

then $h = y - x$ is a feasible descent direction of F at x.

Note For given x $(20.11.1)$ is a system of m linear equations for y. Relation $(20.11.2)$ means that the smallest eigenvalue of $Q_x - Q_y$ is nonnegative. In the important special case $m = 1$, $(20.11.2)$ is reduced to the single quadratic constraint

$$\widehat{x}'Q_{11}\widehat{x} \ge \widehat{y}'Q_{11}\widehat{y}. \qquad(20.11.2a)$$

If $(A(\omega), b(\omega))$ has stochastically independent rows, then $(20.11.2)$ is equivalent to

$$\widehat{x}'Q_{ii}\widehat{x} \ge \widehat{y}'Q_{ii}\widehat{y} \quad \text{for all } i = 1, 2, \ldots, m. \qquad(20.11.2b)$$

In this case solutions y of (20.11) may be obtained by solving for given vector x the convex program

$$
\begin{aligned}
\text{minimize} \quad & \widehat{y}'Q_{i_0 i_0}\widehat{y} \\
\text{subject to} \quad & \widehat{y}'Q_{ii}\widehat{y} \le \widehat{x}'Q_{ii}\widehat{x}, \quad i = 1, 2, \ldots, m \\
& \overline{A}y = \overline{A}x \\
& y \in D,
\end{aligned} \qquad(20.12)
$$

where $1 \le i_0 \le m$ is a fixed integer.

In the general case one has to consider the program

$$
\begin{aligned}
\text{minimize} \quad & \lambda(Q_x - Q_y) \\
\text{subject to} \quad & \overline{A}y - \overline{A}x \\
& y \in D,
\end{aligned} \qquad(20.13)
$$

where $\lambda(Q_x - Q_y)$ denotes the smallest eigenvalue of $Q_x - Q_y$.

20.4   Implementation

20.4.1   Representation of the random matrix $(A(\omega), b(\omega))$

$(A(\omega), b(\omega))$ is defined by

$$(A(\omega), b(\omega)) = (A^0, b^0) + \sum_{j=1}^{r} \omega^j (A^j, b^j). \tag{20.14}$$

where (A^j, b^j), $j = 0, 1, \ldots, r$, are $m \times (n+1)$ matrices to be selected by the user and $\omega^1, \omega^2, \ldots, \omega^r$ are independent normal random variables with mean zero and variance one. A realization $(A(\omega_k), b(\omega_k))$ of $(A(\omega), b(\omega))$ is then represented by

$$(A(\omega_k), b(\omega_k)) = (A^0, b^0) + \sum_{j=1}^{r} \omega_k^j (A^j, b^j),$$

where $\omega_k = (\omega_k^1, \omega_k^2, \ldots, \omega_k^r)$, $k = 0, 1, \ldots$, is a sequence of stochastically independent realizations of the random r-vector $\omega = (\omega^1, \omega^2, \ldots, \omega^r)$ generated by means of a pseudo random generator (converting uniformly distributed pseudo random numbers into normal distributed ones based on the central limit theorem).

20.4.2   Computation of the search directions

We suppose that
$$\text{rank } \overline{A} = \text{rank } A^0 = m < n.$$

The matrix $\overline{A} = (\overline{a}_1, \overline{a}_2, \ldots, \overline{a}_m)$, $\overline{a}_k = k$-th column of $\overline{A}$, must be partitioned by the user into a regular $m \times m$ matrix

$$B = (\overline{a}_{k_1}, \overline{a}_{k_2}, \ldots, \overline{a}_{k_m})$$

and an $m \times (n-m)$ rest matrix

$$E = (\overline{a}_{\kappa_1}, \overline{a}_{\kappa_2}, \ldots, \overline{a}_{\kappa_{n-m}}).$$

The user has then to define the index set

$$\text{INDXA0} = \{k_1, k_2, \ldots, k_m, \kappa_1, \kappa_2, \ldots, \kappa_{n-m}\}.$$

Given the last iteration point x_k, in the subroutine FUNCT a solution y_k of the relations (20.11.1) · (20.11.3) is computed.

At present only the case $D = \mathbb{R}^n$ is implemented. For sake of generality the system of relations (20.11) is solved by means of the program (20.13). However, having a more special situation, the user only has to replace the procedure

(20.13) implemented presently in FUNCT by his own procedure for solving (20.11).

If $y_k \neq x_k$, then $h_k = y_k - x_k$ is a feasible descent direction (see Theorem 3.1) and the next iteration point x_{k+1} is defined by

$$x_{k+1} = x_k + \rho_k(y_k - x_k),$$

where $\rho_k > 0$ is a step size..

If $y_k = x_k$, then FUNCT fails to find a descent direction. Hence, the next iteration point is defined by

$$x_{k+1} = x_k - \rho_k Y_k,$$

where

$$Y_k \in A(\omega_k)' \partial u(A(\omega_k)x_k - b(\omega_k)).$$

20.4.3 Step size

At present the step sizes ρ_k, $k = 0, 1, \ldots$, are defined by

$$\rho_k = \frac{1}{k+1}.$$

For a deterministic step the user may also take $\rho_k = 1$ or $\rho_k = 0.5$.

20.4.4 Loss function u

The following classes of loss functions are implemented:

(a) *Quadratic loss functions*

$$u(z) = c + q'z + z'Wz, z \in \mathbb{R}^m,$$

where c is a fixed number, q denotes an m-vector and W is a positive semidefinite $m \times m$ matrix.

(b) *Polynomial loss function*

$$u(z) = \sum_{j=1}^{m} z_j^{2s}, z = (z_1, \ldots, z_m)' \in \mathbb{R}^m,$$

where s is a fixed integer.

(c) *Sublinear loss function*

$$u(z) = \max_{1 \leq t \leq \rho} f_t'z, z \in \mathbb{R}^m,$$

where $f_1, f_2, \ldots, f_\rho$ are fixed m-vectors.

20.4.5 Stopping criterion

The user has to select a (small) positive number EPS> 0, an integer ITMAX and a number TMAX. The procedure runs until the first of the following conditions is fulfilled:

$$\|x_{k+1} - x_k\| \leq \text{EPS} ,$$
$$k \leq \text{ITMAX} \ (= \text{maximal number of iterations}),$$
$$T \leq \text{TMAX} \ (= \text{maximal computing time}),$$

where $\| \bullet \|$ denotes the Euclidean norm.

Acknowledgment: The FORTRAN code was written by A. Böhme.

References

[1] J. Aitchison and I.R. Dunsmore, *Statistical Prediction Analysis*, Cambridge University Press, 1975.

[2] K.J. Aström, *Introduction to Stochastic Control Theory*, New York-London: Academic Press, 1970.

[3] Yu. Ermoliev, "Stochastic Quasigradient Methods and their Application to System Optimization", *Stochastics* 9(1983), 1–36.

[4] Yu. Ermoliev and A. Gaivornoski, *Stochastic Quasigradient Methods and their Implementation*, IIASA Working Paper, Laxenburg, 1983.

[5] T.S. Ferguson, *Mathematical Statistics*, New York-London: Academic Press, 1967.

[6] J.B. Hiriart-Urruty, *Contributions a la programmation mathematique: Cas deterministe et stochastique*, Thesis, University of Clermont-Ferrand II, 1977.

[7] P. Kall, *Stochastic Linear Programming*, Berlin-Heidelberg-New York: Springer-Verlag, 1976.

[8] H. Kesten, "Accelerated stochastic approximation", *Ann. Math. Statist.* 29(1958), 41–59.

[9] K. Marti and R.-J. Riepl, "Optimale Portefeuilles mit stabil verteilten Renditen", *ZAMM* 57(1977), T337–T339.

[10] K. Marti, *Approximationen stochastischer Optimierungsprobleme*, Königstein/Ts.: Hain, 1979.

[11] K. Marti, "On solutions of stochastic programming problems by descent procedures with stochastic and deterministic directions", *Methods of Op. Res.* 33(1979), 281–293.

[12] K. Marti, "Stochastic linear programs with stable distributed random variables", in Optimization Techniques Part 2, J. Stoer (ed.), Lecture Notes in Control and Information Sciences 7(1978), 76–86.

[13] K. Marti, "Solving stochastic linear programs by semi-stochastic approximation algorithms", in *Recent results in stochastic programming*, P. Kall, A. Prékopa (eds.), Lecture Notes in Economics and Mathematical Systems 179(1980), 191–213.

[14] K. Marti, "On the construction of descent directions in a stochastic program having a discrete distribution", *ZAMM* 64(1984), T336–T338.

[15] K. Marti, "Computation of Descent Directions in Stochastic Optimization Problems with Invariant Distributions", *ZAMM* **65**(1985), 355–378.

[16] K. Marti and E. Fuchs, "Rates of convergence of semi-stochastic approximation procedures for solving stochastic optimization problems". Preprint in: Mitteilungen des Forschungsschwerpunktes Simulation und Optimierung deterministischer und stochastischer dynamischer Systeme. HSBw München, November 1984, to appear in *Optimization*.

[17] K. Marti, "Stochastische Dominanz und stochastische lineare Programme", *Methods of Oper. Res.* **23**(1977), 141–160.

[18] B.T. Poljak and Ya. Z. Tyspkin, "Robust psuedogradient adaptation algorithms", *Automation and Remote Control* **41**(1980), 1404–1410.

[19] L. Schmetterer, "Stochastic approximation", *Proceedings 4th Berkeley Symposium on Mathematical Statistics and Probability*, **Vol. I** (1960), 587–109.

[20] H.W. Sorenson, *Parameter Estimation*, New York-Basel: M. Dekker(1980).

[21] M.T. Wasan, *Stochastic Approximation*, Cambridge University Press(1969).

[22] R. J-B. Wets, "Stochastic Programming: Solution Techniques and Approximation Schemes", in *Mathematical Programming: The State of the Art*, A. Bachem, M. Grötschel, B. Korte (eds.), pp. 566–603, Springer-Verlag 1983.

[23] W.T. Ziemba and R.G. Vickson (eds.): *Stochastic Optimization Models in Finance*, New York-London: Academic Press 1975.

CHAPTER 21

STOCHASTIC INTEGER PROGRAMMING BY DYNAMIC PROGRAMMING

B.J. Lageweg, J.K. Lenstra, A.R. Kan and L. Stougie

Abstract

Stochastic integer programming is a suitable tool for modeling hierarchical decision situations with combinatorial features. In continuation of our work on the design and analysis of heuristics for such problems, we now try to find optimal solutions. Dynamic programming techniques can be used to exploit the structure of two-stage scheduling, bin packing and multiknapsack problems. Numerical results for small instances of these problems are presented.

21.1 Introduction

Stochastic integer programming problems appear to be among the hardest problems in the area of mathematical programming. Most research on these problems has so far concentrated on the design and analysis of *approximation algorithms*. A survey of recent work in this direction, illustrated on the probabilistic analysis of a two-stage scheduling heuristic, can be found elsewhere in this volume [9].

In this chapter, we are interested in *optimization algorithms* for stochastic integer programming. The development of a reasonably efficient general procedure for this purpose seems a tremendous research challenge. Our objective is more modest. We will consider stochastic integer programs of a very special structure. The stochastic parameters will have a discrete distribution with a finite number of points with positive density. Moreover, each realization will lead to a combinatorial optimization problem that is solvable by a dynamic programming routine. The overall stochastic optimization problem will then be solved by a single giant recursion that combines the separate dynamic programming computations for all the individual realizations. This can be done only for problem instances of a relatively small size. Still, our numerical results give valuable insight into the shape of value functions of stochastic integer programming problems.

The following three sections illustrate our approach on two-stage *scheduling*, *bin packing*, and *multiknapsack* problems. In each section, we first formulate the problem in question, then present the dynamic programming algorithm, and finally discuss our numerical results. We note that the computational experience was obtained by improved implementations of the basic recursions, the technical details of which can be found in an extended version of this paper [7].

21.2   Scheduling

21.2.1   Problem Formulation

The two-stage scheduling problem studied in this section was first formulated in [2]. At the aggregate level, one has to decide on the number X of *identical parallel machines* that are to be acquired, while knowing the *cost* c of a single machine, the number n of *jobs* that are to be processed, and the probability distribution of the vector $\omega = (\omega_1, \ldots, \omega_n)$ of their *processing times*. It is assumed that the ω_j are independent and identically distributed random variables with expectation μ. At the detailed level, after X has been determined, a realization $\omega \in \Omega$ of ω becomes known, where Ω denotes the set of all realizations, and one has to decide on a *schedule* in which each machine processes at most one job at a time, job j is processed during an uninterrupted time period of length ω_j $(j = 1, \ldots, n)$ and no job is processed prior to time 0, so as to achieve a minimum value $Y^*(X, \omega)$ of the maximum job completion time. The total cost of the acquisition decision X and the optimal scheduling decision is denoted by $V^*(X, \omega) = cX + Y^*(X, \omega)$.

In the two-stage decision model, the objective is to determine a value $X^* \in \mathbb{N}$ such that the expected total cost is minimized:

$$EV^*(X^*, \omega) = \min_{X \in \mathbb{N}} \{EV^*(X, \omega)\}.$$

In the distribution model, the objective is to determine a function $X^0 : \Omega \to \mathbb{N}$ such that for each $\omega \in \Omega$ the actual total cost is minimized:

$$V^*(X^0(\omega), \omega) = \min_{X \in \mathbb{N}} \{V^*(X, \omega)\}, \qquad \forall \omega \in \Omega.$$

Previous work on this problem concerned the design and analysis of a two-stage heuristic [3]. This heuristic sets the number of machines equal to the value of X that minimizes the *lower bound* $V^{LB}(X) = cX + n\mu/X$ on $EV^*(X, \omega)$ and assigns the jobs to the machines by a *list scheduling* rule. (In our computational experiments, we used the *longest processing time* rule, which puts the jobs on a list in order of nonincreasing processing times and successively assigns the next job on the list to the earliest available machine; this rule has a better worst case performance than arbitrary list scheduling [5].) The relative error of the heuristic tends to 0 as n tends to infinity for various measures of stochastic convergence [3].

21.2.2 Dynamic programming

The second stage scheduling problem of determining $Y^*(X,\omega)$ for given X and ω is NP-hard [4]. We will consider the situation in which the processing times can assume only k distinct values $a_1,\ldots,a_k$, for a fixed value of k. Let us denote by $\omega = [n_1,\ldots,n_k]$ the vector of processing times in which the value a_j occurs n_j times, for $j = 1,\ldots,k$.

One can obtain an optimal schedule on X machines by assigning a certain subset of jobs optimally to $X - 1$ machines and putting the remaining jobs on another machine. This observation leads to the following recurrence relations:

$$Y^*(X,[n_1,\ldots,n_k]) = \min\{\max\{Y^*(X-1,[n_1 - \ell_1,\ldots,n_k - \ell_k]),$$
$$Y^*(1,[\ell_1,\ldots,\ell_k])\}$$
$$|0 \le \ell_j \le n_j(j = 1,\ldots,k)\}(X > 1),$$

$$Y^*(1,[n_1,\ldots,n_k]) = \sum_{j=1}^{k} n_j a_j.$$

Computation of $Y^*(X,\omega)$ by a dynamic programming algorithm based on this recursion requires $O(X \prod_{j=1}^{k} n_j)$ time, which is exponential in k but polynomial for fixed k.

In the more general context of the two-stage scheduling problem, we assume that the processing times have a discrete distribution with k integral values $a_1,\ldots,a_k$ in its support. The independence of the processing times implies that $\omega = [\mathbf{n}_1,\ldots,\mathbf{n}_k]$ has a multinomial distribution. The idea is now to go through the entire recursion once in order to compute $Y^*(X,\omega)$ for all values $X \in \{1,\ldots,n\}$ and for all realizations $\omega \in \Omega$, where Ω is given by

$$\Omega = \{[n_1,\ldots,n_k]|0 \le n_j \le n(j = 1,\ldots,k), n_1 + \cdots + n_k = n\}.$$

The distribution model is then solved by the selection, for each $\omega \in \Omega$, of a value of X that minimizes $V^*(X,\omega) = cX + Y^*(X,\omega)$. The two-stage decision model is solved by the determination of a value of X that minimizes $EV^*(X,\omega) = cX + \sum_{\omega \in \Omega} \Pr\{\omega = \omega\}Y^*(X,\omega)$.

A straightforward application of the above dynamic programming algorithm requires $O(n^k)$ comparisons for each of the $O(n^{k+1})$ pairs (X,ω), and hence $O(n^{2k+1})$ time altogether; the multinomial probabilities are easily computed within this time bound. A more efficient implementation reduces the overall running time to $O(n^2 k)$ [7].

21.2.3 Computational results

The dynamic programming algorithm was coded in PASCAL and run on a CD Cyber 170-750 to solve several instances of the two-stage scheduling problem. The solution of instances with 100 jobs and two possible processing time values or with 50 jobs and three processing time values required about 30 seconds. The values of k considered are admittedly small, but the values of n are realistic and the running times are such that our brute force approach should not be dismissed on grounds of manifest inefficiency.

We illustrate the numerical results on a set of representative instances given by

$$c = 1,$$
$$n = 1, \ldots, 100,$$
$$k = 2, a_1 = 18, a_2 = 14, \Pr\{\omega_j = a_1\} = \Pr\{\omega_j = a_2\} = \tfrac{1}{2} \, (j = 1, \ldots, n).$$

Figure 21.1 shows four functions of the number of jobs:
- the minimal lower bound $\min_X\{V^{LB}(X)\}$ mentioned in Section 21.2.1;
- the minimal expected total cost $EV^*(X^*, \omega)$ (the optimum for the two-stage decision model);
- the expected minimal total cost $EV^*(X^0(\omega), \omega)$ (the optimum for the distribution model, averaged over all realizations);
- the expected approximate total cost obtained by the heuristic mentioned in Section 21.2.1.

Note that the last three functions are defined only for integral n; linear interpolation has been applied to improve the presentation. The distribution model yields slightly better results than the two-stage decision model on average, as expected. A comparison between the optima and the lower and upper bounds confirms that the absolute differences are significant while the relative differences disappear with increasing problem size.

For the case that $n = 100$, Figure 21.2 shows three functions of the first stage decision variable, the number X of machines:
- the lower bound $V^{LB}(X)$;
- the expected total cost $EV^*(X, \omega)$ in case of an optimal second stage decision;
- the expected total cost in case of an approximate second stage decision.

Note that we have interpreted X as a continuous variable: acquisition of a fractional machine costs a fraction of c but yields no benefit at the second stage; the vertical line segments correspond to discontinuities. In spite of the smoothing effect due to averaging over all realizations, both the optimal and the approximate cost functions are highly nonconvex and multimodal. The functions consist of a first stage component, which is linear and increasing, and a second stage component, which is nonconvex and nonincreasing. Addition of the two components can turn the nonconvexities into local minima, and small values of c appear to be most effective in this respect.

21.3 Bin Packing

21.3.1 Problem formulation

The two-stage bin packing problem is formulated as follows. At the aggregate level, one has to decide on the *capacity* Y of *bins*, while knowing the *cost* d of one unit of capacity, the number n of *items* that are to be packed into the bins, and the probability distribution of the vector $\omega = (\omega_1, \ldots, \omega_n)$ of the item *weights*. It is again assumed that the ω_j are independent and identically distributed random variables with expectation μ. At the detailed level, after Y has been determined, a realization $\omega \in \Omega$ of ω becomes known, and one has to decide on a *packing* in which each item is assigned to a bin and the total weight of the items assigned to the same bin does not exceed its capacity Y, so as to achieve a minimum number $X^*(Y, \omega)$ of bins needed. The total cost of the first stage decision Y and the optimal second stage decision is denoted by $W^*(Y, \omega) = dY + X^*(Y, \omega)$.

In the two-stage decision model, the objective is to determine a value $Y^* \in \mathbb{R}_+$ such that

$$EW^*(Y^*, \omega) = \min_{Y \in \mathbb{R}_+} \{EW^*(Y, \omega)\}.$$

In the distribution model, the objective is to determine a function $Y^0 : \omega \to \mathbb{R}_+$ such that

$$W^*(Y^0(\omega), \omega) = \min_{Y \in \mathbb{R}_+} \{W^*(Y, \omega)\}, \qquad \forall \omega \in \Omega.$$

This problem is the symmetric counterpart of the two-stage scheduling problem from the previous section. One can view items as jobs, weights as processing times, bins as machines and their capacity as a job completion deadline, but now the order of the decisions is reversed. In fact, the above cost structure is quite natural in this context. First, a delivery date for the jobs is negotiated, whereby the cost of extending this date by one unit is independent of the number of machines that will turn out to be needed later on.

In analogy to the two-stage scheduling heuristic given at the end of Section 21.2.1, one can consider the following two-stage bin packing heuristic. The bin capacity is set equal to the value of Y that minimizes the *lower bound* $W^{LB}(Y) = dY + n\mu/Y$ on $EW^*(Y, \omega)$, and the items are packed into bins by the *first fit decreasing* rule, i.e., the items are taken in order of nonincreasing weights and each next item is assigned to the first bin that has enough capacity to accommodate it. This heuristic can be shown to have several strong properties of asymptotic optimality [10].

21.3.2 Dynamic programming

The second stage bin packing problem of determining $X^*(Y, \omega)$ for given X and ω is $\mathcal{NP}$-hard [4]. We will again consider the situation in which the stochastic parameters can assume only k values $a_1, \ldots, a_k$, for a fixed k, and write $\omega = [n_1, \ldots, n_k]$ to denote the vector in which the value a_j occurs n_j times, for $j = 1, \ldots, k$.

The following dynamic programming algorithm given in to [6]. Let $C(Y, \omega)$ be the total amount of capacity needed to pack items with weights specified by ω into bins of capacity Y. It is assumed that $C(Y, \omega)$ includes the slack capacity of each bin (which is equal to Y minus the total weight of the items assigned to that bin) except for the slack capacity of the last bin. Thus, if $C(Y, \omega) = XY - \Gamma$ with $X \in \mathbb{Z}_+$ and $0 \leq \Gamma < Y$, then an optimal packing requires X bins and the last bin has a slack capacity of Γ. Let $\Delta(Y, \omega, a)$ be the extra capacity needed when an item with weight a is added to this packing:

$$\Delta(Y, \omega, a) = \begin{cases} a & \text{if } \Gamma \geq a, \\ \Gamma + a & \text{if } \Gamma < a. \end{cases}$$

It is not hard to see that

$$C(Y, [n_1, \ldots, n_k]) = \min_{1 \leq j \leq k : n_j > 0} \{ C(Y, [n_1, \ldots, n_{j-1}, n_j - 1, n_{j+1}, \ldots, n_k]) + \Delta(Y, [n_1, \ldots, n_{j-1}, n_j - 1, n_{j+1}, \ldots, n_k], a_j) \}$$
$$(n_1 + \ldots + n_k > 0),$$

$$C(Y, [0, \ldots, 0]) = 0.$$

We finally have that $X^*(Y, \omega) = \lceil C(Y, \omega)/Y \rceil$.

For the two-stage bin packing problem, we make the same assumptions concerning the distribution of the stochastic parameters as in Section 21.2.2 and apply the same strategy to obtain solutions to both stochastic optimization models. Since the values $a_1, \ldots, a_k$ are integral, there is no loss of generality in considering only integral capacities Y. Let $a_{\max} = \max\{a_1, \ldots, a_k\}$ and note that $1 \leq Y \leq n a_{\max}$. The algorithm requires a fixed number of comparisons for each of the $O(n^{k+1} a_{\max})$ pairs (Y, ω), and hence $O(n^{k+1} a_{\max})$ time altogether. A more efficient implementation reduces the overall running time to $O(n^{k+(1/2)} a_{\max}^{1/2} d^{-(1/2)})$ [7].

Due to the relation between the two-stage scheduling and bin packing problems that was observed above, the $Y^*(X, \omega)$ values from Section 21.2.2 could be used to derive the $X^*(Y, \omega)$ values needed here and vice versa, as long as the set $\{a_1, \ldots, a_k\}$ is the same in both cases. The former recursion has the advantage of requiring strictly polynomial time; the latter one is pseudopolynomial but much faster for small values $a_1, \ldots, a_k$.

21.3.3 Computational results

For the typical problem instance given by

$$d = 1,$$

$$n = 100,$$

$$k = 2, a_1 = 18, a_2 = 14, \Pr\{\omega_j = a_1\} = \Pr\{\omega_j = a_2\} = \tfrac{1}{2}(j = 1, \ldots, n),$$

Figure 21.3 shows three functions of the first stage decision variable, the capacity Y:

- the lower bound $W^{LB}(Y)$;
- the expected total cost $EW^*(Y, \omega)$ in case of an optimal second stage decision;
- the expected total cost in case of an approximate second stage decision.

An investigation of these and other results leads to the same conclusions concerning running time, quality of lower and upper bounds, and the occurrence of multiple local minima as in Section 21.2.3.

21.4 Multiknapsack

21.4.1 Problem formulation

The two-stage multiknapsack problem that we will consider here can be viewed as a *capital budgeting* problem. At the aggregate level, one has to decide on the sizes $X_1, \ldots, X_m$ of m *budgets* that are to be reserved for financing a number of *projects*, while knowing the *cost* c_i of reserving one unit of budget i ($i = 1, \ldots, m$), the *requirement* r_{ij} of project j out of budget i ($i = 1, \ldots, m, j = 1, \ldots, n$), and the probability distribution of the vector $\omega = (\omega_1, \ldots, \omega_n)$ of *revenues* that the projects will yield. It is assumed that all c_i, r_{ij} and ω_j are nonnegative and that the r_{ij} are integral. At the detailed level, after $X = (X_1, \ldots, X_m)$ has been determined, a realization $\omega \in \Omega$ of ω becomes known, and one has to decide on a *selection* S of the projects that maximizes the total revenue $Y^*(X, \omega)$ within the budget constraints:

$$Y^*(X, \omega) = \max_{S \subset \{1, \ldots, n\}} \{\sum_{j \in S} \omega_j \mid \sum_{j \in S} r_{ij} \leq X_i \quad (i = 1, \ldots, m)\}.$$

The total profit of the budgeting decision X and the optimal selection decision is denoted by $Z^*(X, \omega) = -\sum_{i=1}^m c_i X_i + Y^*(X, \omega)$.

In the two-stage decision model, the objective is to determine a vector $X^* \in \mathbf{R}_+^m$ such that

$$EZ^*(X^*, \omega) = \max_{X \in \mathbf{R}_+^m} \{EZ^*(X, \omega)\}.$$

In the distribution model, the objective is to determine a function $X^0 : \Omega \to \mathbb{R}_+^m$ such that

$$Z^*(X^0(\omega), \omega) = \max_{X \in \mathbb{R}_+^m} \{Z^*(X, \omega)\}, \qquad \forall \omega \in \Omega.$$

21.4.2   The distribution model

The knapsack problem, i.e., the second stage problem with $m = 1$, is already NP-hard [4]. Surprisingly, the distribution model is easily solved to optimality. For each $\omega \in \Omega$, the selection $S(\omega)$ of profitable projects is given by $S(\omega) = \{j | \omega_j - \sum_{i=1}^m c_i r_{ij} > 0\}$. The minimum budgets needed to finance these projects are equal to $X_i^0(\omega) = \sum_{j \in S(\omega)} r_{ij} (i = 1, \ldots, m)$, and the corresponding total profit is

$$Z^*(X^0(\omega), \omega) = \sum_{j \in S(\omega)} \left(\omega_j - \sum_{i=1}^m c_i r_{ij} \right), \quad \forall \omega \in \Omega.$$

In the situation that each revenue ω_j can assume only k distinct values, the determination of X^0 requires $O(mn)$ computations for each of k^n realizations ω.

21.4.3   Dynamic programming

The second stage multiknapsack problem is solvable by a classical dynamic programming algorithm from [1]. Let $F_j(X, \omega)$ be the maximum revenue if only the first j projects can be selected, for given budgets $X = (X_1, \ldots, X_m)$ and revenues $\omega = (\omega_1, \ldots, \omega_n)$. An optimal selection is either restricted to the first $j - 1$ projects or includes project j:

$$\begin{aligned} F_j((X_1, \ldots, X_m), \omega) &= \max\{F_{j-1}((X_1, \ldots, X_m), \omega), \\ &\quad F_{j-1}((X_1 - r_{1j}, \ldots, X_m - r_{mj}), \omega) + \omega_j\} \quad (j = 1, \ldots, n), \\ F_0((X_1, \ldots, X_m), \omega) &= \begin{cases} 0 & \text{if } X_1 = \cdots = X_m = 0, \\ -\infty & \text{otherwise.} \end{cases} \end{aligned}$$

Since the requirements r_{ij} are integral, also the budgets X_i can be assumed to be integral. Computation of $Y^*(X, \omega) = F_n(X, \omega)$ requires a single comparison for each of $\prod_{i=1}^m X_i$ vectors $X' \leq X$ at each of n successive stages, and hence $O(n \prod_{i=1}^m X_i)$ time altogether.

For the two-stage multiknapsack problem, we again consider the situation in which each revenue ω_j can assume only k distinct values, for a fixed k. Let $R_i = \sum_{j=1}^n r_{ij}$ and note that $0 \leq X_i \leq R_i (i = 1, \ldots, m)$. At stage j, only the k^j different realizations of $(\omega_1, \ldots, \omega_j)$ need to be distinguished $(j = 1, \ldots, n)$. The algorithm therefore has to consider $O(k^j \prod_{i=1}^m R_i)$ pairs (X, ω) at stage j. Summation over all j yields an $O(k^n \prod_{i=1}^m R_i)$ time bound for the computation of all $Y^*(X, \omega)$ and also for the determination of a budget vector X^* that is optimal in expectation.

21.4.4 Computational results

The dynamic programming algorithm was coded in PASCAL and run on a CD Cyber 170–750 to solve several instances of the two-stage knapsack problem. We set $m = 1$ at the outset and did not attempt to solve proper multiknapsack problems, for which $m \geq 2$. We assumed independence of the revenues ω_j and tried to make the second stage knapsack problem nontrivial by specifying a high correlation between the expected revenue $E\omega_j$ of project j and its budget requirement r_{1j}. The solution of instances with twelve projects and two possible revenue values for each of them required about ten seconds.

For the problem instance given by

$$m = 1, c = 1,$$
$$n = 12, \Pr\{\omega_j = a_{1j}\} = \Pr\{\omega_j = a_{2j}\} = \tfrac{1}{2} \quad (j = 1, \ldots, n),$$

with the values of $r_{1j}, a_{1j}, a_{2j}(j = 1, \ldots, n)$ given in Table 21.1, Figure 21.4 shows the expected total profit $EZ^*((X_1), \omega)$ as a function of the budget size X_1. Note that the profit is shown only for integral X_1; the line segments that start from the points shown with a slope $-c_1$ and that indicate the profit for fractional X_1 have been deleted. Even if we restrict our attention to integral values of X_1, the profit function has many local maxima.

Table 21.1   Knapsack: numerical data

j	1	2	3	4	5	6	7	8	9	10	11	12
r_{1j}	5	2	9	13	10	8	4	7	10	6	4	9
a_{1j}	7	4	12	17	15	12	5	9	14	9	6	11
a_{2j}	3	1	6	11	8	7	1	4	7	7	2	8

References

[1] R.E. Bellman, *Dynamic Programming*, Princeton University Press, Princeton, NJ, 1957.

[2] M.A.H. Dempster, M.L. Fisher, L. Jansen, B.J. Lageweg, J.K. Lenstra, A.H.G. Rinnooy Kan, "Analytical evaluation of hierarchical planning systems", *Oper. Res.* **29**(1981), 707–716.

[3] M.A.H. Dempster, M.L. Fisher, L. Jansen, B.J. Lageweg, J.K. Lenstra, A.H.G. Rinnooy Kan, "Analysis of heuristics for stochastic programming: results for hierarchical scheduling problems", *Math. Oper. Res.* **8**(1983), 525–537.

[4] M.R. Garey and D.S. Johnson, *Computers and Intractability: a Guide to the Theory of NP-Completeness*, Freeman, San Francisco, 1979.

[5] R.L. Graham, E.L. Lawler, J.K. Lenstra, and A.H.G. Rinnooy Kan, "Optimization and approximation in deterministic sequencing and scheduling: a survey", *Ann. Discrete Math.* **5**(1979), 287–326.

[6] M. Held, R.M. Karp, and R. Shareshian, "Assembly-line balancing— dynamic programming with precedence constraints", *Oper. Res.* **11**(1963), 442–459.

[7] B.J. Lageweg, J.K. Lenstra, A.H.G. Rinnooy Kan, and L. Stougie, "Stochastic integer programming by dynamic programming", *Statist. Neerlandica* **39**(1985), 97–113.

[8] J.K. Lenstra, A.H.G. Rinnooy Kan, and L. Stougie, "A framework for the probabilistic analysis of hierarchical planning systems", *Ann. Oper. Res.* **1**(1984), 23–42.

[9] A.H.G. Rinnooy Kan, and L. Stougie, "Stochastic integer programming", this volume.

[10] L. Stougie, *Design and Analysis of Algorithms for Stochastic Integer Programming*, Ph.D. thesis, Centre for Mathematics and Computer Science, Amsterdam, 1985.

PART IV

Applications and Test Problems

CHAPTER 22

FACILITY LOCATION PROBLEM

Yu. Ermoliev

22.1 Introduction

The public provision of urban facilities and services often takes the form of a few central supply points serving a large number of spatially dispersed demand points. These facilities include hospitals, schools, libraries, and emergency provisions such as fire and police services. One of the fundamental features of these systems is the spatial interaction between suppliers and consumers. The need to introduce behavioral patterns more realistic than simply assuming that customers use the nearest facility has been recognized by many authors, among them Coelho and Wilson [4], Hodgson [7], Beaumont [1], and Leonardi [8],[9]. Since the proposed spatial interaction ("gravity") models can be justified both theoretically and empirically, their use in location modeling seems promising.

However, the classical spatial interaction models solve only part of the problem. Although they are based on stochastic assumptions [14], [11], [3] they use only the expected values of the underlying stochastic processes. A natural further step is therefore to introduce the stochastic behavior explicitly, thus allowing for uncertainty in both customer choice and demand knowledge. This was the approach in papers [5],[6]. The aim of this paper is to describe some of the problems arising when such stochastic features are introduced and to discuss the computational feasibility of stochastic quasigradient (SQG) methods. Practical results obtained in [6] are presented for a stochastic problem which deals with the optimal size of school facilities. Real data from Turin, Italy, have been used in the tests and the results are compared to those obtained by other methods. Some results reported on involve objective functions that are not even continuous. The contents of this chapter follows papers [5],[6].

22.2   Statement of the Problem

The simplest formulation of the deterministic facility location problem is as follows: minimize the performance function

$$\sum_{ij}[x_{ij}\ln(x_{ij}) + c_{ij}x_{ij}] \tag{22.1}$$

subject to the constraints

$$\sum_{j=1}^{n} x_{ij} = a_i, i = \overline{1,r}, \tag{22.2}$$

$$x_{ij} \geq 0, \qquad \forall i,j, \tag{22.3}$$

where x_{ij} is an (unknown) expected flow of users from demand location i to facility location $j (i = \overline{1,r}, j = \overline{1,n})$ per unit time; a_i is the total demand (in terms of customers to be served per unit of time) at each demand location i; c_{ij} are the costs of travel between each pair of locations (i,j).

The objective function (22.1) was first introduced into transport planning evaluation by Bregman [2] and Neuburger [13] and extended to location analysis Coelho and Wilson [4]. These authors gave this function an economic interpretation, namely the consumer surplus measure associated with the pattern of consumer trips $\{x_{ij}\}$.

Due to the simple form of the problem (22.1)–(22.3), the closed-form optimal solution is not hard to find:

$$x_{ij} = a_i P_{ij}, x_j = \sum_{i=1}^{r} x_{ij}, \tag{22.4}$$

where

$$P_{ij} = \frac{\exp(-c_{ij})}{\sum_j \exp(-c_{ij})}$$

and x_j is the size of the facility at j. Note that the quantities P_{ij} satisfy the following conditions:

$$\sum_{j=1}^{n} P_{ij} \geq 0, i = \overline{1,r}, j = \overline{1,n} \tag{22.5}$$

Equations (22.4) and (22.5) imply that trips from demand locations to facilities are made according to a very simple interaction rule. The quantity P_{ij} can be interpreted as the probability that a customer living at location i will choose the facility at location j. Then x_{ij} is the expected number of customers traveling between i and j.

It is worth noting that the interpretation of the quantities P_{ij} as probabilities is connected with the theory of probabilistic choice behavior [12]. It has also been shown by Bertuglia and Leonardi [3] that these quantities can be considered as a steady-state distribution of a suitably defined Markov process.

It is now possible to use equation (22.4) as the basis from which to make some generalizations concerning stochasticity. The simplest of these are as follows:

(1) The demand a_i at demand location i is not known in advance; it is a random variable. This assumption is reasonable in many long-term planning applications. For instance, in a high-school location problem the total number of students living in each demand location may change over time and so cannot be known in advance.

(2) Customers living in district i choose their destinations j independently of each other with probability P_{ij}.

These assumptions are embodied in the following model, which assumes that the choices made by the customers are stochastic. Let ε_{ij} be the actual (random) numbers of customers traveling from i to j and define τ_j, the total number of customers attracted to j, as follows:

$$\tau_j = \sum_{i=1}^{r} \varepsilon_{ij}, j = \overline{1, n}.$$

Note also that

$$\sum_{j=1}^{n} \varepsilon_{ij} = a_i, i = \overline{1, r}. \tag{22.6}$$

Let $H_j(y)$ denote the distribution function of τ_j:

$$H_j(y) = P\{\tau_j \leq y\}.$$

The distribution function $H_j(y)$ cannot easily be given in closed form, but random draws of τ_j can be computed using a simple simulation model based on equation (22.6). If x_j is the planned size of the facility at j, then the actual number of τ_j of customers attracted to j may not be equal to x_j. Suppose that a cost

$$\alpha_j^+ (x_j - \tau_j)$$

has to be paid when $x_j \geq \tau_j$ and a cost

$$\alpha_j^- (\tau_j - x_j)$$

has to be paid when $x_j < \tau_j$. We therefore have the cost function

$$f_j(x_j, \tau_j) = \begin{cases} \alpha_j^+ (x_j - \tau_j), & \text{if } x_j \geq \tau_j, \\ \alpha_j^- (\tau_j - x_j), & \text{if } x_j < \tau_j, \end{cases}$$

The resulting stochastic programming problem is then as follows: determine the sizes x_j of the facilities $j = \overline{1,n}$ that minimize the expected cost

$$F(x_1 \ldots x_n) = \sum_{j=1}^{n} Ef_j(x_j, \tau_j)$$

$$= \sum_{j=1}^{n} \left[\alpha_j^+ \int_0^{x_j} (x_j - y) dH_j(y) \right. \tag{22.7}$$

$$\left. + \alpha_j^- \int_{x_j}^{\infty} (y - x_j) dH_j(y) \right]$$

subject to constraints

$$x_j \geq 0, j = \overline{1,n}. \tag{22.8}$$

Note that the objective function contains no spatial interaction embedding term since the behavior of the customer is included in the structure of the probabilities P_{ij}.

Practical problems that lead to the minimization of a function such as equation (22.7) are common in operations research. For example, we could consider a facility allocation problem or a storage inventory control problem where some capacities have to meet random demand and both surpluses and deficits incur penalty costs.

In the special case where $F(x)$ has continuous derivatives, minimization of $F(x)$ by analytical means would lead to the consideration of the partial derivatives

$$\frac{\partial}{\partial x_j} F(x) = \alpha_j^+ \int_0^{x_j} dH_j(y) - \alpha_j^- \int_{x_j}^{\infty} dH_j(y)$$

The solution would then require the determination of $x = (x_1 \ldots x_n)$ such that

$$H_j(x_j) = \frac{\alpha_j^-}{\alpha_j^+ + \alpha_j^-}, \quad j = \overline{1,n}.$$

In general it may not be possible to solve this equation analytically (for instance, if $H_i(y)$ is unknown, as in problem (22.7)–(22.8)).

22.3 The Stochastic Quasigradient Method

The problem (22.7)–(22.8) contains two typical difficulties of stochastic programs (see Chapter 1). First, it is difficult or impossible to compute the exact values of the integrals appearing in (22.7), except for special and well-behaved forms of the distribution functions $H_j(y)$. Actually the functions H_j are defined only by means of a rule for generating random draws by Monte-Carlo-type simulating procedures. Thus, to solve such problems it is necessary to use procedures which do not calculate the exact values of the objective functions. Second, the objective function (22.7) is generally nonsmooth.

This becomes clear after reformulating problem (22.7)–(22.8) as a stochastic minimax problem. It is easy to see that

$$f_j(x_j, \tau_j) = \max\{\alpha_j^+(x_j - \tau_j), \alpha_j^-(\tau_j - x_j)\}.$$

The objective function (7) is therefore

$$F(x) = \sum_{j=1}^{n} E \max\{\alpha_j^+(x_j - \tau_j), \alpha_j^-(\tau_j - x_j)\} \qquad (22.9)$$

Function (22.9) is convex, but in general nonsmooth, since the maximization operator is present under the mathematical expectation sign. The stochastic quasigradient method for this particular stochastic minimax problem works as follows.

Let $x^0 = (x_1^0 \ldots x_n^0)$ be an arbitrary initial approximation and $x^s = (x_1^s \ldots x_n^s)$ be the approximation computed after the s-th iteration. A random observation $\tau^s = (\tau_1^s \ldots \tau_n^s)$ of the vector $\tau = (\tau_1 \ldots \tau_n)$ is obtained by simulation. A new approximation is determined by the rule:

$$x_j^{s+1} = \max\{0, x_j^s - \rho_s \xi_j^s\}, \quad j = \overline{1, n}, \quad s = 0, 1 \ldots \qquad (22.10)$$

where ρ_s is a step multiplier, such that

$$\rho_s \geq 0, \sum_{s=0}^{\infty} \rho_s = \infty, \sum_{s=0}^{\infty} E \rho_s^2 < \infty, \qquad (22.11)$$

and

$$\xi_j^s = \begin{cases} \alpha_j^+, & \text{if } x_j^s \geq \tau_j^s, \\ -\alpha_j^-, & \text{if } x_j^s < \tau_j^s, \end{cases}$$

In principle, the convergence of $\{x^s\}$ will be obtained if the step multipliers ρ_s are chosen so as to satisfy the step conditions (22.11). For the practical construction of the step-size control, these requirements are only of general importance.

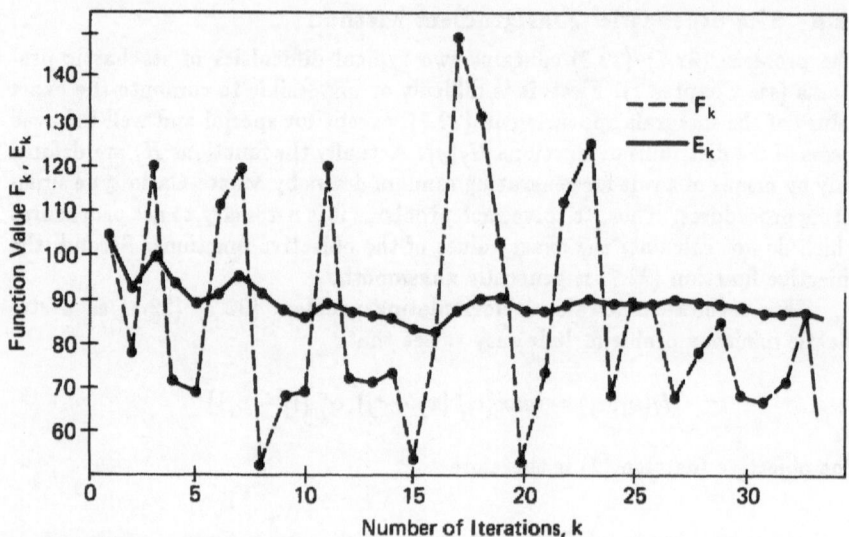

Figure 22.1 The behavior of the sequences $\{F_k\}$ and $\{E_k\}$ as a function of the iteration number.

22.4 Practical Computations

The methods of controlling the step size in stochastic minimization are usually based on keeping the step multipler constant during a number of iterations and then reducing it according to certain rules. In the course of the iterations a succession of function values $F_s = \Sigma_j f_j(x_j^s, r_j^s)$ are observed. Usually these values vary over a wide range. However, the sequence

$$E_k = \frac{1}{k} \sum_{s=0}^{k} F_s = \frac{1}{k} \sum_{s=0}^{k} \sum_{j=1}^{n} f_j(x_j^s, r_j^s) \qquad (22.12)$$

shows a smoother behavior as can be seen in Figure 22.1. Indeed, E_k could be expected to approach a stationary value. One rule of controlling the step size uses on this fact. The method can be summarized as follows:

(1) Choose the initial value ρ^0 for the step multiplier
(2) Using ρ^0 for the step multiplier calculate the value of E_k according to equation (22.12)
(3) When a stationary sequence $\{E_k\}$ is observed, reduce the step multiplier by one half
(4) Go back to step (2) with the new value of step multiplier until no improvement in the test function E_k is observed.

There are some unanswered questions in the procedure outlined above. First, how should the initial step multiplier be chosen? If it is too large, both the sequence $\{E_k\}$ and the iterates x^s will oscillate heavily and no decrease in the objective function will be observed. If the initial step multiplier is too small, the rate of decrease will be very small and perhaps hardly noticeable. From the computational point of view the latter situation is more harmful and should be avoided, while the situation arising from too large a step multiplier is rapidly recognized and hence can be corrected. As a rule of thumb the initial step should be chosen to satisfy

$$\rho \xi_j \approx \mu \bar{x}_j \qquad (22.13)$$

where $\mu \in (0,1)$ and $\bar{x}_j$ is the estimated value for the j-th component of the solution.

The use of step ρ_s also needs further explanations. The ideal way of controlling the procedure would be on-line, where the program continuously plots the values of the sequence $\{E_k\}$ on the screen and where the iterations could be manually interrupted to cut down the step multiplier. This is not always possible and the iterations must be performed in small batches, whereafter the values of E_k are plotted and possible adjustments of the step multipler can take place. A definite way to find the stationary phase of the sequence is to rescale the coordinate axes before plotting the values of a new batch. In this case the stationary phase is in fact recognized as smooth oscillations around a fixed value.

Figure 22.2 shows an example of the behavior of E_k as a function of a iteration number k. The values for coefficients are $\alpha_j^+ = \alpha_j^- = 1.00, j = 1, \ldots, 23, \rho^0 = 1.00$, and the components of the initial estimate and the solution are known to differ by at most five units. Note that the rate of decrease of the sequence $\{E_k\}$ is fast during the first iteration batches but becomes slower as the step size decreases. Hence a crude estimate of the result is obtained after a rather small number of iterations, but for greater accuracies the number of iterations needed grows rapidly.

If rigorously followed, the basic procedure for the step-size control may lead to a slow performance of the algorithm. First, the manual step-size control with many I/O operations requires considerable effort from the person who does the calculations and usually this affects the response time. This happens especially in a time-sharing computer environment where the number of users is large and the average response time is already quite long. Second, the number of iterations needed can be often be significantly reduced.

To overcome the need for numerous manual I/O operations, a simple automatic version of the manual step-size control can be designed. Given three parameters the procedure simulates the behavior of the controlling person and reduces the step multiplier as soon as it observes a stationary or an oscillatory sequence $\{E_k\}$. Let the three input parameters be NB, DIFI, and DIF2. The first parameter NB fixes the batch size, i.e., the iterations will be performed

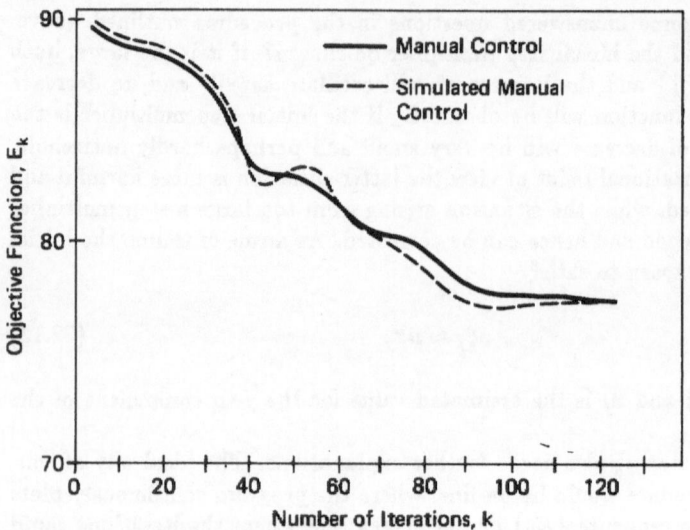

Figure 22.2 The convergence behavior of $\{E_k\}$ in the manual control and simulated manual control cases.

in batches of NB iterations. Let the step multiplier used during the iteration batch be equal to ρ. A test indicator is defined as:

$$d_m = \frac{E_{(m-1)} \cdot NB - E_m \cdot NB}{\sum_{s \in M} \rho^s \|\xi^s\|} \qquad m = 1, \ldots, \qquad (22.14)$$

The procedure then checks the two conditions

$$d_m \leq \text{DIF1} \qquad (22.15a)$$

and

$$\frac{\sum_{s \in M} \Delta^+ E_s}{\max_{s \in M} E_s - \min_{s \in M} E_s} \geq \text{DIF2} \qquad (22.15b)$$

where

$$\Delta^+ E_s = \max(0, E_s - E_{s-1}) \qquad (22.16a)$$

$$M = \{s \mid (m-1) \cdot NB \leq s \leq m \cdot NB\} \qquad (22.16b)$$

In the case when either of these conditions holds the step multiplier is reduced by one half. The first condition (22.15a) tests if the decrease of the sequence in proportion to the step-size used is less that the given limit. The second condition (22.15b) then checks if the sequence is oscillatory. This is done by considering the ratio of the sum of the positive jumps of the sequence $\{E_k\}$ to the maximum change in the sequence that takes place during the iteration batch.

With DIF1 = 0.01 and DIF2 = 0.30 the procedure simulates the manual control very closely (Figure 22.2). Depending on the starting values used for x^0 and ρ^0 sometimes a few more iterations were performed than the manual control would have required, but the total computing time still usually remained smaller than in the case of manual control.

With the aforementioned values for DIF1 and DIF2 the automatic step-size control normally guarantees that the solution is eventually reached, independent of the initial values for x^0 and ρ^0. Often the algorithm can be made faster by using a greater value for DIF1. If for example, DIF1 = 1.00, the use of the control would reduce the step multiplier as soon as the total decrease of the objective function during a batch is less than the total change of the components in that batch. If the solution can initially only be roughly estimated, the number of iterations can be kept of moderate size. This can be done by choosing an initial value for ρ that will reach the solution region in a few iterations and by cutting down the step size as soon as the rate of decrease of the objective function slows down. Using the test indicator d_m of equation (22.14) the program checks if

$$d_m \leq \text{DIF1} \tag{22.17a}$$

or

$$d_m \leq d_m - 1 \tag{22.17b}$$

Instead of E_m, an average of a few neighboring values of E_m can be used to calculate the indicator d_m. If any of conditions (22.17) holds, the step multiplier is cut down by a factor r, which is given as an input.

The effect of the accelerated procedure is seen in Figure 22.3 where the curves correspond to the accelerated step-size control. The reduction coefficient r is 0.5 in both cases, but in the first case, the batch size is 10, in the latter case, 5. DIF1 has now been set to 1.0. It is seen that some decrease in the number of iterations has been obtained in both cases as compared to the situation in Figure 22.2, but the difference is quite small. However, in this example a good estimate of the solution is known in advance and the number of iterations is rather small with any kind of step-size control. Note that if the initial estimate for x is far from the actual solution and a small initial value is used for ρ, then the accelerated procedure may reduce the step too rapidly, and an excessive number of iterations is needed to obtain the solution. As noted earlier, this danger can be normally eliminated by selecting an initial ρ^0 estimate that is too big rather than too small.

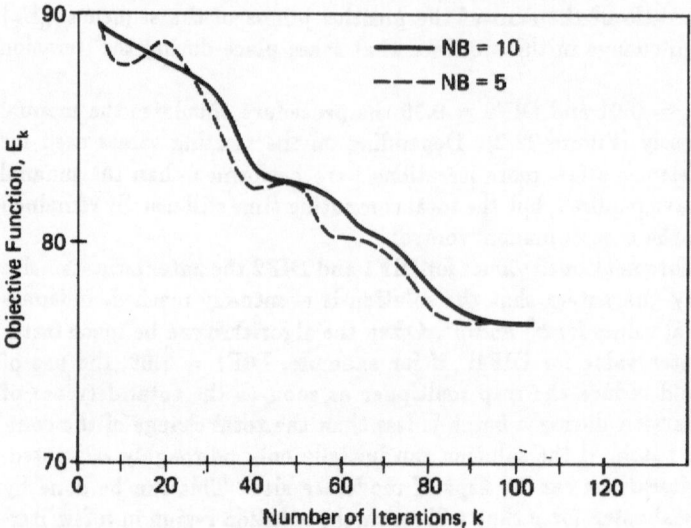

Figure 22.3 The convergence behavior of $\{E_k\}$ in the accelerated step-size control case.

22.5 A Case Study

An example of a resource allocation problem (see [6]) that minimizes costs to meet uncertain demand will be discussed in this section. The problem is a high school location problem in Turin, Italy. The physical setting and the data for this problem are described in Leonardi and Bertuglia [10]. For the purposes of this analysis, Turin is divided into 23 districts, each district being both a demand source and a possible high school facility location. Customers are assumed to behave according to a gravity-type model. For simplicity, travel time is assumed as the only explanatory variable for the choice behavior (some theoretical underpinnings for such models are described in Leonardi [9]).

However, unlike in the standard models, the gravity model will be given here a stochastic interpretation, as suggested in Section 22.2. That is, the relative distribution of students among facilities is a discrete multinomial Bernoulli distribution, rather than as a set of deterministic fractions. in mathematical terms this takes the following form.

Let $a_i, i = 1, \ldots, r$, be the total number of students at point i. The problem is to determine the size x_j of the facilities at points $j, j = 1, \ldots, n$, when it is known that the students at point i choose the facility at point j with probability

$$p_{ij} = \frac{e^{-\lambda c_{ij}}}{\sum_{j=1}^{n} e^{-\lambda c_{ij}}} \qquad (22.18)$$

where λ is a constant and c_{ij} are empirical coefficients that depend on the distance between i and j (in this example: travel times in minutes). The use of (22.18) for the probabilities has theoretical and empirical justifications. Model (22.18) is a simplified form of the logit model discussed in McFadden [11], [12], for example. If the flow of students between i and j is denoted ε_{ij}, the stochastic demand at point j is then

$$\tau_j = \sum_{i=1}^{r} \varepsilon_{ij} \qquad (22.19)$$

while the number of students at point i can be written as

$$a_i = \sum_{j=1}^{n} \varepsilon_{ij} \qquad (22.20)$$

The numbers a_i are now deterministic and given as input. If the unit cost of capacity surplus is α and that of deficit is β and no other costs are considered, then our cost minimization problem is of the type (22.7) with $\alpha_j^+ = \alpha, \alpha_j^- = \beta, j = 1, \ldots, n$.

The ability to generate random realizations, τ_j^S, of the demand vector is essential for applying the quasigradient method. The direct determination of the distribution of τ_j is practically quite difficult in this case. Instead, random vectors can be generated by simulating individual choices of the students according to the probabilities p_{ij} in (22.18).

Table 22.1 shows the solutions obtained for $\alpha = \beta = 1.0$. In this case the solution $x_j = \Sigma_i a_i \cdot p_{ij}$ of a deterministic problem that is based on an entropy approach. The first column in Table 22.1 contains the labels of each district, numbered from 1–23. The second column of Table 22.1 gives the vector $a = (a_1, \ldots, a_{23})$ of total demands in each district; a was also used as the initial estimate for the iteration. Here the original data from Turin have been multiplied by $1/100$. The next three columns show the results originating from the use of different starting values for the iteration. The last column shows the solution based on the deterministic model. In general, a good agreement exists between all the solutions; they are usually within two digits of each other. There are, however, some significant discrepancies. These can be partly explained by the stochastic nature of the convergence and by the flatness of the objective function near the solution. They associate somewhat with the slow convergence of the algorithm as the number of iterations increases.

The discrepancies between the solutions in Table 22.1 can be associated with the shape of the probability densities underlying the probabilities of (22.18). The values that are used for the coefficients c_{ij} are listed in Table 22.6, the value of the constant λ is 0.15. Probability densities can be numerically approximated from this data. Densities for several of the components are drawn in Figure 22.4. The densities are mostly symmetric and strongly peaked. In these cases the stochastic minimization solution, which corresponds to the median of this distribution, and the deterministic solution, which corresponds to

Table 22.1 Optimal location of turin high schools. Solutions obtained for penalty costs $\alpha = \beta = 1.0$.

District	Number of students	$\rho^0 = 1.00$ $NB = 20$	$\rho^0 = 1.00$ $NB = 10$	$\rho^0 = 1.00$ $NB = 5$	Deterministic solution
1	14.0	15.6	17.0	17.8	17.5
2	13.0	12.8	12.8	13.6	13.0
3	15.0	18.6	17.9	17.1	18.7
4	11.0	18.0	18.3	18.9	18.9
5	14.0	17.0	16.4	15.3	16.4
6	14.0	13.0	13.8	14.0	13.7
7	11.0	11.2	10.1	10.0	11.0
8	12.0	10.0	10.0	10.0	10.5
9	12.0	12.9	12.9	13.5	13.2
10	23.0	19.2	19.6	20.1	19.3
11	26.0	25.4	26.7	26.9	26.2
12	23.0	19.9	20.0	19.1	20.3
13	22.0	16.2	16.0	15.5	16.1
14	18.0	15.0	15.6	15.0	15.3
15	14.0	13.9	14.0	14.0	14.3
16	15.0	13.4	13.6	13.8	13.2
17	14.0	13.0	13.0	13.1	12.9
18	14.0	15.0	16.1	15.7	15.8
19	10.0	9.8	10.0	10.2	9.8
20	10.0	10.0	10.9	10.1	10.5
21	5.0	5.0	5.0	5.0	5.1
22	8.0	10.8	11.7	10.8	10.6
23	21.0	16.5	16.1	16.0	16.9

the expected value, should be close to each other. This is in fact demonstrated, for instance, by the facility sizes in districts 8 and 9, where the discrepancies are small. However, for district 1 the density is flat and skew, and the median and expected values are not equal. On the other hand, in the solutions for x_1 the discrepancies are large. The flatness of the densities also explains the large discrepancies between the different solutions obtained from the stochastic minimization procedure.

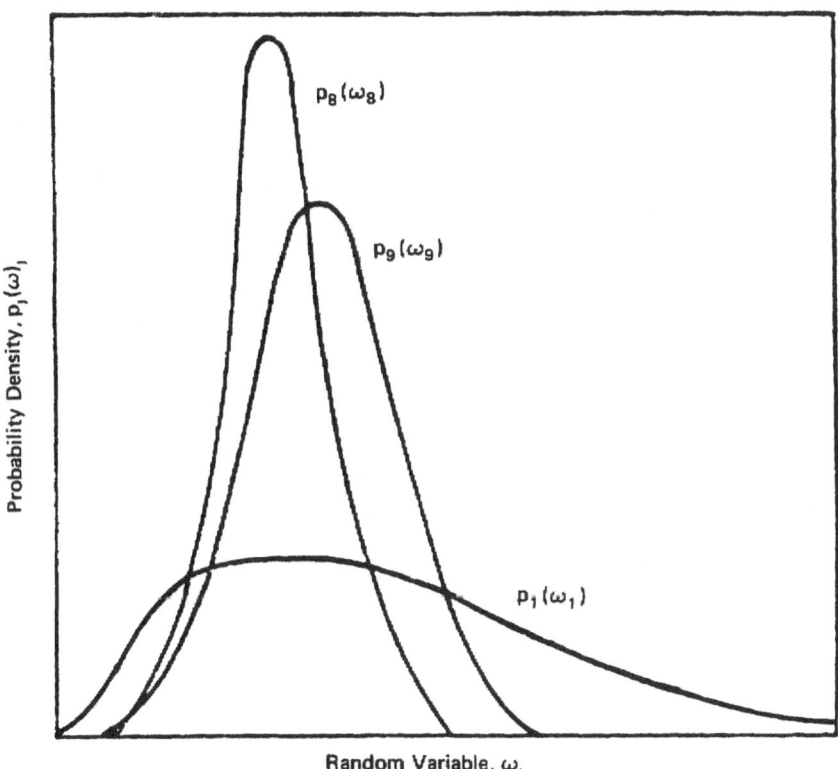

Figure 22.4 The probability densities for random demand τ_j at location $j = 1.8, or\,9$.

In Table 22.2 solutions are presented for cases where α and β differ from each other. As one could expect, the increase in the relative cost of deficit compared to the cost of surplus leads to larger values in the solution vector. If however, the probability density of the corresponding component of τ_j is very peaked, as in the case of $\tau_2 1$, the change in the relative costs does not have any significant influence on the solution.

Table 22.2 Optimal location of Turin high schools. Solutions obtained for different values of penalty costs α and β.

District	$\alpha = 1.00$ $\beta = 1.50$	$\alpha = 1.00$ $\beta = 2.00$	$\alpha = 1.50$ $\beta = 1.00$	$\alpha = 2.00$ $\beta = 1.00$
1	16.9	20.7	12.9	10.9
2	13.9	15.0	11.5	10.8
3	19.7	21.1	17.0	16.0
4	19.6	20.7	17.1	16.0
5	17.5	18.0	15.2	14.7
6	14.6	15.0	13.0	12.0
7	11.2	12.0	10.0	9.4
8	11.1	11.9	9.8	9.1
9	13.6	14.2	12.0	11.0
10	20.4	21.4	18.4	17.6
11	27.4	28.4	25.0	23.2
12	21.9	22.6	19.1	17.2
13	16.7	18.0	14.3	13.9
14	16.0	16.6	14.1	13.4
15	15.0	15.0	14.0	12.5
16	13.8	14.8	12.3	12.0
17	13.2	14.0	12.2	12.0
18	16.7	16.8	14.8	13.9
19	9.9	10.9	9.6	8.8
20	10.0	12.0	9.0	9.0
21	5.0	5.0	5.0	5.0
22	10.9	12.9	8.9	8.5
23	17.4	18.1	15.4	14.3

22.6   A Nonconvex Objective Function

The problem discussed so far lacks some of the main features that are usually considered typical for optimal location problems. For instance, economies of scale, which make location problems nontrivial, are absent in our earlier formulation. In deterministic models, economies of scale are usually introduced by means of fixed charges, to be paid when a facility is established, no matter what the number of attracted customers. This formulation is typical of the well known plant-location problems of operations research. Related ways to introduce scale effects are by means of suitable constraints, as on the total number of facilities or on the minimum feasible size for facilities.

Here the first formulation will be explored. Let a fixed cost γ be defined, to be paid when a facility is established. For simplicity, let us assume that the same value of γ applies to all districts. Then the minimization of the expected cost calls for finding the minimum of the function.

$$G(x) = \sum_{j=1}^{n} \gamma \delta(x_j) + E\left\{ \sum_{j=1}^{n} \max[\alpha(x_j - \tau_j), \beta(\tau_j - x_j)] \right\} \qquad (22.21)$$

where $\delta(x)$ is the unit step function at zero. It is easy to see that with nonnegative $x_j, G(x)$ is not convex and usually has several local minima. The problems of this form are normally treated with mixed integer programming methods. Here we attempt to apply the general idea of stochastic quasigradients to finding the global minimum. Approximating the step function by a logarithmic function, the estimate

$$\xi_j^s = \frac{\gamma}{x_j^s + \varepsilon} + \begin{cases} \alpha & \text{if } x_j^s < \tau_j^s \\ -\beta & \text{if } x_j^s \geq \tau_j^s \end{cases} \qquad (22.22)$$

with ε a small positive constant used in computing the generalized gradient at $x = x^s$. Otherwise the procedure follows the gradient calculated by equation (22.10).

In general, the procedure rapidly find a minimum which is at least local. After that, however, some difficulties arise with the control of the iteration process. In principle, the approximation

$$G_k^1(x) = \gamma \sum_{j=1}^{n} \delta(x_j) + \frac{1}{k} \sum_{s=0}^{k} \sum_{j=1}^{n} \max[\alpha(x_j^s - \tau_j^s), \beta(\tau_j^s - x_j^s)] \qquad (22.23)$$

can be used again to follow the progress of the iterative scheme. Now, however, after a number of iterations the function $G_k^1(x^k)$ may achieve a minimum. On the other hand, some components of the estimation for the generalized gradient as calculated from equation (22.22) may still show a trend toward the origin, where another (at least) local minimum would be found. Note that with a small ε the origin becomes a fixed point for the iteration : if $x^{s0} = 0$ for one s_0, then $x^s = 0$ for all $s > s_0$. To overcome these difficulties, the initial value x^0 should be large enough and the initial step multiplier ρ should be chosen such that the step size is a small fraction of x_j. In this way a fallacious convergence towards zero during the first iterations can be avoided. To assess the behavior of the function $G(x)$ at the various minima, a test function

$$G_k^2(x^k) = \gamma \sum_{j=0}^{n} \frac{x_j^k}{x_j^{k-m}} + \frac{1}{k} \sum_{s=0}^{k} \sum_{j=1}^{n} \max[\alpha(x_j^s - \tau_j^s), \beta(\tau_j^s - x_j^s)] \qquad (22.24)$$

could be used. In this case m is a small integer, the choice of which slightly depends on the relative magnitude of α, β, and γ.

Figure 22.5 shows the behavior of the functions $G_k^1(x^k)$ and $G_k^2(x^k)$ with increasing k for $\alpha = \beta = 0.5, \gamma = 5.0, m = 6$. It is seen that $G_k^2(x^k)$ is monotonically decreasing toward the global minimum while $G_k^1(x^k)$ has two local maxima. Table 22.3 shows the vector x^k at $k = 180$, which corresponds to one local minimum of $G_k^1(x^k)$, and at the end of the iteration $(k = 280)$. It cannot be proved that the solution obtained is the exact solution of the optimization problem. On the other hand, the computational effort that is needed for an estimation by the stochastic quasigradient method is also relatively small when compared to some integer programming methods, for instance.

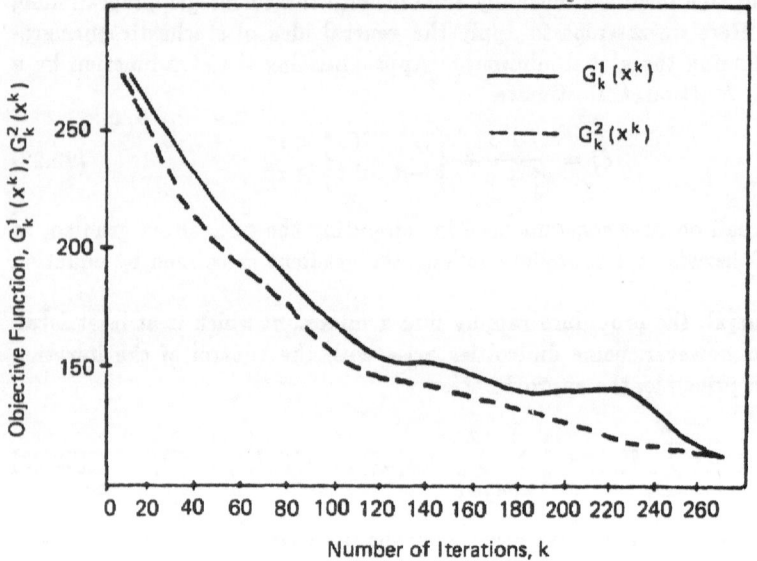

Figure 22.5 The behavior of $G_k^1(x^k)$ and $G_k^2(x^k)$ as a function of k.

The solutions obtained depend mostly on the relative magnitudes of α, β, γ. With increasing fixed costs, γ, more facilities are likely to remain closed. When the β are increased the deficits are more penalized and thus more facilities remain open. Table 22.4 shows results from a sensitivity analysis on the values of α and β. The aim of the analysis is to find which values of α and β will cause the smallest facility (district 21) to disappear from the solution. This will happen almost certainly when β is less than 1.5. However, for a large range of values for x_21, between zero and five, the objective function remains almost constant. Hence, with these parameter values, opening or closing that facility does not have great influence on the value of the objective function.

Table 22.3 Optimal location of Turin high schools. Solutions obtained after 180 and 280 iterations with penalty costs $\alpha = \beta = 0.5$ and fixed charge $\gamma = 5.0$.

District	$k = 180$	$k = 280$
1	—	—
2	—	—
3	4.4	—
4	8.7	—
5	8.3	—
6	—	—
7	—	—
8	—	—
9	—	—
10	16.7	16.1
11	23.2	22.5
12	14.0	13.0
13	7.3	—
14	19.1	—
15	—	—
16	—	—
17	2.7	—
18	8.7	—
19	—	—
20	—	—
21	—	—
22	—	—
23	11.5	10.1

Table 22.5 shows the results of a sensitivity analysis on fixed charge γ. The aim of this analysis is to find the least value of γ leading to a solution with a single facility open.

The fixed charge is fixed and equal to $\gamma = 5.0$.

A few comments are appropriate here on the comparison between the deterministic solutions, as determined in Leonardi and Bertuglia [10], and the solutions obtained with the stochastic quasigradient method. Some general tendencies are shared in common among all solutions, such as the low ranking of district 21 and the high ranking of district 11. The general clusters of open locations show also some similarity. A cluster of central districts (between 1–6), one of the first-ring districts (between 9–18) and a few peripheral districts (usually district 23 only) appear in deterministic solutions as well. However,

Table 22.4 Optimal location of Turin high schools. Results of a sensitivity analysis for changing values of penalty costs α and β. The fixed charge is fixed and equal to $\gamma = 5.0$.

District	$\left.\begin{array}{c}\alpha\\\beta\end{array}\right\} = 1.0$	$\left.\begin{array}{c}\alpha\\\beta\end{array}\right\} = 1.5$	$\left.\begin{array}{c}\alpha\\\beta\end{array}\right\} = 1.75$	$\left.\begin{array}{c}\alpha\\\beta\end{array}\right\} = 2.0$
1	9.2	11.2	11.4	12.9
2	9.8	11.0	11.4	11.4
3	16.0	16.5	17.1	17.7
4	16.4	17.1	17.5	17.3
5	14.0	15.0	15.1	15.2
6	11.1	12.0	12.0	12.1
7	8.1	9.3	10.0	9.7
8	8.3	9.0	9.0	9.0
9	10.0	11.9	11.9	12.0
10	17.0	18.3	18.5	18.7
11	24.7	24.9	24.9	24.9
12	18.0	18.1	18.5	19.0
13	13.8	14.3	14.4	14.5
14	13.8	14.0	14.0	14.0
15	12.4	13.0	13.0	13.0
16	11.6	12.0	12.3	12.4
17	11.7	12.0	12.0	12.0
18	13.9	14.1	14.6	14.9
19	8.4	8.8	9.0	9.0
20	7.0	8.7	9.0	9.2
21	—	—	—	5.0
22	6.4	8.0	8.1	9.0
23	14.7	15.1	15.7	15.8

when one looks at the detailed composition of these clusters, no two of them are the same. Sometimes very striking differences are found, such as the closing or opening of district 1 (the downtown district), which would be difficult to justify to a public authority. The main cause for such a lack of robustness of stochastic methods is the existence of many local minima and many near optimal solutions, with values of the objective function lying within a very narrow range. Of course a deterministic algorithm of an ennumerative nature can still detect small differences, even though it may take a long time. In a stochastic formulation, random fluctuations might well be of the same order of magnitude

Table 22.5 Optimal location of Turin high schools. Results of a sensitivity analysis for changing values of fixed change γ. The penalty costs are fixed and equal to $\gamma = \beta = 1.5$.

District	$\gamma = 10.0$	$\gamma = 15.0$	$\gamma = 20.0$
1	—	—	—
2	—	—	—
3	13.9	—	—
4	14.7	12.3	—
5	12.5	—	—
6	9.9	—	—
7	—	—	—
8	—	—	—
9	8.7	—	—
10	16.2	14.4	—
11	24.5	23.7	19.1
12	18.6	16.0	—
13	14.3	11.2	—
14	13.9	—	—
15	11.8	—	—
16	11.8	—	—
17	11.5	—	—
18	13.1	—	—
19	—	—	—
20	—	—	—
21	—	—	—
22	—	—	—
23	14.1	5.7	—

of the range of the objective function values. This seems to be the case in our examples.

i\j	1	2	3	4	5	6	7	8	9	10	11	12	13	14	15	16	17	18	19	20	21	22	23
1	5	17	13	21	19	19	17	23	29	30	22	32	26	26	27	30	28	13	32	30	50	28	34
2	14	5	14	25	21	26	24	29	18	21	24	32	31	32	36	41	41	28	48	36	50	13	28
3	8	12	5	13	9	17	17	25	18	23	12	20	18	21	28	29	31	22	39	30	51	20	27
4	21	25	14	5	7	22	31	39	21	22	13	14	9	23	30	34	40	31	44	44	61	26	27
5	17	27	10	8	5	16	25	33	30	40	22	22	15	22	48	35	42	30	46	38	57	32	37
6	16	22	18	20	15	5	26	32	35	32	30	31	23	32	36	39	34	27	39	39	56	37	43
7	22	23	19	32	26	25	20	21	31	39	39	39	38	37	38	40	31	18	34	15	49	34	35
8	27	29	26	38	32	30	33	5	37	15	39	46	11	41	45	46	51	17	34	24	37	34	42
9	20	19	20	21	30	34	34	40	5	5	19	29	31	43	48	51	52	38	57	46	62	12	19
10	32	25	24	14	34	39	33	41	15	9	9	20	32	40	48	39	47	39	58	46	58	16	18
11	27	35	14	16	23	30	36	0	32	20	5	7	20	29	38	42	53	34	53	46	64	23	17
12	24	29	23	10	22	35	35	0	31	33	8	20	5	5	29	32	48	38	57	55	67	34	27
13	28	33	17	25	14	25	38	41	46	33	20	29	28	31	15	28	42	33	45	48	64	43	38
14	31	38	21	34	22	12	37	40	48	47	33	45	11	30	29	18	29	24	44	39	70	48	46
15	14	42	32	39	37	31	34	34	51	48	36	30	31	35	23	5	25	24	37	39	73	52	51
16	35	43	31	42	37	33	20	18	38	52	47	52	48	47	44	31	18	18	36	33	71	51	52
17	32	29	24	32	43	28	37	35	58	39	34	44	37	67	39	26	35	18	17	17	57	60	55
18	53	49	40	46	31	40	18	26	46	58	54	57	52	42	71	40	34	18	35	34	74	48	42
19	23	37	36	48	47	40	52	40	63	47	50	55	54	44	45	41	73	59	75	5	58	59	63
20	30	53	54	66	42	60	33	34	11	60	68	73	71	67	50	75	49	37	57	56	54	32	51
21		13	22	26	63	35	35	42	20	17	23	33	36	42	45	50	73	59	75	45	64		68
22		27	26	27	31	41				18	18	28	37	44	50	53	52	39	57	48			30
23					36		35												59			32	5

Table 22.6 The coefficients c_{ij} for probabilities p_{ij}.

22.7 Concluding Remarks

The purpose of this study has been to consider the stochastic quasigradient method for solving a resource allocation problem. The main advantages of the method are undoubtedly its computational simplicity and the small amount of information required—explicit probability distributions are not needed, random observations from a Monte Carlo simulation process will do.

The computational procedure for the basic recursion equation can be written down by using only a few program statements and the storage requirements of the method are minimal. The generation of the random observations, however, may be time-consuming and hence the need for an algorithm made as effluent as possible. The standard step-size control is based on the interactive use of the computer and this normally guarantees that the solution is found after a moderate number of iterations. In this chapter some methods are presented that do not necessarily require continuous control from the person running the program and that often reduce the computation time.

Test are also made for a case where the objective function is nonconvex. In the deterministic formulation, problems of this type lead to integer programming methods that are often slow, unless some special assumptions (like linearity) concerning the objective function and constraints are satisfied. Here the solution is based on the same iteration algorithm as in the convex case. The existence of several local minima may cause some difficulties with the control of the iteration process, but the experience shows that with regard to its simplicity and speed the method can be efficiently applied to obtain good estimates for the solutions of these difficult problems.

The practical results of determining the size of school facilities in Turin were generally seen to be in agreement with the solutions derived by other means although differences in details were found. It is true that, given the special probability structure of equation (22.18), some deterministic algorithms could be used. However, these algorithms do not apply to more general cases, where the stochastic procedure might be advantageous.

References

[1] J.R. Beaumont, "Some issues in the application of mathematical programming in human geography", Wp-256, School of Geography, University of Leeds, 1979.

[2] J.M. Bregman, "Convergence proof of the Shelichovski method for transportation problems", *USSR Computat. Math. Math. Phys.* **7**,1 (1966).

[3] C.S. Bertuglia and G. Leonardi, "Dynamic models for spatial interaction", *Sistemi Urbani* **1**(2)(1979),3–25.

[4] J.D. Coelho and A.G. Wilson, "The optimum location and size of shopping centres", *Regional Studies* **10**(1976), 413-421.

[5] Y. Ermoliev and G. Leonardi, "Some proposals for stochastic facility location models", *Mathematical Modeling* **3**(1982), 407–420.

[6] Y. Ermoliev, G. Leonardi and J. Vira, "The stochastic quasigradient method applied to a facility location problem", WP-81-14, International Institute for Applied Systems Analysis, Laxenburg, Austria, 1981.

[7] M.J. Hodgson, "Towards more realistic allocation in location-allocation models: An interaction approach", *Environment and Planning A* **10**, 1273–1285.

[8] G. Leonardi, "Optimum facility location by accessibility maximizing", *Environment and Planning A* **10**(1978), 1287–1305.

[9] G. Leonardi, "A unifying framework for public facility location problems", WP-80-79, International Institute for Applied Systems Analysis. Laxenburg, Austria (1980) (to be published in *Environment and Planning A.*

[10] G. Leonardi and C.S. Bertuglia, "Optimal High School Location First Results for Turin, Italy", WP-81-5, International Institute for Applied Systems Analysis, Laxenburg, Austria, 1981.

[11] D. McFadden, "The measurement of urban travel demand", *J. Public Economics* **3**(1974), 303–328.

[12] D. McFadden, "Conditional logit analysis of qualitative choice behavior", in *Frontiers in Econometrics*, P. Zarembka, ed., Academic Press, New York, 1973.

[13] H.L.I. Neuburger, "User benefit in the evaluation of transport and land use plans", *J. Transport Economics Policy* **5**(1971), 52–75.

[14] A.G. Wilson, *Entropy in Urban and Regional Modelling*, Pion Press, London, 1970.

CHAPTER 23

LAKE EUTROPHICATION MANAGEMENT: THE LAKE BALATON PROJECT

A.J. King, R.T. Rockafellar,
L. Somlyódy, R.J-B Wets

Abstract

This is a brief overview of a collaborative effort of the Environment and Natural Resources, and the Adaptation and Optimization task forces at IIASA, to design stochastic optimization models for the management of lake eutrophication, and its use in a major case study (Lake Balaton). For further details, consult: Somlyódy [5],[6]; Somlyódy and van Straten [8]; Somlyódy and Wets [9]; Rockafellar and Wets [2]; and King [1].

Lake Balaton (Figure 23.1), one of the largest shallow lakes of the world, which is also the center of the most important recreational area in Hungary, has recently exhibited the unfavorable signs of artificial eutrophication. An impression of the major features of the lake-region system (including phosphorus sources and control alternatives) can be gained from Figure 23.1 (for details, see Somlyódy et al [7]; and Somlyódy and van Straten [8]). Four basins of different water quality can be distinguished in the lake (Figure 23.1) determined by the increasing volumetric nutrient load from east to west (the biologically available phosphorus load, BAP, is about ten times higher in Basin I than in Basin IV). The latter is associated to the asymmetric geometry of the system, namely the smallest western basin drains half of the total watershed, while only 5% of the catchment area belongs to the larger basin.

Based on observations for the period 1971–1982 the average deterioration of water quality of the entire lake is about 10% (in terms of Chlorophyll-a (Chl-a)). According to the OECD classification, the western part of the lake is in a (most advanced) hypertrophic state (which is the result of the large nutrient load), while the eastern portion of it is in an eutrophic stage.

The modeling approach to eutrophication and its management involved 4 major phases (Somlyódy [5]).

1. The description of the dynamics of the lake eutrophication process by a simulation model (LEM) which has two sets of inputs: controllable inputs (mainly artificial nutrient loads) and noncontrollable inputs (meteorological factors, such as temperature, solar radiation, wind, precipitation). The output of the model is the concentrations vector y of a number of water

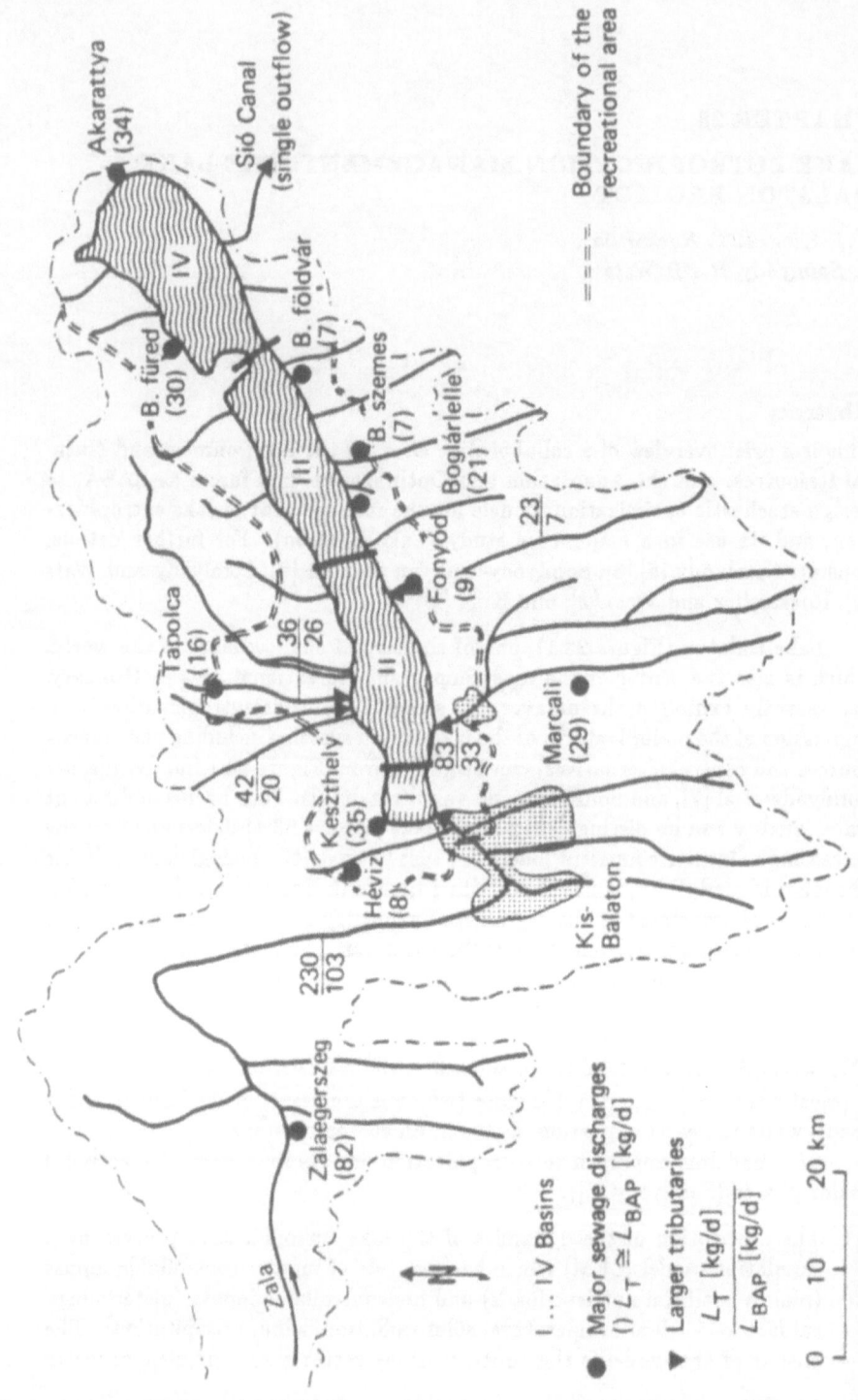

Figure 23.1 Major nutrient sources and control options.

quality components as a function of time (on a daily basis) t, and space $r : y(t, r)$. LEM is calibrated and validated by relying on historical data.

2. Derivation of stochastic inputs and the usage of LEM in a Monte Carlo fashion under systematically changed load conditions resulting in water quality as a stochastic variable: $\tilde{y}(t, r)$. Selection of the indicator for water quality management: for Lake Balaton the annual peak value of (Chl-a) was found to be appropriate. The use of (Chl-a)max. as the indicator allows to eliminate time from the analysis on the level of management.

3. Derivation of the aggregated, stochastic load response model (LEMP) serving the indicator as a function of the load (for Lake Balaton a linear relationship was obtained). Design of a planning type nutrient load model (NLMP) and the incorporation of LEMP and NLMP in a management, optimization model (EMOM).

4. Validation. In the course of this procedure various simplifications and aggregations are made without a quantitative knowledge of the associated errors. Accordingly, the last step in the analysis is validation. That is, the LEM should be run with the "optimal" load scenario (found in the previous step), and the "accurate" and "approximate" solutions generated by the aggregated and nonaggregated versions of LEM can be compared.

The lakes' total P is in an average $315t/yr$ (the BAP load is $170t/yr$); but depending on the hydrologic regime it can reach $550t/yr$. 53% of the load L is carried by tributaries (30% of which is of sewage origin—indirect load, see e.g., the largest city of the region, Zalaegerszeg in Figure 23.1), 17% is associated to direct sewage discharges (the recipient is the lake). Atmospheric pollution is responsible for 8% of the lake's load and the rest comes from direct runoff (urban and agricultural). Tributary load increases from east to west, while the change in the direct sewage load goes in the opposite direction. The sewage contribution (direct and indirect loads) is 30% to P, while it is about 52% to the total biological available load (the load of agricultural origin can be estimated as 47% and 33%, respectively) suggesting the importance of sewage load from the viewpoint of the short term eutrophication control. Figure 23.1 indicates also the loads of sewage discharges and tributaries which were involved in the management optimization model. These cover about 85% of the nutrient load which we consider controllable on the short term (e.g. atmospheric pollution and direct runoff are excluded).

Control alternatives are sewage treatment (upgrading of the biological stage and introduction of P precipitation) and the establishment of prereservoirs as indicated in Figure 23.1 (see e.g. the Kis-Balaton reservoir system planned for a surface area of about 75 km^2).

The nutrient load model for Lake Balaton incorporates control variables associated with control options mentioned. Sewage load was considered deterministic, while tributary load was modeled by the simple relationship.

$$\mathbf{L} = (L_0 + a_1 \mathbf{Q} + \mathbf{L}_\rho)(\xi^- + \xi)$$

where L_0 is the base load (mainly of sewage origin), $\mathbf{Q}$ is the stream flow rate, $\mathbf{L}_\rho$ is the residual, and the variable ξ accounts for the influence of infrequent sampling (ξ^- is the lower bound). The most detailed data set including 25 years of continuous records for Q and 5 years of daily observations for the loads was available for the Zala River (Figure 23.1) draining half of the watershed and representing practically the total load of Basin I. For the Zala River $\mathbf{L}_\rho$ was found to have a normal distribution, while $\mathbf{Q}$ was approached by a lognormal distribution. Tributary load can be controlled by choosing the size of reservoirs (they generally consist of two parts, having separated impacts on dissolved and particulate loads, see Figure 23.1), while the L_0 component can be influenced by sewage treatment. As can be judged from the above equation, sewage treatment affects the expectation of the load, only, while reservoirs affects both expectation and variance (for details see Somlyódy [6]).

The planning type nutrient load model (NLMP) outlined briefly and the linear load response model (LEMP) lead to the affine relation (Somlyódy and Wets [9])

$$\mathbf{y}(x,\omega) = \mathbf{T}(\omega)x - \mathbf{h}(\omega)$$

where $y = (y_1,\ldots,y_4)$ are the water quality indicators in Basins $1,\ldots,4$, the random vector $\mathbf{h}$ incorporates all noncontrollable factors, the x-variables are the control variables and the linear transformation $T(\omega)x$ gives the effect on water quality of the measures taken to control the loads $\mathbf{L}$.

In the formulation of the eutrophication management optimization model (EMOM) the objective must be chosen so as to measure in the most realistic fashion possible the deviations of the indicators from the water quality goals. This led us to a stochastic program with recourse model with associated solution procedure developed by Rockafellar and Wets [3] and implemented by King [1]. We also used a linear programming model, see Somlyódy [6] and Somlyódy and Wets [9] (Section 6) that is based on expectation-variance considerations (for the water quality indicators). In the Lake Balaton case study the results for both this expectation-variance model and the stochastic programming model (5.11) lead to remarkably similar investment decisions. Subsequently, objective functions and results of the two models are briefly discussed.

1. The recourse formulation starts from the following considerations. The model should distinguish between situations that barely violate the desired water quality levels ($\gamma_i, \quad i = 1,\ldots,N$) and those that deviate substantially from these norms. This suggests a formulation of our objective in terms of a penalization that would take into account the observed values of $(y_i(x,w) - \gamma_i)$ for $i = 1,\ldots,4$.

We found that the following class of functions provided a flexible tool for the analysis of these factors. Let $\theta : R \to R_+$ be defined by

$$\theta(\tau) := \begin{cases} 0 & \text{if } \tau \leq 0 \\ \frac{1}{2}\tau^2 & \text{if } 0 \leq \tau \leq 1 \\ \tau - \frac{1}{2} & \text{if } \tau \geq 1 \end{cases}$$

This is a piecewise linear-quadratic-linear function. The penalty functions $(\Psi_i, \quad i = 1, \ldots, N)$ are defined through:

$$\Psi_i(z_i) = q_i e_i \theta(e_i^{-1} z_i) \text{ for } = 1, \ldots, N,$$

where q_i and e_i are positive quantities that allow us to scale each function Ψ_i in terms of slopes and the range of its quadratic component. By varying the parameters e_i and q_i we are able to model a wide range of preference relationships and study the stability of the solution under perturbation of these scaling parameters.

The objective is thus to find a program that in the average minimizes the penalties associated with exceeding the desired concentration levels. This leads to the following formulation of the water quality management problem:

find $x \in R^n$ such that

$$0 \leq x_j \leq r_j, \qquad j = 1, \ldots, n$$

$$\sum_{j=1}^{n} a_{ij} x_j \leq b_i, \qquad i = 1, \ldots, m_1$$

$$\sum_{j=1}^{n} t_{ij}(w) x_j - v_i(w) = h_i(w) \qquad i = 1, \ldots, m_2$$

and $z = \sum_{j=1}^{n} \left(c_j x_j + \dfrac{d_j}{2r_j} x_j^2 \right) + E\{ \sum_{i=1}^{m_2} q_i e_i \theta(e_i^{-1} v_i(w)) \}$ is minimized

to which one refers as *a quadratic stochastic program with simple recourse*; here b_1 is the available budget that we handle as a parameter. For problems of this type, in fact with this application in mind, an algorithm is developed in Rockafellar and Wets [2], and Rockafellar and Wets [3], which relies on the properties of an associated dual problem. In particular it is shown that the following problem:

find $y \in R_+^m$ and $z(\cdot) : \Omega \to R^{m_2}$ measurable such that

$$0 \leq z_i(w) \leq q_i, \qquad i = 1, \ldots, m_2$$

$$u_j = c_j - \sum_{i=1}^{m_1} a_{ij} y_i - E\{ \sum_{i=1}^{m_2} z_i(w) t_{ij}(w) \}, \qquad j = 1, \ldots, n$$

and $\sum_{i=1}^{m_1} y_i b_i - \sum_{i=1}^{m_2} E\{ h_i(w) z_i(w) + \dfrac{e_i}{2q_i} z_i^2(w) \}$

$$- \sum_{j=1}^{n} r_j d_j \theta(d_j^{-1} u_j) \text{ is maximized },$$

is dual to the original problem, provided that for $i = 1, \ldots, m_2$, the e_i and q_i are positive (and that is the case here) and for $j = 1, \ldots, n$, the $d_j > 0$, which is taken care of by a natural perturbation of the objective.

An experimental version of this algorithm that relies on MINOS was implemented at IIASA by A. King (and is available through IIASA as part of a collection of codes for solving stochastic programs), see King [1]. It starts the procedure by solving the deterministic problem with expected values for the coefficients in h and T.

2. As a starting point for the construction of the expectation-variance model, we consider the following objective function:

$$\sum_{i=1}^{N} q_i E\{y_i(x,\cdot) - \gamma_i\}_+^2\}$$

where, as earlier, $y_i(x,w)$ is the water quality indicator characterized by the selected indicator in basin i given the investment program x and the environmental conditions w, γ_i the goal set for basin i and q_i a weighting factor. The objective being quadratic in the area of interest, and the distribution functions $G_i(x,\cdot)$ of the $y_i(x,\cdot)$ not being too far from normal, one should be able to recapture the essence of the effect of this objective function on the decision process by considering just expectations and variances of the $y_i(x,\cdot)$. This observation, and the "soft" character of the management problem, suggest that we could substitute for the original objective

$$\sum_{i=1}^{N} q_i \left(E\{y_i(x,\cdot) - \bar{y}_{oi}\} + \theta\sigma(y_i(x,\cdot) - \bar{y}_{oi}) \right)$$

where θ is a positive scalar (usually between 1 and 2.5), $\bar{y}_{oi} = E\{y_{oi}\}$ is the expected nominal state of basin i, and σ denotes standard deviation,

$$\sigma(y_i(x,\cdot) - \bar{y}_{oi}) = E\{(y_i(x,\cdot) - E\{y_i(x,\cdot)\})^2\}^{\frac{1}{2}}.$$

Since for each $i = 1,\ldots,N$, the y_i are affine (linear plus a constant term) with respect to x, the expression for

$$E\{y_i(x,\cdot) - \bar{y}_{oi}\} = \sum_{j=1}^{n} \mu_{ij}x_j + \mu_{io}$$

as a function of x is easy to obtain from the load equations. The μ_{ij} are the expectations of the coefficients of the x_j and the μ_{io} the expectation of the constant term. Unfortunately the same does not hold for the standard deviation $\sigma(y_i(x,\cdot) - \bar{y}_{oi})$. The nutrient-load model suggest that

$$\sigma(y_i(x,\cdot) - \bar{y}_{oi}) \sim \left(\sum_{\ell} \sigma_{i\ell}^2 x_\ell^2\right)^{\frac{1}{2}}$$

where $\sigma_{i\ell}$ is the part of the standard deviation that can be influenced by the decision variable x_ℓ; for example, the standard deviation of the tributary load.

Cross terms are for all practical purposes irrelevant in this situations since the total load in basin i is essentially the sum of the loads generated by various sources that are independently controlled. This justifies using

$$\sum_{i=1}^{N} q_i \left[\left(\sum_{j=1}^{n} \mu_{ij} x_j \right) + \theta \left(\sum_{j=1}^{n} \sigma_{ij}^2 x_j^2 \right)^{\frac{1}{2}} \right]$$

as an objective for the optimization problem. This function is convex and differentiable on R_+^n except at $x = 0$, and conceivably one could use a nonlinear programming package to solve the optimization problem:

$$\text{find} \quad x \in R^n \text{ such that}$$

$$r_j^- \leq x_j \leq r_j^+ \qquad j = 1, \ldots, n$$

$$\sum_{j=1}^{n} a_{ij} x_j \leq b_i \qquad i = 1, \ldots, m$$

and $z = \sum_{i=1}^{N} q_i [\sum_{j=1}^{n} \mu_{ij} x_j + \theta \left(\sum_{j=1}^{n} \sigma_{ij}^2 x_j^2 \right)^{\frac{1}{2}}]$ is minimized.

One can go one step further in simplifying the problem to be solved, namely by replacing the term.

$$\left(\sum_{j=1}^{n} \sigma_{ij}^2 x_j^2 \right)^{\frac{1}{2}}$$

in the objective, by the linear (inner) approximation

$$\sum_{j=1}^{n} \sigma_{ij} x_j.$$

On each axis of R_+^n, no error is introduced by relying on this linear approximation; otherwise we are over-estimating the effect a certain combination of the $x_j's$ will have on the variance of the concentration levels. Thus, at a given budget level we shall have a tendency to start projects that affect more strongly the variance if we use the linear approximation, and this is actually what we observed in practice. Assuming the cost functions c_j are piecewise linear, we have to solve the *linear* program:

$$\text{find} \quad x \in R^n \text{ such that}$$

$$r_j^- \leq x_j \leq r_j^+, \qquad j = 1, \ldots, n$$

$$\sum_{j=1}^{n} a_{ij} x_j \leq b_i, \qquad i = 1, \ldots, m_1$$

$$\text{and } t = \sum_{i=1}^{N} q_i \sum_{j=1}^{n} (\mu_{ij} + \theta\sigma_{ij})x_j \text{ is minimized.}$$

We refer to this problem as the *(linearized) expectation-variance model.*

We have given only a heuristic "justification" for the use of the expectation-variance model as a management tool. In Section 6 of Somlyódy and Wets [9], this model is also derived from a basic formulation of the management problem that integrates reliability and penalty considerations.

3. Figures 23.2 and 23.3 give a comparison of the results for the recourse and the expectation-variance models when we vary β (the budget level). Statistical parameters (expectation, standard deviation and extremes) of the water quality indicators gained from Monte Carlo procedure are illustrated in Figure 23.2 for the Keszthely basin as a function of the available budget β.

In Figure 23.3, we record the changes in the two major control variables (x_{SN1} and x_{D1}) associated to the treatment plant of Zalaegerszeg and the (second) reed lake segment of the Kis-Balaton system (see Figure 23.1). There is a significant trade-off between these two variables. For decision making purposes, it is important to observe that there are four ranges of possible values of β, in which the solution has different characteristics.

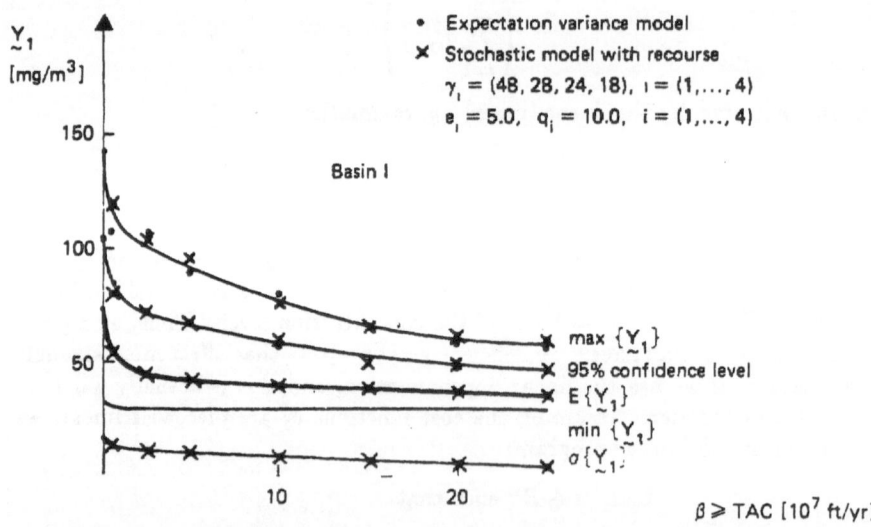

Figure 23.2. Water quality indicator $(\text{Chl} - a)_{\max}$ as a function of the total annual cost.

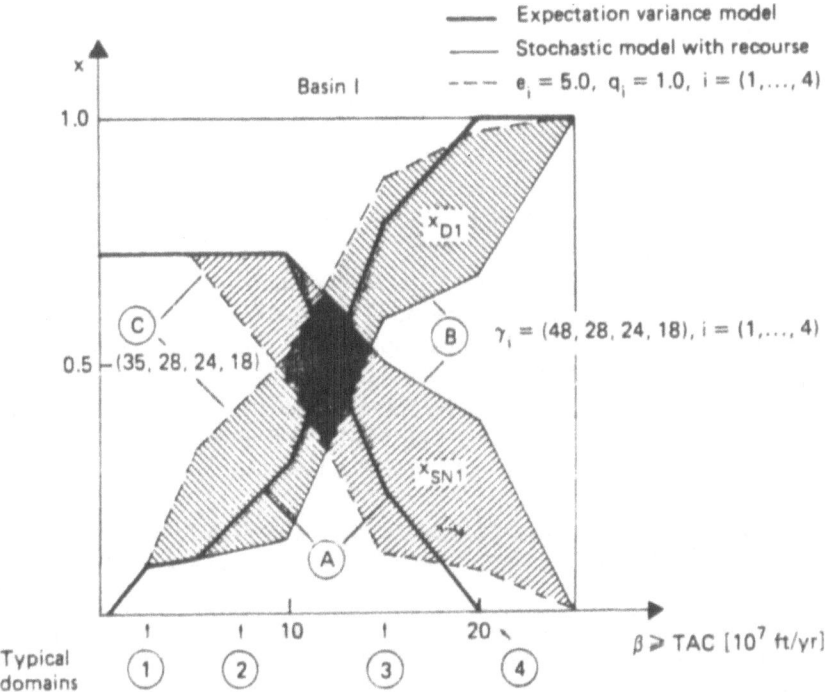

Figure 23.3. Change of major decision variables.

As seen from Figures 23.2-23.3, the two models produce practically the same results in terms of the water quality indicators (including also their distribution). With respect to details there are minor deviations. According to Figure 23.3, the expectation-variance model gives more emphasis to fluctuations in water quality and consequently to reservoir projects, than the stochastic recourse model (see the basic case, B, with the parameters specified), and this is in accordance with the fact that the role of the variance is overstressed in the expectation-variance model.

From this quick comparison of the performances of the two models, we may conclude that the more precise stochastic model validates the use of the expectation-variance model in the case of Lake Balaton.

A more detailed analysis, and further discussion on the role of parameters γ_i, e_i and q_i, and comparison between deterministic models and the stochastic models is given in Section 8 of Somlyódy and Wets [9].

References

[1] A. King, "An Implementation of the Lagrangian Finite-Generation Method", in *Numerical Methods for Stochastic Programming*, Y. Ermoliev and R. Wets (eds.). IIASA Collaborative Volume, Springer Verlag, 1985.

[2] R.T. Rockafellar and R. Wets, "A Dual Solution Procedure for Quadratic Stochastic Programs with Simple Recourse", in *Numerical Methods*, pp.252–265. V. Pereyra and A. Reinoza (eds.). Lecture Notes in Mathematics 1005. Springer Verlag, Berlin, 1983.

[3] R.T. Rockafellar and R. Wets, "A Lagrangian finite Generation Technique for Solving Linear-Quadratic Problem in Stochastic Programming", in *Stochastic Programming: 1984*, A. Prékopa and R. Wets (eds.). *Mathematical Programming Study* (1985).

[4] R.T. Rockafellar and R.J-B. Wets, "Linear-Quadratic Programming Problems with Stochastic Penalties: The Finite Generation Algorithm", WP-85-45, Laxenburg, Austria: International Institute for Applied Systems Analysis, 1985.

[5] L. Somlódy, "A Systems Approach to Eutrophication Management with Application to Lake Balaton", *Water Quality Bulletin* Vol. 9 No.1(1983), 25–37.

[6] L. Somlyódy, "Lake Eutrophication Mangement Models", in *Eutrophication of Shallow Lakes: Modeling and Management. The Lake Balaton Case Study*, pp.207–250. L. Somlyódy, S. Herodek, and J. Fischer (eds.), CP-83-53, Laxenburg, Austria: International Institute for Applied Systems Analysis, 1983.

[7] L. Somlyódy, S. Herodek, and J. Fischer (eds.), "Eutrophication of Shallow Lakes: Modeling and Management", pp.367. *The Lake Balaton Case Study*, CP-83-53, Laxenburg, Austria: International Institute for Applied Systems Analysis, 1983.

[9] L. Somlyódy and G. van Straten (eds.), *Modeling and Managing Shallow Lake Eutrophication with Application to Lake Balaton*. Springer Verlag (in press).

[10] L. Somlyódy and R. Wets, "Stochastic Optimization Models for Lake Eutrophication Management", CP-85-16, Laxenburg, Austria: International Institute for Applied Systems Analysis, 1985.

CHAPTER 24

OPTIMAL INVESTMENTS FOR ELECTRICITY GENERATION: A STOCHASTIC MODEL AND A TEST-PROBLEM

F. V. Louveaux and Y. Smeers

Introduction

In this chapter, we study the problem of optimal investments for electricity generation. We discuss the reasons which justify the use of a multistage stochastic model and present a formulation for such a model. We then propose a two-stage test-problem derived from this model.

24.1 The Problem

Among the various problems related to electricity generation, we consider here the investment problem which consists in finding optimal levels of investment in various types of power plants so as to meet future demands, see Anderson [1]. Three properties of a given power plant i can be singled out in a static analysis: the investment cost c_i, the operating cost q_i and the availability factor a_i which indicates the percent of time the power plant can effectively be operated. Demand for electricity can be considered to be a single product, but the level of demand varies over time. The electricity producers usually represent the demand in terms of a so-called "load-curve" which describes the demand over time in decreasing order of demand level (Figure 24.1). Since we are concerned here with investments over the long run, the load curve we consider is taken over a year.

The load curve can be approximated by a piecewise constant curve (Figure 24.2) with k segments. Let $d_1 = D_1, d_j = D_j - D_{j-1}, \quad j = 2,\ldots,k$ represent the additional power demanded in the so-called "mode j" for a duration T_j. Note that in order to obtain a good approximation of the load curve, it is necessary to consider large values of k. In the static situation, the problem consists in finding the optimal investment for each mode j, i.e. that one which minimizes the total cost of effectively producing 1 MW of electricity during the time T_j.

$$i(j) = \operatorname*{argmin}_{i=1,n}\left\{\frac{c_i + q_i T_j}{a_i}\right\} \tag{24.1}$$

where n is the number of available technologies.

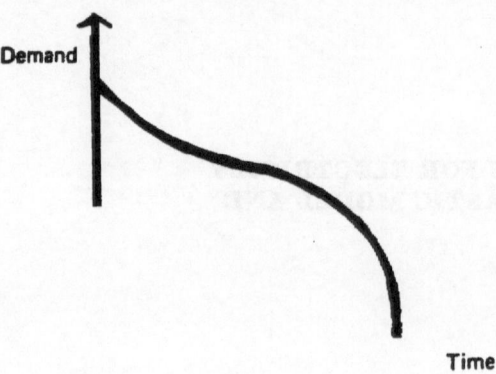

Figure 24.1

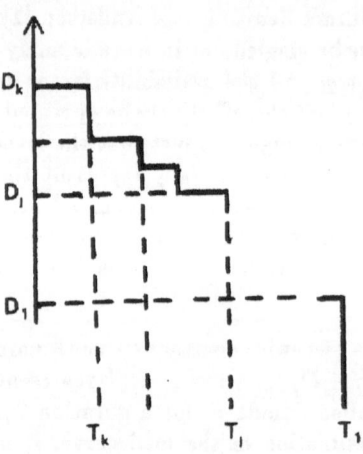

Figure 24.2

The above static model captures one essential feature of the problem namely that base load demand (associated with large values of T_j, i.e. small indices j) is covered by equipment with low operating costs (scaled by availability factor) while peak-load demand (associated with small values of T_j, i.e. large indices j) is covered by equipment with low investment costs (also scaled by their availability factor). For the sake of completeness, note that peak-load equipment should also offer enough flexibility in operations.

24.2 A multistage model

At least four elements justify considering a dynamic or multistage model for the electricity generation investment problem:

- the long-term evolution of equipment costs
- the long-term evolution of the load curve
- the appearance of new technologies
- the obsolescence of presently available equipment.

The equipment costs are influenced by technological progress but also (and for some drastically) by the evolution of fuel costs.

Of significant importance in the evolution of demand is the total energy demanded (the area under the load curve) but also the peak-level D_k which determines the total capacity that should be available to cover demand. The evolution of the load curve is commanded by several factors including the level of activity in industry, energy savings in general as well as electricity producers tariff policy.

The appearance of new technologies depends on the technical and commercial success of research and development while obsolescence of available equipment depends on past decisions and technical life time of equipment.

All these elements together induce that it is no longer optimal to invest only in view of the short-term ordering of equipment given by (24.1) but that a long-term optimal policy should be found.

The following multistage model can be proposed. Let

n = number of technologies available
x_i^t = new capacity made available for technology i at time t
s_i^t = total capacity of i available at time t
a_i = availability factor of i
L_i = life-time of i
g_i^t = existing capacity of i at time t, decided before $t = 1$
d_j^t = maximal power demanded in mode j at time t
T_j^t = duration of mode j at time t
y_{ij}^t = capacity of i effectively used at time t in mode j
c_i^t = unit investment cost for i at time t (on a yearly equivalent basis)
q_i^t = unit production cost for i at time t

The electricity generation N-stage problem is

$$\min_{x,y,s} \sum_{t=1}^{N} \left(\sum_{i=1}^{n} c_i^t \cdot s_i^t + \sum_{i=1}^{n} \sum_{j=1}^{k} q_i^t \cdot T_j^t \cdot y_{ij}^t \right) \tag{24.2}$$

subject to

$$s_i^t = s_i^{t-1} + x_i^t - x_i^{t-L_i} \qquad i = 1, \ldots, n, \quad t = 1, \ldots, N \qquad (24.3)$$

$$\sum_{i=1}^{n} y_{ij}^t = d_j^t \qquad j = 1, \ldots, k, \quad t = 1, \ldots, N \qquad (24.4)$$

$$\sum_{j=1}^{k} y_{ij}^t \le a_i (g_i^t + s_i^t) \qquad i = 1, \ldots, n, \quad t = 1, \ldots, N \qquad (24.5)$$

$$x, y, s \ge 0$$

Decisions in each period t involve new capacities x_i^t made available in each technology and capacities y_{ij}^t operated in each mode for each technology.

Newly decided capacities increase the total capacity s_i^t made available, as given by (24.3) where account is also taken of equipments becoming obsolete after their lifetime. We assume $x_i^\tau = 0$ if $\tau \le 0$, so equation (24.3) only involves newly decided capacities.

By (24.4), the optimal operation of equipments must be chosen in such a way as to meet demand in all modes, using available capacities which by (24.5) depend on capacities g_i^t decided before $t = 1$, newly decided capacities s_i^t and the availability factor.

The objective function (24.2) is the sum of the investment plus maintenance costs and operating costs. Compared to (24.1), availability factors are taken care of in constraints (24.5) and do not need to appear any more in (24.2), the operating costs are exactly the same and are based on operating decisions y_{ij}^t, while the investment annuities and maintenance costs c_i^t apply on the cumulative capacity s_i^t. Placing annuities on the cumulative capacity, instead of charging the full investment cost to the decision x_i^t, simplifies the treatment of end effects and is currently used in many power generation models. It is a special case of the salvage value approach, see e.g. Grinold [3].

24.3 A stochastic model

The same reasons that pleaded for the use of a multistage model can be advocated to motivate resorting to a stochastic model. The evolution of equipment costs, in particular fuel costs, the evolution of total demand, the date of appearance of new technologies, even the lifetime of existing equipments can all be considered truly random. We first present a basic model taking the uncertainty about demand and costs into account leaving the other two aspects for the discussion.

The main difference between the stochastic model and its deterministic case is in the definition of the variables x_i^t and s_i^t. In particular, x_i^t now represents the new capacity of i decided at time t, which becomes available at time $x_i^{t+\Delta_i}$ where Δ_i is the construction delay for equipment i. In other words, to have extra capacity available at time t, it is necessary to decide at $t - \Delta_i$, when less information is available on the evolution of demand and equipment costs. This is especially important since it would be preferable to be able to wait till the last moment to take decisions that would have immediate impact.

Another consequence of delay factors and uncertainty is the fact that the model loses its relatively complete recourse property. This means any choice of investment decisions does not yield a feasible operations policy. To restore the relatively complete recourse property, it is necessary to assume that there exists a technology with high operating costs and zero construction delay. For any period t and any realisation ξ of the random event, an investment is made in that technology, which for simplicity is always supposed to be technology n, if the level of capacity investments in the previous periods is insufficient to cover present demand.

Let

x_i^t = new capacity decided at time t for equipment i, $\quad i = 1, \ldots, n$
s_i^t = total capacity of i available plus in order at time t
n = a technology such that $\Delta_n = 0$
ξ_t = represents the random variable at time t

and the other variables be as before. Then the stochastic model is

$$\min E_\xi \sum_{t=1}^{N} \left(\sum_{i=1}^{n} c_i^t s_i^t + \sum_{i=1}^{n} \sum_{j=1}^{k} q_i^t T_j^t y_{ij}^t \right) \tag{24.6}$$

$$s_i^t = s_i^{t-1} + x_i^t - x_i^{t-L_i} \tag{24.7}$$

$$\sum_{i=1}^{n} y_{ij}^t = d_j^t \tag{24.8}$$

$$\sum_{j=1}^{k} y_{ij}^t \leq a_i (g_i^t + s_i^{t-\Delta_i}) \tag{24.9}$$

$$a_n \left(g_n^t + s_n^{t-1} + x_n^t \right) \geq D_m^t - \sum_{i=1}^{n-1} a_i \left(g_i^t + s_i^{t-\Delta_i} \right) \tag{24.10}$$

$$s, x, y \geq 0$$

The elements forming ξ_t are essentially the demands $(d_1^t, \ldots, d_n^t)$ and the costs (c^t, q^t). The decision vectors (x^t, s^t, y^t) are conditional on the realizations $(\xi_1, \ldots \xi_t)$. The above model has fixed recourse since W and T are fixed and relatively complete recourse thanks to inequality (24.10). In most cases, when periods represent several years, typically five, and N is small enough, equation (24.7) can be simplified into

$$s_i^t = s_i^{t-1} + x_i^t.$$

If one wants to consider the date of appearance of a new technology i to be a random event, the easiest way is to add constraints of the form

$$x_i^t \leq \eta_i^t \cdot u_i$$

where u_i is a fixed upperbound on the investment in any period t, $\eta_i = (\eta_i^1, \ldots, \eta_i^N)$ is a stochastic vector, whose components are zero and one, and such that $\eta_i^{t+1} \geq \eta_i^t, i = 1, \ldots, N-1$. This permits to maintain to the model a fixed recourse structure.

On the other hand, if the availability factors or the life-time are random, then the model no longer possesses the fixed recourse property.

24.4   Techniques of Solution

Techniques of solution used in Louveaux [4] and Louveaux and Smeers [5] to solve (24.6)–(24.10) are based on two observations. First, an accurate approximation of the load curve by a piecewise constant curve, as was done in Section 1, requires the use of many different modes ($k = 20$ to 40, typically). This in fact induces that the size of the model becomes very large. The alternative procedure proposed in [5] is to use a piecewise linear approximation such that a limited number of pieces suffice to adequately describe the load curve. Then, the objective function in (24.6) becomes quadratic in the y_{ij}^t's.

The second observation is that the above model possesses the block-separability property, discussed in [4]. This means that decisions on the operations variables y_{ij}^t, for a given ξ_t, can be taken independently of investment decisions x_i^t of the same period for the same ξ_t, and moreover that the operations variables y_{ij}^t do not influence in any way the choice of subsequent variables x_i^τ for $\tau \geq t+1$. The details of how to handle the special case of the technology n with zero construction delay for which the decision x_n^t can influence y_{nj}^t for the same period are explained in [4]. Using these techniques, problems running over 5 periods and having up to 32 final random realisations have been solved, see [5].

24.5   Test Problem

In this section, we present a two-stage linear version of (24.6)–(24.10) with stochastic right-hand side only, and we discuss the reasons which make this test problem interesting.

24.5.1 The example

The example is a two-stage linear version of (24.6)–(24.10), with 3 operating modes, 4 technologies, one period construction delay for all technologies, and no equipment available, so $g = (0,0,0,0)$. We also assume $d_3 = 2, d_2 = 3$ and $d_1 = \xi$, where ξ can take the value 3, 5 or 7 with probability .3, .4 and .3 respectively. Moreover $T_2 = .6T_1$ and $T_3 = .1T_1$,; we assume $T_1 = 10$. Since $N = 2$ and all equipments have a one period construction lead time, (24.7) reduces to $s_i^t = x_i^t$, so the variables s^t are suppressed from the formulation and the index t can be omitted. The constraint (24.10) takes the simple form $\sum_{i=1}^{4} x_i \geq 12$ where $12 = \max_\xi \xi + d_2 + d_3$.

An upper bound is placed on the budget spent on the first period. The investment costs for the four equipments are $(10, 7, 16, 6)$ respectively. Assuming $T_1 = 10$, the operating costs in mode 1 are $(40, 45, 32, 55)$. Then, if $T_2 = 6$ and $T_3 = 1$, one obtains the following model.

$$z = \min 10x_1 + 7x_2 + 16x_3 + 6x_4 + E_\xi \min(40y_{11} + 45y_{21} + 32y_{31} + 55y_{41}$$
$$+ 24y_{12} + 27y_{22} + 19.2y_{32} + 33y_{42}$$
$$+ 4y_{13} + 4.5y_{23} + 3.2y_{33} + 5.5y_{43})$$

subject to

$$x_1 + x_2 + x_3 + x_4 \geq 12 \qquad y_{11} + y_{12} + y_{10} \leq x_1$$
$$10x_1 + 7x_2 + 16x_3 + 6x_4 \leq 120 \qquad y_{21} + y_{22} + y_{23} \leq x_2$$
$$x \geq 0 \qquad y_{31} + y_{32} + y_{33} \leq x_3$$
$$y_{41} + y_{42} + y_{43} \leq x_4$$
$$y_{11} + y_{21} + y_{31} + y_{41} \geq \xi$$
$$y_{12} + y_{22} + y_{32} + y_{42} \geq 3$$
$$y_{13} + y_{23} + y_{33} + y_{43} \geq 2$$
$$y \geq 0$$

where ξ can take the value 3, 5 or 7 with probability 0.3, 0.4 and 0.3 respectively.

The optimal solution is given by $x_1 = 8/3; x_2 = 4; x_3 = 10/3; x_4 = 2$ with objective value $z = 381.853$. It was obtained by using Birge's NDST3 program, see [2].

24.5.2 Use of the example

One great quality of the above example is that optimal second-stage decisions are easy to derive. This is an interesting feature for the design and verification of a new algorithm or computer code. The same property can also be used to illustrate the advantages of block-separability in multistage programs, see [4].

. We now indicate how the second-stage decisions can be used to obtain one cut of the L-shaped method, see Van Slyke and Wets [6] and Birge [2] for a multistage version.

In the above example, the optimal second-stage decisions, conditional to some realization of ξ, can be obtained by a simple rule, called the "order of merit rule", which states that it is optimal to operate the equipments in the order of increasing operating costs.

To illustrate this, take the example where $\xi = 5$ and $x_1 = 8/3; x_2 = 4; x_3 = 10/3; x_4 = 2$. Following the order of merit rule, the cheapest equipment, namely equipment No. 3, should be used first, i.e. in mode 1, up to the available capacity; since $x_3 = 10/3 \le d_1$, it follows that $y_{31} = x_3$ (this is valid as long as $x_3 \le 5$).

The second cheapest equipment in terms of operating costs is equipment No. 1, hence $y_{11} = 5 - x_3$ and $y_{12} = x_1 - (5 - x_3) = x_1 + x_3 - 5$.

In mode 2, in addition to equipment 1, it is necessary to operate No. 2 as follows: $y_{22} = 3 - (x_1 + x_3 - 5) = 8 - x_1 - x_3$ and finally $y_{23} = 2$. From this, we derive the value of the second stage for $\xi = 5$.

$$Q(x, \xi = 5) = 32x_3 + 40(5 - x_3) + 24(x_1 + x_3 - 5) + 27(8 - x_1 - x_3)$$
$$+ 4.5.2 = 305 - 3x_1 - 11x_3$$

Similarly, for $\xi = 3$, one obtains the second-stage optimal solution

$$y_{31} = 3, y_{32} = x_3 - 3, y_{12} = x_1, y_{22} = 6 - x_1 - x_3, \text{ and } y_{23} = 2,$$

hence the optimal value of the second-stage

$$Q(x, \xi = 3) = 209.4 - 3x_1 - 7.8x_3.$$

Finally, for $\xi = 7$, one optimal solution is

$$y_{31} = x_3, y_{11} = x_1, y_{21} = 7 - x_1 - x_3, y_{22} = 3,$$

$$y_{23} = x_1 + x_2 + x_3 - 10 \text{ and } y_{43} = 12 - x_1 - x_2 - x_3, \text{ so}$$

$$Q(x, \xi = 7) = 417 - 6x_1 - x_2 - 46x_3.$$

Given the probabilities associated to $\xi = 3$, 5, and 7, one obtains

$$Q(x) = E_\xi Q(x_1, \xi) = 309.92 - 3.9x_1 - .3x_2 - 20.54x_3.$$

Hence, the related cut in the L-shaped method of Van Slyke and Wets [6] would be

$$\theta \ge 309.92 - 3.9x_1 - 0.3x_2 - 20.54x_3.$$

References

[1] D. Anderson, "Models for determining least-cost investment in electricity supply", *Bell Journal of Economics and Management Science* **3**(1972), pp. 267–299.

[2] J. Birge, "Decomposition and partitioning methods for multi-stage stochastic linear programs", *Tech. Report 82-6, Dept. of Industrial and Operations Engineering*, University of Michigan (1982), to appear in *Operations Research*.

[3] R.C. Grinold, "Model building techniques for the correction of end effects in multistage convex programs", *Operations Research* **31**(1983), 407–431.

[4] F.V. Louveaux, "Multistage stochastic programs with block-separable recourse", to appear in *Mathematical Programming Study on Recent Advances in Stochastic Programming*, eds. A. Prékopa and R. Wets.

[5] F.V. Louveaux and Y. Smeers, "Stochastic optimization for the introduction of a new energy technology", to appear in *Stochastics*.

[6] R. Van Slyke and R. Wets, "L-shaped linear programs with applications to optimal control and stochastic programming", *SIAM J. Appl. Math.* **17**(1969), 638–663.

CHAPTER 25

SOME APPLICATIONS OF STOCHASTIC OPTIMIZATION METHODS TO THE ELECTRIC POWER SYSTEM

C. Nedeva

25.1   Introduction

The electricity generation, distribution and consumption network is a complex system involving a large number of power sources and consumers, electricity transmission lines, transformers, and so on. The technical and economic characteristics of its components depend on a number of factors: the amount of electricity consumed depends on the introduction of new consumers, on the use of new techniques, on the time of day, the season, etc; the volume of production and the price of electricity depend on the local hydrometeorological conditions, and on the quantities and prices of the available resources etc.

Various types of problems arise in this system: forecasting problems, problems of engineering design, exploitation problems, etc. Most of the resulting mathematical problems are problems of optimization under uncertainty, since it is usually impossible to predict precisely what will happen in the future. A typical design problem is described in the next section.

25.2   Determination of the Optimal Parameters of a Super Conducting Power Cable Line (S.C.P.C.L.)

The aim is to minimize the total cost of the construction, exploitation and support of an S.C.P.C.L. We consider an S.C.P.C.L. of fixed construction with a coaxial disposition of the current-carrying and shielding superconductive elements. This type of construction allows us to express the total cost by means of the following parameters: the dimension x, the nominal tension y and the number z of cables in one line. The cost also depends on a number of factors whose values are determined theoretically or experimentally and not known precisely. The exploitation costs of an S.C.P.C.L. depend on the transmitted power, the non-uniformity and amplitude of the graph of the load, the number of switch-offs, the temperature of the surroundings; and parameters whose values may vary during the operation of the S.C.P.C.L. The construction costs depend on the prices of the materials and the labor costs, which are not known precisely at the time of design—optimistic, pessimistic and most probable values are provided by expert economists.

The parameters whose values are not known will be denoted by $(\omega_1, \ldots, \omega_\ell)$ $= \omega$ (in our problem $\ell \sim 30$), and we shall assume that ω is a random vector with distribution function H. Thus the total cost of the S.C.P.C.L. may be expressed by a very complicated function [1], indeed, so complicated that it is not feasible to discuss it here. We shall denote this function by $f(x, y, z, \omega)$ and record only its essential properties: f is measurable on ω, differentiable and strictly convex on x for every y, z, ω. The variables y and z are discrete and may take a certain (not too large) number of values. The sets of feasible values of y and z will be denoted by Y and Z, respectively.

For given $y \in Y$ and $z \in Z$, the S.C.P.C.L. should possess a steady-state stability margin, which is expressed by the condition

$$\omega_{i_0} x \leq c y^2 z,$$

where c is a known constant and ω_{i_0} is the component of ω corresponding to the transmitted power.

Since the parameters x, y, z should be determined and fixed before the construction and exploitation of the S.C.P.C.L., a reasonable optimization criterion is the "minimization of the mean total cost" with the requirement that the steady-state stability condition is satisfied with sufficiently large probability p_0.

We thus arrive at the following mathematical model: minimize the function

$$F_H(x, y, z) = Ef(x, y, z, \omega) = \int f(x, y, z, t) \, \mathrm{d}H(t) \tag{25.1}$$

subject to

$$P(\omega_{i_0} x \leq c y^2 z) \geq p_0, \tag{25.2}$$

$$a \leq x \leq b, y \in Y, z \in Z. \tag{25.3}$$

Let us assume that the distribution function H is known.

We therefore have to solve a partially-discrete stochastic programming problem. The number of feasible combinations of parameters y and z does not exceed 15, and for fixed $y \in Y$ and $z \in Z$, the minimization problem of the function (25.1) with respect to x subject to (25.2), (25.3) is easily solved by means of a method described below. Enumeration on the discrete parameters y and z is fully acceptable and we shall describe a method for the minimization with respect to x of the function (25.1) subject to (25.2), (25.3), for fixed $y = y_0, z = z_0$. We note first that conditions (25.2) and (25.3) together are equivalent to the condition

$$x \in X^0 = \{x | a \leq x \leq \alpha^*\} \tag{25.4}$$

where $\alpha^* = \max \alpha$ such that $a \leq \alpha \leq b$ and

$$P(c y_0^2 z_0 / \omega_{i_0} < \alpha) \leq 1 - p_0.$$

The distribution function of the random variable ω_{i_0} is known and the problem of finding the right bound α^* is easily solved, for instance, by the golden section method.

The problem (25.1) and (25.4) may be solved by means of stochastic quasi-gradient (SQG) methods. We choose an initial point $x^0 \in X^0$. Suppose we have arrived at a point x^k after k iterations. Then we choose the point ω^k in accordance with the distribution function H and construct the point

$$x^{k+1} = \max[a, \min\{\alpha^*, x^k - \gamma(f(x^k + \frac{1}{k}, y^0, z^0, \omega^k) - f(x^k, y^0, z^0, \omega^k))\}]$$

where $\gamma > 0$ is a constant (in our computations a reasonable choice for γ turns out to be $\gamma = 10^{-5}$).

As a stopping rule we use the inequality

$$\frac{1}{\tau}[\sum_{s=k}^{k+\tau} f(x^s, y^0, z^0, \omega^s) - \sum_{s=k+\tau}^{k+2\tau} f(x^s, y^0, z^0, \omega^s)] \leq 0.001.$$

Optimal parameters were determined for many different sets of input data. The average number of iterations was 510. Table 25.1 gives the main parameters for one particular set of input data.

Table 25.1 The value and type of the main parameters for one set of input data

Parameter	Value(s)/distribution	Type
Length of the S.C.P.C.L.	30 (km)	Fixed
Feasible set for the number of cables in one line (Z)	$\{2,3,4\}$	Fixed
Feasible set for the nominal tension (Y)	$\{10, 20, 40, 60,$ $90, 110\}$ (kv)	Fixed
Transmitted power (ω_{i_0})	sharply normal in $[400,600]$ $E\omega_{i_0} = 500$ (MVA)	Random
Required probability for the steady-state stability condition (p_0)	0.95	Fixed

The optimal solution for the example given above was obtained as follows:

- dimensional parameter $x^* = 0.146$
- number of cables in one line $z^* = 2$
- nominal tension $y^* = 60$ (kv)

The mathematical expectation of the total cost is (approximately): $F = 628.6$ (Lvs/m). The results obtained by this method were compared with some known optimal solutions, and matched them quite closely.

When formulating the mathematical model we assumed that the distribution functions of the random variables are known. However, an analysis of the available information showed that these distributions can be determined only within given classes of distribution functions characterized by some moments or intervals for these moments. Thus, after the unknown parameters have been determined and the values x^*, y^*, z^* have been found, we have to consider how the value of the objective function $F_H(x^*, y^*, z^*)$ changes when the partially known distribution is varied within the given class of distribution functions. This is, to some extent, a problem of sensitivity analysis with respect to those parts of the distribution functions which are only partially known.

In the problem under consideration, certain random variables, such as the transmitted power, the temperature, etc., possess well-defined distribution functions. Other random variables, such as material costs , the nonuniformity and graph of the load, etc., have distribution functions that are only partially known. Thus, the basic problem is to determine the bounds of the objective function as the distribution functions vary between pessimistic and optimistic estimates.

Such problems can be described in formal terms as follows. Let us denote by $\eta \in \Omega$ the group of random variables whose distribution functions are partially known and by $g^0(\eta)$ the value of the objective function as a function of η for fixed optimal parameters. The distribution function H of η belongs to class K, defined in the following way:

$$\int_\Omega g^i(t)\mathrm{d}H(t) \leq a_i, \qquad \ell = 1, \ldots, r$$

$$\int_\Omega \mathrm{d}H(t) = 1$$

with given constants $a_i, i = 1, \ldots, r$. In order to determine the range of possible values of $G^0(H)$ as the distribution H varies in K, we have to solve the extremal problems

$$\min_{H \in K} G^0(H) \tag{25.5}$$

$$\max_{H \in K} G^0(H), \tag{25.6}$$

where

$$G^0(H) = \int_\Omega g^0(t)\mathrm{d}H(t).$$

Numerical methods for solving such problems are given in [3]. Since the extremal distributions are not of importance, we can use the so-called dual approach, when under rather general assumptions

$$\max_{H \in K} G^0(H) = \min_{u \in U^+} \left\{ \sum_{k=1}^r a_u^k u_k + \max_{v \in \Omega} \left[g^0(v) - \sum_{k=1}^r u_k g^k(v) \right] \right\},$$

where

$$U^+ = \{ u \in R^r \,|\, u_i \geq 0, \qquad i = 1, \ldots, r \}.$$

An analogous proposition holds for problem (25.5). We can then use the stochastic procedure described below (see [5]).

We start by choosing points $u^0 \in U^+$, $v^0 \in \Omega$, and suppose that after s iterations we have arrived at (u^s, v^s). Then we generate a point $\tilde{v}^s$ in accordance with the uniform measure on Ω, and determine

$$v^{s+1} = \begin{cases} v^s, & \text{if } g^0(v^s) - \sum_{k=1}^r u_k g^k(v^s) \le g^0(\tilde{v}^s) - \sum_{i=1}^r u_k g^k(\tilde{v}^s) \\ \tilde{v}^s & \text{otherwise} \end{cases}$$

we then compute

$$u_i^{s+1} = \max\{0, u_i^s - \frac{1}{s}(a_i - g^i(v^{s+1}))\}, \qquad i = 1, \ldots, r.$$

When implementing this method we used the inequality

$$|\frac{1}{r} \sum_{s=k}^{k+r} \sum_{i=1}^r u_i^s (a_i - g^i(v^{s+1}))| \le 0.01$$

as a stopping criterion. The computational results obtained by this method showed a great variation in the degree to which the value of the objective function depends on the distribution functions. This is demonstrated in Table 25.2 and Table 25.3.

Table 25.2 Some experimentally determined parameters.

Random variable Class	K_1 of distribution function
Critical density of the flow at temperature $3 - 4.2 \mathrm{roman}K(\omega_1)$	$\omega_1 \in [4 \times 10^{10}, 8 \times 10^{10}]$ $E\omega_1 \in [4.95 \times 10^{10}, 5.05 \times 10^{10}]$
Relation of the expansion coefficient to the solidity coefficient (for the strengthening) (ω_2)	$\omega_2 \in [0.90, 1.40]$ $E\omega_2 \in [1.095, 1.105]$
Relation of the expansion coefficient to the solidity coefficient (for the system of shielding flow tubes) (ω_3)	$\omega_3 \in [1.05, 1.25]$ $E\omega_3 \in [1.195, 1.205]$

Table 25.3 The main economic parameters.

Meaning of the random variables	Class GW of distribution functions
Coefficient of the price of material (for the cold zone) (ω_{25})	$\omega_{24} \in [0.3375, 0.5625]$ $E\omega_{24} \in [0.40, 0.50]$
Coefficient of the price of installation (for the cold zone) (ω_{26})	$\omega_{26} \in [0.135, 0.165]$ $E\omega_{26} \in [0.145, 0.155]$
Coefficient of the price of material (for the cryogenic-covering) (ω_{27})	$\omega_{27} \in [0.3375, 0.5625]$ $E\omega_{27} \in [0.40, 0.50]$
Coefficient of the price of installation (for the cryogenic-covering) (ω_{28})	$\omega_{28} \in [0.135, 0.165]$ $E\omega_{28} \in [0.145, 0.155]$
Price of the refrigerator stations (ω_{29})	$\omega_{29} \in [9, 11](mln.lv.)$ $E\omega_{29} \in [9.7, 10.3]$

We divided the set of the partially known distributions into two subsets K_1 and K_2. The minimal (maximal) value of the objective function with respect to distributions of class K_1 was $F^{\min} = 627$. ($F^{\max} = 642.7$), and the result was obtained after 1243 (1406) iterations. This result shows that the value of the objective function does not depend strongly on the choice of distributions for these experimentally determined parameters. Therefore their following specification is not important.

The minimal value of the objective function with respect to distributions of class K_2 is $F^{\min} = 623.9$ (obtained after 2432 iterations), while the maximal value is $F^{\max} = 5967.8$. This result, which shows that the value of the objective function depends strongly on the "pessimistic" bounds, calls for additional input from experts.

25.3 Optimization of the Electricity Generating Stations

A specific feature of the above problem is the abundance of inexact input data. We shall make use of this characteristic in stochastic programming methods which we shall use to solve some exploitation problems in electricity generation.

The problem can be briefly formulated as follows: determine the active and the reactive powers of the electricity generating stations (the power is usually expressed as a complex number $x = x' + Ix''$ and x' and x'' are called "active" and "reactive" power, respectively) so that the price of the electric power produced is minimized subject to the following conditions:

- total production is equal to total consumption,
- the resulting power flow is technically feasible.

Let us denote the active power of consumer i, by S'_i , its reactive power by S''_i, $i = 1, \ldots, p$, and suppose that they are random variables with known distribution functions. For the stations we shall use x'_i and x''_i to denote the active and the reactive powers, respectively, which must be in the intervals $[\alpha'_i, \beta'_i], [\alpha''_i, \beta''_i]$, $i = 1, \ldots, q$. The cost of one unit of electrical power produced at the station i is $c_i, i = 1, \ldots, q$. For every node j (power station or consumer) an interval $[\underline{u}_j, \overline{u}_j]$, $j = 1, \ldots, n, n = p+q$ for the voltage modulation is given. We shall take the active and reactive powers of the stations as control variables. Other control variables could include the transformation coefficients for some lines, the reactive powers of certain consumers, etc.—these do not influence the basic structure of the problem, but make its description more complicated.

We use the following mathematical model to determine the vector $x' = (x'_1, \ldots, x'_q)$: minimize

$$L(x') = \sum_{i=1}^{q} c_i x'_i$$

subject to

$$\sum_{i=1}^{q} x'_i = \sum_{i=1}^{p} S'_i,$$

$$\alpha'_i \leq x'_i \leq beta'_i, \qquad i = 1, \ldots, q.$$

This simple linear programming problem possesses an explicit solution $\bar{x}' = x'(\sigma')$, where $\sigma' = \sum_{i=1}^{p} S'_i$. Even when a quadratic objective function is used (instead of a linear one), the solution may be simply expressed by the values of the random variable σ'.

The values of the reactive powers—vector $x'' = (x''_1, \ldots, x''_q)$—have to be determined so that a technically feasible power flow exists. Let us explain this condition. If the values of the active and the reactive powers of all nodes are appropriately assigned, in the nodes of the system definite voltages u_j, $j = 1, \ldots, n$, arise. Mathematically this is expressed by the fact that a nonlinear complex system called system of nonlinear equations of the power flow possesses a solution $u = (u_1, \ldots, u_n)$

$$\sum_{j=1}^{n} a_{ij} u_j = \frac{S'_i + IS''_i}{\hat{u}_i} + b_i, \qquad i = 1, \ldots, p,$$

$$\sum_{j=1}^{n} a_{p+ij} u_{p+j} = \frac{x'_i(\sigma') + Ix''_i}{\hat{u}_{p+i}} + b_{p+i}, \qquad i = 1, \ldots, q.$$

Here $\{a_{ij}\}, i = 1, \ldots, n, j = 1, \ldots, n$—the admittance matrices which include also complex constants. By I, the imaginary unit, and by $\hat{u}_i$, the conjugate number of u_i are denoted. This system consisting of $2n$ equations for $2n$ unknown we shall denote by

$$h^j(x'', y, \omega) = 0, \qquad j = 1, \ldots, 2n, \qquad (25.7)$$

where ω is a random vector including the consumers powers and also the vector $x'(\sigma')$ and $y = (y_1, \ldots, y_n, y_{n+1}, \ldots, y_{2n})$ is the vector composed by components of the voltages of the nodes.

For fixed values of the components of the vector ω a vector x'' has to be found such that the following condition to be satisfied

$$y \in Y = \{y \in R^{2n} | \underline{u}_j \leq y_j \leq \overline{u}_j, \quad j = 1, \ldots, n, 0 \leq y_{n+j} \leq 2\pi, j = 1, \ldots, n\}.$$

Since this problem must be solved in real time, it is convenient to apply a parametrization of the solution: the solution will be searched as *a priori* given vector-function $x''(\omega) = x''(v, \omega)$, which depend on the random vector ω and on the unknown vector $v \in R^m$ that has to be determined. For convenience let us denote

$$h^i(x'(v, \omega), y, \omega) = g^i(v, y, \omega), \qquad i = 1, \ldots, 2n,$$

$$f(v, y, \omega) = \max_i |g^i(v, y, \omega)|.$$

We state the following problem for the vector v: minimize the function

$$F(v) = E \min_{y \in Y} f(v, y, \omega)$$

subject to

$$v \in V = \{v \in R^n | P\{\alpha_i'' \le x''(v,\omega) \le \beta_i''\} = 1, \qquad i = 1,\ldots,q\}.$$

Usually functions $x''(v,\omega)$ are chosen as linear functions, for example $x_i''(v,\omega) = v_i\sigma''$, $i = 1,\ldots,q$, where $\sigma'' = \sum_{i=1}^p S_i''$. In this case the set V is a parallelogram.

$$V = \{v \in R^a | \underline{v}_i = \alpha_i'' / \max_\omega \sigma'' \le v_i \le \beta_i'' / \min_\omega \sigma'' = \overline{v}_i, \qquad i = 1,\ldots,q\}.$$

Now we shall describe a numerical method for solving the problem with such parametrization. The method is of the stochastic ε-quasigradient type methods (see [6]).

Let $v^0 \in V$ be an initial point and let after s iterations, we have arrived at v^s. Then we choose the observations $S_i'(s), S_i''(s), i = 1,\ldots,p$, in accordance with their distribution, compute the vector $x'(\sigma_s')$ as a solution of the linear programming problem described above for $\sigma' = \sigma_s' = \sum_{i=1}^p S_i'(s)$, and also compute $\sigma_s'' = \sum_{i=1}^p S_i''(s)$. Thus we have determined the vector ω^s, composed by $S_i'(s), S_i''(s), i = 1,\ldots,p$, and $x'(\sigma_s')$.

Then we determine a vector $y^s = y^s(v^s,\omega^s)$ such that

$$f(v^s, y^s, \omega^s) \le \min_{y \in Y} f(v^s, y, \omega^s) + \varepsilon_s, \varepsilon_s > 0,$$

and define

$$i^s = \operatorname*{argmax}_i |g^i(v^s, y^s, \omega^s)|,$$

$$\delta_s = \operatorname{sign} g^{i^s}(v^s, y^s, \omega^s).$$

We compute the point

$$v_i^{s+1} = \max\{\underline{v}_i, \min[\overline{v}_i, v_i^s - \rho_s \delta_s g_{v_i}^{i^s}(v^s, y^s, \omega^s)]\}, \qquad i = 1,\ldots,q,$$

where $\rho_s > 0$ is the stepsize and $g_v^{i^s}(v^s, y^s, \omega^s)$ is the gradient of $g^{i^s}(v, y, \omega)$.

The determination of a vector $y^s = y^s(v^s, \omega^s)$ when v^s, ω^s are given, is a well-known problem in the electroenergetics, the so-called 'problem of the power flow'. For its approximate solution, numerous methods of nondifferentiable optimization can be used. As a termination criteria, the following inequality has been applied

$$\frac{1}{r} \sum_{s=k}^{k+r} f(v^s, y^s, \omega^s) \le 0.5.$$

Example: This example illustrates the computational results for a network with 6 nodes, 3 power station, and 3 consumers $(p = 3, q = 3, n = 6)$. The active and reactive powers of the consumers are supposed to be normal distributed and may take values which are not more than 20% less or greater than their mathematical expectations. As input data for the powers of the consumer we give only the values of the expectations. The active and the reactive powers of the power stations have to be determined (as we have described above) and intervals for their volumes are given. At the end, for every node an interval for the nominal tension of the voltage is given.

Input data for the nodes.

No.	Type	Active power	Reactive power	Nominal Tension
1	consumer	300 (MW) ±20%	150 (MVAR) ±20%	[210, 230] (KV)
2	consumer	150 (MW) ±20%	100 (MVAR) ±20%	[205, 230] (KV)
3	consumer	150 (MW) ±20%	50 (MVAR) ±20%	[210, 230] (KV)
4	station	[0, 300] (MW)	[0, 200] (MVAR)	[390, 410] (KV)
5	station	[100, 500] (MW)	[0, 300] (MVAR)	[205, 230] (KV)
6	station	[100, 500] (MW)	[0, 300] (MVAR)	[390, 410] (KV)

The input data for the electro-transmission line consists of the admittance matrices of the lines. We shall not describe all these complex numbers and only note that 8 branches (lines) are assumed.

The computational results was obtained after 109 iterations (18 sec., when the computer ES-1040 is used). The parameters v_1, v_2, v_3, of the linear parametrization

$$(x_i'' = v_i \sigma'', \quad i = 1, 2, 3, \quad \sigma'' = S_1'' + S_2'' + S_3'')$$

were determined as follows

$$v_1 = 0.38, v_2 = 0.89, v_3 = 0.53.$$

The value of the objective function is

$$F\{v\} = 0.4805.$$

This result shows the average value of the maximal "nonbalance" in the system of the power flow (25.7), when the reactive powers of the stations are chosen in accordance to the parameterization low described above. Such result is fully satisfactory from the technical point of view.

References

[1] E. Blinkov and A. Savoliev, *Tehniko-ekonomiceski rascet i optimisacia SPK*, Moscow, 1977 (in Russian).

[2] Y.M. Ermoliev and Z.V. Nekrylova, "The methods of stochastic subgradients and its applications", Notes of seminar on Theory of Optimal Solutions, Academy of Sciences of the USSR, Kiev, 1967.

[3] Y. Ermoliev, "Method for stochastic programming in randomized strategies", *Kibernetica* 1(1970).

[4] Y. Ermoliev and C. Nedeva, "Stochastic optimization problems with partially known distribution functions", CP-82-60, International Institute for Applied Systems Analysis, Laxenburg, Austria, 1982.

[5] Y. Ermoliev and A. Gaivoronski, "A stochastic algorithm for minimax problems", CP-82-88, Laxenburg, Austria: International Institute for Applied Systems Analysis, Laxenburg, Austria, 1982.

CHAPTER 26

POWER GENERATION PLANNING WITH UNCERTAIN DEMAND

O. Janssens de Bisthoven, P. Schuchewytsch and Y. Smeers

Abstract

We consider a multistage stochastic version of the power generation planning problem and present a solution technique for tackling it. The model can include uncertainties in the cost and demand parameters as well as in the technology matrix; it embeds the classical LOLP reliability constraints. The solution method is a mixture of decomposition and cutting plane techniques. Because of the complexity of this type of problem compared to the more classical LP formulation, we provide a discussion of its practical relevance on the basis of a case study.

26.1. Introduction

Power generation planning consists of finding the mix of new production capacities that will satisfy the future electric demand at minimal investment and operations cost. The problem has given rise to many mathematical programming formulations that would be too long to recall here (see [1] for some references). In its most usual form (see the classical paper by Anderson [2]) the model is formulated as the following linear program

$$\text{minimize} \quad \sum_{\tau=1}^{T}(K_\tau y_\tau + c_\tau x_\tau) \qquad (26.1)$$

$$\text{subject to} \quad \sum_{\tau=1}^{T} A_{t\tau} y_\tau + B_t x_t = a_t, \quad t = 1,\ldots,T$$

$$C_t x_t = b_t, \quad t = 1,\ldots,T$$

$$\sum_{\tau=1}^{T} D_{t\tau} y_\tau \le d_t, \quad t = 1,\ldots,T \qquad (26.2)$$

that we interpret as follows. y and x are respectively the vectors of investment and operations variables. The objective function evaluates the present value of the capacity expansion and exploitation costs over the horizon. The first constraint provides a linkage between the operations and investment variables, it expresses the fact that the exploitation is limited by the existing capacities.

The second constraint summarizes technical restrictions on the operations of the plants (e.g. lack of flexibility of the nuclear plants) and the satisfaction of the demand. Special attention must be given to inequalities (26.2) which are introduced as surrogates of a reliability criterion. In their most common form they express that the total installed capacity must be larger than the peak demand plus some margin.

While models of the type (26.1)–(26.2) are usually sufficient for long term scenario studies, some authors ([1], [3]) have introduced more refined tools where the linear inequalities (26.2) are replaced by a true reliability criterion.

$$F_t(y_1, \ldots, y_t) \le d_t \qquad (26.2')$$

which, in one of its common forms, expresses that the probability of not being able to satisfy the peak demand cannot be larger than some amount d_t. This criterion, usually referred to as the loss of load probability (LOLP) makes the new model (26.1)–(26.2') considerably more difficult to solve than its linearized counterpart (26.1)–(26.2). Other versions of the problem which use slightly different reliability criteria (loss of energy probability (LOEP)) are equally difficult. Bender's decomposition has been proposed as a natural way to tackle these more complex problems.

We consider in this paper the treatment of a stochastic version of (26.1)–(26.2) where uncertainties can appear in the cost coefficients, the demand parameters and the technology matrix. Problems of this type are of immediate interest these days where parameters such as investment and fuel costs, demand or availabilities of certain plants are typically uncertain.

This extended version of (26.1)–(26.2') can be stated as a multistage stochastic program with recourse (see [4], [5]). In order to stick to the solution procedure adopted in this paper we shall immediately define the extensive form of the deterministic equivalent of the problem.

Let ET designate an event tree of depth T, Π_i is the probability of node i and $A(i)$ the set of its ancestors (including i itself) in ET. We consider the following multistage linear program

$$\text{minimize} \quad \sum_{i \in ET} \Pi_i (K_i y_i + c_i x_i) \qquad (26.3)$$

$$\text{subject to} \quad \sum_{j \in A(i)} A_{ij} y_j + B_i x_i = a_i, \quad i \in ET$$

$$C_i x_i = b_1, \quad i \in ET$$

$$F_i(y_j, j \in A(i)) \le d_i, \quad i \in ET \qquad (26.4)$$

The model (26.3)–(26.4) will usually be quite large and hence difficult to solve; it is handled in this paper by a mixture of decomposition and cutting plane techniques which is discussed in Section 26.2. The current implementation of the method is presented in Section 26.3. The last part of the paper discusses the relevance of the approach compared to the more classical deterministic models. This is done in the context of a study of the commissioning of new nuclear capacities in Belgium in 1984.

26.2   Methodological Aspects

This section is devoted to an intuitive discussion of the method adopted for solving the problem (26.3)–(26.4). Our aim here is more to motivate the general approach than to provide a rigorous treatment of it (see [6] for an exposition and a convergence proof of the mixed decomposition/cutting plane algorithm used). Throughout the paper, the discussion will be illustrated with the help of the event tree given in Figure 26.1.

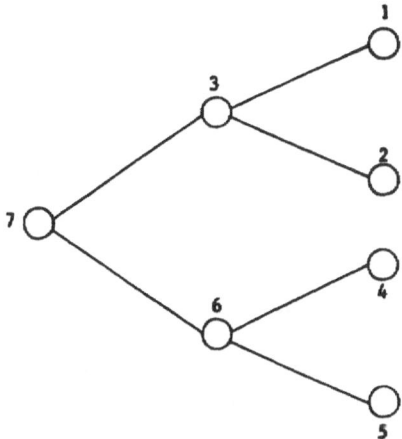

Figure 26.1   Illustrative event tree

Consider the linear programming problem consisting of the set of relations (26.3) only. It has a lower triangular block structure which in the case of our example is represented on Figure 26.2.

Various algorithms exist for taking advantage of this property of the matrix. We shall in this paper rely on the extension of decomposition [7] and nested decomposition [8] proposed by Kallio and Porteus [9] for arborescent linear programs. By definition the program

$$\text{Min} \sum_{\ell=1}^{N} e_\ell x_\ell$$

$$\sum_{\ell=1}^{N} B_{k\ell} x_\ell = f_k, \qquad k = 1, \ldots, N$$

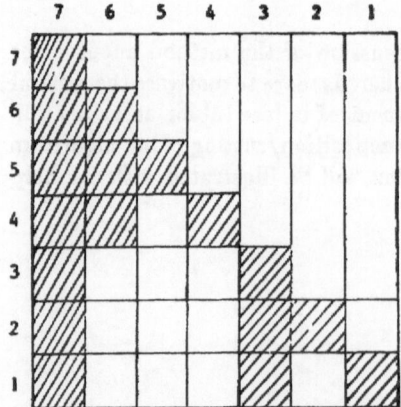

Figure 26.2 Block structure of the matrix

is arborescent if there exists an arborescence having nodes 1 to N and such that $B_{k\ell} \neq 0$ implies the existence of a directed path from k to ℓ. As we shall see both the primal and the dual of problem (26.3) can be looked at as arborescent programs. The implementation of the method is described in [10], the following summary of the principle of the approach will suffice for our purpose in this paper. We consider an arborescent matrix as illustrated in Figure 26.3. Decomposition proceeds by breaking the original model into a set of nested masters and subproblems according to the structure of the matrix. Referring to Figure 26.3 the original problem, noted 7 consisted of a coupling block and two linked blocks noted 3 and 6; each of these latters has the same structure as the original model, namely a coupling constraint set and two linked matrices.

In the decomposition algorithm (see Figure 26.4) the global problem 7 will be replaced by a master problem (noted 7) that will receive proposals from its subproblems 3 and 6 and to which it will transfer prices. Because of the nested block structure each of the subproblems can itself be replaced by a master problem (also noted 3 and 6 respectively) which receives proposals from its own subproblems (subproblems 1, 2 for the master 3 and subproblems 4, 5 for the master 6) and returns price signals.

Particular cases of this general decomposition method arise when the matrix reduces to a single block angular structure ([7]) or, when each master only has a single subproblem ([8]).

Arborescent linear programming can be applied in different ways to model (26.3). Working directly on the primal problem, an exploitable structure is the

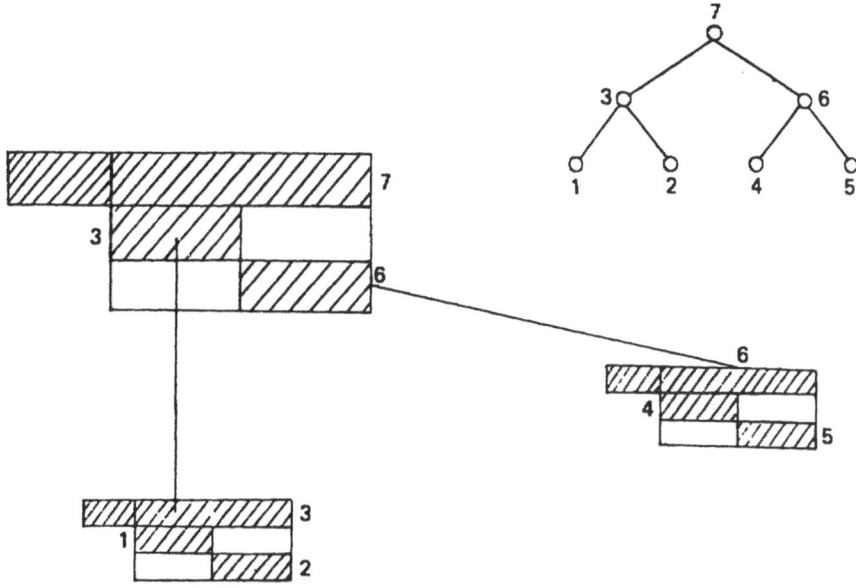

Note: We have assumed that the tree describing the matrix is identical to the event tree of Figure 26.1. This is by no means necessary but will help in later discussions.

Figure 26.3 Nested block angular structure

one indicated in Figure 26.5. It is easy to see that this corresponds to taking advantage of the sole multitemporal aspect of the problem. From the point of vie of the data structure, this implies that the size of the subproblem in a given time period is determined by the number of nodes in that period. This is admittedly embarrassing in a stochastic program where the number of terminal nodes can quickly become large.

In contrast, working on the dual permits a much higher degree of decomposition. The structure of the dual matrix is given in Figure 26.6 which also shows its nested block angular structure. The size of the subproblem is then entirely determined by the size of each block in the matrix. This is a much more favorable situation and it is this structure that we shall exploit here.

We now turn to the handling of the reliability constraint. It is most common in power generation planning to characterize a plant by its rated power U and its availability factor p (see [2]). Leaving aside, for the time being, the fact that we are dealing with continuous capacity variables and not multiples of the rated power, the loss of load probability in a node i of the event tree can be defined as follows. Let ξ_i be the demand of electricity in node i, ξ_i is a random variable whose distribution is entirely determined by the load duration

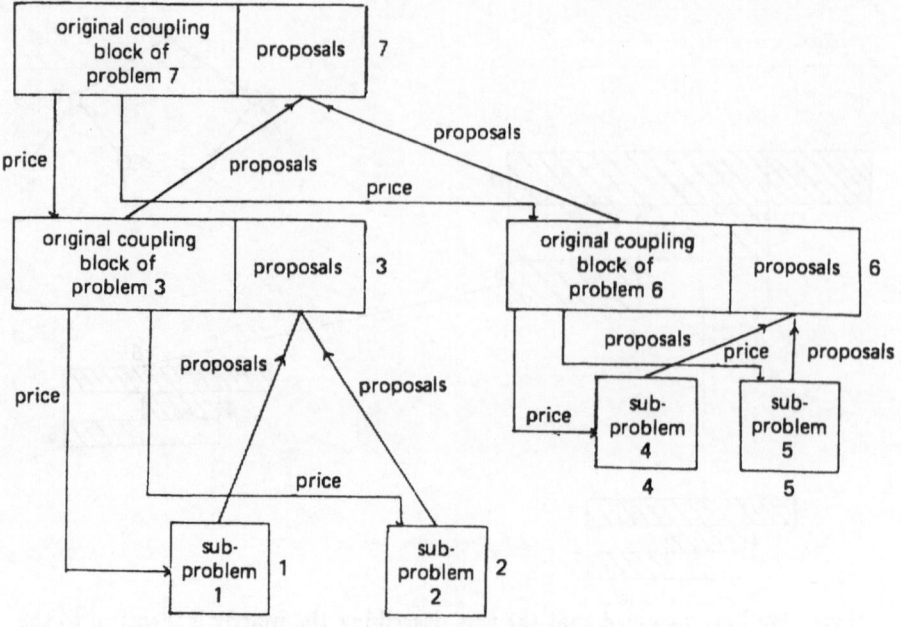

Figure 26.4   Transfer of information between subproblems

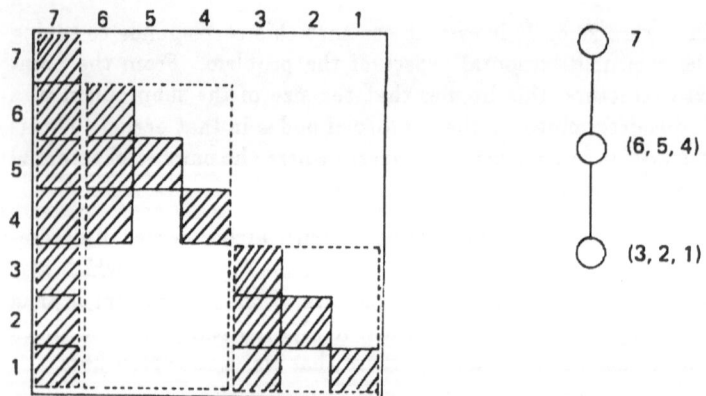

Figure 26.5   Nested block structure obtained by working on the primal problem

curve. Let S_i be the set of plants existing in node i and $\{y_j, j \in A(i)\}$ the vector of installed capacities (that we take as integer variables in the course of this discussion), we can define for each plant s, the random variable η_{si} equal

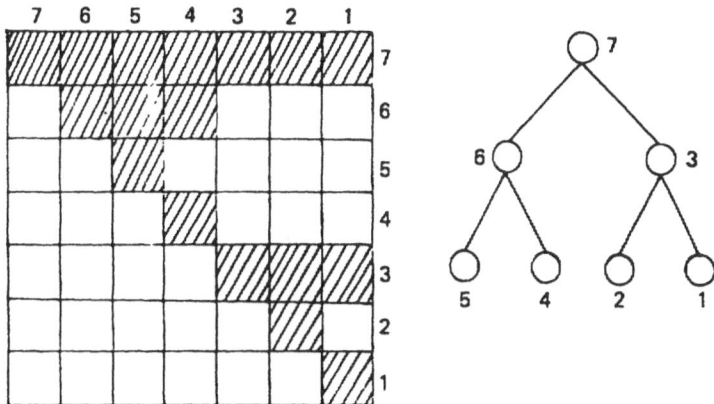

Figure 26.6 Matrix structure of the dual problem

to the available capacity of plant type s in node i. The reliability criterion in node i can then be written as

$$\Pr \left[\sum_{s \in S_i} \eta_{si} \leq \xi_i | U_{si}, (y_j, j \in A(i)) \right] \leq d_i \qquad (26.5)$$

which is a chance constraint. Besides very special cases, it is impossible to write a deterministic equivalent of (26.5) which is, in any case, already difficult to evaluate numerically (see [11], [12], [13]) for examples of numerical methods). The inclusion of reliability constraints in planning models has mainly been done through Benders' decomposition ([2], [3]); we shall follow a similar approach but reason instead in terms of cutting planes.

Let $(\Delta_i, i \in ET)$ be the vector of (exogenously determined) capacities to be scrapped. Starting with the solution $(y_i, i \in ET)$ of problem (26.3), that is without reliability constraints, one can define the available capacity z_i at node i as

$$z_i = \sum_{j \in A(i)} y_j - \sum_{j \in A(i)} \Delta_j.$$

Strictly speaking, the loss of load probability is only defined for values of z_i that are multiples of the rated capacities; let $[z_i]$ be the vector derived from z_i by rounding down the capacities to multiples of the commercial powers and δ_i be defined as

$$\delta_{is} = \frac{F_i([z_i]) - F_i([z_i] + e_s U_s)}{U_s}, \qquad s \in S_i$$

where e_s is the s-th unit vector. δ_{is} can be seen as the decrease of the loss of load probability resulting from a unitary investment in plant s. If the reliability criterion is not satisfied at $[z_i]$ we add the constraint

$$F_i([z_i]) + \sum_{s \in S_i} \delta_{is}(z_{is} - [z_{is}]) \leq d_i. \qquad (26.6)$$

This amounts to replacing the initial reliability constraint by some inner linearization.

Because the storage of a linear program and its manipulation by the revised simple ι method are essentially column oriented, the addition of a cut is not a natural operation in most commercial codes. This is a fortiori so if the solution technique is based on column generation such as in decomposition or nested decomposition. In contrast, the addition of a cut in the primal becomes the addition of a column when working in the dual and can thus be nicely inserted in the decomposition algorithm. The implementation of the combination of these techniques is discussed in the following section.

26.3   Implementation

Stochastic optimization although introduced at the very beginning of mathematical programming does not seem to be in widespread use. This may be due not only to the lack of specialized codes capable of dealing with these problems but also to the fact that stochastic models seem, at least in our experience, more difficult to formulate (event trees are more complex to arrive at than scenarios) and to generate (commercial matrix generators such as OMNI [14] do not permit easy manipulation of trees). It thus seems essential in order to implement the approach discussed in the preceding section to leave the maximal possible freedom to the user and in particular to refrain from imposing him constraints originating from the solution procedure. The following approach has thus been adopted. In a first state the user writes the extensive form of his model in the MPS format using standard matrix generation techniques. A program transforms this version of the primal model into an MPS representation of the dual. A third program rearranges the input of the dual in a form suitable for the decomposition code. The fourth stage is the optimization itself; the last one, the report writer, is essentially missing in the current implementation but should be developed in the future. We briefly review these different stages.

26.3.1   Problem generation

While standards exist for defining two stage stochastic programs [15], the case of multistage models remains largely untouched. We have assumed in this work that the modeler directly constructs the extensive form of the deterministic equivalent of his problem in MPS format using a commercial matrix generator. We allow him the most general formulation of a linear programming problem, namely

$$\text{Min } c^t x$$
$$r \leq Ax \leq s \tag{26.7}$$
$$\ell \leq x \leq u$$

which contains ranges on the constraints and bounds on the variables. In order to allow for subsequent treatment, it is required that row and column names corresponding to a given node have as their two last characters the identificator of the node; the current implementation supposes that the nodes of the tree are numbered in postorder; this constraint can however be relaxed easily.

26.3.2 Construction of the dual problem

The dual of (26.7) is written as

$$\begin{aligned}
\text{minimize} \quad & -s^t y - r^t z - u^t v - \ell^t w \\
\text{subject to} \quad & A^t y + A^t z + v + w = c \\
& y, v \leq 0 \\
& z, w \geq 0
\end{aligned} \qquad (26.8)$$

and is constructed automatically from the MPS input file of the primal problem (MAGENOUT file in the OMNI system). The formulation is rather unusual to the extent that it involves nonpositive variables (y and z) as well as the more common nonnegative variables (z and w). It can be justified as follows: numerical elements in MPS format files are represented in twelve character fields, one character being used for specifying the sign of the number. Our version of the dual problem can be defined through an MPS file which contains the same numerical elements as those of the primal problem and hence does not require any change of sign (no change of sign is required in the constraints and the minus signs of the objective function can be generated using the facilities of the MPS software), this permits keeping the s, r, u and ℓ with their original sign in the dual MPS file. Besides the time gained by not having to change sign, this construction leads to a dual which is numerically fully equivalent to the original problem.

26.3.3 Rearrangement of the MPS input file

This rearrangement is specific to the decomposition code used. It is discussed in detail in [10].

26.3.4 Optimization

The main features of the decomposition code are discussed in [10] and will not be recalled here. The interaction between this code and the reliability criterion is represented on Figure 26.7. This part of the implementation is currently far from optimal; the following discussion will help clarify the issue.

The decomposition code is fed with the reorganized MPS file of the dual problem (see section above). Because decomposition methods provide feasible dual solution every time a cycle with a bounded subproblem is completed, it is possible to extract the dual solution when convergence has almost been reached and to evaluate the corresponding capacities in each node of the event tree. The reliability criterion is then evaluated everywhere or for a subset of the nodes where the user feels that the LOLP is most likely to be violated; additional cuts are generated when necessary.

The approach, although simple in principle present several challenging features that have not been handled most satisfactory now. We briefly report on these in the following.

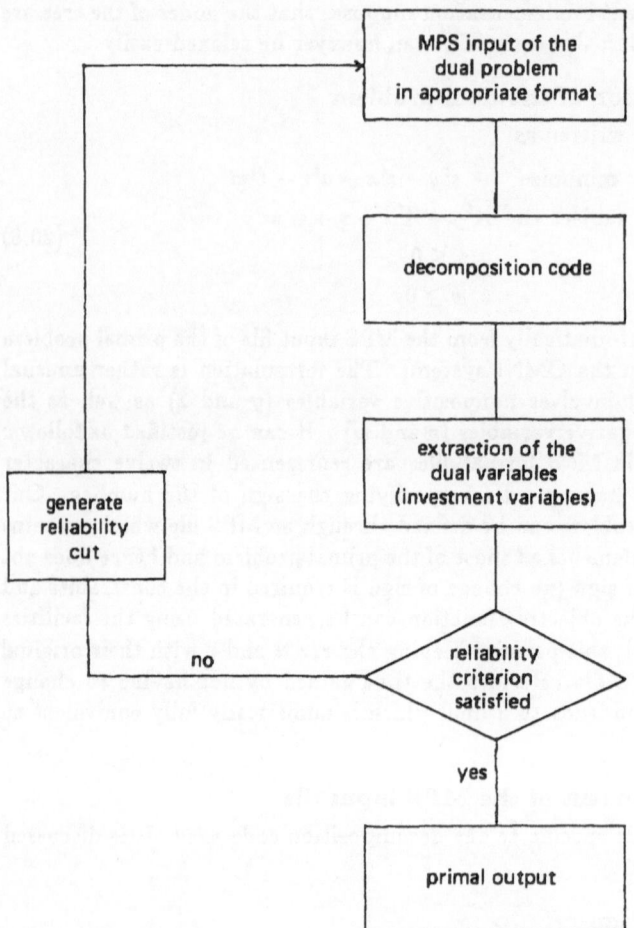

Figure 26.7 Interaction between the decomposition code and the reliability criterion

Computation of the reliability criterion

The evaluation of the loss of load probability is a costly operation and it is out of question to restart it from scratch at each evaluation of the reliability criterion. The cumulant method introduced in [12] and [13] provides an elegant solution to this problem. The cumulants (see [16] for the definition of this notion) of the different plants and of the load duration curves at each node i can be computed once for all at the outset of the study; the evaluation of the reliability criterion is then drastically reduced afterwards.

Insertion of reliability cuts

As mentioned above, a cut becomes a column in the dual and can be added relatively easily in the input of the dual. Some elementary restart procedures have been included in the current decomposition code which allow one not to go through the whole optimization from scratch. Although these cuts could ideally be directly added to the internal representation of the different subproblems as new columns, this approach has however not yet been explored and the cuts are now included through the MPS file.

Report generation

The question of the connection of the decomposition code with a commercial report writer has not been explored yet. At this stage the dual variables of the subproblems (the primal variables of the original problem) are directly extracted from the internal representation of the solution of the subproblems and constitute the output. This implies that the report writer must be written in a general purpose high level language (such as PL/1 with MPSX/370).

26.4   A Case Study

This machinery is rather complex, at least compared to the direct use of a commercial linear programming code. It is thus important, before resorting to the stochastic programming approach, to evaluate the additional insight that it can bring into the decision process. The following discussion is taken from a study of the commissioning of new nuclear plants in Belgium in early 1984. The general decision context is discussed in [17]; we focus here more on the numerical results. Consider the event tree of Figure 26.8 where the probabilities of the different scenarios are indicated at the right of the corresponding terminal nodes.

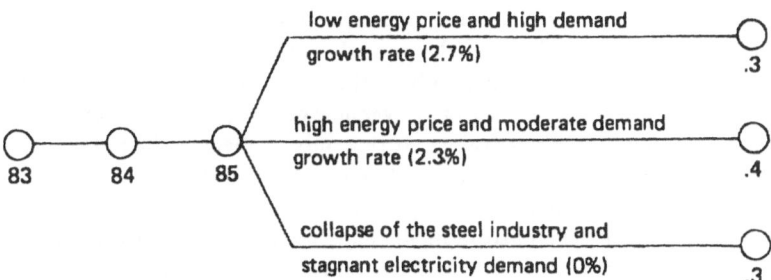

Figure 26.8   Event tree relative to the commissioning of new nuclear plants in Belgium in 1984

The tree has been constructed by a governmental agency and taken as such. It models a process where investment decisions must be taken in 1984 and 1985 without knowing the future demand. This latter is supposed to be revealed in 1985 and later decisions taken with perfect foresight from that period on. Relevant to the use of stochastic programming is the existence of two relatively similar scenarios (2.7 and 2.3%) together with a more contrasted one (0%). Dropping the last scenario would probably make the stochastic model useless; having more contrast in the two first evolutions would increase its interest.

The discussion will focus on the size of the model and the impact of both the uncertain demand growth rate and the reliability criterion. We shall conclude by some quantitative evaluation of the stochastic programming approach and comments on its implications for policy analysis.

Size of the model

A first criticism against the use of the preceding machinery arises from the present capabilities of commercial codes. It can indeed be claimed that, given the existing possibilities of these codes, it is simply not reasonable to set up models that require more computational resources. In order to assess this argument, consider Table 26.1 which reports the capacities of nuclear plants coming on line in 1994 and 1995 with deterministic models where the horizon is limited to 1995 and 2000 respectively.

Table 26.1 1993 and 1994 nuclear capacities with deterministic models of different horizons (capacities coming on lines (in MW))

	1993		1994	
	horizon 1995	horizon 2000	horizon 1995	horizon 2000
2.7%	508	1972	575	442
2.3%	1065	1681	405	412

Both versions of the model deal with end effects by assuming that the salvage values of the plants at the end of the horizon is equal to the discounted sum over the rest of the technical life of the annual values of the investment cost. The difference of the results clearly points to the importance of recurring to the longer term horizon model. A stochastic version of the problem limited to a 1996 horizon has about 9000 constraints. It certainly challenges the possibilities of commercial codes such as MPSX/370 but can be handled by them. The 2000 horizon model has about 20,000 constraints and cannot be handled by MPSX/370. The longer term horizon model appears necessary, but commercial software would have difficulties (or find it impossible) to solve its stochastic version.

Uncertain demand growth rate and reliability constraint

The handling of the reliability criterion certainly adds to the complexity of the methodology Because LOLP constraints are rarely treated explicitly in long term power planning models, one may question their usefulness in this more complex set up. Table 26.2 reports the results of a pure scenario analysis and of the stochastic model when the horizon is limited to 96 (the shortened version has been selected to reduce computer costs) with and without accounting for the LOLP constraints. Perfect foresight induces the immediate commissioning of new nuclear plants in the scenario approach, which results in the satisfaction of the LOLP criterion (except in the 2.7% case where gas turbines are required for reliability purposes). In contrast, the uncertainty about the growth rate first postpones investment decisions which however remain of nuclear type. The obtained generation system, however, violates the reliability constraints in 1994 and 1995. The role of the LOLP constraint appears in the last three rows of Table 26.2. While gas turbines are again coming on line as soon as 1990 in the 2.7% scenario, coal fired plants are introduced for reliability purposes in 1994 and 1995. This new effect justifies considering the LOLP constraint in stochastic model.

Table 26.2. Comparison of the scenario and stochastic approaches under different LOLP constraints (capacities coming on lines (in MW))

		1990	1993	1994	1995	Remarks
Scenario	2.7%	·	508	575	·	All investments are nuclear. Gas turbines
	2.3%	·	1065	405	·	are introduced in 1990 with the
	0%	·	·	·	·	2.7 scenario
Stochastic without LOLP	2.7%	·	·	·	856	All investments
	2.3%	·	·	·	1170	are nuclear
	0%	·	·	·	·	plants
Stochastic with LOLP	2.7%	·	117	330	938	coal
		155	·	·	·	gas turbine
	2.3%	·	·	·	1042	nuclear
		·	735	171	·	coal
	0%	·	·	·	·	

Valuation of the stochastic programming approach

We now consider the 2001 horizon model and evaluate two criteria usually found in the literature in relation to stochastic models. The value of information [18] compares the expected cost obtained in the deterministic scenario studies and the objective function value of the stochastic model. It corresponds to the value of perfect forecast. The value of the stochastic solution [18] evaluates the gain brought about by acting according to the solution of the stochastic program. Supposing a certain behaviour of the decision maker (for instance selecting a mean value approach in the first periods) it compares the cost resulting from that behaviour to the one associated with the solution of the stochastic program. Taking the value of information first, one finds that the average cost of the scenario models amounts to $7\ 656\ 10^6$ \$ (of year 1982) while the cost of the stochastic programming model is $7\ 714\ 10^6$ \$. Although this may look like a negligible difference in percentage, it is certainly important when considered in marginal terms. Because the generation system remains basically unchanged until 1994 (we can neglect the additional gas turbine capacities of 2.7% scenario which are only introduced for reliability purpose and are not exploited) the cost differences must be related to the eight years of the period 1994–2001 which, after proper discounting operations, amounts to $25.6\ 10^6$ \$/year.

The situation is more striking for the value of the stochastic solution. Taking the average of the deterministic solution as the initial decision we end up with an infeasible stochastic programming approach. This corresponds to an infinite value of the stochastic solution. This result can be explained as follows; two policy constraints are implemented in the zero growth scenario which have to do with particular features of the Belgium situation; one requires an additional consumption of national coal in case of the collapse of the steel industry; the other one imposes a minimum level of operations to the new nuclear plants. Admittedly these constraints have little economic sense; they have however a lot of political relevance and formalize concerns often expressed in the public opinion. Together they render the operations of the power sector in 90 infeasible in the 0% growth with the investments resulting from the mean value approach. This is admittedly an extreme case (which does not appear in the 96 horizon model) it however shows the utility of the stochastic programming approach with respect to the more classical scenario approach.

Policy implication

The commissioning of new nuclear plants in Belgium has been delayed from 1981 to 1984 when a small participation to a French station (~ 450 MW) was decided. The discussions during those three years have mainly concentrated on demand forecasts and on whether, because of the current uncertainties, one should not defer any immediate decision. The scenario approach, with its first stage decision depending drastically on the assumptions, has been relatively difficult to use in that context. In contrast, the stochastic programming approach, because it immediately deals with the whole set of scenario answers the question of whether it is better to wait until additional information is available.

26.5 Conclusion

The present uncertainties that prevade the economic environment of the utilities make the sole ise of classical deterministic power generation planning models difficult to justify. In particular the scenario approach, whatever its usefulness for exploring the impact of uncertainties on present decisions, can prove useless when the solutions are too much different for equally plausible scenarios. Stochastic programming has long been proposed as a natural way to tackle the problem. We present an implementation of the approach and show that it is both computationally feasible and practically relevant.

References

[1] J.A. Bloom, "Solving an electricity generation capacity expansion problem by generalized Benders' decomposition", *Operations Research* **31**(1)(1983), 84–100.

[2] D. Anderson, "Models for determining least-cost investments in electricity supply", *Bell Journal of Economics* (spring)(1972), 267–299.

[3] G. Coté and M. Laughton, "Stochastic Production Costing in Generation Planning: A Large-Scale Mixed Integer Model", *Mathematical Programming Study* **20**(1982).

[4] P. Olsen, "Multistage stochastic program with recourse: The equivalent deterministic problem", *SIAM Journal on Control and Optimization* **14**(1976), 495–517.

[5] P. Olsen, "When is a multistage stochastic programming problem well defined?", *SIAM Journal on Control and Optimization* **14**(1976), 518–527.

[6] O. Janssens de Bisthoven and Y. Smeers. "An algorithm for solving multistage power generation planning models with recourse" (in preparation).

[7] G.B. Dantzig and P. Wolf, "Decomposition principles for linear programs", *Operations Research* **8**(1960), 101–111.

[8] J.K. Ho and A.S. Manne, "Nested decomposition for dynamic models", *Mathematical Programming* **6**(1974), 121–140.

[9] M. Kallio and E.L. Porteus, "Decomposition of arborescent linear programs", *Mathematical Programming* **13**(1977), 348–356.

[10] O. Janssens de Bisthoven, E. Loute and Y. Smeers, "An implementation of Kallio and Porteus algorithm for arborescent linear programs" (in preparation).

[11] H. Baleriaux, E. Jamoulle and F. Linard de Guertechin, "Simulation de l'exploitation d'un parc de machines thermiques de production d'électricité couplé à des stations de pompage", *Revue E* **5**(1967), 225–245.

[12] J.P. Stremel, R.T. Jenkins, R.A. Babb and W.D. Bayles, "Production costing using the cumulant method of representing the equivalent load curve", *I.E.E.E. Transactions on Power Apparatus and Systems*, Vol. PAS-99, No. 5, Sept/Oct(1980).

[13] J.P. Stremel, "Production costing for long-range generation expansion planning studies", *I.E.E.E. Transactions on Power Apparatus and Systems*, Vol. PAS-101, No. 3.

[14] Haverly Systems Inc., *OMNI Linear Programming System*, User's manual, version 1983.

[15] J. Edwards, J. Birge, A.J. King and L. Nazareth, "Standardized input formats for stochastic programs with recourse", mimeo, IIASA, Laxenburg, Austria, 1984.

[16] H. Cramer, *Mathematical Methods of Statistics*, Princeton University Press, 1974.

[17] O. Janssens de Bisthoven, P. Schuchewytsch and Y. Smeers, "Dealing with uncertain demand in power generation planning", in: *Models in the Decision Making Process: Planning under Uncertainty*, Proceedings of the Second IMACS Symposium on Energy Modelling and Simulation, Brookhaven, National Laboratories.

[18] J.R. Birge, "The value of the stochastic solution in stochastic linear programs", *Mathematical Programming* **24**(1982), 314–325

CHAPTER 27

EXHAUSTIBLE RESOURCE MODELS WITH UNCERTAIN RETURNS FROM EXPLORATION INVESTMENT

J.R. Birge

Abstract

Exhaustible resource models that do not consider exploration investment have typically low values of perfect information and sometimes even optimal myopic policies. In this paper, we add exploration and capacity investment and allow the returns from exploration to be stochastic. We show that, in this model, the stochastic program solution may be quite valuable and that myopic policies are far from optimal.

27.1 Introduction

Exhaustible resource models have been studied by a number of authors. Hotelling [3] initially formulated a model that demonstrated that the market price of an exhaustible resource grows exponentially as it is depleted. Nordhaus [7] introduced the idea of a "backstop" technology to this model. The result is the Hotelling-Nordhaus model in which a finite resource is used until its production cost exceeds that of the inexhaustible backstop technology. The backstop technology is then introduced and the two technologies are never used simultaneously.

Manne [5] and Manne and Richels [6] use the Hotelling-Nordhaus model in their analysis of the effect of the uncertainty of the introduction date of the fast breeder reactor. They formulate a stochastic linear program and solve it to find the expected value of perfect information (EVPI). Their results indicate that the expected value of perfect information in this model is low and that, therefore, deterministic problem solutions provide close approximations to the solution of the stochastic problem.

Chao [2] presents an analytical justification for the observations of Manne and Richels. He formulates a mathematical program for the Hotelling-Nordhaus model. Under certain assumptions that include a demand that is independent of price, Chao shows that a myopic policy of using the most inexpensive available technology first is optimal. He also introduces a price responsive demand function to his model and again shows that the EVPI is low.

In this paper, we expand upon Chao's model by allowing exploration investment that could yield additional resource supplies. The amount of increase in the supply per unit of investment is however uncertain. We show that the

EVPI and the value of the stochastic solution (VSS) (Birge [1]) can be large when this type of uncertainty is included. We give examples illustrating these observations.

27.2 The Basic Model

Our results concern two measures of the effect of uncertainty in stochastic programs, the expected value of perfect information and the value of the stochastic solution. We present these measures in the context of two-stage stochastic programs with recourse. We first formulate the deterministic program

$$\text{minimize} \quad \varphi(x, \xi) = cx + \min[qy | Wy = \xi + Tx, y \geq 0]$$
$$\text{subject to} \quad Ax = b, x \geq 0 \tag{27.1}$$

where the vectors $c \in \mathbb{R}^n$, $q \in \mathbb{R}^n$, and $b \in \mathbb{R}^m$ are known, the m_2-vector ξ is a random vector defined on the probability space $(\Xi, \mathcal{F}, F)$, and A, W, and T are correspondingly dimensioned known real-valued matrices. A decision vector $\overline{x}(\hat{\xi})$ obtained in Program 27.1 represents an optimal first period decision given a realization $\hat{\xi}$ of the random vector.

If an optimal first period decision is taken for all possible realizations of the random vector, then we obtain in expected value the "wait-and-see" (WS) solution value (Madansky [4]), where

$$\text{WS} = E_\xi[\min_x \varphi(x, \xi)].$$

The stochastic program with recourse (Wets [8]) involves optimizing after taking the expected value. We write the value of this program as

$$\text{RP} = \min_x E_\xi[\varphi(x, \xi)].$$

For $E(\xi) = \overline{\xi}$, we obtain a third value that is the expectation of the expected value (EEV) solution $\overline{x}(\overline{\xi})$ that is optimal in (27.1) for $\xi = \overline{\xi}$. This quantity is

$$\text{EEV} = E_\xi[\varphi(\overline{x}(\overline{\xi}), \xi)].$$

The effects of uncertainty are measured by differences among WS, RP, and EEV. The expected value of perfect information represents the amount one is willing to spend in gaining information about the stochastic variables. It is calculated as

$$\text{EVPI} = \text{WS} - \text{RP}.$$

The value of the stochastic solution, on the other hand, measures the additional value of solving the stochastic program over solving the deterministic expected value problem. We define

$$\text{VSS} = \text{EEV} - \text{RP}.$$

In the discussion below, we describe VSS and EVPI in the context of an exhaustible resource model originally due to Chao.

Chao's basic model is a linear program to determine an optimal dynamic production schedule to minimize the present value of the cost of satisfying an increasing sequence of demand requirements over time. The demand may be satisfied by any of $m-1$ substitutable technologies, each using one distinct finite resource, and by one backstop technology with no resource limit. The resulting linear program is

$$\text{minimize} \quad \sum_{i=1}^{m}\sum_{t=1}^{\infty}\beta^t c_i y_{it} + \sum_{i=1}^{m}\sum_{t=0}^{T}\beta^t k_i x_{it}$$

$$\text{subject to} \quad \sum_{t=0}^{\infty} y_{it} \leq R_i, \quad i=1,\ldots,m;$$

$$\sum_{i=1}^{m} y_{it} = D_t, \quad t=1,\ldots,T; \tag{27.2}$$

$$y_{i,t+1} = y_{it} + \sum_{s=0}^{\infty}(\delta_s - \delta_{s-1})x_{i,t-s}, \quad t=0,1,\ldots,$$

$$y_{it} \geq 0; x_{it} \geq 0; \quad t=0,1,\ldots; i=1,\ldots,m;$$

where y_{it} is the amount of period t demand, D_t, satisfied by resource i at time t, x_{it} is the amount of resource i committed at t to be extracted later, c_i is the current cost of technology i, k_i is the capital cost of i, β is the discount factor, δ_t is the extraction rate, and R_i is the initial availability of the resource used by technology i. It is assumed that y_{io} and x_{it} are known for $i=1,\ldots,n$ and for $t=0,-1,\ldots$, and that $y_{i0} = \sum_{t=0}^{\infty}\delta_{-t}x_{it}$. It is also assumed that $D_1 \leq D_2 \leq \ldots \leq D_{T-1} \leq D_T$.

Chao defines γ as the capital recovery factor for the standard time profile where $\gamma = 1/(\sum_{s=0}^{\infty}\beta^s\delta_s)$ and lets d_t be the demand for new resource commitments where $D_t = sum_{s=0}^{\infty}\delta_s d_{t-s}$. The result it that (27.1) can be rewritten as

$$\text{minimize} \quad \sum_{i=1}^{m}\sum_{t=0}^{T}(k_i + c_i/\gamma)\beta^t x_{it}$$

$$\text{subject to} \quad \sum_{t=0}^{T} x_{it} \leq R_i - \sum_{t=-1}^{-\infty}(\sum_{s=-t}^{-\infty}\delta_s)x_{it}, \quad i=1,\ldots,m;$$

$$\sum_{i=1}^{m} x_{it} = d_t, \quad t=0,\ldots,T; \tag{27.3}$$

$$x_{it} \geq 0, \quad i=1,\ldots,m; \text{ and all } t.$$

Chao uses Program 27.3 to derive his results on myopic solutions. He shows that the corresponding transportation problem can be solved optimally by the

Northwest Corner Rule if the resource costs $k_i + c_i/\gamma$ are arranged in increasing cost order within each period.

The result leads to an expected value of perfect information of zero because the WS solution is the same as the RP solution. It also yields a VSS of zero because the EEV value is the same as RP when myopic solutions are optimal.

Chao introduces price-responsive demands to the basic model in (27.3) and obtains a nonlinear programming model that does not have myopic optimal decisions. He computes an upper bound on the EVPI and shows that distant future uncertainties and low price elasticities lead to a small EVPI. In the next section, we introduce investment uncertainty into the basic model and show that this may lead to a significant EVPI and VSS.

27.3   A Model with Uncertain Exploration Returns

We assume that R_i in Program 27.3 represents the amount of resource i that is known to be available at time 0. This amount can be increased by exploration investment, but the amount of the increase is uncertain. We also assume that there is a capacity limit L_i on the amount of a resource which may be committed at time 0. This amount may also be increased by investment in new capacity and that return is assumed known with certainty. The stochastic linear program derived from (27.3) is then

$$\text{minimize} \quad \sum_{i=1}^{m}(k_1 + c_i/\gamma)x_{i0} + \sum_{i=1}^{m} d_i u_{i0} + \sum_{i=1}^{m} g_i v_{i0} + \qquad (27.4.0)$$

$$\sum_{t=1}^{T}\sum_{i=1}^{m}\sum_{j=1}^{K_t} p_t^j \beta^t \{(k_i + c_i/\gamma)x_{it}^j + d_i u_{it}^j + g_i v_{it}^j\}$$

$$\text{subject to} \quad x_{it}^j \leq R_i + \sum_{s=0}^{t-1} \alpha_{is}^{a(j)} u_{is}^{a(j)} - \sum_{s=0}^{t-1} x_{is}^{a(j)} \qquad (27.4.1)$$

$$i = 1,\ldots,m; t = 0,\ldots,T; j = 1,\ldots,K_t;$$

$$x_{it}^j \leq L_i + \sum_{s=0}^{t-1} v_{is}^{a(j)}, \qquad (27.4.2)$$

$$i = 1,\ldots,m; t = 0,\ldots,T; j = 1,\ldots,K_t;$$

$$\sum_{i=1}^{m} x_{it}^j = d_t; \quad t = 0,\ldots,T; j = 1,\ldots,K_t;$$

$$x_{it}^j \geq 0, \quad i = 1,\ldots,m; t = 0,\ldots,T; j = 1,\ldots,K_t; \quad (27.4.4)$$

where d_i is the cost of one unit of exploration for resource i, u_{it}^j is the amount of exploration, g_i is the cost of capital investment in resource i, v_{it}^j is the amount of that investment, p_t^j is the probability of scenario j at time t, K_t is the number of scenarios at time t, and α_{it}^j is the return per unit of exploration for resource i

under scenario j. Each scenario j is preceded by *ancestor* scenarios in previous periods which are designated by $a(j)$.

The stochastic nature of Program 27.4 is contained only in the return on exploration investment, α_{it}^j. In general, these values may vary continuously, but the discrete formulation in (27.4) is used for simplicity. This program involves a stochastic technology matrix, but it may be formulated with stochastic right-hand sides by defining new variables w_{it}^ℓ, $\ell \geq 0$, such that

$$u_{i,t-1}^{a(j)} = \sum_{\ell=1} to\mathcal{L}_i^t w_{it}^\ell, \tag{27.5}$$

and

$$x_{it}^j \leq R_{i,t-1}^{a(j)} + \sum_{\ell=1}^{\mathcal{L}_i^t} \alpha_{it}^\ell w_{it}^\ell - x_{i,t-1}^{a(j)}, \tag{27.6}$$

where $R_{i,t-1}^{a(j)}$ is the availability of resource i in period $t-1$, there are $\mathcal{L}_i^t$ different values of $\alpha_{i,t-1}$, and $w_{it}^\ell \leq 0$ for all ℓ except for $\ell = \ell^j$ such that $\alpha_{it}^{\ell j} = \alpha_{i,t+1}^{a(j)}$. The upper bound on $w_{it}^{\ell j}$ is sufficiently large to allow any investment level. The stochastic right-hand side problem is then formed by substituting (27.5), (27.6), and a constraint where R_{it}^j is set equal to the right-hand side of (27.6), for Constraint 27.4.1 in Program 27.4.

In the deterministic version of (27.4), the investment decisions may skip from investment in one resource to another according to the values of α_{it}^j. This is due to the basic property of the linear program in which extreme point values correspond to investments in single resources. The solution of (27.4) allows for many more combinations of alternative investment decisions and, hence, provides for hedging against other possibilities. This hedging characteristic yields a positive VSS for many cases and the value of knowing the investment return yields a positive EVPI. An example of these occurrences appear in the next section.

27.4   Example

We consider a two period problem to demonstrate the potential effect of investment uncertainty. In this example, we consider three technologies. The first technology uses a resource in which investment return is highly variable. The second technology corresponds to a resource in which investment in additional capacity results in certain returns. The third technology is an infinitely available backstop. The data for the model are in Table 27.1.

Table 27.1   Model Input Data

Resources	Current Cost	Initial Availability
Res 1	5.0	25.0
Res 2	10.0	10.0
Backstop	16.7	$+\infty$

Investment	Cost	Return
Res 1 - Good Luck	1.0	1.0
Bad Luck	1.0	0.1
Res 2	1.0	1.0

Periods	Demand
First	15.0
Second	25.0

Scenarios	Probability
Good Luck	0.5
Bad Luck	0.5

Discount Factor	$\beta = 0.6$

The only uncertainty in this model is in the return for Resource 1 exploration investment. Resource 2 investment can be interpreted as building additional capacity. This model can be formulated as a stochastic linear program with recourse and with uncertainty in the right-hand side by using constraints as in (27.5) and (27.6). In this case, we obtain the following two-stage stochastic linear program in which x represents first period decisions and y represents second period decisions.

minimize $z = 5x_1 + 10x_2 + 16.7x_3 + x_4 + x_5 + E_\xi[3y_5 + 6y_6 + 10y_7]$

subject to $x_1 \leq 25$

$x_2 \leq 10$

$x_1 + x_2 + x_3 \geq 15$

$-x_1 + y_1 + .1y_3 + y_4 = 0$

$x_4 - y_3 - y_4 = 0$

$-x_2 + x_5 + y_2 = 0$

$y_4 \leq \xi$

$y_1 + y_5 \leq 25$

$y_2 + y_6 \leq 10$

$y_5 + y_6 + y_7 \geq 25,$

$$x_1, \ldots, x_5 \geq 0, y_1, \ldots, y_7 \geq 0,$$

(27.7)

where $P\{\xi = 0\} = 0.5$ and $P\{\xi = 10\} = 0.5$. In this program, x_1, x_2, and x_3 represent commitments of the resources, x_4 and x_5 are investment variables, y_1 and y_2 represent the net changes in resource availabilities, y_3 and y_4 represent the amount of new Resource 1 availability obtained through investment, and y_5, y_6, and y_7 represent commitments in the second period.

The alternatives to Program 27.7 are to solve deterministic models that assume good luck, bad luck, a mean value with $\xi = \overline{\xi} = 5$, or a single myopic solution. For each of these solutions, we obtain the expectation of the two period costs after using the first period solution obtained by these deterministic problems (as in finding the EEV). These values are

Scenario	Deterministic Value	Expectation Value
Good Luck	175.0	196.5
Bad Luck	200.0	200.0
Mean	185.0	200.75
Myopic	215.0	215.0

These values can be compared to the value of the stochastic program (27.7), which is 192.5.

We can then obtain the information values, EVPI and VSS. The expected value of perfect information is

$$EVPI = RP - WS = 192.5 - 187.5 = 5.0.$$

The value of the stochastic solution is

$$VSS = EEV - RP = 200.75 - 192.5 = 8.5.$$

The value of the stochastic solution relative to the myopic, or no investment, solution is also of interest. It is $215.0 - 192.5 = 22.5$.

The difference between the EVPI and VSS values demonstrates how these quantities reflect different values of uncertainty. The EVPI is lower than the VSS because the RP solution can fairly adequately hedge against either of the future outcomes. In the RP solution, there is investment in both Resource 1 and Resource 2 capacity ($x_4 = 10$ and $x_5 = 4$) so that no backstop usage is necessary in either scenario. The mean value solution, however, only involves investment in Resource 1 so that the backstop must be used in the bad luck scenario. This leads to a higher VSS than EVPI and shows the merit of using the stochastic program solution.

Investment in two resources is unique to the stochastic program solution. Any deterministic scenario only involves investment in one resource. This again shows the utility of the stochastic program. It is able to blend the deterministic solutions so that the decision maker does not have to decide between two completely different solutions.

We also note that the addition of investment has a significant effect on the value relative to the myopic solution. If no investment is allowed then the myopic solution would be optimal, and the backstop would necessarily be used to satisfy five units of demand in the second period. An exhaustible resource model with investment therefore clearly must consider future scenarios, and the solution of an equivalent stochastic program can have significant advantages over the solution of a deterministic expected value problem.

References

[1] J.R. Birge, "The value of the stochastic solution in stochastic linear programs with fixed recourse", *Mathematical Programming* **24**(1982), 314–325.

[2] H.P. Chao, "Exhaustible resource models: the value of information", *Operations Research* **29**(1981), 903–923.

[3] H. Hotelling, "The economics of exhaustible resources", *J. Political Econ.* **39**(1931), 173–175.

[4] A. Madansky, "Inequalities for stochastic linear programming problems", *Management Science* **6**(1960), 197–204.

[5] A.S. Manne, "Waiting for the breeder", in: Review of Economic Studies Symposium, (1974), 47–65.

[6] A.S. Manne and R.G. Richels, "A decision analysis of the U.S. breeder reactor program", *Energy* **3**(1978), 747–767.

[7] W.D. Nordhaus, "The allocation of energy resources", *Brookings Papers on Economic Activity* **3**(1973), 529–576.

[8] R.J-B. Wets, "Stochastic programs with fixed recourse: the equivalent deterministic program", *SIAM Review* **16**(1974), 309–339.

CHAPTER 28

A TWO-STAGE STOCHASTIC FACILITY-LOCATION PROBLEM WITH TIME-DEPENDENT SUPPLY

S.W. Wallace

Abstract

A stochastic facility-location problem with recourse is solved by the L-shaped decomposition method. The purpose is to find which plants, from a set of potential plants, should be opened. The supply is random and varying over time.

To each potential plant is attached a fixed cost. The decomposition results in a stochastic transportation problem and an NP-hard problem with quasi-concave objection function and linear constraints.

28.1 Introduction

We are concerned with the following problem: A set of supply ponts is given, each point having a supply that varies over the year. The supply points in general have their supply peaks at different times. The supply is random.

A set of potential demand points is also given. We want to establish which of them should be kept/built and which should be closed/not built. For the existing ones we also consider the possibility of increasing their capacities. To each potential demand point is attached a fixed cost depending on the capacity of the demand point, which also is to be determined.

Due to the variation of supply over time, we will divide the year into T time periods. Clearly we cannot expect the capacity at the demand points to be fully utilized in all time periods. Still the fixed cost will be the same in all periods, namely the one given by the amount received in the most intensive period.

The problem is motivated by a problem from the Norwegian fish meal and fish oil industry. The supply points represent fishing grounds for which the quotas are stochastic and variable throughout the year. The demand points are potential plants (see Section 28.7 for further details). References [15] and [16] also give background information.

Transportation costs are given between all pairs of supply/potential demand points. If handling costs differ among the plants, they must be included in the transportation costs, and not in the fixed costs of the plants. The formulation is as follows:

$$\min_x E_\omega \{\min_y \sum_i \sum_j \sum_t c_{ij} y_{ij}^t\} + \sum_{j=1}^{M2-1} \tilde{h}_j(x_j) \qquad (28.1)$$

$$\text{subject to} \quad \sum_{j=1}^{M2} y_{ij}^t = S_i^t(\omega) \qquad t = 1,\ldots, \quad Ti = 2,\ldots,M1 \qquad (28.2)$$

$$-\sum_{i=2}^{M1} y_{ij}^t \geq -x_j \qquad t = 1,\ldots,T, \quad j = 1,\ldots,M2-1 \quad (28.3)$$

$$0 \leq x_j \leq d_j \qquad j = 1,\ldots,M2-1 \qquad (28.4)$$

$$y_{ij}^t \geq 0 \qquad (28.5)$$

where

y_{ij}^t equals the number of loads from supply point i to demand point j in time period t. (We relax the natural integrality requirement.)

c_{ij} equals the cost per load sent from supply point i to demand point j.

x_j equals the capacity in loads per time period for demand point j.

$\tilde{h}_j(x_j)$ equals the fixed cost attached to demand point j as a function of x_j.

$S_i^t(\omega)$ equals the uncertain amount supplied at supply point i in time period t.

Note that demand point $M2$ has infinite capacity, i.e. it represents a recourse action such as sending to a second rate market or dumping. Therefore we will assume that $c_{iM2} > c_{ij}$ for $j = 1,\ldots,M2-1$. Clearly the problem is always feasible.

The requirements (28.4) might be dropped, depending on the situation in which the method is used.

The function $\tilde{h}_j(x_j)$ is assumed to be quasi-concave; in practice we will assume the following form

$$\tilde{h}_j(x_j) = \begin{cases} 0 & \text{if } x_j = 0 \\ H_j + h_j x_j & \text{if } x_j > 0, H_j \geq 0 \end{cases}$$

Although we are concerned about several time periods, our problem does not belong to "dynamic facility location" problems or "multiperiod capacity expansion" problems. An important reason for this is that although both problems (usually) operate with T time periods (T finite), our problem does not end here, but rather starts in period 1 again. In dynamic location problems (see e.g. [5], [9]), however, time T marks the end of the time horizon. Therefore, the time of investment is important due to present value considerations. In our case, investments are done at the start of period one and capacity kept at that level throughout time.

Our problem will therefore, to a certain extent, belong to the one-period facility location problem, although that period is divided into T subperiods.

The complications here are naturally due to the stochastic supply, but also:

1. The size of the plants are variables, thereby spoiling the network structure of the constraints.

2. The discontinuous (quasi-concave) objective function.

3. The variable part of the fixed cost of a plant cannot be included in the transportation costs because we have more than one time period.

Problem 1 will be attacked through decomposition, problem 2 through enumeration of extreme points or a series of linear programs.

Problem 3 complicates the decomposition since the master problem of the decomposition must determine not only which plants to open, but also their sizes.

We will use the L-shaped decomposition of the problem, outlined in [13] and [18]. This amounts to writing the problem in the following form:

$$\text{minimize} \quad \sum_{j=1}^{M2-1} \tilde{h}_j(x_j) + \theta$$

$$\text{subject to} \quad \underline{Q}(x) \leq \theta$$
$$0 \leq x_j \leq d_j \quad j = 1, \ldots, M2-1$$

where $\underline{Q}(x)$ is defined as $\underline{Q}(x) = E_\omega Q(x, \omega)$ and $Q(x, \omega)$ is given by

$$Q(x,\omega) = \inf\{\sum_i \sum_j \sum_t c_{ij} y_{ij}^t \mid \underline{N}' y \begin{pmatrix} = \\ \geq \end{pmatrix} b', y \geq 0\} \tag{28.6}$$

where $\underline{N}'$ is the coefficient matrix for the y's in (2) and (3) and

$$b' = \begin{pmatrix} S(\omega) \\ -x \end{pmatrix} \text{ is appropriately sorted to fit } \underline{N}'.$$

The L-shaped algorithm is a "tight" cutting plane algorithm that in general allows for both feasibility and optimality cuts. Since our subproblem (i.e. to find $\underline{Q}(x)$ for x given) is always feasible, we will only need optimality cuts.

A method very similar to the L-shaped is an outer approximation using nonstochastic tenders, see [12]. We will see later that due to some separability properties in our problem, these two methods are equivalent.

The L-shaped algorithm can be viewed as a version of Benders' decomposition [1], as applied to L-shaped structured problems.

Note that the solution to our problem is a set of variables x_j, $j = 1,\ldots,$ $M2 - 1$, and *not* a set of variables x_j and y_{ij}. The problem is a so-called two-stage stochastic optimization problem or stochastic program with recourse. This means that first the decision-maker must determine the x_i's on the basis of only the distribution of the supply. Then after the realization of the supply, the short run (second stage) recourse variables y_{ij} are determined. When solving (28.6), we therefore (in general) will get different y's for the different realizations of ω, while e.g. [6] get a solution consisting of both y_{ij} and x_j. So even though these problems (i.e. ours and [6]) may look similar, their nature is significantly different.

28.2   Determination of $\underline{Q}(x)$, the Subproblem

$$Q(x) = E_\omega Q(x,\omega) \text{ where}$$

$$Q(x,\omega) = \inf\{\sum_i \sum_j \sum_t c_{ij} y_{ij}^t | \underline{N}' y \begin{pmatrix} = \\ \geq \end{pmatrix} b', y \geq 0\} \text{ and} \tag{28.6}$$

$$b' = \begin{pmatrix} S(\omega) \\ -x \end{pmatrix} \text{ appropriately sorted to fit } \underline{N}'.$$

If we write this in more detail we get $Q(x,\omega) =$

$$\text{minimize} \quad \sum_i \sum_j \sum_t c_{ij} y_{ij}^t \tag{28.7}$$

$$\text{subject to} \quad \sum_{j=1}^{M2} y_{ij}^t = S_i^t(\omega) \qquad t = 1,\ldots,T, i = 2,\ldots,M1 \tag{28.8}$$

$$-\sum_{i=2}^{M1} y_{ij}^t \geq -x_j \qquad t = 1,\ldots,T, j = 1,\ldots,M2-1 \tag{28.9}$$

$$y_{ij}^t \geq 0$$

For x fixed (28.7), (28.8) and (28.9) is separable in T subproblems, so

$$Q(x,\omega) = \sum_t Q^t(x,\omega) \text{ where}$$

$$Q^t(x,\omega) = \inf\{\sum_i \sum_j c_{ij} y_{ij} | \underline{N} y \begin{pmatrix} = \\ \geq \end{pmatrix} b^t, y \geq 0\} \qquad (28.10)$$

where $\underline{N}$ and b^t are the coefficient matrix and right hand side of the system

$$\sum_{j=1}^{M2} y_{ij} = S_i^t(\omega) \quad i = 2,\ldots,m1 \qquad (28.11)$$

$$-\sum_{i=2}^{M1} y_{ij} \geq -x_j \quad j = 1,\ldots,M2-1 \qquad (28.12)$$

The constraints of the subproblem, namely (28.11) and (28.12) are not written in standard transportation format. We therefore introduce a dummy supply point, supply point 1, and let $c_{1j} = 0$ for all j.

Furthermore we change the inequality signs in (28.12) to equalities and let

$$S_1^t(\omega) = \max\left\{0, \sum_{j=1}^{M2-1} x_j - \sum_{i=2}^{m1} S_i^t(\omega)\right\}$$

$$x_{M2}^t(\omega) = \max\left\{0, \sum_{i=2}^{M1} S_i^t(\omega) - \sum_{j=1}^{M2-1} x_j\right\}$$

Thereby we get (leaving out the indices ω and t)

$$\text{minimize} \quad \sum_{i=2}^{M1} \sum_{j=1}^{M2} c_{ij} y_{ij}$$

$$\text{subject to} \quad \sum_{j=1}^{M2} y_{ij} = S_i \quad i = 2, \dots, M1 \tag{28.11'}$$

$$\sum_{i=2}^{M1} y_{ij} = -x_j \quad j = 1, \dots, M2 \tag{28.12'}$$

$$y_{ij} \geq 0$$

Since the constraints of a transportation network are linearly dependent, we have omitted the equation for supply point 1.

The dual of this is

$$\text{maximize} \quad \sum_{i=2}^{M1} S_i \pi(s_i) - \sum_{j=1}^{M2} x_j \pi(x_j)$$

$$\text{subject to} \quad \pi(s_i) - \pi(x_j) \leq c_{ij} \quad i = 2, \dots, M1, j = 1, \dots, M2$$

$$-\pi(x_j) \leq c_{1j} = 0 \quad j = 1, \dots, M2$$

$$\pi(s_i), \pi(x_j) \text{ unrestricted in sign.}$$

But these constraints can be rewritten as

$$\pi(s_i) - \pi(x_j) \leq c_{ij} \quad i = 2, \dots, M1, j = 1, \dots, M2$$
$$\pi(x_j) \geq 0 \quad j = 1, \dots, M2 \tag{28.13}$$

$$\pi(s_i) \text{ unrestricted in sign.}$$

If we take the dual once more we get:

$$\min \sum_{i=2}^{M1} \sum_{j=1}^{M2} c_{ij} y_{ij}$$

$$\sum_{j=1}^{M2} y_{ij} = S_i \quad i = 2, \dots, M1 \tag{28.11''}$$

$$-\sum_{i=2}^{M1} y_{ij} \geq -x_j \quad j = 1, \dots, M2 \tag{28.12''}$$

$$y_{ij} \geq 0$$

Except for the constraint for $j = M2$ this is equal to (28.11) and (28.12) but since $c_{iM2} > c_{ij}$ for $j = 1, \dots, M2 - 1$, we know that leaving out the inequality for $j = M2$ in (28.12″) will not alter the solution. Therefore solving problem (28.11′) and (28.12′) instead of (28.11) and (28.12) give the correct dual variables.

Alternatively we can say that since relaxing constraint $M2$ in (28.12″) to $(x_{M2} + \varepsilon)$ will not make any difference (no flow will be moved from any of the other demand nodes), $\pi(x_{M2}) = 0$. By putting $\pi(x_{M2}) = 0$ into (28.13) and then taking the dual, we will get (28.11) and (28.12). The conclusion is therefore:

Remark: By introducing a dummy supply node into (28.11) and (28.12), making sure that supply equals demand and letting all inequalities be equalities, we get a transportation problem for which the dual variables coincide with those of (28.11) and (28.12).

From now on, we will use formulation (28.11′) and (28.12′). We will call the coefficient matrix $\underline{N}$ and the right hand side b although the number of rows have increased by one.

Assuming that ω has a finite number K of possible outcomes, with p_k the probability of outcome k, $\underline{Q}(x)$ becomes

$$\underline{Q}(x) = \sum_{k=1}^{K} p_k \sum_{t=1}^{T} Q^t(x, \omega_k)$$

where $Q^t(x, \omega_k)$ now is

$$\inf \left\{ \sum_i \sum_j c_{ij} y_{ij} \middle| \underline{N}y = b_k^t, y \geq 0 \right\} \tag{28.10'}$$

The dual of (28.10′) is given by

$$P^t(x, \omega_k) = \sup\{\pi b^t | \pi \underline{N} \leq c\} \tag{28.14}$$

Let y_0 be the optimal solution to (28.10′) and π_0 the optimal solution to (28.14). Then $\pi_0 b^t = c y_0$, i.e. $P^t(x, \omega) = Q^t(x, \omega)$.

So therefore

$$\underline{Q}(x) = \sum_{k=1}^{K} p_k \sum_{t=1}^{T} P^t(x, \omega_k) \tag{28.15}$$

If $\underline{N}_0$ is the optimal basis, we now that $\pi_0 = c_0 \underline{N}_0^{-1}$, i.e. it is not a function of the right hand side b^t. Therefore if b_1^t and b_2^t both have the same optimal basis, they have equal dual variables, which can be utilized in (28.15) by bunching all possible right hand sides of (28.15) which have the same optimal basis. (There is a total of $T \cdot K$ right hand sides). How this can efficiently be done is explained in [17]. Let $\pi_k^t(s_i)$ be the optimal value of the variable in (28.14) that corresponds to supply point i and $\pi_k^t(x_j)$ the same for the demand point j.

Following formula (2.20) of [18], we get the following optimality cut if a given x is not optimal.

$$V x + \theta \geq \nu \tag{28.16}$$

where

$$V = \sum_{k=1}^{K} p_k \sum_{t=1}^{T} \pi_k^t(x) \tag{28.17}$$

and

$$\nu = \sum_{k=1}^{K} p_k \sum_{t=1}^{T} \pi_k^t(s) S^t(\omega_k)$$

If $p_k = \frac{1}{K}$, which often will be quite reasonable, then

$$V = \frac{1}{K} \sum_{k=1}^{K} \sum_{t=1}^{T} \pi_k^t(x) \tag{28.16'}$$

$$\nu = \frac{1}{K} \sum_{k=1}^{K} \sum_{t=1}^{T} \pi_k^t(s) S^t(\omega_k) \tag{28.17'}$$

in which case (28.16) can be written as

$$V' x + K\theta \geq \nu'$$

with all coefficients integer provided c_{ij} is integer.

We have already shown that $\pi(x_j) \geq 0$ for all j. It is also easy to demonstrate that $\pi(s_i)$ is greater or equal to zero. Note that since $c_{ij} \geq 0$ for all i and j and since we are minimizing, (28.11) and (28.12) can be rewritten as

$$\sum_{j=1}^{M2} y_{ij} \geq S_i^t(\omega) \quad i = 2, \ldots, M1$$

$$-\sum_{i=2}^{M1} y_{ij} \geq -x_j \quad j = 1, \ldots, M2 - 1$$

This is clearly true since we always will try to send as little as possible, forcing equality in the supply constraints.

The dual of this is (using the objective function (28.7)):

$$\text{maximize} \quad \sum_{i=2}^{M1} S_i^t(\omega) \pi^t(s_i) - \sum_{j=1}^{M2} x_j \pi^t(x_j)$$

$$\text{subject to} \quad \pi^t(s_i) - \pi^t(x_j) \leq c_{ij}$$

$$\pi^t(s_i) \geq 0$$

$$\pi^t(x_j) \geq 0$$

$$\pi^t(s_1) = 0$$

From this it follows that all elements in V are nonnegative and ν is positive in (28.16). Furthermore $\underline{Q}(x) = \nu - Vx$.

A disadvantage of this method is that even if the number of used demand points is low, we solve a transportation problem of full size. Other approaches, such as [6] avoid this, but if we were to follow these methods we would loose other advantages, such as the efficiency of the dual decomposition outlined in [17].

We then turn to:

28.3   The Equivalent Deterministic Program

If ω has a finite number of outcomes, $\underline{Q}(x)$, the subproblem, will be polyhedral in x, [13]. Thereby (28.1)–(28.5) can be written equivalently as

$$\text{minimize} \quad \sum \tilde{h}_j(x_j) + \theta \qquad\qquad (28.19)$$

$$\text{subject to} \quad V_s x + \theta \geq \nu_s \quad s = 1, \ldots, R \qquad (28.20)$$

$$0 \leq x \leq d$$

We call (28.19)–(28.20) the equivalent deterministic program. It has the same solution as (28.1)–(28.5). The constraints $V_s x + \theta \geq \nu_s$ are of the form (28.16) generated by the subproblem.

We will next define the relaxed deterministic problem as

$$\text{minimize} \quad \sum_j \tilde{h}_j(x_j) + \theta \qquad\qquad (28.19)$$

$$\text{subject to} \quad V_s x + \theta \geq \nu_s \quad s = 1, \ldots, r \qquad (28.21)$$

$$0 \leq x \leq d$$

The relaxed deterministic program is a part of an iteration in order to approximate the equivalent deterministic program. The iteration can be stated as follows: Pick an arbitrary (reasonable) x^0. Solve the subproblem, i.e. find $\underline{Q}(x^0)$. Determine an optimality cut of the type (28.16), and solve (28.19), (28.21) setting $r = 1$ and $i = 1$.

Let x^i, θ^i be the optimal solution fo (28.19), (28.21). Then find $\underline{Q}(x^i)$. If $\underline{Q}(x^i) \leq \theta^i$ stop, otherwise construct a new optimality cut of type (28.16), increase r and i by one and resolve.

The program can also be started by a number of (intelligent) guesses x^i, $i = 1, \ldots, r$ if such are available. This fits the idea of nonstochastic tenders in [12].

Since (28.19) is assumed to be quasi-concave, we know that the solution to the relaxed deterministic problem can be found in one of the extreme points of (28.21). We will therefore use an idea presented in [10], although the algorithm as such can be found in [4] and [14]. This is an exact method. In the next section we will present a heuristic approach.

The algorithm is based on the following propositions:

Proposition 1. *The k-th best extreme point in the set of feasible solutions to an LP will always be the neighbor of one of the* $(k-1)$ *best.*

Proposition 2. *Given the graph G with one node for each extreme point and one arc for each pair of neighbor extreme points. Then there is a path from any node in the graph to the root node (representing the optimal solution) on which the objective function is nonincreasing.*

If the purpose (as in [14]) is to find the k best extreme points in the set of feasible solutions, the algorithm goes as follows, based on Proposition 1. (Assuming here that degeneracy does not occur, this only to make the presentation simpler.)

(i) Find the optimal solution. Let $t = 1$.

(ii) Find the $(t+1)$-st best extreme point as the neighbor of one of the t best.

(iii) Increase t by one. If $t = k$, stop. Otherwise go to (ii).

In order to sove step (ii), one will usually store some information about all the neighbors of the t best extreme points. Otherwise the amount of work will be too large. Therefore in step (iii), before returning to step (ii) one must calculate the appropriate information about those neighbors of extreme point t that have not already been found.

If the algorithm proceeds as explained above, it easily follows from Proposition 1 and 2.

Proposition 3. *The value of the objective function for all nodes created in step (iii) will be at least as high as for node t.*

We now show how by relying on the above proposition and the preceding algorithm, we can solve our problem which has a quasi-concave objective function.

Assume as before that

$$\tilde{h}_j(x_j) = \begin{cases} 0 & \text{if } x_j = 0 \\ H_j + h_j x_j & \text{if } x_j > 0. \end{cases}$$

Then as first step of the algorithm solve:

$$\begin{aligned} \text{minimize} \quad & \sum h_j x_j \\ \text{subject to} \quad & V_s x + \theta \geq \nu_s \quad s = 1, \dots, r \\ & 0 \leq x \leq d \end{aligned} \tag{28.22}$$

An optimal solution to this is easy to find, and it is called extreme point 1. The variables z_1 and w_1, defined below are associated with this solution

$$z_1 = \sum_j h_j x_{1j}$$

and

$$w_1 = \sum_{B_1} H_j$$

where B_1 is the set of all *positive* basic variables for extreme point 1.

A variable WMIN is given as a lower bound on w. If no a priori information exist, WMIN $= 0$ can always be used.

Let FOPT $= z_1 + w_1$, OPTEX $= 1$. Find all the neighbors of node one for which $z_j <$ FOPT $-$ WMIN . Put them into a list D sorted after increasing values of z_j.

In a systematic way we now examine the rest of the extreme points of (28.22). Assume at some step that we have found the k best extreme points, (i.e. according to the objective function of (28.22)). Assume

$$z_t + w_t = \min_{j=1,\dots,k} \{z_j + w_j\}$$

such that FOPT $= z_t + w_t$ and OPTEX $= t$.

Next pick the first node in the list D. If D is empty, OPTEX is the optimal extreme point. Otherwise this node represents extreme point $(k+1)$. Check if $z_{k+1} + w_{k+1} <$ FOPT . If that is true, let OPTEX $= k+1$ and FOPT $= z_{k+1} + w_{k+1}$, and delete from the list D all nodes with $z_j >$ FOPT $-$ WMIN . Increase k by one and repeat.

Clearly, a good approximation of WMIN is crucial for the speed of convergence.

With linear objective functions, one would always expect that the last cut generated will be binding in the next iteration. With functions like ours, this will in general *not* be the case. The following way of rewriting the relaxed deterministic problem is therefore *not* valid.

$$\begin{aligned}
\text{minimize} \quad & \sum \tilde{h}_j(x_j) + \theta \\
\text{subject to} \quad & V_s x + \theta \geq \nu_s \quad s = 1,\dots,r-1 \\
& V_r x + \theta \geq \nu_r \\
& 0 \leq x \leq d
\end{aligned}$$

The main disadvantage of this method so far is that we will expect x to change little from one step to the next. Therefore the optimal x from the previous step is likely to be a good guess. The problem is, however, that by using this x as a starting point (using a few dual simplex steps) we have no stopping criterion, although the optimal solution might be just a few pivots away. We therefore suggest the following approach.

Take the x from the previous step. It will represent a primal infeasible but dual feasible solution. Use a dual method to find a primal feasible solution. The number of steps will probably be rather low. Denote this extreme point by 0 (zero). Find w_0, z_0 and let FOPT $= z_0 + w_0$ and OPTEX $= 0$. Then start the main procedure.

The advantage is now that provided extreme point 0 is a good guess (which most likely is not the case for extreme point 1) the number of nodes needed to find the optimal solution will decrease since the check for whether or not nodes in the list D can be deleted is likely to be more powerful. This idea has been tried, and the results are outlined in Section 28.8.

It is possible for the number of cuts of the type (28.10) to become very large. But due to a result by Murty [11] we only need to keep a maximum of $M1 + M2 - 1$ (the number of rows in the node-arc incidence matrix). Theoretically we can, therefore, in each step drop all nonbinding constraints. Our experience, however, is that such an approach is extremely difficult, due to the unstructured behavior of the quasi-concave objective function. Even dropping only one hyperplane is difficult, since the hyperplane with the largest slack in one iteration, easily becomes binding in the next. We return to this in Section 28.8.

As an alternative to the extreme point enumeration we present another method which must be viewed as heuristic.

28.4   A Heuristic Approach to Solve the Relaxed Program

As we will outline further in a later section, the method described in the previous section for solving the relaxed deterministic problem is not very efficient for this specific problem.

In this section we therefore present a heuristic approach based on cardinality constraints, [8]. Very loosely, the idea can be expressed as follows:

Idea. The relaxed deterministic problem (28.19), (28.21) will for reasonable values of h_j and H_j be unimodal in k, the number of plants.

We will return to the problem of determining when unimodality is present, but assume so far that this is actually the case.

The advantage of this approach is that we can solve a relaxed deterministic problem using the cost function $\sum h_j x_j$ instead of the much more complicated $\sum \tilde{h}_j(x_j)$.

The algorithm which is based on a series of LP's can take on two different forms depending on which of the following questions we ask:

- What is the best structure given that we use exactly k plants?
- What is the best structure given that we use no more than k plants?

The first of these questions can be answered by checking all possibilities of k plants, but always using an extra cut such that the LP-code only finds a feasible solution if the current combination of k plants is the best so far (since it is faster to find infeasibility than optimality of a feasible problem). The extra cut is made from the coefficients of the objective function.

The second question above can, provided $H_j = \overline{H}$, be solved using a cardinality constrained LP, see [8]. Both methods are exponential.

Provided we solve a series of LP's, these are basically two approaches:

- Solve the problem sequentially until we find a k such that $\sum \tilde{h}_j(x_{k+1,j}) > \sum \tilde{h}_j(x_{kj})$ where x_{kj} is the size of the j-th potential plant given that we have k plants.
- Use a golden section search on k.

If the cardinality constrained LP is used, we can solve sequentially with the extra constraint:

$$N(x) \leq k \tag{28.23}$$

where $N(x)$ is the number of open plants, until (28.23) is not binding. Then the solution found in the previous step is optimal. Or we can do a bisection on k. A bisection on k will need a maximum of $\log_2(M2 - 1)$ steps.

28.5 Complexity of the Relaxed Equivalent Problem

In this section we show that the equivalent deterministic problem (28.19), (28.21) is NP-hard. For a detailed treatment of NP-problems, see [7]. Problem (28.19), (28.21) is clearly equivalent to the following mixed zero-one integer programming (IP) problem. Given h, H, V, b and D as nonnegative rational matrices and vectors, find

$$
\begin{aligned}
\text{minimize} \quad & hx + Hy + \theta \\
\text{subject to} \quad & Vx + 1\theta \geq b \\
& x \leq Dy \\
& y \in \{0,1\}^{M2-1} \qquad X \\
& x \in Q_+^{M2-1} \\
& \theta \in Q
\end{aligned}
$$

The recognition version of the above mixed zero-one IP called (MIP) is

"Does $hx + Hy + \theta \leq M$ and $(x, y, \theta) \in X$ have a solution?"

We will show the following theorem.

Theorem. (MIP) *is* NP-*complete.*

Proof: First we show that (MIP) is in the class NP. Any feasible solution to (MIP) will have $y \in \{0,1\}^{M2-1}$. The remaining components are determined by an LP in the original coefficients. Since LP is in the class P which is a subset of NP the result follows.

We complete the proof by showing that there exists an NP-complete problem that transforms polynomially into (MIP).

The zero-one IP

"Given A and b, does $Ay \geq b$, $y \in \{0,1\}^{M2-1}$ have a solution?"

is NP-complete even if A and b are restricted to nonnegative entries. Let an arbitrary instance of zero-one IP be given by $\underline{A}$ and $\underline{b}$. We show how to construct in polynomial time an instance I of (IMP) such that the zero-one IP has a solution if and only if I has a solution.

Let $h = H = 0$, $V = A$, $D = I$ and $M = 0$. It is immediate that if the zero-one IP has the solution y^* then I has the solution $\hat{x} = \hat{y} = y^*$ and $\hat{\theta} = 0$. Conversely, if I has a solution $\hat{x}$, $\hat{y}$, $\hat{\theta}$ then $A\hat{y} \geq A\hat{x} \geq b - 1\theta \geq b$. Hence $y^* = \hat{y}$ is a solution to the zero-one IP. Q.E.D.

This result justifies the use of exponential algorithms to solve the relaxed deterministic problem.

28.6   An Alternative Approach

In the previous section we outlined a heuristic method for solving the relaxed deterministic problem. Based on the idea of unimodality another heuristic approach is reasonable to try. The idea is:

Idea: The two-stage stochastic facility-location problem (28.1)–(28.5) is unimodal in k, the number of plants.

As for the relaxed deterministic problem we can again either perform:

  – bisection by adding (28.23)
  – linear search by adding (28.23)
  – golden search by adding

$$N(x) = k \qquad\qquad (28.24)$$

  – linear search by adding (28.24)

In each of these cases we will decompose (28.1)–(28.5) and (28.23) or (28.24) as explained in the previous sections into a relaxed deterministic problem and a stochastic transportation problem. But now the relaxed deterministic problem only has to be solved for one value of k.

Note that also here $H_j = \bar{H}$ is necessary if the cardinality-constrained LP is to be used.

By this method we have moved the iteration on k from an inner loop to an outer loop. It is not clear to us which approach is best. The approach in this section, however, has the advantage that for each k not found to be optimal we get information about the optimal structure for that specific value of k. (With (28.23) this is only true for k smaller than the optimal value.)

We then turn to the problem of determining when the problem (28.1)–(28.5) is unimodal in k, the number of plants.

If unimodality with respect to minimization is not present, we must have a situation like Figure 28.1.

In the following we will assume that $H_j = \bar{H}$ and $h_j = \bar{h}_j$ i.e. all potential plants have the same cost structure. Let x_k be the optimal plant structure with k plants and d_k the total expected transportation costs associated with it.

Mathematically lack of unimodality means that

$$k\bar{H} + \bar{h}\sum_j x_{kj} + d_k \geq (k-1)\bar{H} + \bar{h}\sum_j x_{k-1,j} + d_{k-1}$$

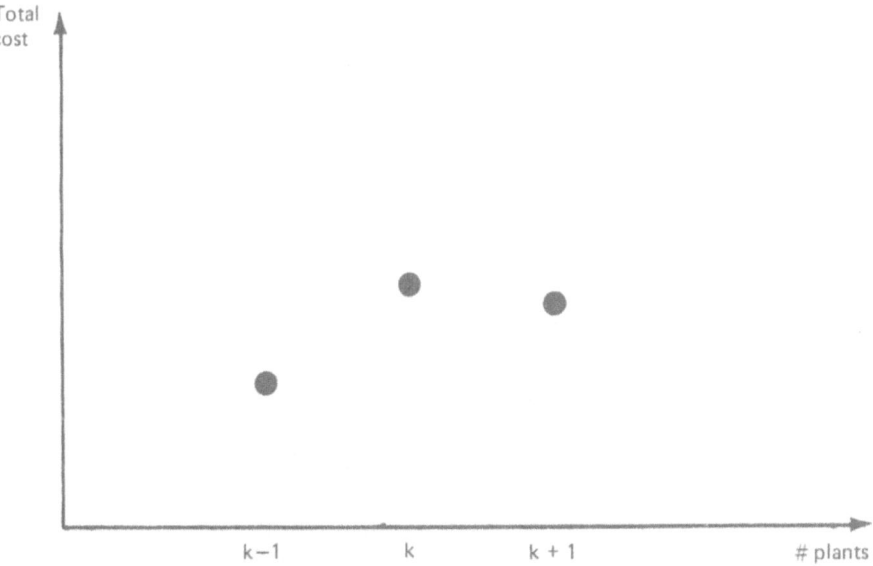

Figure 28.1 A situation violating unimodality with respect to minimization.

$$k\bar{H} + h\sum_{j} x_{kj} + d_k \geq (k+1)\bar{H} + h\sum_{j} x_{k+1,j} + d_{k+1}$$

We add to get

$$h\sum_{j} x_{kj} + d_k \geq \frac{1}{2}[h\sum_{j}(x_{k-1,j} + x_{k+1,j}) + d_{k-1} + d_{k+1}] \tag{28.25}$$

Following [2, p. 204], inequality (28.25) is the exact definition of concavity over integers. Let $f_k = h\sum_{j} x_{kj} + d_k$. We then get

Proposition 3. *The problem (28.1)–(28.5) is unimodal in* k*, the number of plants, provided* f_k *is strictly convex over the integers* $[1, M2 - 1]$.

The above proposition is clearly not necessary. There are certain concavity situations which are acceptable. We will not go into details here, just note the following:

- If sequential search on k is performed, we only need to require that the global minimum is the left-most local minimum.
- With bisection or golden search concavity in f_k can be acceptable, but (28.1)–(28.5) must be unimodal.

28.7 Example

The purpose of this example is to investigate the following question:

"If the Norwegian fish meal and fish oil industry were to be established today, what would the plant structure in southern Norway be, provided we assume there are enough vessels available?"

The reason for asking such a question is the structure of today's industry. All the plants we have today are both small and old. Compared to e.g. Denmark, even our largest plant is small, and the largest Norwegian plant (which is in northern Norway) is almost twice as large as the second largest plant.

Since many of the existing plants are very old, the action of building a new plant (of the size suggested in this report) will not be very different from rebuilding one of the old plants.

What is wrong with this approach, however, is that in the short run the fixed cost of an existing plant is lower than assumed here since the alternative value of it is most often close to zero. But in the long run, one will not reinvest the amount needed to maintain the plant unless the profit is as good as elsewhere. Therefore the approach can be considered appropriate at least in the long run.

Fishing grounds

We have assumed 5 different fisheries, taking place at 14 fishing grounds. Position, quotas and fishing seasons are based on the situation over the last few years. The 5 fisheries are given in the table below.

Table 28.1 Expected values and standard deviation for quotas, fishing seasons and positions for the 5 fisheries used in this article. Quotas are measured in hectoliters.

Fishery	Expected Quota	Standard Dev. Quota	Starts in Weeks	Duration in Weeks	Position (+=East)	
Mackerel (Scomber scombrus)	250.000	10.000	29	2	63.6	- 0.5
Blue whiting (Gladus poutassou)	410.000	75.000	11	2	59.4	-18.0
	820.000	150.000	13	4	59.4	-10.2
	410.000	75.000	17	2	60.0	- 4.0
	410.000	75.000	19	2	61.5	- 1.0
Sprat (Clupea Sprattus)	170.000	75.000	1	1.5	53.7	3.5
	400.000	100.000	45	5	55.0	1.0
Sand-eel	75.000	15.000	12	14	60.0	3.0
	225.000	45.000	12	14	57.5	5.0
	75.000	15.000	12	14	56.0	4.0
	125.000	25.000	12	14	54.3	1.5
Norway pout	80.000	15.000	1	50	60.6	0.5
	1200.000	200.000	1	50	59.5	3.5
	320.000	45.000	1	50	57.8	5.6

Fish meal plants

We have assumed 11 potential plants along the coast of western Norway. The table below shows their positions.

Table 28.2 Region and position for 11 potential plants of southern Norway.

Position	Position	
	Latitude (°N)	(°E)
Sunnmøre	62.5	5.5
Nordfjord	62.0	5.0
Sunnfjord	61.5	5.2
Ytre-Sogn	61.1	5.0
Nordhordaland	60.8	5.0
Bergen	60.3	5.3
Sunnhordland	59.8	5.1
Nordrogaland	59.4	5.3
Stavanger	59.0	5.7
Flekkefjord/Egersund	58.4	6.3
Lindesnes	58.0	7.5

Fixed costs for the plants

The fixed costs for a plant consist in general of two parts:

(1) The cost related to maintenance in order to keep the plant as new. This should include what is needed to update the equipment technologically.

(2) Alternative cost for the capital bound in the plant.

(2) will be different depending on whether we consider an old plant or will build a new one. If there is no alternative use of an old plant, the alternative cost will be zero.

In this report we only consider building new plants, i.e. we consider the problem: What would we do if we were to establish the Norwegian fish-meal industry today. Based on data presented in [15] we have found the following linear approximation of the sum of (1) and (2) above.

$$\text{FIX} = 1.84 + 0.18x$$

where x is the capacity measured in number of loads (each 5000 hl) per month. The cost is measured in millions of NOK.

Transportation costs

The transportation costs per load are calculated between each pair of fishing grounds and plants. The vessels' use of fuel is found according to formulas presented in [3] on the basis of a chosen speed and vessel size.

The cost of bringing one load to the demand point representing the recourse action of going to Denmark is calculated as follows:

1. As pure transportation cost, use the most expensive within the country, i.e. $\max_j c_{ij}$.
2. Add a fixed cost per hl. This cost is meant to reflect the loss due to the fact that Denmark gets the profit from processing and selling the fish.

We have used data from a simulation reported in [16], but we have raised the world market prices for fish meal and fish oil to NOK 3.30/kg and NOK 2.71/kg, respectively. Furthermore, we have set the alternative value of labor to zero, to reflect that the alternative jobs do not exist in rural Norway. This gives us a loss of NOK 32 per hl when the fish is sent to Denmark.

Calculating the quotas

The quota for a certain fishery at a given position is found as follows.

First a total quota is found from a normal distribution with expectation and variance as given in Table 28.1. If the quota is smaller than $\mu - 2\sigma$, it is set equal to this value. If the quota is larger than $\mu + 2\sigma$, it is set equal to that value.

The quota is in the input distributed over a set of time period. $a_{it} = 0.5$ means that 50 percent of the quota of fishery i will be caught in time period t, i.e. we expect to catch $\mu_i a_{it}$. The amount allocated to fishery i in time period t is then drawn for a normal distribution with expectation $\mu_i a_{it}$ and variance $b_{it}\mu_i a_{it}$, again avoiding outliers as above. Note that this value is independent of the actual quota found above. b_{it} gives the standard deviation as a fraction of $a_{it}\mu_i$.

In this way we might experience both fisheries that finish before planned and fisheries that run out of time.

This process is repeated a number of times to obtain several right hand sides.

Results

With the data given, the solution turns out to be that only one plant should be constructed. The formal solution is to have one plant in the Stavanger area. Clearly this solution, being robust toward changes in quotas, might be nonrobust toward changed positions of the fisheries. We should therefore conclude as follows.

One plant with a capacity of 46.500 hl/day should be built somewhere between Haugesund and Egersund.

A plant of this size will be almost 3 times as large as the largest existing plant in Norway.

We would again like to point out that this is a long-run result. In the short run the alternative value of an existing plant is much lower than the 7 percent of the "new value" we have used, so in the short run many of the existing plants should be kept. In the long run, however, one will not reinvest in these old plants (since there are better alternatives). The result above, given the input, is therefore the long run goal, which is stable toward changes in quotas.

It must be admitted that this result is rather surprising. Below we stress some shortcomings of the model, and show in which direction they would move the solution.

(a) Aspects strengthening the one-plant solution
   - We have not assumed any economy of scale in the variable part of the fixed cost. Hence if we let

$$\tilde{h}_i(x_i) = H_i + h_i(x_i)$$

   where $h_i(x_i)$ is concave, the tendency towards one plant would be strengthened.
   - We have not been able to model the fact that continuous production is advantageous (as reported in [15]), and that a low number of plants means a high level of continuity.
   - Changes in positions of fishing ground could be such that the spread decreases, strengthening the one-plant solution.

(b) Aspects weakening the one-plant solution.
   - The fishing grounds could be more spread than assumed.
   - The alternative value of labor could differ between the potential sites of the plants, making it cheap to establish several plants at low cost sites. (But still we easily get a one-plant solution at a low cost site.)
   - The H_i's can be overestimated.

28.8 Computational Experiences

We shall start by reporting our experience with the method used to solve the subproblem, i.e. the stochastic transportation model. The method used is as mentioned outlined in [17], where we reported good computational experiences. The idea is to decompose the requirement space of the transportation problem into a set of polyhedral cones. These cones represent all possible optimal bases (whichever right hand side is used) and they are all dual feasible. Going from one cone to its neighbor is equivalent to taking a dual step of the simplex algorithm.

As long as the problem at hand is relatively small, all the cones can be generated, and the method is extremely efficient. If, however, the problem is large (as ours with 15 supply points and 12 demand points) the number of cones is so huge that one can only create some of them. With unimodal distributions, our clear advice in [17] was to create cones in "circles" around the one containing the expected values of the uncertain right hand sides, since these cones in some sense are large. The example treated in this report, however, does not subsume this unimodality condition. Therefore, although we created 4000 cones, only a few percent of the right hand sides fell into these cones (with an optimal solution with a larger number of plants, the number could probably have been better, since "the expected value of the right hand side" was assumed to be "all plants open").

Each of the 996 transportation problems solved in the subproblem took approximately 0.4 second CPU time, which is reasonable for a 15×12 system. This was despite the fact that for almost all the right hand sides the dual method used to find the optimal solution if none of the 4000 cones were optimal, had to be called. Most of the time was used in this subroutine.

Note that this does not mean that we could as well have dropped the dual decomposition. The cones still represent a set of very efficient dual steps.

There are two important questions when solving the relaxed deterministic problem. One is which method to use once the problem is established. The other one, which to us seems to be extremely important and difficult, is which hyperplanes to drop. Despite the result of Murty [11], dropping all nonbinding hyperplanes is not practical at all. On the other hand, one must limit the number of hyperplanes to keep a manageable problem. The reason for the problem is the unstructured behavior of the objective function. The example below shows how the extreme points can be ordered for a very simple example.

If we solve this problem, we find the optimal solution $x_1 = 10$, $x_2 = 0$. A cut is therefore created, forcing (10,0) out of the feasible region. If a hyperplane is to be dropped on the basis of the largest slack, we will drop the plane going through (0,8) and (1,4). The next optimization will then bring us to (0,5). We therefore see that the hyperplane we dropped was the only one that would make the objective function decrease instead of increase. The hyperplane through (0,8), will again be added, and we must drop another hyperplane. With the given rule, the newly created hyperplane that removed (10,0) from the feasible

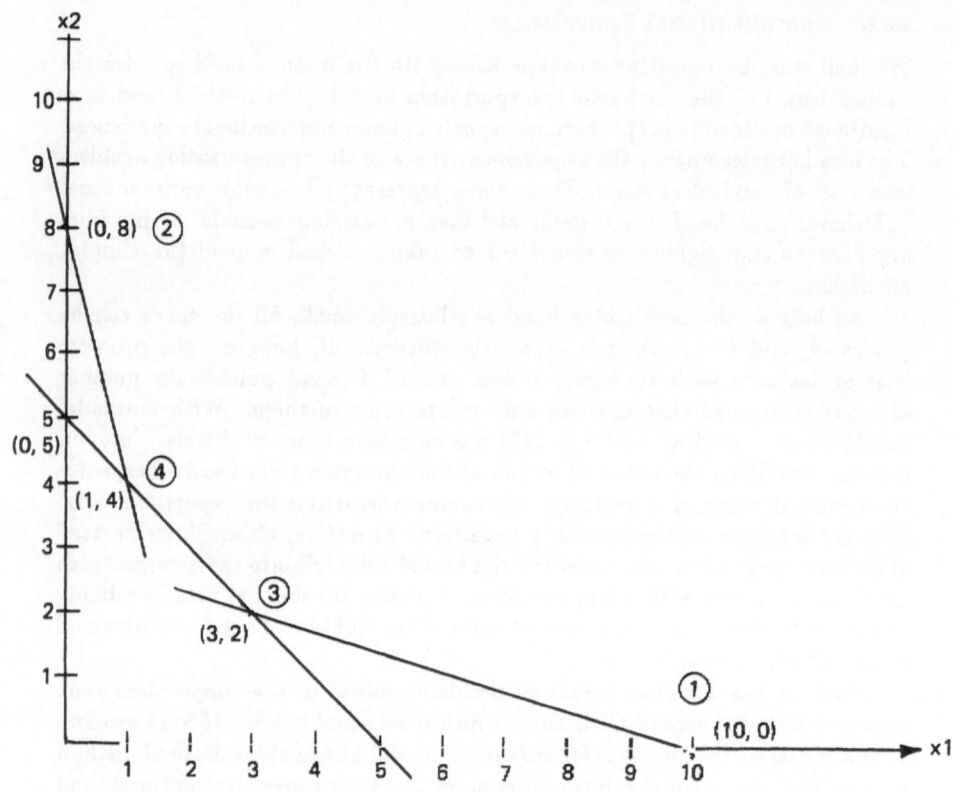

Figure 28.2.   Ordering of extreme points when $H_1 = 10$, $H_2 = 8$, $h_1 = 1$ and $h_2 = 2$.

region will be dropped. We have entered a situation where we alternate between (0,5) and (10,0).

Measures can be taken to avoid this lack of convergence, e.g., redoing the dropping if the objective function does not increase. The example, however, very well illustrates the problems inherent in this kind of objective function.

Very closely related to the problem of dropping hyperplanes is which method to use to solve the relaxed deterministic problem. The reason is that the two methods we have outlined (extreme point enumeration and solving a series of LP's) react differently with respect to increases in the number of constraints.

Instead of giving a general description of these methods, we will only outline our experience as a result of the example in the previous section, where the optimal number of plants was low.

The problem had 13 variables and 13 slacks. When employing the extreme point enumeration technique we tried to drop hyperplanes when the number of them exceeded 13. It turned out that with these 26 variables and 13 constraints it took literally hours to solve one iteration. This is not mainly due to the complexity of extreme point enumeration, but because all hyperplanes were almost parallel as the example in Figure 28.3 shows.

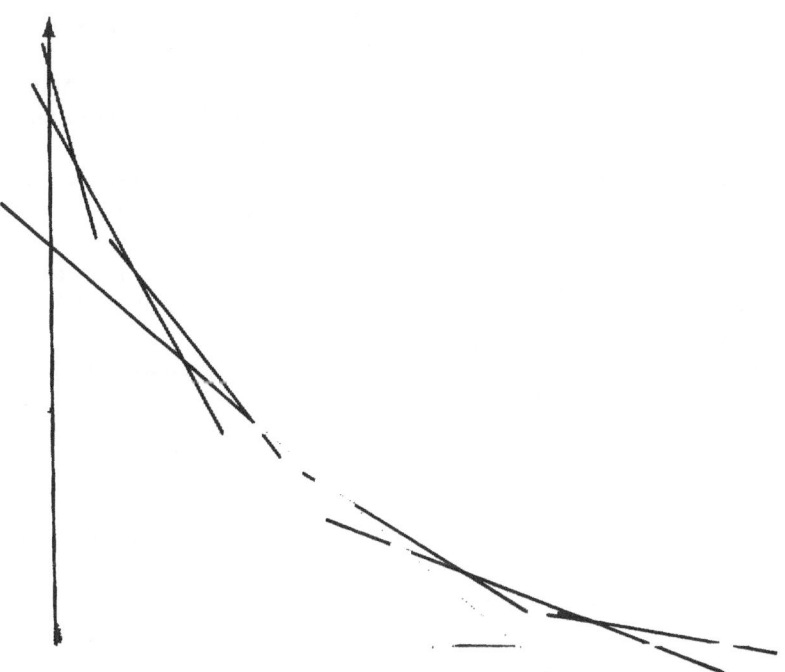

Figure 28.3. Example showing how the hyperplanes tend to become almost parallel as the iteration proceeds.

Therefore the costs $\sum h_j x_j$, on which the enumeration is based, are almost the same in the vast majority of extreme points. Thus the procedure that deletes extreme points from the list of extreme points to be examined is almost without any power, i.e. we tend to examine almost all extreme points in the polyhedron.

Note that the problems outlined above do not exclude extreme point enumeration methods when the polyhedron has a more normal form, because then the deletion of extreme points from the list is likely to be much more powerful.

It is reasonable to believe that the enumeration would have worked better if the optimal number of plants had been higher, since that would have tended to obtain fewer parallel hyperplanes.

The method improved a little when we started by defining a node 0 in order to strengthen the deletion procedure, but not very much, although we put some effort into getting a good node 0.

We also tested the method based on unimodality. Since the number of plants in the solution was low, this method was very efficient. It converged very fast even when we let the number of constraints increase to 26. With 26 constraints the main iteration converged, so we did not have to drop any hyperplanes.

If the optimal number of plants had been around $\frac{M^2-1}{2}$ this method would clearly not be very efficient since an exponential number of LP's would have had to be solved in each main iteration.

We have not tested Holm's cardinality constrained method [8], but it should be tried since it is likely to be quite efficient here even though it is also an exponential method.

We have also not yet examined the possibility of using an outer iteration scheme on k, the number of plants.

Acknowledgements

I would like to thank Roger Wets for introducing me to the area of stochastic programming and for suggesting the main ideas of this report. I would also like to thank my colleague Bengt Aspvall for his help with the proof in Section 28.5 showing that the relaxed deterministic problem is NP-hard.

References

[1] J.F. Benders, "Partitioning procedures for solving mixed-variables programming problems", *Num. Math.* **4**(1962), 238–252.

[2] E.V. Denardo, *Dynamic Programming*, Prentice-Hall, Englewood Cliffs, NJ, 1982.

[3] T. Digernes, "Fraktsatser for føringsordningen i ringnoetflåten", Report No. 662.5-1-1, Fiskeriteknologisk forskningsinstitutt, Trondheim, Norway, 1981.

[4] M.E. Dyer and L.G. Proll, "An algorithm for determining all extreme points of a convex polytope", *Math. Progr.* **12**(1977), 81–96.

[5] D. Erlenkotter, "A comparative study of approaches to dynamic location problems", *European J. Operational Res.* **6**(1981), 133–143.

[6] P.M. Franca and H.P.L. Luna, "Solving stochastic transportation location problems by generalized Benders decomposition", *Transportation Science* **16**(1982), 113–126.

[7] M.R. Garey and D.S. Johnson, *Computers and interactability. A guide to the theory of NP-completeness*, W.H. Freeman and Company, San Francisco, CA, 1979.

[8] S. Holm, "An algorithm for the cardinality constrained LP", in: *Proceedings from Nordic Symposium on Linear Complementarily Problems and Related Areas*, T.Larson and P.A. Smeds (eds.), Linköping Institute of Technology, Sweden, January 1983, pp. 60–71.

[9] A.S. Manne (ed.), *Investments for Capacity Expansion: Size, Location and Time Phasing*, The MIT Press, Cambridge, MA, 1967.

[10] K.G. Murty, "Solving the fixed charge problem by ranking the extreme points", *Oper. Res.* 16(1968), 268–279.

[11] K.G. Murty, "Linear programming under uncertainty: A basic property of the optimal solution", *Z. Warscheinlichkeitstheorie Verw. Geb.* 10(1968), 284–288.

[12] L. Nazareth and R. Wets, "Algorithms for stochastic programs: The case of nonstochastic tenders", IIASA Working Paper, WP-83-5, Laxenburg, Austria, 1983.

[13] R.M. Van Slyke and R. Wets, "*L*-shaped linear programs with applications to optimal control and stochastic programming", *SIAM J. Appl. Math.* 17(1969), 638–663.

[14] S.W. Wallace, "Enumeration algorithm in linear programming", Report in informatics No. 3, University of Bergen, Norway, 1982.

[15] S.W. Wallace, "Økonomisk beskrivelese av sildemelindustrien", CMI-report no. 822550-1, Chr. Michelsen Institute, Bergen, Norway, 1982.

[16] S.W. Wallace, "Kapasitetsreduksjoner i sildenœringen", CMI-report no. 822550-6, Chr. Michelsen Institute, Bergen, Norway, 1982.

[17] S.W. Wallace, "Decomposing the requirement space of a transportation problem into polyhedral cones", CMI-report no. 832555-8, Chr. Michelsen Institute, Bergen, Norway, 1983. To appear in *Math. Progr. Study*

[18] R. Wets, "Stochastic Programming: Solution techniques and approximation schemes", in *Mathematical Programming: The State of the Art*, eds. A. Bachem, M. Grötschel and B. Korte, Springer-Verlag, Berlin 1983, pp. 566–603.

CHAPTER 29

SOME TEST PROBLEMS FOR STOCHASTIC NONLINEAR MULTISTAGE PROGRAMS

X. de Groote, M.C. Noël and Y. Smeers

29.1 Introduction

Few algorithms exist for handling multistage nonlinear programming problems with recourse. It is thus reasonable to provide, at this stage, test problems that can be handled without sophisticated implementations but still offer a sufficiently broad range of complexity.

We propose in this paper different economic growth models that can be used for testing algorithms for nonlinear multistage programming models. All problems are variations of the nonlinear part of Manne's energy economy model ETA-MACRO ([2], [3]).

This set of test problems offers the following advantages:

(i) The models are quite simple in terms of rows and variables in each period and for each event. They have been benchmarked for the case of the European Community and thus, provide in some sense a set of (very much related) realistic problems.

(ii) The models are ranked in order of increasing complexity. This is to be meant not only in terms of the number of rows and equations, number of periods or number of events, but also with respect to the nonlinearities that they contain and the modeling of the recourse that they imply.

(iii) The data required by the models are reduced to a minimum. We provide in this paper all details necessary for setting up the problems.

The paper is organized as follows. Section 2 describes the deterministic versions of the models. Section 3 provides the required data, analyzes the numerical behavior of the different deterministic modes and introduces the construction of the stochastic versions of the problems. Section 4 is more specific to our implementation in the sense that it gives the dual of the different models; this could be relevant for other algorithms that use primal and dual information. Finally, the numerical results are discussed in the last section.

29.2   The Test Problems: Deterministic Forms

The problems considered in this paper have been constructed from the energy-economy model ETA-MACRO developed by Manne and his coauthors since 1977 ([1]; see [2] and [3] for more recent developments). ETA-MACRO assumes a two sector representation of the economy. The energy sector is described by process analysis while the rest of the economy is represented by a production function. We consider two simplified versions of ETA-MACRO; in the first one, noted B, (basic), the representation of the energy sector is reduced to the production of electric and nonelectric energy, each of them by a single activity. The second simplified version of the model, noted E (electricity), recognizes both a capital and operations variable for the production of electricity (the production of nonelectric energy being still represented by a single operations variable). Besides the fact that they lead to models with different number of constraints and variables, these problems also present variations of formulation that are interesting from the point of view of stochastic programming. In particular, the long construction time assumed for the stock of capital in power generation reduces the recourse possibilities of the energy sectors.

ETA-MACRO is formulated as a putty-clay model; perfect malleability is assumed for the new capital stock while the production structure of the old capital stock is fixed. An alternative approach, which is less realistic, is to suppose perfect malleability of the whole capital stock. The distinction, which is quite important from the point of view of economic modeling is also relevant in the context of stochastic programming where the putty-clay model offers less recourse than the putty-putty one. Each of the two models B and E will be considered a putty-clay (PC) and putty-putty (PP) version. We thus present a total of four models, which all deal with the same system but correspond to different degrees of realism, number of constraints and variables, and numerical difficulties.

We now describe these models in more details.

Model A (Basic Putty-Clay)

The output of the economy in period t, Y_t, is decomposed into a contribution due to the existing capital stock and an additional YN_t due to the capacity becoming available in t. Following ETA-MACRO the contribution YN_t is constructed as

$$YN_t = \left[a \left(EN_t^{\tilde{\beta}} NN_t^{1-\tilde{\beta}} \right)^{\tilde{\rho}} + b \left(KN_t^{\tilde{\alpha}} LN_t^{1-\tilde{\alpha}} \right)^{\tilde{\rho}} \right]^{\frac{1}{\tilde{\rho}}} \tag{A.1}$$

where EN_t, NN_t and LN_t are the inputs in electric energy, nonelectric energy and labor consumed by the new capital stock KN_t.

The total output of the economy and its consumption of capital, labor, electric and nonelectric energy are then given by the relations:

$$Y_t - \lambda Y_{t-1} - YN_t = 0 \tag{A.2}$$
$$K_t - \lambda K_{t-1} - KN_t = 0 \tag{A.3}$$
$$E_t - \lambda E_{t-1} - EN_t = 0 \tag{A.4}$$
$$N_t - \lambda N_{t-1} - NN_t = 0 \tag{A.5}$$

where λ is the decay rate of the existing capital stock over one period (usually several years).

These equations are written for all t from 2 to the end of the horizon T.

In order to link KN_t to the investments, we introduce the additional relations for $t = 2, \ldots, T$,

$$K_t - \lambda K_{t-1} - \alpha I_t - \beta I_{t-1} = 0 \tag{A.6}$$

where I_t and I_{t-1} are the investments in the current and preceding period; α and β are their respective contribution to the capital stock of period t.

As in ETA-MACRO the global output of the economy is allocated to private consumption, investments and the input of electric and nonelectric energy; this is expressed as:

$$C_t + I_t + pe_t E_t + pn_t N_t - Y_t = 0 \tag{A.7}$$

where pe_t and pn_t are respectively the unitary input of electric and nonelectric energy in period t.

In contrast with ETA-MACRO, we do not disaggregate the expressions $pe_t E_t$ and $pn_t N_t$ into their components as in an energy model. Needless to say this could be done later in order to investigate the behavior of larger stochastic models; such an extension would however go beyond the scope of this paper.

We conclude the description of this first model by giving the objective function and the terminal condition. As in ETA-MACRO, it is assumed that the system is geared by a multitemporal utility function:

$$\sum_{t=1}^{T-1} \rho^t \log C_t + \frac{\rho^T}{1-\rho} \log C_T \tag{A.8}$$

where the first $T - 1$ terms deal with private consumption during the beginning of the horizon; the last term accounts for end effects as discussed below.

End effects are dealt with by assuming that the economy is growing at a rate g after the horizon; investments in period T must be sufficient to guarantee this growth of the stock of capital after accounting for equipment decay. This is expressed by the relation:

$$-I_T + (1 + g - \lambda^A)K_T = 0. \tag{A.9}$$

where λ^A is the annual decay rate.

The model is operated on a horizon decomposed in five year periods. Data for the European Community have been adapted from Rogner et al. ([5]) and are given in the next section with a discussion of the initial conditions.

The second model (B - PP) supposes a putty-putty description of the economy; the capital stock is homogeneous and perfectly malleable in each period. This eliminates the need for distinguishing between new and old capital stock. The model is then written as follows: the global output of the economy Y_t is given as

$$Y_t = \left[a \left(E_t^{\tilde{\beta}} N_t^{1 - \tilde{\beta}} \right)^{\tilde{\rho}} = b \left(K_t^{\tilde{\alpha}} L_t^{1 - \tilde{\alpha}} \right)^{\tilde{\rho}} \right]^{\frac{1}{\tilde{\rho}}}. \tag{B.1}$$

Because the capital stock is completely malleable, we only need to describe its accumulation through time; this is done in the constraint:

$$K_t - \lambda K_{t-1} - \alpha I_t - \beta I_{t-1} = \lambda. \tag{B.2}$$

The total output of the economy is similarly allocated between private consumption, investments and input for energy; this leads to an equation identical to (A.7) that we note (B.3); the objective function (A.8) and terminal condition (A.9) are similarly unchanged and become the objective function (B.4) and terminal condition (B.5) of the new model. This is summarized below:

$$C_t + I_t + pe_t E_t + pn_t N_t - Y_t = 0 \tag{B.3}$$

$$\sum_{t=1}^{T-1} \rho^t \log C_t + \frac{\rho^T}{1 - \rho} \log C_T \tag{B.4}$$

$$-I_T + (1 + g - \lambda^A)K_T = 0. \tag{B.5}$$

As will be discussed later, it is interesting to consider a version of the model where the power generation plants are represented with their construction lead time. We accordingly introduce a new capital stock for power generation and disaggregate the input of electric energy its the fuel and investment components. Taking up the putty-clay model first (E - PC), we maintain equation (A.1) that gives the contribution of the new capital stock KN_t to the gross output of the economy:

$$YN_t = \left[a \left(EN_t^{\tilde{\beta}} NN_t^{1-\tilde{\beta}} \right)^{\tilde{\rho}} + b \left(KN_t^{\tilde{\alpha}} LN_t^{1-\tilde{\alpha}} \right)^{\tilde{\rho}} \right]^{\frac{1}{\tilde{\rho}}}. \tag{C.1}$$

The total output of the economy and its consumption of capital, labor, electric and nonelectric energy are given as in model A:

$$Y_t - \lambda Y_{t-1} - YN_t = 0 \tag{C.2}$$

$$K_t - \lambda K_{t-1} - KN_t = 0 \tag{C.3}$$

$$E_t - \lambda E_{t-1} - EN_t = 0 \tag{C.4}$$

$$N_t - \lambda N_{t-1} - NN_t = 0 \tag{C.5}$$

$$K_t - \lambda K_{t-1} - \alpha I_t - \beta I_{t-1} = 0. i \tag{C.6}$$

New relations are introduced however to describe the evolution of the power generation system and the production of electricity:

$$KE_t - \lambda_E KE_{t-1} - \alpha_E IE_t - \beta_E IE_{t-1} = 0 \quad t = 2, \dots, T \tag{C.7}$$

$$E_t - d_E KE_t \le 0 \quad t - 2, \dots, T. \tag{C.8}$$

The first relation describes the accumulation of the capital stock in the power generation sector (using a particular decay factor λ_E over the period) while (C.8) relates the production of electric energy to the installed capacity through a utilization rate d_E.

The allocation of the gross output of the economy is somewhat modified in order to account for the new representation of the power sector:

$$C_t + I_t + ce_t IE_t + \overline{pe}_t E_t + pn_t N_t - Y_t = 0 \tag{C.9}$$

where ce_t is the input in monetary units of the reference year of a unitary investment in the power generation sector; $\overline{pe}_t$ is the average fuel cost of the power generation sector.

The objective function and the terminal condition for the nonelectric capital stock are identical to those of model A:

$$\sum_{t=1}^{T-1} \rho^t \log C_t + \frac{\rho^T}{1-\rho} \log C_T \tag{C.10}$$

$$-I_T + (1 + g - \lambda^A) K_T = 0. \tag{C.11}$$

We introduce the last terminal condition for the capital stock of the electricity sector:

$$-IE_T + (1 + g - \lambda_E^A) KE_T = 0 \tag{C.12}$$

where λ_E^A is the annual decay factor of the power sector.

MODEL D (E-PP)

The last model is the putty-putty version of model C. The assumptions underlying its construction have already been discussed in the context of model B.
We only list its relations.

The global output of the economy is given by relation (B.1) and the evolution of the nonelectricity part of the capital stock by the expression (B.2)

$$Y_t = \left[a \left(E_t^{\tilde{\beta}} N_t^{1-\tilde{\beta}} \right)^{\tilde{\rho}} + b \left(K_t^{\tilde{\alpha}} L_t^{1-\tilde{\alpha}} \right)^{\tilde{\rho}} \right]^{\frac{1}{\tilde{\rho}}} \qquad (D.1)$$

$$K_t - \lambda K_{t-1} - \alpha I_t - \beta I_{t-1}. \qquad (D.2)$$

The accumulation of the power generation capacity and the relation between
the existing capacity and the production of electric energy are given by:

$$KE_t - \lambda_E KE_{t-1} - \alpha_E IE_t - \beta_E IE_{t-1} = 0 \, t = 1, \dots, T \qquad (D.3)$$

$$E_t - d_E KE_t \le 0, t = 2, \dots, T. \qquad (D.4)$$

The rest of the model consists of the allocation of the gross output of the
economy, the objective function and the end-effect conditions. Those are the
same as in model C:

$$C_t + I_t + ce_t IE_t + \overline{pe}_t E_t + pn_t N_t - Y_t = 0 \qquad (D.5)$$

$$\sum_{t=1}^{T-1} \rho^t \log C_t + \frac{\rho^T}{1-\rho} \log C_T \qquad (D.6)$$

$$-I_T + \left(1 + g - \lambda^A \right) K_T = 0 \qquad (D.7)$$

$$-IE_T + \left(1 + g - \lambda_E^A \right) KE_T = 0 \qquad (D.8)$$

In order to ease the manipulation of these models, the different relations are
listed in Table 29.1 with an indication of the model where they appear.

Table 29.1

Equations	Use in the different models			
$YN_t = \left[a \left(EN_t^{\tilde{\beta}} NN_t^{1-\tilde{\beta}} \right)^{\tilde{\rho}} + b \left(KN_t^{\tilde{\alpha}} LN_t^{1-\tilde{\alpha}} \right)^{\tilde{\rho}} \right]^{\frac{1}{\tilde{\rho}}}$	A.1	C.1		
$Y_t - \lambda Y_{t-1} - YN_t = 0$	A.2	C.2		
$K_t - \lambda K_{t-1} - KN_t = 0$	A.3	C.3		
$E_t - \lambda E_{t-1} - EN_t = 0$	A.4	C.4		
$N_t - \lambda N_{t-1} - NN_t = 0$	A.5	C.5		
$K_t - \lambda K_{t-1} - \alpha I_t - \beta I_{t-1} = 0$	A.6	B.2	C.6	D.2
$C_t + I_t + pe_t E_t + pn_t N_t - Y_t = 0$	A.7	B.3		
$-I_T + (1 + g - \lambda^A) K_T = 0$	A.9	B.5	C.11	D.7
$KE_t - \lambda_E KE_{t-1} - \alpha_E IE_t - \beta_E IE_{t-1} = 0$			C.7	D.3
$E_t - d_E + KE_t \leq 0$			C.8	D.4
$C_t + I_t + ce_t IE_t + \overline{pe}_t E_t + pn_t N_t - Y_t = 0$			C.9	D.5
$-IE_T + (1 + g - \lambda_E^A) KE_T = 0$			C.12	D.8
$Y_t = \left[a \left(E_t^{\tilde{\beta}} N_t^{1-\tilde{\beta}} \right)^{\tilde{\rho}} + b \left(k_t^{\tilde{\alpha}} L_t^{1-\tilde{\alpha}} \right)^{\tilde{\rho}} \right]^{\frac{1}{\tilde{\rho}}}$		B.1		D.1
Objective: $\sum_{t=1}^{T-1} \rho^t \log C_t + \frac{\rho^T}{1-\rho} \log C_T$	A.8	B.4	C.10	D.6

29.3   The Test Problems

Different stochastic programs can be constructed on the basis of the models of section 2. Our test problems arise from considering the following economic situation. In the present oil glut, it is expected that energy prices will remain weak for some time, with the possible consequence that exploration activity may decrease in the near future; this could lead to a renewed dependence on OPEC with possibly a new tightening of the market in the mid-nineties. Our test problems are attempts to formalize the question of whether the economy should adapt right now to possible price increases in the future or should wait until they occur. In order to model that problem we shall assume that the evolution of the oil prices over the horizon is random and that it can be represented in extensive form by a binary tree. The two branches originating from a node respectively correspond to high and low price increases during the period. The tree is rooted in year 1980 and extends over 5, 7 or 9 five-year periods, depending on the problem. The price growths are 0 and 4 occurring with an even probability. This corresponds to a median growth rate of 2IEW ([7]). This evolution is the only random element of the model; all other factors are supposed to be perfectly known; they can be described as follows.

The initial values of the capital stock (electric KE_1 and nonelectric K_1) are given as well as the consumption of electric (E_1) and nonelectric N_1 energy and the gross output of the economy (Y_1) in the first period. Also known are the initial investments I_1 and IE_1. All these are given in Table 29.2 with the price and cost assumptions.

The values of the other coefficients result from plain assumptions or from equilibrium conditions. The evolution of labor $(L$ and $LN)$ is exogenous in each period; its growth rate is given in Table 29.3 with various parameters appearing in the production function and the constraints. The values have been selected from ([4]), ([5]) and ([6]). The remaining coefficients in the model are obtained from a benchmark at some equilibrium year. Consider the putty-putty model first.

The first order condition,

$$\frac{\partial Y}{\partial N} = Y^{1-\tilde{\rho}} E^{\tilde{\beta}\tilde{\rho}} N^{\tilde{\rho}(1-\tilde{\beta})-1}(1-\tilde{\beta})a = pn,$$

allows one to compute the coefficient a if all other values are known. b can then be derived by difference from:

$$b = \frac{Y^{\tilde{\rho}} - a\left(E^{\tilde{\beta}}N^{1-\tilde{\beta}}\right)^{\tilde{\rho}}}{K^{\tilde{\alpha}\tilde{\rho}}}.$$

These calculations were done for the year 78 and the results summarized in Table 29.4. The coefficients derived for a putty-putty production function remain valid under a putty-clay assumption, if we assume that the consumption of the different factors is increasing at the same rate throughout.

No other numerical information is needed to specify the problems.

Table 29.2 Initial values: base year 1980

Variable	Notation	Value	Unit
Capital stock	K_1	5982.	10^9 \$75
Electric Energy	E_1	1.059	10^{12} KWH
Nonelectric Energy	N_1	21.28	quads
Gross output	Y_1	1658.	10^9 \$ 75
Price of electric energy	pe_1	25.	10^9 \$75/10^{12} KWH
Price of non-electric energy	pn_1	3.5	10^9 \$ 75/quad
Labor	L_1	1.07743	See table 4
Fuel cost of power generation sector	$\overline{pe}_1$	9.7945	10^9 \$75/10^{12} KWH
Cost of unitary investment in power generation sector	ce_1	.630853	10^9 \$75/10^6 KW
Investment in power generation sector	IE_1	25.596	10^6 KW
Capital stock in power generation sector	KE_1	25.596	10^6 KW

Table 29.3 Main coefficients of the model

Electricity value share: $\quad \tilde{\beta} = .39$
Capital value share: $\quad \tilde{\alpha} = .33$
Elasticity of substitution: $\quad \sigma = .388$
Derived value of $\tilde{\rho}$: $\quad \tilde{\rho} = \frac{\sigma - 1}{\sigma} = -1.58$
Annual decay rate of the capital stock: $\quad \lambda^A = .96$
Annual decay rate in the power sector: $\quad \lambda_E^A = .967$
Coefficients describing the accumulation
of capital: $\quad \alpha = 3; \beta = 2$
Coefficients describing the accumulation
of capital in the power sector: $\quad \alpha_E = 0; \beta_E = 5$
Social discount rate: $\quad \delta = .06$
Derived value of ρ: $\quad \rho = \left(\frac{1}{1+\delta}\right)^5 = .747$
Capacity utilization in the power sector: $\quad d_E = .0052$ (in GWH/KW)

period	1	2	3	4	5	6	7	8	and others
annual growth rate	3.4	3	3	2.7	2.7	2.5	2.5	2	

Table 29.4 Evaluation of the coefficients a and b Data for the reference equilibrium year 1978.

$$Y = 1553 \ 10^9 \ \$75$$
$$E = 1.02 \ 10^{12} \ \text{KWH}$$
$$K = 5668 \ 10^9 \ \$75$$
$$N = 21.63 \ \text{quads}$$
$$L = 1.0 \ (\text{by definition})$$
$$pn = 1.53 \ 10^9 \ \$75/\text{quads}$$

Values of a and b

$$a = .63 \ 10^{-5}$$
$$b = .80 \ 10^{-3}$$

Value of L_{80}

$$L_{80} = \left\{ \left[\frac{Y_{80}^{\tilde{\rho}} - a \left(E_{80}^{\tilde{\beta}} N_{80}^{1-\tilde{\beta}} \right)^{\tilde{\rho}}}{b} \right]^{1/\tilde{\rho}} \frac{1}{K_{80}^{\tilde{\alpha}}} \right\}^{1/(1-\tilde{\alpha})} = 1.07743.$$

29.4 The Dual Problems

The test problems introduced in the preceding sections have all been handled by applying nested decomposition on their dual. The methodological approach has been presented at length in ([8]) and ([9]) and we shall not return to it here. This section will state the extensive form of the deterministic equivalents of the test problems, present their duals and discuss some of their features.

Extensive forms are conveniently presented by referring to event trees. Let TR be this tree and i be one of its nodes. $S(i)$ is the set of successors of i and $P(i)$ its predecessor in TR. $D(i)$ is the depth of node i and π_i its probability; the root of the tree is noted l and the set of the terminal nodes L.

The writing of the dual problems involves a profit function Y^* defined as:

$$Y^* (\Pi_K, \Pi_E, \Pi_N) = \max -\Pi_K K - PI_E E - \Pi_N N + Y(K, E, N)$$

where PI_K, Π_E and Π_N are positive scalars; Y is the production function where the value of L has been fixed. Because Y is differentiable, the profit function can be computed by finding the solution vectors:

$$k^* = K(\Pi_K, \Pi_E, \Pi_N)$$
$$E^* = E(\Pi_K, \Pi_E, \Pi_N)$$
$$N^* = N(\Pi_K, \Pi_E, \Pi_N)$$

of the inverted demand system

$$\frac{\partial Y}{\partial K} = \Pi_K$$

$$\frac{\partial Y}{\partial K} = \Pi_E$$

$$\frac{\partial Y}{\partial N} = \Pi_N$$

and substituting through.

Although K^*, E^* and N^* can be computed analytically, the derivation of an explicit expression of the profit function appears cumbersome. We shall thus always evaluate $Y^*(\Pi_K, \Pi_E, \Pi_N)$ by numerical substitution and will note it as:

$$Y^*(\Pi_K, \Pi_E, \Pi_N) = -PI_K K^* - \Pi_E E^* - \Pi_N N^* + Y(K^*, E^*, N^*)$$

in the rest of the paper.

Because of the conjugacy condition, we also have:

$$\frac{\partial Y^*}{\partial \Pi_K} = -K^*$$

$$\frac{\partial Y^*}{\partial \Pi_E} = -E^*$$

$$\frac{\partial Y^*}{\partial \Pi_N} = -N^*$$

which gives the gradient of the profit function. We can now proceed to state the problems.

Basic Putty-Clay Model

Primal Problem

$$\text{minimize} \quad \sum_{\substack{i \in TR \\ i \notin L}} \pi_i \rho^{D(i)} \log C_i + \sum_{i \in L} \pi_i \frac{\rho^{D(i)}}{1-\rho} \log C_i$$

subject to
$$C_i + I_i + pe_i + E_i + pn_i N_i - Y_i \leq \quad U_{i1} \quad i \in TR$$
$$Y_i - \lambda Y_{P(i)} - YN_i \leq 0 \quad u_{i2} \quad i \in TR, i \neq 1$$
$$KN_i - K_i + \lambda K_{P(i)} \leq 0 \quad u_{i3} \quad i \in TR, i \neq 1$$
$$EN_i - E_i + \lambda E_{P(i)} \leq 0 \quad u_{i4} \quad i \in TR, i \neq 1$$
$$NN_i - N_i + \lambda N_{P(i)} \leq 0 \quad u_{i5} \quad i \in TR, i \neq 1$$
$$K_i - \lambda K_{P(i)} - \alpha I_i - \beta I_{P(i)} \leq 0 \quad u_{i6} \quad i \in TR, i \neq 1$$
$$-I_i + (1+g-\lambda^A)K_i \leq 0 \quad u_{i7} \quad i \in L$$

Dual Problem

minimize $\displaystyle\sum_{\substack{i\in TR \\ i\notin L}} -\rho^{D(i)}\pi_i\left[1+\log\frac{u_{i1}}{\pi_i\rho^{D(i)}}\right]$

$\displaystyle\sum_{i\notin L} -\frac{\rho^{D(i)}\pi_i}{1-\rho}\left[1+\log\frac{1-\rho}{\pi_i\rho^{D(i)}}u_{i1}\right] - [I_1+pe_1E_1+pn_1N_1-Y_1]u_{11}$

$\displaystyle +(\beta I_1+\lambda K_1)\left(\sum_{i\in S(1)}u_{i6}\right) - \lambda N_1\left(\sum_{i\in S(1)}u_{i5}\right)$

$\displaystyle -\lambda N_1\left(\sum_{i\in S(1)}u_{i4}\right) - \lambda K_1\left(\sum_{i\in S(1)}u_{i3}\right) + \lambda Y_1\left(\sum_{i\in S(1)}u_{i2}\right)$

$\displaystyle +\sum_{\substack{i\in TR \\ i\neq 1}} [u_{i2}YN_i - u_{i3}KN_i^* - u_{i4}EN_i^* - u_{i5}NN_i^*]$

subject to $\displaystyle u_{i1}-\alpha u_{i6}-\beta\sum_{j\in o(i)} u_{j6}\geq 0 \quad i\in TR, i\neq 1, i\notin L$

$\displaystyle pe_i u_{i1}-u_{i4}\lambda\sum_{j\in o(i)} u_{j4}\geq 0 \quad i\in TR, i\neq 1, i\notin L$

$\displaystyle pn_i u_{i1}-u_{i5}\lambda\sum_{j\in o(i)} u_{j5}\geq 0 \quad i\in TR, i\neq 1, i\notin L$

$\displaystyle u_{i1}-u_{i2}\lambda\sum_{j\in o(i)} u_{j2}\geq 0 \quad i\in TR, i\neq 1, i\notin L$

$\displaystyle u_{i6}\lambda\sum_{j\in o(i)} u_{j6}-u_{i3}+\lambda\sum_{j\in S(i)} u_{j3}\geq 0 \quad i\in TR, i\neq 1, i\notin L$

$u_{i1}-\alpha u_{i6}-u_{i7}\geq 0 \quad i\in L$

$pe_i u_{i1}-u_{i4}\geq 0 \quad i\in L$

$pn_i u_{i1}-u_{i5}\geq 0 \quad i\in L$

$-u_{i1}-u_{i2}\geq 0 \quad i\in L$

$u_{i6}-u_{i3}+(1+g-\lambda^A)u_{i7}\geq 0 \quad i\in L$

where KN_i^*, EN_i^*, and NN_i^* are computed using the prices $\frac{u_{i3}}{u_{i2}}, \frac{u_{i4}}{u_{i2}}$, and $\frac{u_{i5}}{u_{i2}}$ respectively.

Basic Putty-Putty Model
Primal Problem

$$\text{minimize} \quad \sum_{\substack{i \in \text{TR} \\ i \notin L}} \pi_i \rho^{D(i)} \log C_i + \sum_{i \in L} \pi_i \frac{\rho^{D(i)}}{1-\rho} \log C_i$$

$$\text{subject to} \quad C_i + I_i + pe_i E_i + pn_i N_i - Y_i \leq 0 \quad u_{i1} \quad i \in \text{TR}$$

$$k_i - \lambda K_{P(i)} - \alpha I_i - \beta I_{P(i)} \leq 0 \quad u_{i2} \quad i \in \text{TR}, i \neq 1$$

$$-I_i + (1 + g - \lambda A) K_i \leq 0 \quad u_{i3} \quad i \in L$$

Dual Problem

$$\text{maximize} \quad \sum_{\substack{i \in TR \\ i \notin L}} \pi_i \rho^{D(i)} \pi_i \left[1 + \log \frac{u_{i1}}{\pi_i \rho^{D(i)}} \right]$$

$$+ \sum_{i \in L} - \frac{\rho^{D(i)} \pi_i}{1-\rho} \left[1 + \log \left(\frac{1-\rho}{\pi_i \rho^{D(i)}} u_{i1} \right) \right]$$

$$- [I_1 + pi_1 E_1 + pn_1 N_1 - Y_1] u_{11} + \beta I_1 + \lambda K_1 \left(\sum_{i \in S(1)} u_{i2} \right)$$

$$+ \sum_{\substack{i \in TR \\ i \neq 1}} u_{i1} \left[Y_i - \frac{d_i}{u_{i1}} K_i^* - pe_i E_i^* - pn_i N_i^* \right]$$

$$\text{subject to} \quad u_{i1} - \alpha u_{i2} - \beta \sum_{j \in S(i)} u_{j2} \geq 0 \quad i \in TR, i \neq 1, i \notin L$$

$$d_i - u_{i2} + \lambda \sum_{j \in S(i)} u_{j2} = 0 \quad i \in TR, i \neq 1, i \notin L$$

$$u_{i1} - \alpha u_{i2} - u_{i3} \geq 0 \quad i \in L$$

$$d_i - u_{i2} - (1 + g - \lambda^A) u_{i3} = 0 \quad i \in L$$

where K_i^*, E_i^* and N_i^* are computed using the prices $\frac{d_i}{u_{i1}}$, pe_i and pn_i.

Electricity Putty-Clay Model
Primal Problem

maximize $\displaystyle\sum_{\substack{i\in TR \\ i\notin L}} \pi_i \rho^{D(i)} \log C_i + \sum_{i\in L} \pi_i \frac{\rho^{D(i)}}{1-\rho} \log C_i$

subject to $\quad C_i + I_i + cd_i IE_i + \overline{pe}_i E_i + pn_i N_i - Y_i \le 0 \qquad u_{i1} \quad i \in TR$

$Y_i - \lambda Y_{P(i)} - Y N_i \le 0 \qquad u_{i2} \quad i \in TR, i \ne 1$

$KN_i - K_i + \lambda K_{P(i)} \le 0 \qquad u_{i3} \quad i \in TR, i \ne 1$

$EN_i - E_i + \lambda E_{P(i)} \le 0 \qquad u_{i4} \quad i \in TR, i \ne 1$

$NN_i - N_i + \lambda N_{P(i)} \le 0 \qquad u_{i5} \quad i \in TR, i \ne 1$

$K_i - \lambda K_{P(i)} - \alpha I_i - \beta I_{P(i)} \le 0 \qquad u_{i6} \quad i \in TR, i \ne 1$

$E_i - d_E KE_i \le 0 \qquad u_{i7} \quad i \in TR, i \ne 1$

$KE_i - \lambda_E KE_{P(i)} - \alpha_E IE_i - \beta_E IE_{P(i)} \le 0 \qquad u_{i8} \quad i \in TR, i \ne 1$

$-I_i + (1 + g - \lambda^A) K_i \le 0 \qquad u_{i9} \quad i \in L$

$-IE_i + (1 + g - \lambda_E^A) KE_i \le 0 \qquad u_{i10} \quad i \in L$

Dual Problem

minimize
$$\sum_{\substack{i \in TR \\ i \notin L}} -\rho^{D(i)} \pi \left[1 + \log \frac{u_{i1}}{\pi_i \rho^{D(i)}} \right]$$

$$+ \sum_{i \in L} -\frac{\rho^{D(i)} \pi_i}{1 - \rho} \left[1 + \log \left(\frac{1 - \rho}{\pi_i \rho^{D(i)}} u_{i1} \right) \right]$$

$$- [I_1 + ce_1 IE_1 + \overline{pe}_1 E_1 + pn_1 N_1 - Y_1] u_{11} + (\beta I_1 + \lambda K_1) \left(\sum_{i \in S(1)} u_{i6} \right)$$

$$- \lambda N_1 \left(\sum_{c \in S(1)} u_{i5} \right) - \lambda E_1 \left(\sum_{i \in S(1)} u_{i4} \right) - \lambda K_1 \left(\sum_{i \in S(1)} u_{i3} \right)$$

$$+ \lambda Y_1 \left(\sum_{i \in S(1)} u_{i2} \right) + (\beta_E IE_1 + \lambda_E KE_1) \left(\sum_{i \in S(1)} u_{i8} \right)$$

$$+ \sum_{\substack{i \in TR \\ i \neq 1}} [u_{i2} Y N_i - u_{i3} K N_i^* - u_{i4} E N_i^* - u_{i5} N N_i^*]$$

subject to
$$cd_i u_{i1} - \alpha_e u_{i8} - \beta_j \sum_{j \in S(i)} u_{j8} \geq 0 \qquad i \in TR, i \neq 1, i \notin L$$

$$u_{i8} - d_E u_{i7} - \lambda_E \sum_{j \in S(i)} u_{j8} \geq 0 \qquad i \in TR, i \neq 1, i \notin L$$

$$u_{i1} - \alpha u_{i6} - \beta \sum_{j \in S(i)} u_{j6} \geq 0 \qquad i \in TR, i \neq 1, i \notin L$$

$$\overline{pe}_i u_{i1} - u_{i4} + \lambda \sum_{j \in S(i)} u_{j4} + u_{i7} \leq 0 \qquad i \in TR, i \neq 1, i \notin L$$

$$pn_i u_{i1} - u_{i5} + \lambda \sum j \in S(i) u_{j5} \geq 0 \qquad i \in TR, i \neq 1, i \notin L$$

$$- u_{i1} + u_{i2} - \lambda \sum_{j \in S(i)} u_{j2} \geq 0 \qquad i \in TR, i \neq 1, i \notin L$$

$$u_{i3} + \lambda \sum_{j \in S(i)} u_{j3} + u_{i6} - \lambda \sum_{j \in S(i)} u_{j6} \geq 0 \qquad i \in TR, i \neq 1, i \notin L$$

$$ce_i u_{i1} - \alpha_E u_{i8} - u_{i10} \geq 0 \qquad i \in L$$

$$u_{i8} - d_E u_{i7} + (1 + g - \lambda_E^A) u_{i10} \geq 0 \qquad i \in L$$

$$u_{i1} - \alpha u_{i6} - u_{i9} \geq 0 \qquad i \in L$$

$$pe_i u_{i1} - u_{i4} + u_{i7} \geq 0 \qquad i \in L$$

$$pn_i u_{i1} - u_{i5} \geq 0 \qquad i \in L$$

$$u_{i1} + u_{i2} \geq 0 \qquad i \in L$$

$$- u_{i3} + u_{i6} + (1 + g - \lambda^A) u_{i9} \geq 0 \qquad i \in L$$

where KN_i^*, EN_i^* and NN_i^* are computed using the prices $\frac{u_{i3}}{u_{i2}}$, $\frac{u_{i4}}{u_{i2}}$ and $\frac{u_{i5}}{u_{i2}}$, respectively.

Electricity Putty-putty Model
Primal Problem

maximize $\displaystyle\sum_{\substack{i \in TR \\ i \notin L}} \pi_i \rho^{D(i)} \log C_i + \sum_{i \in L} \pi \frac{\rho^{D(i)}}{1 - \rho} \log C_i$

subject to
$C_i + I_i + ce_i IE_i + \overline{pe}_i E_i + pn_i N_i - Y_i \leq 0 \quad u_{i1} \quad i \in TR$

$K_i - \lambda K + P(i) - \alpha I_i - \beta I_{P(i)} \leq 0 \quad u_{i2} \quad i \in TR, i \neq 1$

$E_i - d_E KE_i \leq 0 \quad u_{i3} \quad i \in TR, i \neq 1$

$KE_i - \lambda_E KE_{P(i)} - \alpha_E IE_i - \beta_E IE_{P(i)} \leq 0 \quad u_{i4} \quad i \in TR, i \neq 1$

$-I_i + (1 + g - \lambda^A) K_i \leq 0 \quad u_{i5} \quad i \in L$

$-IE_i + (1 + g - \lambda_E^A) KE_i \leq 0 \quad u_{i6} \quad i \in L$

Dual Problem

$$\text{minimize} \quad \sum_{\substack{i \in TR \\ i \notin L}} -\rho^{D(i)}\pi_i \left[1 + \log\left(\frac{u_{i1}}{\pi_i \rho^{D(i)}}u_{i1}\right)\right]$$

$$+ \sum_{i \in L} -\frac{\rho^{D(i)}\pi_i}{1-\rho}\left[1 + \log\left(\frac{1-\rho}{\pi_i \rho^{D(i)}}u_{i1}\right)\right]$$

$$[I_1 + ce_1 IE_1 + \overline{pe}_1 E_1 + pn_1 N_1 - Y_1]u_{11}$$

$$+ (\beta I_1 + \lambda K_1)\left(\sum_{j \in S(1)} u_{j2}\right)$$

$$+ (\beta_E IE_1 + \lambda_E KE_1)\left(\sum_{j \in S(1)} u_{j4}\right)$$

$$+ \sum_{\substack{i \in TR \\ i \neq 1}} u_{i1}\left[y_i - \frac{d_i}{u_{i1}}K_i^* - \left(\overline{pe}_i + \frac{u_{i3}}{u_{i1}}\right)\tilde{E}_i^* - pn_i N_i^*\right]$$

subject to
$$u_{i1} - \alpha u_{i2} - \beta \sum_{j \in S(i)} u_{j2} \geq 0 \quad i \in TR, i \neq 1, i \notin L$$

$$d_i - u_{i2} + \lambda \sum_{j \in S(i)} u_j 2 = 0 \quad i \in TR, i \neq 1, i \notin L$$

$$-d_E u_{i3} + u_{i4} - \lambda_E \sum_{j \in S(i)} u_{j4} \geq 0 \quad i \in TR, i \neq 1, i \notin L$$

$$ce_i u_{i1} - \alpha_E u_{i4}\beta_E \sum_{j \in S(i)} u_{j4} \geq 0 \quad i \in TR, i \neq 1, i \notin L$$

$$u_{i1} - \alpha u_{i2} - u_{i5} \geq 0 \quad i \in L$$

$$d_i - u_{i2} - (1 + g - \lambda^A)u_{i5} = 0 \quad i \in L$$

$$-d_E u_{i3} + u_{i4} + (1 + g - \lambda_E^A)u_{i6} \geq 0 \quad i \in L$$

$$cd_i u_{i1} - \alpha_E u_{i4} - u_{i6} \geq 0 \quad i \in L$$

where K_i^*, E_i^* and N_i^* are computed using the prices $\frac{d_i}{u_{i1}}$, $\left(\overline{pe}_i + \frac{u_{i3}}{u_{i1}}\right)$ and pn_i, respectively.

29.5   Numerical Experiments

This section presents some of the results obtained by applying nested decomposition on the test problems. Recall here that we deal with four models that we run on horizons of 5, 7 and 9 periods. The size of the resulting problems given in Table 29.5.

Nested decomposition proceeds by a sequence of cycles and can be stopped when the relative error between the current objective function value and the lower bound generated by the algorithm is sufficiently small or when no proposition is generated at some cycle. Table 29.6 reports the overall convergence properties of the method.

Although these results certainly appear reasonable if one considers the size of the problems, one should keep in mind that they do not give a complete overview of the method. This is illustrated by considering the evolution, through the algorithm, of the objective function and of some of the variables of the problem BPP with seven periods (see Table 29.7). It can be seen that while the objective function converges rather quickly (cycle 10), all the cycles are necessary in order to achieve convergence of the primal variables.

It is clearly impossible to list the complete optimal solution of those different problems. In order to provide some references for future numerical experiments, we report in the end of this section the optimal solution of some of the variables for the four first periods. The numbering of the nodes is as given in Figure 29.1.

Table 29.5   Size of the Test Problems

	Primal Problem			Dual Problem		
	Linear Constraints	Nonlinear Constraints	Variables	Constraints	Linear Variables	Nonlinear Variables
B-PP-5	47	30	151	60	46	61
B-PP-7	191	126	631	252	190	253
B-PP-9	767	510	2551	1020	766	1021
E-PP-5	123	30	211	120	92	91
E-PP-7	5078	126	883	504	380	379
E-PP-9	2043	510	3571	2040	1532	1531
B-PC-5	167	30	271	150	46	151
B-PC-7	695	126	1135	630	190	631
B-PC-9	2807	510	4591	2550	766	2551
E-PC-5	243	30	331	210	122	151
E-PC-7	1011	126	1387	882	506	631
E-PC-9	4083	510	5611	3570	2042	2551

Table 29.6 Stopping Condition

	BPP	BPC	EPP	EPC
5 periods deterministic	18 0.0	12 $.45\ 10^{-7}$	27 $.46\ 10^{-7}$	22 $.9\ 10^{-7}$
5 periods stochastic	18 $.46\ 10^{-7}$	12 $.46\ 10^{-7}$	28 $.9\ 10^{-7}$	23 $.46\ 10^{-7}$
7 periods deterministic	21 $.46\ 10^{-7}$	14 $.45\ 10^{-7}$	35 $.45\ 10^{-7}$	32 $.9\ 10^{-7}$
7 periods stochastic	20 $.9\ 10^{-7}$	14 $.45\ 10^{-7}$	40 $.9\ 10^{-7}$	31 $0.\ 10^{-7}$
9 periods deterministic	25 $.9\ 10^{-7}$	16 $.45\ 10^{-7}$	41 $.9\ 10^{-7}$	37 $.45\ 10^{-7}$
9 periods stochastic	20 $.8\ 10^{-5}$	13 $.9\ 10^{-7}$	31 $.9\ 10^{-5}$	22 $2.9\ 10^{-4}$

N.B. The first number refers to the number of cycles, the second to the relative error.

Table 29.7 Convergence of some of the Variables. Problem BPP 7 Periods.

Cycle	Objective Function Value	Lower Bound	Investment in a Node of Period 2	Investment in a Node of Period 4
	22.	22.		
10	034908	021512	235.7	362.7
11	029445	024762	206.2	387.6
12	027981	024917	185.5	391.3
13	026600	024993	220.0	392.7
14	025326	025033	197.6	389.3
15	025193	025049	204.1	394.3
16	025113	025071	207.0	393.0
17	025096	025074	203.1	392.6
18	025082	025074	204.4	393.0
19	025078	025075	205.2	393.6

Table 29.8 Optimal Objective Value

Problems	Objective Value
BPP 5 Periods	21.874876
BPP 7 Periods	22.025077
BPP 9 Periods	22.100000
BPC 5 Periods	21.852174
BPC 7 Periods	22.004176
BPC 9 Periods	22.182961
EPP 5 Periods	21.894914
EPP 7 Periods	22.044848
EPP 9 Periods	22.119711
EPC 5 Periods	21.871362
EPC 7 Periods	22.023133
EPC 9 Periods	22.102913

Table 29.9 BPP 5 Periods: Optimal Solution

Nodes	K (10^9\$75)	Y (10^9\$75)	N (quads)	E (10^{12} KWH)	I (10^9\$75)	C (10^9\$75)
1	5982.0	1658.0	21.28	1.059	340.0	1217.0
2	6169.1	1859.5	16.12	1.443	203.8	1563.1
3	6336.6	2071.2	17.96	1.607	299.6	1668.6
4	6909.3	2353.1	20.40	1.826	381.1	1854.9
5	6878.3	2323.7	17.97	1.957	370.8	1847.4
6	6321.9	2064.4	15.83	1.724	294.7	1659.2
7	6865.1	2342.2	17.96	1.956	373.6	1843.2
8	6831.4	2332.0	15.81	2.095	362.4	1835.3
9	6153.3	1853.2	14.21	1.547	198.6	1555.4
10	6301.2	2062.2	15.81	1.722	295.6	1656.2
11	6857.3	2341.3	17.95	1.955	376.1	1839.9
12	6821.5	1330.9	15.81	2.094	364.2	1832.5
13	6275.5	2053.8	13.93	1.845	287.0	1648.5
14	6796.5	2328.1	15.79	2.092	368.5	1325.5
15	6759.8	2317.1	13.89	2.240	356.3	1817.2

Table 29.10 BPP 7 Periods: Optimal Solution

Nodes	K (10⁹$75)	Y (10⁹$75)	N (quads)	E (10¹² KWH)	I (10⁹$75)	C (10⁹$75)
1	5982.0	1658.0	21.28	1.059	340.0	1217.0
2	6173.1	1859.9	16.12	1.443	205.2	1562.2
3	6357.6	2073.5	17.98	1.609	304.6	1665.7
4	6973.9	2360.3	20.46	1.832	393.6	1849.3
5	6950.3	2351.7	18.03	1.964	385.8	1840.1
6	6341.5	2066.5	15.84	1.726	299.2	1656.7
7	6932.6	2349.8	18.02	1.962	387.8	1836.2
8	6909.5	2340.8	15.87	2.103	380.1	1825.9
9	6156.5	1853.5	14.21	1.548	199.6	1554.7
10	6317.1	2063.9	15.82	1.723	299.3	1654.1
11	6911.8	2347.4	18.00	1.960	387.5	1834.3
12	6892.1	2338.8	15.86	2.101	380.9	1823.2
13	6299.7	2056.4	13.94	1.847	293.5	1644.5
14	6873.1	2336.7	15.84	2.099	383.1	1819.0
15	6847.3	2326.9	13.95	2.249	374.6	1808.2

Table 29.11 BPP 9 Periods: Optimal Solution

Nodes	K (10⁹$75)	Y (10⁹$75)	N (quads)	E (10¹² KWH)	I (10⁹$75)	C (10⁹$75)
1	5982.0	1658.0	21.28	1.059	340.0	1217.0
2	6151.4	1857.7	16.11	1.442	198.0	1567.3
3	6345.1	2072.2	17.97	1.608	311.2	1657.9
4	699.3	2362.5	20.48	1.833	399.1	1845.9
5	6975.4	2354.5	18.05	1.966	393.1	1835.4
6	6321.9	2064.4	15.83	1.734	303.4	1650.5
7	6960.5	2352.9	18.04	1.965	399.6	1827.3
8	6912.0	2341.0	15.87	2.103	383.5	1822.8
9	6153.9	1853.3	14.21	1.547	198.8	1555.3
10	6322.7	2064.5	15.83	1.724	302.5	1651.6
11	6912.4	2347.5	18.00	1.960	384.09	187.8
12	6919.4	2341.9	15.88	2.104	386.4	1820.6
13	6297.7	2056.2	13.94	1.847	294.1	1643.7
14	6873.9	2336.8	15.84	2.099	383.5	1818.7
15	6844.7	2326.6	13.95	2.249	373.8	1808.6

Table 29.12   BPC 5 Periods: Optimal Solution

Nodes	K (10^9\$75)	Y (10^9\$75)	N (quads)	E (10^{12} KWH)	I (10^9\$75)	C (10^9\$75)
1	5982.0	1658.0	21.28	1.059	340.0	1217.0
2	6482.1	1895.1	21.68	1.307	308.2	1478.4
3	7034.7	2144.4	22.51	1.551	377.7	1649.2
4	7702.2	2436.4	24.06	1.813	403.6	1903.3
5	7684.6	2432.6	23.37	1.851	397.8	1889.0
6	7021.7	2141.3	21.93	1.585	373.3	1635.9
7	7670.0	2429.6	22.89	1.879	399.4	1885.8
8	6751.1	2425.5	22.29	1.920	393.1	1869.9
9	6449.8	1892.3	21.16	1.338	304.1	1464.7
10	7007.8	2138.6	21.50	1.610	374.8	1632.0
11	7660.5	2427.6	22.54	1.899	399.0	1885.2
12	7641.6	2423.5	29.94	1.940	392.7	1868.7
13	6993.5	2135.3	20.99	1.646	370.0	1615.4
14	7625.6	2420.3	21.52	1.969	394.4	1865.2
15	7605.2	2415.9	20.98	2.013	38716	1845.7

Table 29.13   BPC 7 Periods: Optimal Solution

Nodes	K (10^9\$75)	Y (10^9\$75)	N (quads)	E (10^{12} KWH)	I (10^9\$75)	C (10^9\$75)
1	5982.0	1658.0	21.28	1.059	340.0	1217.0
2	6491.0	1896.1	21.66	1.309	311.2	1476.3
3	7036.7	2144.3	22.41	1.557	373.9	1653.1
4	7755.7	2441.7	23.79	1.836	423.4	1889.1
5	7740.5	2438.2	23.13	1.875	418.4	1874.4
6	7023.4	2141.2	21.83	1.592	369.5	1639.0
7	7725.5	2435.2	22.66	1.903	420.0	1871.1
8	7708.2	2431.2	22.07	1.945	414.2	1854.0
9	6479.1	1893.3	21.14	1.341	307.2	1462.6
10	7009.7	2138.5	21.41	1.617	370.8	1636.1
11	7714.9	2433.0	22.31	1.924	419.2	1870.0
12	7697.8	2429.1	21.73	1.966	413.5	1853.8
13	6995.2	2135.1	20.90	1.654	366.0	1619.5
14	7681.0	2425.8	21.31	1.995	415.1	1850.3
15	7662.5	2421.5	20.79	2.040	409.0	1830.5

Table 29.14 BPC 9 Periods: Optimal Solution

Nodes	K (10^9\$75)	Y (10^9\$75)	N (quads)	E (10^{12} KWH)	I (10^9\$75)	C (10^9\$75)
1	5982.0	1658.0	21.28	1.059	340.0	1217.0
2	6564.8	1904.0	21.68	1.318	335.7	1459.4
3	7265.4	2169.1	22.49	1.584	413.7	1637.1
4	8282.3	2498.2	24.01	1.896	510.3	1856.5
5	8258.5	2493.8	23.33	1.938	502.3	1843.6
6	7247.6	2165.6	21.91	1.620	407.8	1624.0
7	8237.2	2490.2	22.85	1.966	504.1	1839.6
8	8211.2	2485.4	22.25	2.010	495.4	1824.5
9	6550.2	1901.0	21.16	1.350	330.9	1446.2
10	7230.5	2162.5	21.48	1.645	409.3	1620.6
11	8223.8	2487.7	22.50	1.987	503.2	1839.0
12	8197.9	2483.0	21.90	2.031	494.6	1824.1
13	7210.8	2158.6	20.97	1.683	402.7	1605.2
14	8174.5	2479.0	21.47	2.061	496.5	1819.7
15	8146.0	2473.8	20.94	2.108	487.0	1802.1

Table 29.15 EPP 5 Periods: Optimal Solution

Nodes	K (10^9\$75)	Y (10^9\$75)	N (quads)	E (10^{12} KWH)	I (10^9\$75)	C (10^9\$75)	KE (10^6 KW)	IE (10^6 KW)
1	5982.0	1658.0	21.28	1.059	340.0	1217.0	201.48	25.59
2	6173.6	1861.5	15.73	1.567	205.3	1576.4	298.05	14.84
3	6350.3	2074.0	17.64	1.712	302.0	1682.9	325.76	16.90
4	6918.7	2355.0	20.21	1.889	379.0	1873.6	359.46	20.86
5	6881.6	2343.1	18.16	1.889	366.6	1864.2	359.46	26.04
6	6317.9	2063.9	15.86	1.712	291.1	1674.3	325.76	22.45
7	6857.9	2342.4	17.74	2.035	374.7	1858.4	387.20	21.89
8	6815.2	2329.4	15.93	2.035	360.5	1848.8	387.20	27.96
9	6147.6	1852.9	14.15	1.567	196.7	1567.5	298.05	20.85
10	6304.4	2064.4	15.43	1.870	299.5	1670.7	355.83	16.25
11	6861.1	2342.4	17.82	2.006	373.9	1858.4	281.58	23.06
12	6818.4	2329.3	16.01	2.005	359.7	1848.7	281.58	29.09
13	6270.9	2053.7	13.87	1.870	288.3	1661.2	355.83	22.09
14	6798.1	2329.1	15.64	2.159	369.4	1842.5	410.78	23.76
15	6752.1	2315.1	14.04	2.159	354.1	1832.1	410.78	30.50

Table 29.16 EPP 7 Periods: Optimal Solution

Nodes	K $(10^9\$75)$	Y $(10^9\$75)$	N (quads)	E $(10^{12}$ KWH$)$	I $(10^9\$75)$	C $(10^9\$75)$	KE $(10^6$ KW$)$	IE $(10^6$ KW$)$
1	5982.0	1658.0	21.28	1.059	340.0	1217.0	201.48	25.59
2	6174.8	1861.6	15.74	1.567	205.7	1575.1	299.04	16.28
3	6369.0	2076.5	17.54	1.750	307.6	1686.8	332.95	15.09
4	6986.1	2360.5	20.28	1.892	392.6	1867.1	356.46	21.05
5	6953.2	2351.1	18.23	1.893	381.6	1856.2	356.48	27.03
6	6339.6	2066.7	15.78	1.752	297.8	1671.3	332.95	21.09
7	6935.5	2351.9	17.83	2.031	390.2	1852.3	386.51	19.91
8	6898.8	2338.6	16.03	2.032	378.0	1841.0	386.51	26.52
9	6149.0	1853.0	14.15	1.567	197.1	1566.9	298.04	21.26
10	6318.3	2066.0	15.42	1.882	303.4	1668.0	357.98	16.69
11	6927.3	2350.0	17.84	2.026	389.6	1851.7	385.62	20.43
12	6893.2	2377.9	16.03	2.027	378.2	1839.9	385.62	26.80
13	6287.3	2055.6	13.86	1.882	293.1	1657.2	357.98	23.84
14	6876.0	2339.5	15.60	2.214	387.8	1835.5	421.35	20.13
15	6834.4	2325.0	14.01	2.215	373.9	1923.5	421.35	27.61

Table 29.17 EPP 9 Periods: Optimal Solution

Nodes	K $(10^9\$75)$	Y $(10^9\$75)$	N (quads)	E $(10^{12}$ KWH$)$	I $(10^9\$75)$	C $(10^9\$75)$	KE $(10^6$ KW$)$	IE $(10^6$ KW$)$
1	5982.0	1658.0	21.28	1.059	340.0	1217.0	201.48	25.59
2	6159.2	1860.0	15.73	1.552	200.6	1574.4	298.05	23.24
3	6376.1	2079.3	17.04	1.933	317.6	1679.0	367.76	6.57
4	6999.4	2362.8	20.59	1.804	388.4	1870.2	343.23	22.90
5	6967.2	2351.3	18.51	1.804	377.3	1858.8	343.23	29.08
6	6343.9	2069.4	15.31	1.933	306.9	1668.9	367.76	14.94
7	6956.0	2352.9	17.92	2.008	389.8	1852.7	385.11	22.87
8	6915.4	2340.3	16.06	2.024	376.3	1842.7	385.11	29.06
9	6165.2	1854.7	14.16	1.570	202.6	1566.0	298.05	16.54
10	6323.7	2065.1	15.74	1.757	297.2	1673.2	334.26	16.54
11	6900.5	2345.6	18.10	1.918	383.3	1848.5	364.84	28.46
12	6879.1	2334.8	16.28	1.918	376.2	1838.6	364.84	26.05
13	6297.1	2054.9	14.15	1.757	288.3	1661.2	334.26	23.55
14	6864.6	2335.8	15.83	2.101	384.5	1833.2	399.90	24.60
15	6825.3	2322.4	14.22	2.102	371.4	1918.3	399.90	35.76

Table 29.18 EPC 5 Periods: Optimal Solution

Nodes	K ($10^9$$75)	Y ($10^9$$75)	N (quads)	E (10^{12} KWH)	I ($10^9$$75)	C ($10^9$$75)	KE (10^6 KW)	IE (10^6 KW)
1	5982.0	1658.0	21.28	1.059	340.0	1217.0	201.48	25.59
2	6490.3	1897.2	21.43	1.410	310.9	1489.7	298.05	12.36
3	7044.0	2146.6	22.27	1.647	376.7	1663.4	313.38	19.62
4	7709.5	2438.3	23.82	1.906	404.2	1918.4	362.61	21.72
5	7688.8	2433.6	23.23	1.906	397.3	1904.0	362.61	23.45
6	7027.0	2142.6	21.78	1.647	371.0	1649.4	313.38	21.09
7	7671.5	2430.5	22.73	1.945	399.9	1900.0	369.96	23.47
8	7649.1	2425.5	22.21	1.944	392.5	1883.1	369.96	25.07
9	6477.7	1894.5	20.94	1.448	306.7	1475.0	298.05	14.89
10	7016.6	2140.8	21.29	1.714	373.8	1646.2	326.05	21.26
11	7668.4	2429.8	22.32	2.005	399.9	1900.5	381.53	23.26
12	7645.7	2424.7	21.80	2.004	392.3	1883.8	381.53	25.25
13	6998.0	2136.5	20.85	1.714	367.6	1629.9	326.05	22.57
14	7627.2	2421.3	21.37	2.039	395.4	1879.5	388.08	24.94
15	7602.8	2415.8	20.91	2.040	387.2	1859.8	388.08	26.78

Table 29.19 EPC 7 Periods: Optimal Solution

Nodes	K ($10^9$$75)	Y ($10^9$$75)	N (quads)	E (10^{12} KWH)	I ($10^9$$75)	C ($10^9$$75)	KE (10^6 KW)	IE (10^6 KW)
1	5982.0	1658.0	21.28	1.059	340.0	1217.0	201.48	25.59
2	6500.7	1898.4	21.39	1.424	314.4	1487.1	298.05	12.92
3	7047.6	2146.8	22.15	1.662	372.8	1667.7	316.20	19.73
4	7763.0	2443.5	23.58	1.921	423.7	1904.8	365.54	21.76
5	7743.4	2438.9	23.01	1.921	417.1	1890.0	365.54	23.78
6	7030.1	2142.7	21.66	1.662	366.9	1653.7	316.20	21.55
7	7728.3	2436.2	22.50	1.969	420.8	1885.4	374.64	23.71
8	7706.1	2431.1	22.00	1.969	413.4	1868.4	374.64	25.56
9	6488.5	1895.7	20.90	1.463	310.4	1472.5	298.05	15.13
10	7019.2	2140.8	21.21	1.720	369.3	1650.9	327.24	21.17
11	7720.5	2434.6	22.14	2.008	419.5	1886.2	382.06	23.68
12	7699.3	2429.6	21.64	2.008	412.5	1869.3	382.06	25.46
13	7000.4	2136.4	20.77	1.720	363.1	1634.4	327.24	22.95
14	7682.6	2426.6	21.19	2.055	416.2	1864.4	390.97	25.54
15	7659.2	2421.1	20.75	2.055	408.3	1844.7	390.97	27.29

Table 29.20 EPC 9 Periods: Optimal Solution

Nodes	K (10⁹$75)	Y (10⁹$75)	N (quads)	E (10¹²KWH)	I (10⁹$75)	C (10⁹$75)	KE (10⁶ KW)	IE (10⁶ KW)
1	5982.0	1658.0	21.28	1.059	340.0	1217.0	201.48	25.59
2	6488.6	1897.1	21.37	1.424	310.4	1475.3	298.05	36.05
3	7094.5	2154.1	21.73	1.869	394.4	1665.3	431.81	0.00
4	7836.4	2450.3	23.88	1.917	421.0	1915.8	364.49	17.72
5	7811.2	2448.8	23.24	1.916	412.6	1901.0	364.49	21.42
6	7036.8	2145.4	21.33	1.823	375.2	1643.3	431.81	28.81
7	7788.5	2446.1	21.93	2.256	433.5	1895.8	508.55	2.09
8	7778.4	2443.1	21.39	2.311	430.1	1863.4	508.55	25.48
9	6496.9	1896.7	20.90	1.469	313.1	1471.6	298.05	13.70
10	7028.1	2141.0	21.31	1.682	368.2	1653.6	320.08	19.07
11	7734.5	2434.6	22.35	1.921	422.5	1881.3	365.52	26.68
12	7725.6	2430.9	21.84	1.921	419.5	1865.9	365.52	21.36
13	7015.9	2137.6	20.87	1.684	364.1	1629.9	320.08	30.00
14	7759.9	2436.6	21.01	2.197	437.0	1855.2	420.18	22.14
15	7715.8	2429.1	20.57	2.208	422.3	1833.9	420.18	34.31

References

[1] A.S. Manne, "ETA-MACRO: A Mode of Energy-Economy Interactions", Research Project 1014, Department of Operations Research, Stanford University, Stanford, California 94305, 1977.

[2] A.S. Manne, "ETA-MACRO: A User's Guide", EA-1724 Research Progress 1014, Department of Operations Research, Stanford University, Stanford, California 94305, 1981.

[3] A.S. Manne, M.A. Beltrano, T.F. Rutherford, A.N. Svoronos and T.F. Wilson, "ETA-MACRO: A Progress Report", Research Project 1014, Department of Operations Research, Stanford University, Stanford, California 94305, 1983.

[4] A.S. Man and T. Sira, "European Energy Supplies and Demands: A Long-Term Perspective", International Energy Program, Discussion Paper Stanford University, Stanford, California 94305, 1980.

[5] H.H. Rogner, "A Long-Term Macroeconomic Equilibrium Model for the European Community", RR-82-13, International Institute for Applied Systems Analysis, Laxenburg, 1982.

[6] W. Sassis, A. Holzl, H.H. Rogner, and L.Schrattenholzer, "Fueling Europe in the Future. The Long Term Energy Problem in the EC Countries: Alternative R&D Strategies", International Institute for Applied Systems Analysis, RR-83-9/EUR 8421-EN, 1983.

[7] A.S. Manne and L. Schrattenholzer, "International Energy Workshop: A summary of the 1983 Poll Responses", International Institute for Applied Systems Analysis, 1983.

[8] M.C. Noël and Y. Smeers, "On the Use of Nested Decomposition for Solving Nonlinear Multistage Stochastic Programs", paper presented in 1983 at the IFIP Workshop on Stochastic Programming: Algorithms and Applications, Gargnano.

[9] M.C. Noël and Y. Smeers, "Nested Decomposition of Multistage Nonlinear Programs with Recourse", paper submitted to *Mathematical Programming*.

[24] C. M. Chen, "Comments on 'Grade of Service of a Multiple-Scan Queuing System with a Single Server'," *IEEE Transactions on Communications*, COM-28, No. 3, 1980.

[25] Willemain, and ..., *Time of Decomposition of Multidimensional Propagation with Dispersion*, ...

CHAPTER 30

STOCHASTIC PROGRAMMING PROBLEMS: EXAMPLES FROM THE LITERATURE

A.J. King

Introduction

This is a small collection of problems which have appeared in the stochastic programming literature over the past two and a half decades. The intention guiding the choices was to provide a number of test problems with solutions for researchers who are developing and testing algorithms. Of course anyone can jot down a stochastic linear program. This collection seeks to provide the researcher with a variety of formulations, some classical and some new and as yet unsolved, as templates from which many test problems can be generated.

The problems range from classical chance-constrained and simple recourse models to dynamic models with both chance-constrained and general recourse examples. There are some unfortunate omissions. Chief among these are Kallberg, White and Ziemba's financial planning model, Prékopa's et al STABIL model, Somlyody and Wets' Lake Balaton model, and Kall and Keller's collection of general recourse problems. For these we give references together with a brief description and classification below; these models were excluded for reasons of lack of published data and/or lack of space for presentation herein.

Finally a disclaimer: None of the solutions has been checked for accuracy. There have been many opportunities in the process of collecting and publishing these problems for errors to creep in and replicate. Beware!

We would be grateful for any corrections to these problems, and for any additions that may be proposed by the readers for future editions of this collection.

In addition some of these problems will be distributed in the special format for recourse problems (as described in this volume) on a computer tape to be distributed in early 1985 by the International Institute for Applied Systems Analysis.

AIRCRAFT ALLOCATION PROBLEM
Reference

G. Dantzig: *Linear Programming and Extensions*, Princeton University Press, 1963, pp. 572-597.

This is the classic example of a stochastic program with simple recourse.

An airline wishes to allocate airplanes of various types among its routes to satisfy an uncertain passenger demand, in such a way as to minimize operating costs plus the lost revenue from passengers turned away.

This problem will be available on the stochastic programming computer tape distributed by IIASA.

Stochastic program with simple recourse.

Choose $x_j (j = 1, \ldots, 17)$ to *minimize*

$$\sum_{j=1}^{17} c_j x_j + E\left\{\sum_{k=1}^{5} q_k \mathbf{v}_k^-\right\}$$

subject to

$$x_1 + x_2 + x_3 + x_4 + x_5 \leq b_1$$
$$x_6 + x_7 + x_8 + x_9 \leq b_2$$
$$x_{10} + x_{11} + x_{12} \leq b_3$$
$$x_{13} + x_{14} + x_{15} + x_{16} + x_{17} \leq b_4$$
$$x_j \geq 0 \quad j = 1, \ldots, 17$$
$$\mathbf{v}_k^+ \geq 0, \quad \mathbf{v}_k^- \geq 0$$
$$\mathbf{v}_1^+ - \mathbf{v}_1^- = t_1 x_1 + t_{13} x_{13} - b_1$$
$$\mathbf{v}_2^+ - \mathbf{v}_2^- = t_2 x_2 + t_6 x_6 + t_{10} x_{10} + t_{14} x_{14} - \mathbf{h}_2$$
$$\mathbf{v}_3^+ - \mathbf{v}_3^- = t_3 x_3 + t_7 x_7 + t_{15} x_{15} - \mathbf{h}_3$$
$$\mathbf{v}_4^+ - \mathbf{v}_4^- = t_4 x_4 + t_8 x_8 + t_{11} x_{11} + t_{16} x_{16} - \mathbf{h}_4$$
$$\mathbf{v}_5^+ - \mathbf{v}_5^- = t_5 x_5 + t_9 x_9 + t_{12} x_{12} + t_{17} x_{17} - \mathbf{h}_5$$

$x_1, \ldots, x_5$:	type 1 aircraft assigned to routes $1, \ldots, 5$
$x_6, \ldots, x_9$:	type 2 aircraft assigned to routes $2, \ldots, 5$
x_{10}, x_{11}, x_{12}:	type 3 aircraft assigned to routes 2, 4, 5
$x_{13}, \ldots, x_{17}$:	type 4 aircraft assigned to routes $1, \ldots, 5$
b_i:	number of aircraft available of type $i = 1, \ldots, 4$
c_j:	cost of operating aircraft/route $j = 1, \ldots, 17$
q_k:	revenue lost per passenger turned away on route $k = 1, \ldots, 5$
$\mathbf{v}_k^+$:	empty seats on route k
$\mathbf{v}_k^-$:	passengers turned away on route k
t_j:	passenger capacity on aircraft/route j
$\mathbf{h}_k$:	passenger demand for route k.

Data:

$c =$ [18, 21, 18, 16, 10, 15, 16, 14, 9, 10, 9, 6, 17, 16, 17, 15, 10]

$q =$ [13, 13, 7, 7, 1]

$b =$ [10, 19, 25, 15]

$t =$ [16, 15, 28, 23, 81, 10, 14, 15, 57, 5, 7, 29, 9, 11, 22, 17, 55]

$\mathbf{h}_k$ are discretely distributed as follows

$\mathbf{h}_1 \sim$ [200, 220, 250, 270, 300] w.p. (0.2, 0.05, 0.35, 0.2, 0.2)

$\mathbf{h}_2 \sim$ [50, 150] w.p. (0.3, 0.7)

$\mathbf{h}_3 \sim$ [140, 160, 180, 200, 220] w.p. (0.1, 0.2, 0.4, 0.2, 0.1)

$\mathbf{h}_4 \sim$ [10, 50, 80, 100, 340] w.p. (0.2, 0.2, 0.3, 0.2, 0.1)

$\mathbf{h}_5 =$ [580, 600, 620] w.p. (0.1, 0.8, 0.1)

Solution:

Calculated to one decimal place accuracy

Aircraft Type	1	2	3	4
Route				
1	$x_1 = 10$	*	*	$x_{13} = 7.4$
2	$x_2 = 0$	$x_6 = 12.8$	$x_{10} = 4.3$	$x_{14} = 0.0$
3	$x_3 = 0$	$x_7 = 0.9$	*	$x_{15} = 7.6$
4	$x_4 = 0$	$x_8 = 5.3$	$x_{11} = 0$	$x_{16} = 0$
5	$x_5 = 0$	$x_9 = 0$	$x_{12} = 20.7$	$x_{17} = 0$

CLEEF'S TEST PROBLEM

Reference

H.J. Cleef, "A solution procedure for the two-stage stochastic program with simple recourse." *Zeitschrift für O.R.*, **25** (1981) 1-13.

Stochastic program with simple recourse

choose $x_j (j = 1, \ldots, 16)$ to *minimize*

$$\sum_{j=1}^{16} c_j x_j + E\{\sum_{k=1}^{6} (q_k^+ \mathbf{v}_k^+ + q_k^- \mathbf{v}_k^-)\}$$

subject to:

$$\sum_{j=1}^{16} a_{ij} x_j = b_i \qquad i = 1, 2, 3$$

$$\sum_{j=1}^{16} t_{kj} x_j + \mathbf{v}_k^+ - \mathbf{v}_k^- = \mathbf{h}_k \qquad k = 1, \ldots, 6$$

$$x_j \geq 0, \mathbf{v}_k^+ \geq 0, \mathbf{v}_k^- \geq 0$$

Data

h   is discretely distributed as follows:

$h_1 \sim$ [3, 4, 5, 6, 7, 8, 10, 11, 12, 13, 14, 15] w.p.(.05, .08, .1, .1, .1, .1, .2, .1, .05, .05, .04, .03)

$h_2 \sim$ [5, 6, 6.5, 7.0, 7.5] w.p.(.08, .12, .4, .2, .2)

$h_3 \sim$ [0, 1.0, 2.0, 2.5, 3.0, 3.25, 3.5, 3.75, 4.0, 4.25, 4.5, 4.75] w.p.(.06, .06, .06, .06, .06, .06, .2, .18, .06, .08, .06, .06)

$h_4 \sim$ [-2.0, -1.0, 1.0, 2.0] w.p.(.13, .5, .25, .12)

$h_5 \sim$ [0.0, 1.0, 2.0, 2.5, 3.0, 6.0, 7.0, 10.0, 12.0] w.p.(.08, .12, .1, .1, .2, .1, .1, .12, .08)

$h_6 \sim$ [8.0, 24.0] w.p. (.5, .5)

Solution

$x_1 = 0$	$x_7 = 0.108757541$	$x_{13} = 0.955190541$
$x_2 = 0$	$x_8 = 0$	$x_{14} = 0.301132174$
$x_3 = 0.521568144$	$x_9 = 0$	$x_{15} = 0$
$x_4 = 0$	$x_{10} = 0.250267079$	$x_{16} = 0$
$x_5 = 0$	$x_{11} = 0.392736833$	
$x_6 = 0.203628174$	$x_{12} = 0.351412474$	

KALL AND KELLER'S COMPLETE RECOURSE PROBLEMS

Reference

E. Keller, "GENSLP: A program for generating input for stochastic linear programs with complete fixed recourse". Manuscript, Institut für Operations Research der Universität Zürich, Zürich, CH-8006.

This is a computer program which generates random general recourse problems, and is available from computer tape to be distributed by IIASA. The program was written by E. Keller under the direction of P. Kall at the Institute for Operations Research, University of Zürich.

The format of the general recourse problem they pose is as follows: choose $x \in \mathbf{R}^n$ to *minimize*

$$c^T x + E\{\min_{y} q^T y\}$$

subject to

$$x \geq 0, \qquad y \geq 0$$
$$Ax = b$$
$$\mathbf{T}x + W\mathbf{y} = \mathbf{h}$$

where A, b, c, q, W are deterministic matrices or vectors of appropriate dimension, and $\mathbf{T}, \mathbf{h}$ are random of the following form:

$$\mathbf{T} = T^0 + \tau_1 T^1 + \ldots + \tau_k T^k$$
$$\mathbf{h} = h^0 + \tau_1 h^1 + \ldots + \tau_k h^k$$

Here $(\tau_1, \ldots, \tau_k)$ is a random vector and $(T^0, \ldots, T^k)(h^0, \ldots, h^k)$ are fixed.

Keller's computer program generates the data $A, b, c, q, W, h^0, \ldots, h^k$, $T^0, \ldots, T^k$ randomly with appropriate checks for feasibility of A and W (for complete recourse). P. Kall has conducted a series of tests using the problems generated by GENSLP, to appear in 1985.

PROJECT SCHEDULING PROBLEM

Reference

H.J. Cleef, W. Gaul, "Project scheduling via stochastic programming", *Math. Operationsforsch. Statist., Ser. Optimization* **13**(1982), 449-468.

We are given a directed-graph representation of a project where arcs represent activities and nodes represent points at which choices between various activities must be made. The problem is to choose the length of completion time for each activity so that the total time consumed is less than some prespecified limit and the project cost is minimized. Activity completion times are subject to lower and upper bounds, and costs increase as completion time is lowered. In some cases the activity completion times are only estimates, and there are recourse costs once the true completion time is known: if completion time is too short there are costs associated with obtaining additional resources; if completion time is too generous then there are gains associated with resources freed to work elsewhere. This problem will be available on the IIASA stochastic programming computer tape.

Stochastic program with simple recourse

Choose $x_j (j = 1, \ldots, 25)$ to *minimize*

$$2805.0 - \sum_{j=1}^{25} c_j x_j + E\{\sum_{j=1}^{25} [q_j^+ y_j^+ + q_j^- y_j^-]\}$$

subject to

$$x_j \leq \sum_{k=1}^{15} e_{jk} \pi_k \qquad j = 1, \ldots, 25$$

$$\pi_{15} - \pi_1 \leq \gamma$$

$$x_j + y_j^+ - y_j^- = \xi_j \qquad j = 1, \ldots, 25$$

$$\ell_j \leq x_j \leq u_j; \quad y_j^+, y_j^- \geq 0 \quad j = 1, \ldots, 25$$

x_j: the scheduled length of time to complete task $j = 1, \ldots, 25$

e_{jk}:   the node-arc incidence matrix

π_k: the scheduled time at which decision node $k = 1, \ldots, 15$ is reached

γ: specified total time for project

c_j: cost of completing scheduled task j in one unit less time

ℓ_j, u_j: lower/upper bounds on time to complete task j

ξ_j: actual project completion time for task j

y_j^+:   excess time scheduled over actual time for completion of task j

y_j^-:   deficit of scheduled time over actual time for task j

q_j^+:   per unit value of excess time for task j

q_j^-:   per unit value of deficit time for task j

Data

node-arc incidence matrix e_{jk}:

Nodes

Arcs	1	2	3	4	5	6	7	8	9	10	11	12	13	14	15
1	-1	1	0	0	0	0	0	0	0	0	0	0	0	0	0
2	-1	0	1	0	0	0	0	0	0	0	0	0	0	0	0
3	0	-1	1	0	0	0	0	0	0	0	0	0	0	0	0
4	0	-1	0	0	0	0	1	0	0	0	0	0	0	0	0
5	0	-1	0	0	0	0	0	0	0	0	0	1	0	0	0
6	0	0	-1	1	0	0	0	0	0	0	0	0	0	0	0
7	0	0	-1	0	1	0	0	0	0	0	0	0	0	0	0
8	0	0	-1	0	0	1	0	0	0	0	0	0	0	0	0
9	0	0	0	-1	0	0	0	0	0	0	1	0	0	0	0
10	0	0	0	-1	0	0	0	0	0	0	0	1	0	0	0
11	0	0	0	0	-1	0	0	1	0	0	0	0	0	0	0
12	0	0	0	0	0	-1	0	0	0	0	0	1	0	0	0
13	0	0	0	0	0	-1	0	0	0	0	0	0	0	1	0
14	0	0	0	0	0	0	-1	0	0	0	0	1	0	0	0
15	0	0	0	0	0	0	-1	0	0	0	0	0	1	0	0
16	0	0	0	0	0	0	0	-1	1	0	0	0	0	0	0
17	0	0	0	0	0	0	0	0	-1	0	0	1	0	0	0
18	0	0	0	0	0	0	0	0	-1	0	0	0	1	0	0
19	0	0	0	0	0	0	0	0	0	-1	1	0	0	0	0
20	0	0	0	0	0	0	0	0	0	-1	0	1	0	0	0
21	0	0	0	0	0	0	0	0	0	0	-1	1	0	0	0
22	0	0	0	0	0	0	0	0	0	0	-1	0	0	1	0
23	0	0	0	0	0	0	0	0	0	0	0	-1	0	0	1
24	0	0	0	0	0	0	0	0	0	0	0	0	-1	0	1
25	0	0	0	0	0	0	0	0	0	0	0	0	0	-1	1

c: $[-2, -5, -3, -10, 0, -9, -7, -4, 8, -1, 12, 14, -3, 18, -1, -9, 10, -4, 0, -9, -11, -3, -5, 1, 1]$

ℓ: $[1, 1, 2, 1, 0, 3, 2, 3, 6, 1, 5, 3, 1, 6, 3, 1, 6, 2, 0, 1, 2, 3, 1, 2, 3]$

u: $[31, 31, 31, 31, 0, 31, 31, 31, 12, 31, 13, 12, 31, 24, 31, 31, 18, 31, 0, 31, 31, 31, 31, 31, 31]$

γ: 30

q^+: $[4, 6, 5, 12, 0, 12, 9, 6, 0, 6, 0, 0, 14, 0, 5, 13, 0, 7, 0, 12, 13, 6, 7, 3, 3]$

q^-: $[-3, -5, -4, -11, 0, -10, -7, -5, 0, -2, 0, 0, -4, 0, -3, -10, 0, -5, 0, -10, -12, -4, -6, -2, -2]$

Note: the completion times for arcs $j = 5, 9, 11, 12, 14, 17, 19$ are deterministic, hence the recourse penalties are 0.

The random completion times are independent, discretely distributed:

ξ	value		probability			
1	3,5,10,13,20	0.2,	0.3,	0.3,	0.15,	0.05
2	3,5,10,13,20	0.2,	0.3,	0.3,	0.15,	0.05
3	4,6,8,10,12	0.15,	0.25,	0.25,	0.2,	0.15
4	2,3,5,6,7	0.1,	0.2,	0.5,	0.1,	0.1
5	deterministic(dummy)	-				
6	6,9,15,20,25	0.175,	0.55,	0.2,	0.025,	0.05
7	6,7,8,12,18	0.15,	0.075,	0.3,	0.3,	0.175
8	6,9,15,20,25	0.175,	0.55,	0.2,	0.025,	0.05
9	deterministic	-				
10	2,3,5,6,7	0.1,	0.2,	0.5,	0.1,	0.1
11	deterministic	-				
12	deterministic	-				
13	3,5,10,13,20	0.2,	0.3,	0.3,	0.15,	0.05
14	deterministic	-				
15	6,9,15,20,25	0.175,	0.55,	0.2,	0.025,	0.05
16	3,5,10,13,20	0.2,	0.3,	0.3,	0.15,	0.05
17	deterministic	-				
18	4,6,8,10,12	0.15,	0.25,	0.25,	0.2,	0.15
19	deterministic(dummy)	-				
20	2,3,5,6,7	0.1,	0.2,	0.5,	0.1,	0.1
21	6,7,8,12,18	0.175,	0.55,	0.2,	0.025,	0.05
22	6,9,15,20,25	0.175,	0.55,	0.2,	0.025,	0.05
23	2,3,5,6,7	0.1,	0.2,	0.5,	0.1,	0.1
24	4,6,8,10,12	0.15,	0.25,	0.25,	0.2,	0.15
25	6,9,15,20,25	0.175,	0.55,	0.2,	0.025,	0.05

Solution

optimal value: 2208

optimal solution:

$\pi_1 = 0$	$\pi_6 = 12$	$\pi_{11} = 18$
$\pi_2 = 1$	$\pi_7 = 3$	$\pi_{12} = 29$
$\pi_3 = 3$	$\pi_8 = 18$	$\pi_{13} = 28$
$\pi_4 = 6$	$\pi_9 = 19$	$\pi_{14} = 27$
$\pi_5 = 5$	$\pi_{10} = 18$	$\pi_{15} = 30$

FINANCIAL PLANNING MODEL

Reference

J.G. Kallberg, R.W. White, W.T. Ziemba, "Short term financial planning under uncertainty". *Management Science* **28** 6 1982, 670–682.

A multiperiod simple recourse model with discrete probability models. Many variations solved, however some data is lacking in this article.

The description may be found in the reference; a brief sketch follows.

A firm must adjust its portfolio of short term assets and liabilities to minimize the net cost of cash surpluses and deficits over a fixed planning horizon. Uncertainties arise in the firm's need for cash, as well as in certain transaction costs. (These are modeled as discrete random variables.) The format is as follows:

choose $x_{j,t} (j = 1, \ldots, 14, t = 1, \ldots, 4)$ to *minimize*:

$$\sum_{t=1}^{4} \sum_{j=1}^{14} c_{j,t} x_{j,t} + E\{\sum_{t=1}^{4} \sum_{\ell=1}^{3} (q_{\ell,t}^{+} y_{\ell,t}^{+} + q_{\ell,t}^{-} y_{\ell,t}^{-})\}$$

$$+ \sum_{j=1}^{14} (d_{j}^{+} v_{j}^{+} + d_{j}^{-} v_{j}^{-})$$

subject to:

$$x_{j,t} \geq 0, v_{j}^{+} \geq 0, v_{j}^{-} \geq 0, y_{\ell,t}^{+} \geq 0, y_{\ell,t}^{-} \geq 0$$

$$\sum_{j=1}^{14} a_{i,j,t} x_{j,t} \leq b_{i,t} \qquad i = 1, \ldots, 7, \quad t = 1, \ldots, 4$$

$$\sum_{j=1}^{14} e_{j,t} x_{j,t} + \sum_{j=1}^{14} f_{j,t-1} x_{j,t-1} = g_t \qquad t = 1, \ldots, 4$$

$$y_{\ell,t}^{+} - y_{\ell,t}^{-} = \sum_{j=1}^{14} t_{\ell,t} x_{j,t} - \xi_{\ell,t} \qquad \ell = 1, 2, 3, \quad t = 1, \ldots, 4$$

$$v_{j}^{+} - v_{j}^{-} = x_{j,4} - \overline{v}_{j} \qquad j = 1, \ldots, 14$$

PRODUCT MIX PROBLEM

This problem is a stochastic version of a linear program derived in [1; p. 50]. It is an example of the use of a random technology matrix. The random version of this problem is due to R.J-B. Wets.

A furniture shop has 6000 man-hours available in the carpentry shop and 4000 man-hours in the finishing shop per period. All employees are on salary, however, and the actual man-hours available are assumed to be normally distributed random variables with deficits resulting from employee absences and surpluses due to voluntary overtime. There are four classes of products each consuming a certain number of man-hours in carpentry and finishing; the actual time consumed is assumed to be a uniformly distributed random variable. Each product earns a certain profit per item, and the shop has the option to purchase casual labor from outside. Note that the cost of the salaried labor is fixed, and thus does not enter the problem.

Stochastic program with simple recourse.

Choose $x_j (j = 1, \ldots, 4)$ to *maximize*

$$\sum_{j=1}^{4} c_j x_j - E\left\{\sum_{k=1}^{2} q_k \mathbf{v}_k^+\right\}$$

subject to

$$x_j \geq 0 \quad j = 1, \ldots, 4$$

$$\mathbf{v}_k^+ \geq 0, \quad \mathbf{v}_k^- \geq 0 \quad k = 1, 2$$

$$\mathbf{v}_k^+ - \mathbf{v}_k^- = \sum_{j=1}^{4} \mathbf{t}_{kj} x_j - \mathbf{h}_k \quad k = 1, 2$$

x_j: amount of product j produced

c_j: profit per unit of product j

$\mathbf{v}_k^+$:   hours of casual labor required of type k

q_k: cost per hour for casual labor of type k

$\mathbf{t}_{kj}$:   hours required of type k to produce product j

$\mathbf{h}_k$:   hours of salaried labor of type k available.

Data:

$c =$ [12, 20, 18, 40]

$q =$ [5, 10]

$\mathbf{h}_1$ Normal, mean 6000, st. dev. 100

$\mathbf{h}_k$ Normal, mean 4000, st. dev. 50

$\mathbf{t}_{11} \sim$ $U[3.5, 4.5]$, $\mathbf{t}_{12} \sim U[8, 10]$, $\mathbf{t}_{13} \sim U[6, 8]$, $\mathbf{t}_{14} \sim U[9, 11]$

$\mathbf{t}_{21} \sim$ $U[0.8, 1.2]$, $\mathbf{t}_{22} \sim U[0.8, 1.2]$, $\mathbf{t}_{23} \sim U[2.5, 3.5]$, $\mathbf{t}_{24} \sim U[36, 44]$

Solution

This problem has been solved using a technique developed in [2] with computer codes developed at IIASA (see [3]). Here are results for a run where the random measures we approximated by empirical measures derived from Monte Carlo simulations. The accuracy is to within a duality gap of 0.1%.

Number of samples:	1028
Solution:	$x_1 = 1384.80$
	$x_2 = 0.0$
	$x_3 = 0.0$
	$x_4 = 55.5370$
Optimal value:	17690.54

Next we solve the problem where the measures are replaced by a discretization based on conditional expectations. This is the "lower-bounding" scheme described in [4] and implemented in [3]; of course we obtain an upper bound here because this problem is a maximization. Here is the actual discretization of the measures. We divide the range of each random variable ξ into 4 intervals, I_1, I_2, I_3, I_4 of equal probability $p = \frac{1}{4}$ and calculate $\xi_k = E\{\xi | I_k\}, k = 1, \ldots, 4$. Then each random variable ξ is approximated by the discrete random variable taking values $\xi_1, \xi_2, \xi_3, \xi_4$ each with probability $\frac{1}{4}$:

$h_1 = [5872.9331, 5967.49, 6032.51, 6127.0669]$ w.p.$(0.25, 0.25, 0.25, 0.25)$

$h_2 = [3936.4666, 3983.7450, 4016.2550, 4063.5334]$ w.p.$(0.25, 0.25, 0.25, 0.25)$

$t_{11} = [3.625, 3.875, 4.125, 4.375]$ w.p.$(0.25, 0.25, 0.25, 0.25)$

$t_{12} = [8.25, 8.75, 9.25, 9.75]$ w.p.$(0.25, 0.25, 0.25, 0.25)$

$t_{13} = [6.25, 6.75, 7.25, 7.75]$ w.p.$(0.25, 0.25, 0.25, 0.25)$

$t_{14} = [9.25, 9.75, 10.25, 10.75]$ w.p.$(0.25, 0.25, 0.25, 0.25)$

$t_{21} = [0.85, 0.95, 1.05, 1.15]$ w.p.$(0.25, 0.25, 0.25, 0.25)$

$t_{22} = [0.85, 0.95, 1.05, 1.15]$ w.p.$(0.25, 0.25, 0.25, 0.25)$

$t_{23} = [2.625, 2.875, 3.125, 3.375]$ w.p.$(0.25, 0.25, 0.25, 0.25)$

$t_{24} = [37.0, 39.0, 41.0, 43.0]$ w.p.$(0.25, 0.25, 0.25, 0.25)$

The solution, again accurate to within 0.1% duality gap, is:

x_1	=	1377.26
x_2	=	0.0
x_3	=	0.0
x_3	=	55.8027

Optimal value: 17715.03

This problem will be available on the stochastic programming computer tape to be distributed by IIASA.

Reference

1. G. Dantzig: *Linear Programming and Extensions*, Princeton University Press, 1963.
2. R.T. Rockafellar and R.J-B. Wets, "A Lagrangian finite generation technique for solving linear-quadratic problems in stochastic programming," in A. Prékopa and R. J-B. Wets, *Stochastic Programming 1984* Mathematical Programming Study, North Holland (1985)
3. A.J. King, "An implementation of the Lagrangian finite generation method," this volume.
4. J. Birge and R.J-B. Wets, "Designing approximation schemes for stochastic optimization problems", in A. Prékopa and R. J-B. Wets, *Stochastic Programming 1984* Mathematical Programming Study, North Holland (1985)

LAKE BALATON MODEL

Reference

L. Somlyody and R.J-B. Wets, "Stochastic models for lake eutrophication management". Collaborative Paper CP-85- , International Institute for Applied Systems Analysis, Laxenburg, Austria (1985).

The problem is to choose an optimal level of investments in sewage treatment facilities so that expected deviations of pollutant concentration levels are minimized. The form of the problem is as follows:

choose $x_j (j = 1, \ldots, 54)$ to *minimize*

$$E\{\sum_{k=1}^{4} e_k q_k \theta[e_k^{-1}(\mathbf{v}_k - \bar{v}_k)]\}$$

subject to
$$0 \leq x_j \leq 1 \qquad\qquad j = 1, \ldots, 54$$
$$\sum_{j=1}^{54} a_{ij} x_j \leq b_i \qquad\quad i = 1, \ldots, 35$$
$$\mathbf{v}_k = \mathbf{h}_k - \sum_{j=1}^{54} \mathbf{t}_{kj} x_j \quad k = 1, \ldots, 4$$

where $\theta(\cdot)$ is a linear-quadratic penalty function

$$\theta(\tau) = \begin{cases} 1/2\tau^2 & \text{if } 0 \leq \tau \leq 1 \\ \tau - 1/2 & \text{if } \tau \geq 1 \\ 0 & \text{if } \tau \leq 0. \end{cases}$$

This problem has been solved using the Lagrange finite generation technique, see [1] and [2]. Details of the problem are available from the above authors.

1. R.T. Rockafellar and R.J-B. Wets, "A Lagrangian finite generation technique for solving linear-quadratic problems in stochastic programming". In A. Prékopa and R. J-B. Wets *Stochastic Programming 1984* Mathematical Programming Study, North Holland 1985.

2. A.J. King, "An implementation of the Lagrangian finite generation technique", this volume.

MULTIPERIOD PRODUCTION PLANNING

Reference

R.J. Peters, K. Boskma, and H.A.E. Kuper, "Stochastic programming in production planning: a case with non-simple recourse," *Statistica Neerlandica* 31 (1977), 113-126.

A factory must decide upon a production schedule and plan increases/decreases in production activity over several periods in order to meet a random demand for its mix of products. There are costs associated with changes in activity from one period to the next. The factory may engage in recourse activities, buying product to cover shortages and storing product to cover surpluses. A surplus in one period is carried over to the next period thus imparting a general recourse feature to this problem.

Multistage general recourse problem

Choose x, u to *minimize*

$$\sum_{t=1}^{4}\Big[\sum_{j=1}^{2}c_j x_{j,t} + \sum_{i=1}^{3}d_i u_{i,t}\Big] + \sum_{t=1}^{3}\Big[e^{+}w_t^{+} + e^{-}w_t^{-}\Big]$$

$$+E\Big\{\min s_y \sum_{t=1}^{4}\sum_{j=1}^{2}[q_{j,t}^{+}\mathbf{y}_{j,t}^{+} + q_{j,t}^{-}\mathbf{y}_{j,t}^{-}]\Big\}$$

subject to:

$$x_{j,t} \geq 0 \qquad j = 1,2 \quad t = 1,\ldots,4$$

$$u_{i,t} \geq 0 \qquad i = 1,2,3, \quad t = 1,\ldots,4$$

$$w_t^{+}, w_t^{-} \geq 0 \quad t = 1,\ldots,4$$

$$\sum_{j=1}^{2} a_{ij} x_{j,t} - u_{i,t} \leq b_{i,t} \qquad i = 1,2,3, \quad t = 1,\ldots,4$$

$$u_{i,t} \leq f_{i,t} \qquad i = 1,2,3, \quad t = 1,\ldots,4$$

$$\sum_{j=1}^{2} a_{3j}[x_{j,t+1} - x_{j,t}] = w_t^{+} - w_t^{-} \qquad t = 1,2,3$$

$$\mathbf{y}_{j,t}^{+} \geq 0, \quad \mathbf{y}_{j,t}^{-} \geq 0 \quad j = 1,2, \quad t = 1,\ldots,4$$

$$x_{j,1} + \mathbf{y}_{j,1}^{+} - \mathbf{y}_{j,1}^{-} = \boldsymbol{\xi}_{j,1} \quad j = 1,2$$

$$x_{j,t} + \mathbf{y}_{j,t-1}^{-} + \mathbf{y}_{j,t}^{+} - \mathbf{y}_{j,t}^{-} = \boldsymbol{\xi}_{j,t}, \quad j = 1,2, \quad t = 2,3,4$$

$x_{j,t}$: amount of product $j = 1,2$ produced in period $t = 1,\ldots,4$

$\boldsymbol{\xi}_{j,t}$: demand for product j in period t

$\mathbf{y}_{j,t}^{+}$: amount of deficit product j purchased in period t

$\mathbf{y}_{j,t}^{-}$: amount of surplus product j stored in period t

c_j: cost of producing product j

$q_{j,t}^{+}, q_{j,t}^{-}$: cost of deficit/surplus product j in period t

$u_{i,t}$: extra capacity of production activity $i = 1,2,3$ used in period t

$b_{i,t}$: normal capacity of production activity i in period t

$f_{i,t}$: maximum expansion of capacity for activity i in period t

d_i: cost of extra capacity of production activity in i

w_t^{+}, w_t^{-}: change in utilization of production activity 3 from period $t = 1,2,3$ to period $t+1 = 2,3,4$

e^{+}, e^{-}: cost of change of production activity 3.

Data:

		$t = 1$	2	3	4
b:	$j = 1$	4000	4000	4000	3500
	2	3000	3000	2500	3000
	3	4500	4500	3750	3500
f:	$j = 1$	400	400	400	350
	2	300	300	250	350
	3	450	450	375	350

a: $\begin{pmatrix} 4 & 5 \\ 3 & 6 \\ 3 & 7 \end{pmatrix}$ c: $(100, 150)$

d: $(15, 20, 10)$ $e^+ = 20, \quad e^- = 15$

		$t = 1$	2	3	4
q^-:	$i = 1$	25	25	25	100
	2	30	30	30	150
q^+:	$i = 1$	400	400	400	400
	2	450	450	450	450

The demands $\xi_{j,t}$ $j = 1, 2$ $t = 1, \cdots, 4$ are independent normal (mean, standard deviation):

Period	Product 1	Product 2
1	$(300, 45)$	$(500, 75)$
2	$(320, 45)$	$(500, 75)$
3	$(440, 45)$	$(500, 75)$
4	$(480, 45)$	$(600, 75)$

Solution:

		$t = 1$	2	3	4
x:	$j = 1$	341.35	304.56	493.22	401.96
	$j = 2$	560.85	576.62	377.91	377.73
u:	$i = 1$	170.60	115.80	7.32	27.95
	$i = 2$	0.	0.	0.	0.
	$i = 3$	450.0	450.0	375.0	350.0
w^+:		0.	0.	0.	
w^-:		0.	825.	275.	

MANPOWER PLANNING

Reference

E.P.C. Kao, M. Queyranne: *Aggregation in a Two-Stage Stochastic Program for Manpower Planning in the Service Sector*, Working Paper, Center for Health Management, University of Houston, 1981.

An employer must decide upon a base level of regular staff at various skill levels. The recourse actions available are regular staff overtime or outside temporary help in order to meet unknown demand for services at minimum cost.

Multistage general recourse problem.

Choose $x_j (j = 1, 2, 3)$ to *minimize*

$$\sum_{j=1}^{3} c_j x_j + \sum_{t=1}^{12} E \left\{ \min_{y,z} \sum_{j=1}^{3} [q_j \mathbf{y}_{j,t} + r_j \mathbf{z}_{j,t}] \right\}$$

subject to: $x_j \geq 0$

$\mathbf{y}_{j,t} \geq 0, \quad \mathbf{z}_{j,t} \geq 0$

$$\sum_{j=1}^{3} [\mathbf{y}_{j,t} + \mathbf{z}_{j,t}] \geq \xi_t - \alpha_t \sum_{j=1}^{3} x_j \quad t = 1, \ldots, 12$$

$\mathbf{y}_{j,t} \leq 0.2 \, \alpha_t x_j \quad j = 1, 2, 3, \quad t = 1, \ldots, 12$

$\gamma_{j-1} [x_{j-1} + \mathbf{y}_{j-1,t} + \mathbf{z}_{j-1,t}] - [x_j + \mathbf{y}_{j,t} + \mathbf{z}_{j,t}] \geq 0 \quad j = 1, 2, 3 \quad t = 1, \ldots, 12$

x_j:	base level of regular staff at skill level $j = 1, 2, 3$
$\mathbf{y}_{j,t}$:	amount of overtime help
$\mathbf{z}_{j,t}$:	amount of temporary help
c_j:	cost of regular staff at skill level $j = 1, 2, 3$
q_j:	cost of overtime
r_j:	cost of temporary
ξ_t:	demand for services
α_t:	anticipated absentee rate for regular staff at time $t = 1, \ldots, 12$
γ_{j-1}:	ratio of amount of skill level j per amount of skill level $j - 1$ required.

Data

$c =$ $[7.03, 4.53, 3.44]$
$q =$ $[9.59, 6.18, 4.69]$
$r =$ $[11.70, 9.95, 5.78]$
$\alpha =$ $[.8943, .8917, .8948, .9086, .9032, .8842, .8513, .8798, .8871, .9043, .8606, .8341]$

$\gamma =$ $[0.6, 0.2]$

the demands ξ_t, $t = 1, \ldots, 12$ are independent $N(\overline{\mu}_t, \overline{\sigma}_t^2)$ where $\overline{\sigma}_t^2 \triangleq \ell \overline{\mu}_t$.
$\overline{\mu} =$ $[11975, 11740, 12169, 13132, 13525, 12598, 13503, 14168, 12602, 11807, 11334, 10410]$

Solution

The problem is not solved. Instead the author has worked out upper and lower bounds for the objective function corresponding to various values of ℓ (i.e. changing the variance of the random demand).

ℓ	Upper Bound UB	Lower Bound LB	Difference $\Delta = UB - LB$	Relative Gap $(\Delta/LB) \times 100\%$
0	852,230	846,706	5524	0.65
1	852,997	846,287	6690	0.79
10	855,539	851,458	4081	0.48
30	859,505	854,706	4799	0.56

FLOOD CONTROL PROBLEM

Reference

A. Prékopa, T. Szántai: "Flood control reservoir system design using stochastic programming." *Mathematical Programming Study* **9**, North-Holland, 1978, 138-151.

The object is to choose the optimal size of reservoirs placed at certain (fixed) locations in order to control flooding due to random stream inputs. The criterion is to find the lowest cost solution which controls floods a given percentage of the time. The probability model for the stream flows is multivariate normal and gamma.

Chance constrained problem

Choose $x_j (j = 1, \ldots, 5)$ to *minimize*:

$$\sum_{j=1}^{5} c_j x_j$$

subject to

$$0 \leq x_j \leq u_j \qquad j = 1, \ldots, 5$$

$$P[\sum_{j=1}^{5} t_{kj} x_j \geq \sum_{i=1}^{5} t_{ki} \xi_i, k = 1, \ldots, 9] \geq p$$

x_j: capacity of reservoir $j = 1, \ldots, 5$
u_j: upper bound on capacity of reservoir j
c_j: cost per unit capacity of reservoir j
ξ_i: streamflow for tributary i, $i = 1, \ldots, 5$

(The system of conditions $T\xi \leq Tx$ is equivalent to the condition that the net flow volume be less than the capacity of the furthest downstream reservoir.)

Data

$$u = [1.0, 1.0, 1.0, 2.0, 3.0]$$
$$c = [0.4, 0.5, 0.6, 1.2, 1.8]$$

$$T = \begin{bmatrix} 0 & 0 & 0 & 0 & 1 \\ 0 & 0 & 0 & 1 & 1 \\ 1 & 0 & 0 & 1 & 1 \\ 0 & 1 & 0 & 1 & 1 \\ 0 & 0 & 1 & 1 & 1 \\ 1 & 1 & 0 & 1 & 1 \\ 1 & 0 & 1 & 1 & 1 \\ 0 & 1 & 1 & 1 & 1 \\ 1 & 1 & 1 & 1 & 1 \end{bmatrix}$$

For the random variables ξ_i $i = 1, \ldots, 5$ we specify:

	means	st. dev.
ξ_1	0.8	0.2
ξ_2	1.5	0.3
ξ_3	1.2	0.6
ξ_4	0.5	0.4
ξ_5	0.7	0.3

Solution

The problem was solved using three different correlation matrices for the random variables $\xi_1, \ldots, \xi_5$, (but with the same means and variances given above):

$$R_1 = \begin{pmatrix} 1.0 & 0.0 & 0.6 & 0.4 & 0.0 \\ 0.0 & 1.0 & 0.5 & 0.3 & 0.3 \\ 0.6 & 0.5 & 1.0 & 0.7 & 0.6 \\ 0.4 & 0.3 & 0.7 & 1.0 & 0.4 \\ 0.0 & 0.3 & 0.6 & 0.4 & 1.0 \end{pmatrix},$$

$$R_2 = \begin{pmatrix} 1.0 & -0.5 & 0.0 & 0.3 & -0.5 \\ -0.5 & 1.0 & -0.8 & 0.0 & 0.2 \\ 0.0 & -0.8 & 1.0 & 0.0 & 0.3 \\ 0.3 & 0.0 & 0.0 & 1.0 & 0.0 \\ -0.5 & 0.2 & 0.3 & 0.0 & 1.0 \end{pmatrix},$$

$R_3 = E$, the identity matrix,

and two probability levels $p = 0.8, p = 0.9$. The following table gives the solutions:

Numerical Results

Type of distribution	Correlation matrix	Probability level	X_1	X_2	X_3	X_4	X_5	Objective function	Computing time*
Multivariate gamma	R_1	p=0.8	0.807	1	1	1.356	1.412	5.591	00:52:657
		p=0.9	0.751	1	1	1.976	1.398	6.289	00:35:688
	R_3	p=0.8	1	1	1	1.539	1.193	5.494	00:16:785
		p=0.9	1	1	1	1.268	1.848	6.348	00:11:343
Multivariate normal	R_1	p=0.8	0.796	1	1	1.591	1.383	5.816	01:03:444
		p=0.9	0.998	1	1	1.885	1.524	1.6505	00:25:126
	R_2	p=0.8	0.906	1	1	1.351	1.371	5.551	00:58:078
		p=0.9	0.833	1	1	1.239	1.830	6.214	00:51:426
	R_3	p=0.8	1	1	1	1.226	1.431	5.547	00:43:461
		p=0.9	1	1	1	1.650	1.374	5.953	00:57:478

*Time in minutes/seconds/milliseconds.

In the multivariate gamma case for R_1 we have

$$\xi_1 = \tfrac{1}{20}(y_1 + y_2 + y_3),$$
$$\xi_2 = \tfrac{3}{50}(y_4 + y_5 + y_6 + y_7),$$
$$\xi_3 = \tfrac{3}{10}(y_1 + y_4 + y_5 + y_8 + y_9 + y_{10} + y_{11}),$$
$$\xi_4 = \tfrac{8}{25}(y_2 + y_6 + y_8 + y_9 + y_{12}),$$
$$\xi_5 = \tfrac{9}{70}(y_4 + y_8 + y_{10} + y_{13}),$$

where $y_1, \ldots, y_{13}$ have standard gamma distributions with the following parameters:

$$\vartheta_1 = 0.576 \qquad \vartheta_6 = 0.225 \qquad \vartheta_{10} = 0.850$$
$$\vartheta_2 = 0.160 \qquad \vartheta_7 = 23.875 \qquad \vartheta_{11} = 2.055$$
$$\vartheta_3 = 15.264 \qquad \vartheta_8 = 0.140 \qquad \vartheta_{12} = 0.758$$
$$\vartheta_4 = 0.315 \qquad \vartheta_9 = 0.280 \qquad \vartheta_{13} = 4.940$$
$$\vartheta_5 = 0.585$$

STABIL

Reference

(1) A. Prékopa, S. Ganczer, I. Deak, K. Patyi, "The STABIL stochastic programming model and its experimental application to the electrical energy sector of the Hungarian economy.", in M. Dempster, ed., *Stochastic Programming*, Academic Press, 1980, 369-385.

(2) A. Prékopa, S. Ganczer, I. Deak, K. Patyi, "A STABIL sztohasztikus programozási modell és annak kísérleti alkadmazása ā Magyar villamosenergia-iparra," *Alkalmazott Matematikai Lapok* **1**(1975) 3-22 (in Hungarian)

A large-scale chance constrained model with multi-variate normal and gamma distributions. A description of the model is in (2), with an edited version in (1). The format of the problem is as follows:

choose $x_j (j = 1, \ldots, 52)$ to *minimize*

$$\sum_{j=1}^{52} c_j x_j$$

subject to:

$$x_j \geq 0$$

$$\sum_{j=1}^{52} a_{ij} x_j \geq b_i \qquad i = 5, \ldots, 110$$

$$P\left\{ \sum_{j=1}^{52} a_{ij} x_j \geq b_i + \sigma_i \xi_i, \quad i = 1, 2, 3, 4 \right\} \geq p.$$

GAS DELIVERY PROBLEM

Reference

J.-M. Guldmann: "Supply, storage, and service reliability decisions by gas distribution utilities: a chance-constrained approach." *Management Science* **29**(8)(1983), 884-906.

A gas delivery company has two options for gas supply: to purchase from a pipeline, or to withdraw gas from its own storage field. The demand for gas consumption is assumed random. The company must make three types of decisions: to decide the maximum monthly contract (which commits the pipeline company to allocate this capacity to the delivery company), to decide to increment its storage capacity, and to decide the actual monthly supply request from the pipeline company on a month to month basis. The contract decision and the storage capacity increment decision are made once at the beginning of each year. Any monthly surplus or deficit of pipeline supply vs. gas consumption is stored or withdrawn respectively from the gas delivery company's own storage field. The delivery company's objective is to meet its consumption demand at minimum cost, subject to feasibility constraints on the operation of its gas storage facility.

Chance constrained non-linear version.

Choose $x_1, \ldots, x_{12}, \quad y, \quad z$ to *minimize*

$$\sum_{t=1}^{12} c_j x_j + c_0 \quad E\left\{\sum_{t=1}^{12} |x_t - \xi_t|\right\} + c_{13} y + c_{14} z$$

subject to

$$x_t \geq 0 \qquad t = 1, \ldots, 12$$
$$y \geq x_t \qquad t = 1, \ldots, 12$$
$$z \geq 0$$

$$P \left\{ \begin{array}{ll} (x_t - \xi_t) - a_1 \sum_{s=1}^{t-1}(x_s - \xi_s) - a_2 z \leq b_1 \\ -(x_t - \xi_t) - a_3 \sum_{s=1}^{j-1}(x_s - \xi_s) - a_4 z \leq b_2 \\ \sum_{s=1}^{t}(x_s - \xi_s) - a_5 z \leq b_3 \\ \sum_{s=1}^{t}(x_s - \xi_s) \geq 0 \end{array} \middle| t = 1, \ldots, 12 \right\} \geq p$$

x_t: gas ordered from pipeline in month $t = 1, \ldots, 12$
y: contract capacity per month
z: gas storage capacity increment
ξ_t: actual gas consumed in month $t = 1, \ldots, 12$ (random)
c_t: cost per unit of gas ordered in month $t = 1, \ldots, 12$
c_0:   cost of transferring one unit gas into or out of storage facility
c_{13}:   cost of contract capacity
c_{14}:   cost of storage facility capacity increment

Data.

$$c_1, \ldots, c_7 = 1202.4 \qquad c_8, \ldots, c_{12} = 1299.3 \qquad c_0 = 33.23$$
$$c_{13} = 392.0 \qquad\qquad c_{14} = 57.0$$

$$a_1 = -0.078 \qquad\qquad a_2 = 0.8 \qquad\qquad b_1 = 118075.2$$
$$a_3 = 0.15 \qquad\qquad a_4 = 0.049 \qquad\qquad b_2 = 7232.1$$
$$\qquad\qquad\qquad\qquad\quad a_5 = 0.41 \qquad\qquad b_3 = 60513.5$$

ξ_t $(t = 1, \ldots, 12)$ are independent, normal:

$$\xi_t = 14{,}900 + 36.583 \eta_t$$

where η_t are normal as follows:

	mean	st. dev.		mean	st. dev.
η_1	506.6	90.5	η_7	371.6	91.1
η_2	248.2	88.3	η_8	712.6	85.6
η_3	50.5	28.8	η_9	1071.6	145.8
η_4	11.0	9.4	η_{10}	1207.7	129.5
η_5	18.9	14.1	η_{11}	1046.3	115.2
η_6	120.5	42.1	η_{12}	892.5	125.4

Solution.

The author did not solve the given problem. The nonlinear term presents difficulties. T.. cope with this difficulty he specifies:

$$P \left\{ \begin{array}{ll} x_t \geq \xi_t & t = 1, \ldots, 7 \\ x_t \leq \xi_t & t = 8, \ldots, 12 \end{array} \right\} \geq 0.999$$

in which case the nonlinear term is approximated by:

$$c_0 \left[\sum_{t=1}^{7} (x_t - E\{\xi_t\}) + \sum_{t=8}^{12} (E\{\xi_t\} - x_t) \right].$$

He then obtains the following results:

Reliability

Month (t)		99%	95%	90%	85%	80%
		Monthly Supply x_t (MMCF)				
April	(1)	46,680	46,680	46,680	46,680	46,680
May	(2)	44,471	36,909	36,909	43,153	42,954
June	(3)	40,571	39,016	37,871	36,615	36,017
July	(4)	38,695	36,846	35,568	34,260	33,590
August	(5)	30,291	33,319	31,499	24,959	24,156
September	(6)	25,472	25,472	25,472	25,472	25,472
October	(7)	41,820	41,820	41,820	41,820	41,820
November	(8)	28,442	28,442	28,442	28,442	28,442
December	(9)	32,764	32,764	32,764	32,764	32,764
January	(10)	40,132	40,132	40,132	40,132	40,132
February	(11)	36,326	36,326	36,326	36,326	36,326
March	(12)	29,197	29,197	29,197	29,197	29,197
Total		434,861	426,923	422,680	419,820	417,550
		Minimum Storage Capacity (MMCF)				
		360,138	324,158	304,927	291,981	281,677
		Expected Purchases and Storage Operation Costs ($1000's)				
		564,093	554,284	549,042	545,509	542,704

The minimum storage capacity is (in our notation) $z + 147594$. We refer the reader to the reference for more details of the author's solution.

INDEX

Springer Series in Computational Mathematics

Editorial Board: R. L. Graham, J. Stoer, R. Varga

Computational Mathematics is a series of outstanding books and monographs which study the applications of computing in numerical analysis, optimization, control theory, combinatorics, applied function theory, and applied functional analysis. The connecting link among these various disciplines will be the use of high-speed computers as a powerful tool. The following list of topics best describes the aims of *Computational mathematics:* finite element methods, multigrade methods, partial differential equations, multivariate splines and applications, numerical solutions of ordinary differential equations, numerical methods of optimal control, nonlinear programming, simulation techniques, software packages for quadrature, and p.d.e. solvers.

Computational Mathematics is directed towards mathematicians and appliers of mathematical techniques in disciplines such as engineering, computer science, economics, operations research and physics.

Volume 1

R. Piessens, E. de Doncker-Kapenga, C. W. Überhuber, D. K. Kahaner

QUADPACK

A Subroutine Package for Automatic Integration

1983. 26 figures. VII, 301 pages.
ISBN 3-540-12553-1

Contents: Introduction. – Theoretical Background. – Algorithm Descriptions. – Guidelines for the Use of QUADPACK. – Special Applications of QUAD-PACK. – Implementation Notes and Routine Listings. – References.

Volume 2

J. R. Rice, R. F. Boisvert

Solving Elliptic Problems Using ELLPACK

1985. 53 figures. X, 497 pages. ISBN 3-540-90910-9

Contents: The ELLPACK System. – The ELLPACK Modules. – Performance Evaluation. – Contributor's Guide. – System Programming Guide. – Appendices. – Index.

Springer-Verlag
Berlin Heidelberg New York
London Paris Tokyo

Volume 3

N. Z. Shor

Minimization Methods for Non-Differentiable Functions

Translated from the Russian by K. C. Kiwiel, A. Ruszczyński

1985. VIII, 162 pages. ISBN 3-540-12763-1

Contents: Introduction. – Special Classes of Nondifferentiable Functions and Generalizations of the Concept of the Gradient. – The Subgradient Method. – Gradient-type Methods with Space Dilation. – Applications of Methods for Nonsmooth Optimization to the Solution of Mathematical Programming Problems. – Concluding Remarks. – References. – Subject Index.

Volume 4

W. Hackbusch

Multi-Grid Methods and Applications

1985. 43 figures, 48 tables. XIV, 377 pages.
ISBN 3-540-12761-5

Contents: Preliminaries. – Introductory Model Problem. – General Two-Grid Method. – General Multi-Grid Iteration. – Nested Iteration Technique. – Convergence of the Two-Grid Iteration. – Convergence of the Multi-Grid Iteration. – Fourier Analysis. – Nonlinear Multi-Grid Methods. – Singular Perturbation Problems. – Elliptic Systems. – Eigenvalue Problems and Singular Equations. – Continuation Techniques. – Extrapolation and Defect Correction Techniques. – Local Techniques. – The Multi-Grid Method of the Second Kind. – Bibliography. – Subject Index.

Volume 5

V. Girault, P-A. Raviart

Finite Element Methods for Navier-Stokes Equations

Theory and Algorithms

1986. 21 figures. X, 374 pages. ISBN 3-540-15796-4

Contents: Mathematical Foundation of the Stokes Problem. – Numerical Solution of the Stokes Problem in the Primitive Variables. – Incompressible Mixed Finite Element Methods for Solving the Stokes Problem. – Theory and Approximation of the Navier-Stokes Problem. – References. – Index of Mathematical Symbols. – Subject Index.

Springer

Springer